Lecture Notes in Computer Science 3506

Commenced Publication in 1973
Founding and Former Series Editors:
Gerhard Goos, Juris Hartmanis, and Jan van Leeuwen

Lecture Notes in Computer Science 3576

Commenced Publication in 1973
Founding and Former Series Editors:
Gerhard Goos, Juris Hartmanis, and Jan van Leeuwen

Editorial Board

David Hutchison
Lancaster University, UK
Takeo Kanade
Carnegie Mellon University, Pittsburgh, PA, USA
Josef Kittler
University of Surrey, Guildford, UK
Jon M. Kleinberg
Cornell University, Ithaca, NY, USA
Friedemann Mattern
ETH Zurich, Switzerland
John C. Mitchell
Stanford University, CA, USA
Moni Naor
Weizmann Institute of Science, Rehovot, Israel
Oscar Nierstrasz
University of Bern, Switzerland
C. Pandu Rangan
Indian Institute of Technology, Madras, India
Bernhard Steffen
University of Dortmund, Germany
Madhu Sudan
Massachusetts Institute of Technology, MA, USA
Demetri Terzopoulos
New York University, NY, USA
Doug Tygar
University of California, Berkeley, CA, USA
Moshe Y. Vardi
Rice University, Houston, TX, USA
Gerhard Weikum
Max-Planck Institute of Computer Science, Saarbruecken, Germany

Choonsik Park   Seongtaek Chee (Eds.)

Information Security and Cryptology – ICISC 2004

7th International Conference
Seoul, Korea, December 2-3, 2004
Revised Selected Papers

 Springer

Volume Editors

Choonsik Park
Seongtaek Chee
NSRI (National Security Research Institute)
161 Gajeong-dong, Yuseong-gu, Daejeon, Korea
E-mail: {csp,chee}@etri.re.kr

Library of Congress Control Number: 2005926827

CR Subject Classification (1998): E.3, G.2.1, D.4.6, K.6.5, F.2.1, C.2, J.1

ISSN   0302-9743
ISBN-10   3-540-26226-1 Springer Berlin Heidelberg New York
ISBN-13   978-3-540-26226-8 Springer Berlin Heidelberg New York

Springer is a part of Springer Science+Business Media

springeronline.com

© Springer-Verlag Berlin Heidelberg 2005
Printed in Germany

Typesetting: Camera-ready by author, data conversion by Scientific Publishing Services, Chennai, India
Printed on acid-free paper   SPIN: 11496618   06/3142   5 4 3 2 1 0

Preface

The 7th International Conference on Information Security and Cryptology was organized by the Korea Institute of Information Security and Cryptology (KIISC) and was sponsored by the Ministry of Information and Communication of Korea.

The conference received 194 submissions, and the Program Committee selected 34 of these for presentation. The conference program included two invited lectures. Mike Reiter spoke on "Security by, and for, Converged Mobile Devices." And Frank Stajano spoke on "Security for Ubiquitous Computing." We would like to first thank the many researchers from all over the world who submitted their work to this conference. An electronic submission process was available. The submission review process had two phases. In the first phase, Program Committee members compiled reports (assisted at their discretion by subreferees of their choice, but without interaction with other Program Committee members) and entered them, via a Web interface, into the Web Review software. We would like to thank the developers, Bart Preneel, Wim Moreau, and Joris Claessens. Without the Web Review system, the whole review process would not have been possible. In the second phase, Program Committee members used the software to browse each other's reports, and discuss and update their own reports. We are extremely grateful to the Program Committee members for their enormous investment of time, effort, and adrenaline in the difficult and delicate process of review and selection.

We are most grateful to Dr. Jin Hong and Dr. Aaram Yun from NSRI (National Security Research Institute, Korea). Skillfully and patiently, they carried the main load of background work of the Program Co-chairs, in particular in setting up the submission and review servers, providing technical help to the authors and committee members, and in the preparation of this proceedings.

February 2005   Choonsik Park and Seongtaek Chee

Organization

General Chair

Pil Joong Lee   POSTECH, Korea

Program Committee Co-chairs

Choonsik Park   NSRI, Korea
Seongtaek Chee   NSRI, Korea

Program Committee

Alex Biryukov	Katholieke Universiteit Leuven, Belgium
Daniel Bleichenbacher	Bell Laboratories, USA
Kyo Il Chung	ETRI, Korea
Robert Deng	Singapore Management University, Singapore
Gene Itkis	Boston University, USA
Thomas Johansson	Lund University, Sweden
Antoine Joux	DCSSI Crypto Lab, France
Toshinobu Kaneko	Tokyo University of Science, Japan
Hyoung Joong Kim	Kangwon National University, Korea
Myung-Hwan Kim	Seoul National University, Korea
Seungjoo Kim	Sungkyunkwan University, Korea
Yongdae Kim	University of Minnesota, USA
Dong Hoon Lee	Korea University, Korea
Arjen K. Lenstra	Citibank, USA and Eindhoven University of Technology, The Netherlands
Masahiro Mambo	Tohoku University, Japan
Tsutomu Matsumoto	Yokohama National University, Japan
SangJae Moon	Kyungpook National University, Korea
David Naccache	Gemplus Card International, France
Phong Q. Nguyen	CNRS/École Normale Supérieure, France
Tatsuaki Okamoto	NTT Labs, Japan
Josef Pieprzyk	Macquarie University, Australia
David Pointcheval	École Normale Supérieure, France
Vincent Rijmen	IAIK, Graz University of Technology, Austria, and Cryptomathic

Matt Robshaw	Royal Holloway, University of London, UK
Jae-Cheol Ryou	Chungnam National University, Korea
Kouichi Sakurai	Kyushu University, Japan
Palash Sarkar	Indian Statistical Institute, India
William Whyte	NTRU Cryptosystems, USA
Sung-Ming Yen	National Central University, Taiwan, ROC
Moti Yung	Columbia University, USA
Yuliang Zheng	University of North Carolina at Charlotte, USA

Organizing Committee Chair

Ji Hong Kim	Semyung University, Korea

Organizing Committee

Young Sub Koo	Ministry of Information and Communication, Korea
Jae Sung Kim	KISA, Korea
Jeong Nyeo Kim	Electronics and Telecommunications Research Institute, Korea
Gwang Soo Rhee	Sookmyung Women's University, Korea
Kye Sang Lee	Dongeui University, Korea
Young Ho Park	Sejong Cyber University, Korea
Heekuck Oh	Hanyang University, Korea
Chang Han Kim	Semyung University, Korea
Kang Bin Yim	Soonchunhyang University, Korea
Jae Cheol Ha	Korea Nazarene University, Korea

External Reviewers

Michel Abdalla	Dario Catalano	Christophe Clavier
Seigo Arita	Fabienne Cathala	Scott Contini
Roberto Avanzi	Julien Cathalo	Jean-Sébastien Coron
Yoo-Jin Baek	Byungki Cha	Nora Dabbous-Costa
Claude Barral	Sanjit Chatterjee	Tanmoy Das
Olivier Benoît	Chien-Ning Chen	Christophe De Cannière
Carine Boursier-Guesdon	Jung Hee Cheon	Benne de Weger
An Braeken	Benoît Chevallier-Mames	Jean-François Dhem
Éric Brier	Eunsung Cho	Xuhua Ding
Julien Brouchier	Hamid Choukri	Christophe Doche

Håkan Englund
Pierre-Alain Fouque
Jacques Fournier
Steven Galbraith
Karine Gandolfi-Villegas
Pierre Girard
Benoît Gonzalvo
Louis Granboulan
Pascal Guterman
Jaime Gutierrez
Dong-Guk Han
Helena Handschuh
Martin Hell
Seokhie Hong
Nick Hopper
Yoshiaki Hori
Nick Howgrave-Graham
Chao-Chih Hsu
Misuk Huh
Kenji Imamoto
Marc Joye
Ju-Sung Kang
Vishal Kher
Jinhae Kim
Sangwook Kim
Jaehyung Ko
Xeno Kovah
Jin Kwak
Soonhak Kwon
Tanja Lange
Joseph Lano
Younggyo Lee

Hyang-Sook Lee
Insok Lee
Jeong Hyun Lee
Yingjiu Li
Hsi-Chung Lin
Chun-Shien Lu
Philip Mackenzie
Karthikeyan Mahadevan
Subhamoy Maitra
Stefan Mangard
Keith Martin
Gwenaëlle Martinet
Alexander Maximov
Robert McNerney
Frédéric Muller
Sean Murphy
Junghyun Nam
Khánh Quốc Nguyễn
Masayuki Numao
Katsuyuki Okeya
Francis Olivier
Ivan Osipkov
Elisabeth Oswald
Béatrice Péirani
Pascal Paillier
Je Hong Park
S.Y. Park
Kenny Paterson
Mireille Pauliac
Duong Hieu Phan
Stéphanie Porte
Anne-Marie Praden

Bart Preneel
Kyung-Hyune Rhee
Ludovic Rousseau
Scott Russell
Mark Shaneck
Kyungah Shim
Jun-Bum Shin
Wook Shin
Taizo Shirai
Igor Shparlinski
Stéphane Socié
Kyungho Son
Paul Souradyuti
Martijn Stam
Ron Steinfeld
Po-Chyi Su
Hung-Min Sun
Toshihiro Tabata
Tsuyoshi Takagi
Kouhei Tatara
Michael Tunstall
Lionel Victor
Guilin Wang
Peng Wang
Huaxiong Wang
Claire Whelan
Christopher Wolf
Hongjun Wu
Yongdong Wu
Yanjiang Yang
Huafei Zhu

Table of Contents

PKI and Related Implementation

Digital Signature

Elliptic Curve Cryptosystem

Provable Security and Primitives

Network Security

Steganography

Biometrics

Security by, and for, Converged Mobile Devices

Mike Reiter

CyLab. Electrical & Computer Engineering,
Carnegie Mellon University, USA
`reiter@cmu.edu`

Abstract. Inheriting the vast mobile phone market, converged mobile devices ("smartphones") are poised to become the first ubiquitous personal computing platform. In this talk we detail our vision of the smartphone as a universal access control device—replacing physical keys, access tokens, etc.—and describe our efforts to address some of the technical challenges that stand in the way of this vision. Our discussion will focus on: techniques to prevent the misuse of a stolen device; novel user interfaces that aid in the secure use of such a device; and the design of an access control framework for the variety of authorization scenarios that such a device must accommodate. We also describe our efforts to deploy this technology in a testbed on the Carnegie Mellon campus.

C. Park and S. Chee (Eds.): ICISC 2004, LNCS 3506, p. 1, 2005.
© Springer-Verlag Berlin Heidelberg 2005

Security for Ubiquitous Computing

Frank Stajano

Computer Laboratory, University of Cambridge, UK
fms27@cam.ac.uk

Abstract. Ubiquitous computing, over a decade in the making, has finally graduated from whacky buzzword through fashionable research topic to something that is definitely and inevitably happening. This will mean revolutionary changes in the way computing affects our society: changes of the same magnitude and scope as those brought about by the World Wide Web. When throw-away computing capabilities are embedded in shoes, drink cans and postage stamps, security and privacy take on entirely new meanings. Programmers, engineers and system designers will have to learn to think in new ways. Ubiquitous computing is not just a wireless version of the Internet with a thousand times more computers, and it would be a naive mistake to imagine that the traditional security solutions for distributed systems will scale to the new scenario. Authentication, authorization, and even concepts as fundamental as ownership require thorough rethinking. At a higher level still, even goals and policies must be revised. One question we should keep asking is simply "Security for whom?" The owner of a device, for example, is no longer necessarily the party whose interests the device will attempt to safeguard. Ubiquitous computing is happening and will affect everyone. By itself it will never be "secure" (whatever this means) if not for the dedicated efforts of people like us who actually do the work. We are the ones who can make the difference. So, before focusing on the implementation details, let's have a serious look at the big picture.

C. Park and S. Chee (Eds.): ICISC 2004, LNCS 3506, p. 2, 2005.
© Springer-Verlag Berlin Heidelberg 2005

Algebraic Attacks on Combiners
with Memory and Several Outputs*

Nicolas T. Courtois

Axalto Cryptographic Research & Advanced Security, 36-38 rue de la Princesse,
BP 45, F-78430 Louveciennes Cedex, France
courtois@minrank.org

Abstract. Algebraic attacks on stream ciphers [14] recover the key by
solving an overdefined system of multivariate equations. Such attacks
can break many LFSR-based stream ciphers, when the output is obtained by a Boolean function, see [14, 15, 16]. Recently this approach has
been successfully extended also to combiners with memory, provided the
number of memory bits is small, see [1, 16, 2]. In [2] it is shown that, for
ciphers built with LFSRs and an arbitrary combiner using a subset of
k LFSR state bits, and with l inner state/memory bits, a polynomial
attack always do exist when k and l are fixed. Yet this attack becomes
very quickly impractical: already when k and l exceed about 4.

In this paper we give a simpler proof of this result from [2], and prove
a more general theorem. We show that much faster algebraic attacks
exist for any cipher that (in order to be fast) outputs several bits at a
time. In practice our result substantially reduces the complexity of the
best attack known on four well known constructions of stream ciphers
when the number of outputs is increased. We present interesting attacks
on modified versions of Snow, E0, LILI-128 and Turing ciphers.

Keywords: algebraic cryptanalysis, LFSR-based stream ciphers,
Boolean functions, combiners with memory, LILI-128, Turing cipher,
Snow, E0.

Note: An extended version is available at eprint.iacr.org/2003/125/.

1   Introduction

In this paper we study LFSR-based stream ciphers. In such ciphers there is an
inner state updated by an iterated linear function, and a stateful or stateless
nonlinear combiner that produces the output, given the inner state of the first
(linear) part. Our goal is to extend the recent very powerful and very general
algebraic attacks on stream ciphers to the case of combiners with several outputs.
Such constructions appear naturally if we want to construct ciphers being fast
in practice.

---

* Work supported by the French Ministry of Research RNRT Project "X-CRYPT".

C. Park and S. Chee (Eds.): ICISC 2004, LNCS 3506, pp. 3–20, 2005.

Up till recently, for stateless combiners - using a Boolean function - most general attacks known were so called correlation attacks, see for example [29, 23, 11, 8]. In order to resist such attacks, many authors focused on proposing Boolean functions that will have no good linear approximation and that will be correlation immune with regard to a subset of several input bits, see for example [11]. Unfortunately there is a tradeoff between these two properties. One of the proposed remedies is to use a stateful combiner. This idea is used in the Bluetooth wireless protocol cipher E0 [6]. Yet the simplicity of E0 made it vulnerable to advanced correlation attacks [25] and other attacks [2, 1, 16].

Recently the scope of application of the correlation attacks have been extended to consider higher degree correlation attacks with respect to non-linear low degree multivariate functions, or in other words, allowing to exploit low degree approximations [14]. The paper [14], proposes an algebraic approach to the cryptanalysis of stream ciphers. It will reduce the problem of key recovery, to solving an overdefined system of algebraic equations. Following [14] and [15], all LFSR-based stream ciphers are (potentially) vulnerable to algebraic attacks. The argument says that, if by some method, we are able to deduce from the output bit(s), only one multivariate equation of low degree in the LFSR state bits, then the same can (probably) be done for many other states. Each equation remains also linear with respect to any other LFSR state, and given many keystream bits, we inevitably obtain a very overdefined system of equations (i.e. many equations). Then we may apply the XL algorithm from Eurocrypt 2000 [35], adapted for this purpose in [14], or the simple linearization as in [15, 5], to efficiently solve the system.

In the paper [15], the scope of algebraic attacks is substantially extended, by showing new non-trivial methods to obtain low degree equations, that are not low degree approximations. This gives attacks that are not correlation attacks anymore, and are purely algebraic attacks on stream ciphers. The key ingredient is a simple but very powerful method to reduce the degree of the equations: instead of considering outputs as functions of inputs, one should rather study **algebraic relations** between the input and output bits. They turn out to have a substantially lower degree. The general idea of using multivariate algebraic relations in cryptanalysis of various public and secret key cryptosystems is not new and have been proposed (for very different purposes) by Patarin'95 [33], Jakobsen'98 [26], Courtois [13, 17, 18], and recently by Courtois-Pieprzyk in attempt to break AES [19].

In stream ciphers, this type of attacks have been proposed first in [15] and turn out to be quite powerful. In most cases, as already explained, due to the recursive structure of the cipher, finding just one such multivariate relation will give a polynomial attack on a stream cipher. Very surprisingly, this "multivariate relation" attack [15], extends also to combiners with memory, in particular when the number of possible inner states is small. This can be seen as an algebraic counterpart of previous results by Meier, Staffelbach and Golic on correlation attacks on combiners with one or a few memory bits [30, 24]. For algebraic attacks, the possibility of eliminating memory bits has been first suggested by Courtois and Meier in [15]. The heuristics of [15] only says that such attacks may exist, and exhibits also a counter-example for which the current method will fail to find a useful multivariate relation that would lead to an attack (cf. Section 7 of [15]). Yet, considering relations that imply potentially many output bits, seems very promising, except that finding useful relations becomes a hard problem (how to know which outputs will be used in the relation ?). The first attack of this type for a realistic cipher E0, has been found by careful elimination by hand, done by Armknecht [1]. A substantial speed-up is achieved with "Fast Algebraic Attacks" [16, 3, 28].

Even more surprisingly, Krause and Armknecht have recently proven a Theorem, to the effect that for **any** combiner with k inputs and l bits of memory, an algebraic attack of this type will always exist [2]. More precisely, they show that required multivariate relations do always exist with degree at most $\lceil k(l+1)/2 \rceil$. It generalises an earlier theorem due to Courtois and Meier, giving degree $\lceil k/2 \rceil$ for $l = 0$, published in [15].

With this bound on the degree from [2], starting from about $l = 4$ memory bits algebraic attacks will quickly become quite impractical. In this paper we will give a new, much simpler proof of this theorem, and we will present a **much more general** theorem, for combiners that use several outputs instead of one. For correlation attacks, this issue has been studied in [38, 8]. For algebraic attacks, we will show that having several outputs allows to substantially lower the degree of the relations, which in turn will dramatically decrease the complexity of an algebraic attack on most LFSR-based stream ciphers. Our new theorem will also give new results for combiners without memory (i.e. using just Boolean functions).

2 Notation

We consider stream ciphers in which there is a state with a linear feedback function (for example composed of one or several LFSRs). Let $K = (K_0, \ldots, K_{n-1})$ be an n-bit secret key. Let $s = K$ be the initial state of the LFSR or the linear part of the cipher. At each clock $t = 0, 1, 2, \ldots$, the new state of the linear part is computed as $s \leftarrow L(s)$, with L being some multivariate linear transformation, for example corresponding to the connection polynomial of an LFSR, a combination of several parallel LFSRs, or a linear cellular automaton. We assume that L is public.

Only k out of n bits of the linear part of the cipher are used in the next part of the cipher called the combiner. The combiner has k inputs, m outputs and l internal memory bits. At each clock $t = 0, 1, 2, \ldots$, the combiner outputs m bits $y_0^{(t)}, \ldots, y_{m-1}^{(t)}$, for $t = 0, 1, 2, \ldots$. These output bits depend deterministically on the k input bits $x_0^{(t)}, \ldots, x_{k-1}^{(t)}$ and on internal memory bits that before and at the time t are $a_0^{(t-1)}, \ldots, a_{l-1}^{(t-1)}$. In all generality, the second component is described as a pair of functions $F = (F_1, F_2) : GF(2)^{n+l} \to GF(2)^{m+l}$, that given the current state and the input, compute the next state and the output:

$$F : \begin{cases} (y_0^{(t)}, \ldots, y_{m-1}^{(t)}) = F_1(x_0^{(t)}, \ldots, x_{k-1}^{(t)}, a_0^{(t-1)}, \ldots, a_{l-1}^{(t-1)}) \\ (a_0^{(t)}, \ldots, a_{l-1}^{(t)}) = F_2(x_0^{(t)}, \ldots, x_{k-1}^{(t)}, a_0^{(t-1)}, \ldots, a_{l-1}^{(t-1)}) \end{cases}$$

The initial inner state is $a^{(-1)}$, exists before $t = 0$, and can be anything (it is and remains unknown, the goal of the attacks being to eliminate all the monomials in the a_i).

3   Algebraic Attacks on Stream Ciphers

This Section summarizes the general idea of algebraic attacks on stream ciphers from [14], greatly extended and developed in [15].

We recall that the linear part of our cipher (a combination of one or several binary LFSRs) is composed of n bits $s_0, \ldots, s_{n-1}$. At the beginning $s = K$ (the initial LFSR state) and at each clock of the cipher, it is updated as $s \leftarrow L(s)$, with L being some known multivariate linear transformation. The general algebraic attack on such stream ciphers, following closely [15] or [16], works as follows:

- Find (by some method that is very different for each cipher) one (at least, but one is enough) multivariate relation Q of low degree d between the LFSR state bits and some M following outputs, for example:

$$Q(s_0, s_1, \ldots, s_{n-1}, \ y^{(0)}, \ldots, y^{(M-1)}) = 0$$

- The same equation will apply to all consecutive windows of M states

$$Q([L^t(K)]_0, [L^t(K)]_1, \ldots, [L^t(K)]_{n-1}, \ y^{(t)}, \ldots, y^{(t+M-1)}) = 0$$

- The $y^{(t)}, \ldots, y^{(t+M-1)}$ are replaced by their values known from the observed output of the cipher.
- For each keystream bit, we get a multivariate equation of degree k in the x_i.
- Due to the linearity of L, for any t, the degree of these equations is still d.
- Given many keystream bits, we inevitably obtain a very overdefined system of equations (i.e. great many multivariate equations of degree d in the K_i).
- Then we may apply the XL algorithm from Eurocrypt 2000 [35], adapted to equations of degree higher than 2 in [14]. Better results should be obtained with modern Gröbner bases techniques, such as the F5 algorithm [22].
- If we dispose of a sufficient amount of keystream, (which is frequently not very big, see [15]), the XL algorithm is not necessary and may be replaced by the so called linearization method that is particularly simple. There are about $T \approx \binom{n}{d}$ monomials of degree $\leq d$ in the n variables K_i (assuming $d \leq n/2$). We consider each of these monomials as a new variable V_j. Given about $\binom{n}{d} + M$ keystream bits, and therefore $R = \binom{n}{d}$ equations on successive windows of M bits, we get a system of $R \geq T$ linear equations with $T = \binom{n}{d}$ variables V_i that can be easily solved by Gaussian elimination on a linear system of size T.

- In theory, the Gaussian elimination takes time T^ω with $\omega \leq 2.376$ [12]. However the fastest practical algorithm we are aware of, is Strassen's algorithm [37] that requires about $7 \cdot T^{log_2 7}$ operations. Since our basic operations are over $GF(2)$, we expect that a careful bitslice implementation of this algorithm on a modern CPU can handle 64 such operations in one single CPU clock. Thus, in all numerical complexity results given in this paper we will give $7/64 \cdot T^{log_2 7}$ as an estimation of the number of CPU clocks necessary in the attack.

4 The Proof Method

Our general Theorem 5.1, given later, considers arbitrary combiners with k input bits, l memory bits, and m output bits and shows the existence of equations of some degree that lead to an algebraic attack. It generalises the main result of [2] for arbitrary combiners with one output, i.e. with $m = 1$, which in turn generalises a result obtained in [15] for memoryless combiners with single output, i.e. for $m = 1$ and $l = 0$. Our proof technique is very different than in [2] and is very similar to one used in [15].

In this section, in order to illustrate the simplicity of our proof technique, we will first prove the following theorem for combiners with $m = 1$ and $l = 1$, that is in fact a special case of both our general Theorem 5.1 given later, and of the main theorem of [2].

Theorem 4.1 (Special Case of Krause-Armknecht Theorem).
Let F be an arbitrary fixed circuit/component with k binary inputs x_i, one bit of memory a, and one output y. (The output and the next state of the memory bit a, depend in an arbitrary way (but deterministically) on the k inputs and the previous memory bit.)

Then, given $M = 2$ consecutive states $(t, t+1)$, there is a multivariate equation R of degree k in the $x_j^{(i)}$, that relates only the input and the output bits, **without** any of the inner state/memory bits $a^{(t-1)}, a^{(t)}$, as follows:

$$R\left(x_0^{(t)}, \ldots, x_{k-1}^{(t)}; \ x_0^{(t+1)}, \ldots, x_{k-1}^{(t+1)}; \ y^{(t)}, y^{(t+1)}\right) = 0.$$

Remark: In this and later theorems, we will only limit the degree of the equations in the $x_j^{(i)}$. The degree in the $y_j^{(i)}$ is not important, as in an attack these values will be fixed.

Proof: We consider $2k$ variables as follows: $x_0^{(t)}, \ldots, x_{k-1}^{(t)}, x_0^{(t+1)}, \ldots, x_{k-1}^{(t+1)}$. We know that the following two memory bits $a^{(t)}$ and $a^{(t+1)}$ and the two outputs $y^{(t)}, y^{(t+1)}$, do depend only on these $2k$ variables, plus additionally on the bit $a^{(t-1)}$ present in memory at the beginning. Thus, the four values $a^{(t)}$, $a^{(t+1)}$, $y^{(t)}$ and $y^{(t+1)}$, do depend deterministically only on the $2k + 1$ variables $x_0^{(t)}, \ldots, x_{k-1}^{(t+1)}$ and $a^{(t-1)}$. This is summarised on the following picture:

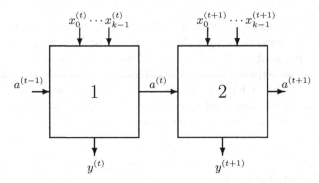

Fig. 1. Two successive applications of a combiner with k inputs, 1 output and 1 memory bit

We define the following set of monomials A: we consider all the monomials of degree up to k in the following $2k$ variables: the $x_i^{(t)}$ together with the $x_j^{(t+1)}$. The size of A is exactly $\sum_{i=0}^{k} \binom{2k}{i} = 2^{2k-1} + \frac{1}{2}\binom{2k}{k}$, which is strictly greater than 2^{2k-1}.

Now we will create the following matrix:

- Lines are all the possible values for $x_0^{(t)}, \ldots, x_{k-1}^{(t)}, x_0^{(t+1)}, \ldots, x_{k-1}^{(t+1)}$ and for the memory bit $a^{(t-1)}$. There are 2^{2k+1} lines.
- The columns correspond to products of successive monomials of A, multiplied by any out of the 4 possible monomials in the two variables $y^{(t)}, y^{(t+1)}$. There are $4 \cdot |A| = 2^{2k+1} + 2\binom{2k}{k} > 2^{2k+1}$ columns.
- Each entry in the matrix is the value $\in \{0,1\}$ of the column monomial in the case corresponding to the current line.

The number of columns is strictly greater than the number of lines. Therefore one column must be a linear combination of other columns. Since columns are products of monomials, and all the cases are treated, this gives a multivariate equation, true with probability 1, for all possible entries and whatever is the initial value of $a^{(t-1)}$. By construction, it does not involve memory bits $a^{(i)}$. This ends the proof of Theorem 4.1.   □

Remark 1: It can be seen that there are at least $2\binom{2k}{k} - 1$ such equations, which could greatly reduce the keystream requirements of some attacks. For simplicity we do not exploit this in the present paper.

Remark 2: In the extended version of this paper, we give another proof of this Theorem, in which the result is stronger and there is much less monomials present.

5   New General Result on Combiners with Memory

Similarly we prove our main result that extends the main theorem of [2].

Theorem 5.1 (Our Key Theorem).
Let F be an arbitrary fixed circuit/component with k binary inputs x_i, l bits of memory a_i, and m outputs y_i. Let d and M be two integers such that:

$$2^{Mm} \cdot \sum_{i=0}^{d} \binom{Mk}{i} > 2^{Mk+l} \qquad \textbf{(KE)}$$

Then, considering M consecutive steps/states $(t, \ldots t + M - 1)$, there is a multivariate equation (and relation) R of degree d in the $x_j^{(i)}$, relating[1] the input and the output bits for these states

$$R\left(x_0^{(t)}, \ldots, x_{k-1}^{(t)}, \ldots, x_0^{(t+M-1)}, \ldots, x_{k-1}^{(t+M-1)};\right.$$

$$\left. y_0^{(t)}, \ldots, y_{m-1}^{(t)}, \ldots y_0^{(t+M-1)}, \ldots, y_{m-1}^{(t+M-1)}\right) = 0.$$

Proof: Our proof is very similar as in the special case above (Theorem 4.1), and also gives a new, much simpler proof of the original (but less general) result of [2].

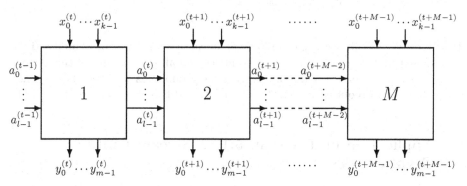

Fig. 2. M successive applications of a combiner with k inputs, m outputs and l bits of memory

We start with the following (cf. Fig. 2):

- We have $M \cdot m$ output bits: $y_0^{(t)}, \ldots, y_{m-1}^{(t)}; \ldots; y_0^{(t+M-1)}, \ldots, y_{m-1}^{(t+M-1)}$
- The total of $M \cdot k$ input bits, $x_0^{(t)}, \ldots, x_{k-1}^{(t)}; \ldots; x_0^{(t+M-1)}, \ldots, x_{k-1}^{(t+M-1)}$.
- We have l initial memory bits, $a_0^{(t-1)}, \ldots, a_{l-1}^{(t-1)}$.
- In all we have $l + Mk$ input variables. The memory bits for second and following inner states, $a_j^{(t+i)}$, $0 < i < M$ do depend only on these $l + Mk$ variables.
- Thus, for our M consecutive steps/states $t, \ldots, t + M - 1$, all the outputs $y_j^{(t+i)}$, $i < M$ do depend deterministically only on the $l + Mk$ variables listed above.

---

[1] Again, **without** any of the inner state/memory bits $a_j^{(i)}$.

We define the following set of monomials A: we consider all the monomials of degree up to d in all the Mk variables $x_i^{(t+i)}$. The size of A is exactly $\sum_{i=0}^{d} \binom{Mk}{i}$. Now we will create the following matrix:

- Lines are all the possibilities for the $l + Mk$ input variables. There are 2^{Mk+l} lines.
- The columns are all products of monomials of A, multiplied by any of the possible monomials in the $y_j^{(t+i)}$. There are $2^{Mm} \cdot |A| = 2^{Mm} \cdot \sum_{i=0}^{d} \binom{Mk}{i}$ columns.
- Each entry in the matrix is the value $\in \{0,1\}$ of the column monomial in the case corresponding to the current line.

The key argument is the same as before. The number of columns in our matrix should be strictly greater than the number of lines, and the requirement to achieve this, is precisely our previous assumption:

$$2^{Mm} \cdot |A| = 2^{Mm} \cdot \sum_{i=0}^{d} \binom{Mk}{i} > 2^{Mk+l}$$

Therefore we get at least one non-trivial linear combination of columns (i.e. monomials) that is zero, for all possible entries and all possible initial (memory) states. This multivariate equation is (with the monomials we have chosen) exactly of the form required by our Theorem 5.1, and this ends the proof.   □

6   Application of Theorem 5.1 to Stream Cipher Cryptanalysis

Theorem 5.1 and other results of this paper, allow to find equations and execute the algebraic attack described in Section 3. In some cases this Theorem would work even when $d = 0$, when other variables are such that (KE) holds, but the equations of degree 0 in the $x_j^{(i)}$ will only contain the $y_j^{(i)}$, and cannot be used to recover the secret key of a cipher (though can probably be exploited to predict the future keystream). For simplicity, in this paper we will always apply Theorem 5.1 for $d \geq 1$.

6.1   The Complexity of the Attacks Based on Theorem 5.1

Our algebraic attack on stream ciphers has two main steps:

Step 1. Find the equations by Gaussian reduction on the matrix given in the proof of the Theorem. This step requires about $2^{\omega(Mk+l)}$ computations.

Step 2. From the Step 1. for each keystream bit, we get one equation of degree d in the $x_j^{(i)}$ (with $d \geq 1$). The $x_j^{(i)}$ are known linear combinations of the key bits K_i and these equations are also of degree d in the key bits. When the $x_j^{(i)}$ are replaced by their actual values obtained from the keystream, we get multivariate equations that only contain monomials of degree d in the key bits K_i. Then, given about $T = \binom{n}{d}$ keystream bits, we solve these equations by linearization in about $T^\omega \approx 2^{\omega d \log n}$ computations.

In some cases (when M is small), the complexity of the first step may be negligible compared to the second step (cf. Section 7.5 and examples in Table 2). In some cases the complexity of the first step may always be very large (examples in Table 3). In other cases there will be a tradeoff between the complexity of the two steps, see Section 7.6.

Remark: The complexity of replacing the $x_j^{(i)}$ in the equations of Step 1., by the relevant (known) linear combinations of the key bits K_i (cf. Section 3) has been neglected for simplicity (it can be seen to be smaller than the maximum of complexities given above).

6.2   Important Remark

It is important to understand that, in general, this Theorem 5.1 does not show that the algebraic attack will always work. There are some (very special) cases in which it will not work as well as expected from our Theorem 5.1. We will see this on an example.

Assume that we have a component that has $m = 10$ outputs, and we artificially add 10 more outputs computed as some 10 Boolean functions of the "real" outputs:

$$(y_{10}, \ldots, y_{19}) = (F_{10}(y_0, \ldots, y_9), \ldots, F_{19}(y_0, \ldots, y_9)).$$

Now we have (in theory) $l = 20$, and from the formula (KE) we see easily that in most cases our Theorem 5.1 will give for $m = 20$ equations of substantially lower degree than for $m = 10$. These equations are real (their existence is proven). Yet these equations will not be useful in an attack. For example there will be equations such as $y_{10} = F_{10}(y_0, \ldots, y_9)$, and a great many of derived equations: different linear combinations of these equations multiplied by many different monomials. All these equations are in a sense "artificial" and unfortunately they will all reduce to 0 later in the attack, after when the $y_0, \ldots, y_{19}$ are replaced by their values obtained from the output of the cipher.

This example shows that in some very special cases, the algebraic attack will probably not work for the degree given by our Theorem 5.1. Yet, it will probably work perfectly well for the degree corresponding to the "real" value of $l = 10$. It is conjectured that when the output bits are fully independent and not related by some algebraic relation, and if the output takes all the possible 2^m values, the attack should always work, for **every** equation obtained from the above Theorem 5.1. Moreover, in practice, the difference between the number of lines and the number of columns, in the matrix (the one we generated to prove the theorem) will be big, and there will be not only one but, (for example) thousands of equations obtained. The chances that the attack would not work for all of them, are negligible.

7   How to Choose Parameters in Theorem 5.1

In the previous Section 6 we showed that it is straightforward to use Theorem 5.1 to design an algebraic attack on stream ciphers following Section 3. Another question is to choose parameters in such a way that the complexity of the attack will be optimal. For this we need to study the behaviour of the key inequality (KE): $2^{Mm} \cdot \sum_{i=0}^{d} \binom{Mk}{i} > 2^{Mk+l}$.

In order to minimise the complexity of Step 2. of the attack (cf. Section 6.1) we simply need to choose M that gives the smallest possible d. Yet, as we will see later (in particular when $m \geq k$, cf. Section 7.6) things are not always as simple to optimise the Step 1.

7.1   Asymptotic Behaviour of (KE) and Theorem 5.1

In order choose (M, d) that satisfy (KE): $2^{Mm} \cdot \sum_{i=0}^{d} \binom{Mk}{i} > 2^{Mk+l}$ we have two cases:

A. If $m < k$, when $M \to \infty$ we have no hope to satisfy the key inequality (KE). In this case we conjecture that the best attack (and the smallest degree d) will be achieved taking M as small as possible (or close to it). This case is studied in Section 7.5.

B. If $m \geq k$ then when $M \to \infty$ we can always satisfy the key inequality (KE). In this case we should take M as big as possible, but not too big because the complexity to find the equations required by the attack (Step 1. cf. Section 6.1) could become bigger than the complexity of the attack itself (Step 2.). This case is studied in Section 7.6.

Remark: For the (less general) theorem from [2], there is only the case A., because $m = 1$.

7.2   Necessary Condition for (KE) and Theorem 5.1

We want to solve (KE) given the values m and l. Since one always has $\sum_{i=0}^{d} \binom{Mk}{i} \leq 2^{Mk}$, we cannot have $Mm \leq l$, and this gives **a necessary condition** $Mm > l$, hence $Mm \geq l+1$ which gives

$$M \geq \lceil (l+1)/m \rceil. \qquad (C)$$

7.3   Sufficient Conditions for (KE) and Theorem 5.1

Conversely, it is easy to see that, each time $M \geq \lceil (l+1)/m \rceil$, we have $Mm \geq l+1$, and the formula (KE) will be satisfied for some $d \leq Mk$.

Sufficient Condition 1: For any given values m and l, and for any $M \geq \lceil (l+1)/m \rceil$, the formula (KE) will be satisfied by some d being at most $d \leq Mk$.

When the minimum $M = \lceil (l+1)/m \rceil$ is chosen, we can use $d = k \cdot \lceil (l+1)/m \rceil$, but in fact one can do better. A smaller d can be achieved for this same (minimal)

M. Indeed, since M is an integer, the minimal value of M does not imply that we need to take a maximal value for d. From (KE) we get the following condition:

$$\sum_{i=0}^{d} \binom{Mk}{i} > 2^{Mk} \cdot 2^{l-m \cdot \lceil (l+1)/m \rceil}$$

It can be seen that $d = \lceil kM/2 \rceil = \lceil k \lceil (l+1)/m \rceil /2 \rceil$ is always sufficient. Indeed we always have:

$$\sum_{i=0}^{d} \binom{Mk}{i} > 2^{Mk}/2.$$

And we also always have:

$$\frac{1}{2} \geq 2^{l-m \cdot \lceil (l+1)/m \rceil}.$$

Sufficient Condition 2: From the above, we get immediately the following Theorem:

Theorem 7.4 (Generalised Krause-Armknecht Theorem).
Let F be an arbitrary fixed circuit/component with k binary inputs, l bits of memory, and m outputs. Then, considering $M = \lceil (l+1)/m \rceil$ consecutive steps/states $(t, \ldots t + M - 1)$, there is a multivariate relation, involving only the input bits (the $x_j^{(i)}$) and the output bits (the $y_j^{(i)}$) for these states, and with degree $\lceil kM/2 \rceil = \lceil k \lceil (l+1)/m \rceil /2 \rceil$ in the $x_j^{(i)}$.

Remark: If we put $m = 1$ in this Theorem 7.4 (1 output bit), we obtain exactly the main result of [2]. This in turn generalises the theorem given in [15], which is exactly the above result with $m = 1$ and $l = 0$, i.e. the case of Boolean functions that are memoryless combiners with 1 output bit.

7.5   How to Use Theorem 5.1 When $m < k$

All the remarks above are true both for $m < k$ and for $m \geq k$, however we expect that (cf. Section 7.1) choosing the smallest possible M should be optimal (or close to optimal) only when $m < k$. In some cases, the choice of Theorem 7.4 above: $M = \lceil (l+1)/m \rceil$ and $d = \lceil kM/2 \rceil$ will be optimal for Theorem 5.1. However in most cases, there will be a non-zero difference between $M = \lceil (l+1)/m \rceil = 1$ and $(l+1)/m$ that will imply that $\frac{1}{2} \gg 2^{l-m \cdot \lceil (l+1)/m \rceil}$ in the derivation of Theorem 7.4 above. In such cases, it seems that the best method[2] is to take still $M = \lceil (l+1)/m \rceil$ (or very close to this) and try to the lowest d that satisfies the key requirement of Theorem 5.1 which is $2^{Mm} \sum_{i=0}^{d} \binom{Mk}{i} > 2^{Mk+l}$.

---

[2] Again when $m < k$, if in similar case $m \geq k$, it could be even better to increase M, cf. Section 7.6.

The Complexity of the Attacks Based on Theorem 7.4

Let $d = \lceil k \lceil (l+1)/m \rceil /2 \rceil$ be the degree obtained in Theorem 7.4. Following Section 6.1, the complexity of the first step of the attack (to find the equations) will be about $2^{\omega(Mk+l)} = 2^{\omega(k\lceil (l+1)/m \rceil + l)}$ and this is roughly $\left(2^{\omega(d/2+l)}\right)$. For the second step the complexity will be about $\binom{n}{d}^{\omega} \approx n^{d\omega}$ (see Section 3). Though this d is not always the best degree we will get and use in an attack, we expect that when $m < k$ the complexity of the first step of the attack will frequently be substantially smaller than for the second step (cf. examples in Table 2).

7.6   How to Use Theorem 5.1 When $m \geq k$

If $m \geq k$, then when $M \to \infty$ we can always satisfy the key inequality (KE).

$$2^{Mm} \cdot \sum_{i=0}^{d} \binom{Mk}{i} > 2^{Mk+l} \qquad (KE)$$

This fact is obvious when $m > k$ and still true when $m = k$, because then it is sufficient to take $M = \lceil (l+1)/k \rceil$ and $d = Mk$. (Remark: here M cannot be smaller than $\lceil (l+1)/k \rceil$ because following Section 7.2, $M \geq \lceil (l+1)/m \rceil$ and here it is equal to $\lceil (l+1)/k \rceil$.)

It can be seen that in all cases when $m \geq k$, when $M \to \infty$, then d may be an arbitrarily small integer > 0 (i.e. we will even get $d = 1$ when M is large enough).

In practice, we should take M as big as possible, but not too big because the complexity to find the equations (Step 1 of the attack) will become too big: it is following Section 6.1 about $2^{\omega(Mk+l)}$ computations. (While Step 2. requires about $\binom{n}{d}^{\omega} \approx 2^{\omega \log_2 nd}/d!$)

In order to get the best attack, we need to minimise $2^{\omega(Mk+l)} + 2^{\omega \log_2 nd}/d!$ under the condition $\binom{Mk}{d} > 2^{M(k-m)+l}$. The behaviour of these complexities is not simple, because $M \geq \lceil (l+1)/m \rceil$ and must be an integer. Our experience shows that sometimes $M = \lceil (l+1)/m \rceil$ is optimal, sometimes it isn't. Sometimes the best attack will be when both complexities are about equal, sometimes the first step will always take much more time than the second step (even for the minimal $M = \lceil (l+1)/m \rceil$). Some relevant examples are given in Table 3 and Table 2.

7.7   Summary or How to Design the Best Algebraic Attack

In order to find the fastest attack with Theorem 5.1, we recommend to proceed as follows:

- First we try to apply Theorem 7.4, and get a (working) solution (M, d).
- Then with same M, take the lowest d such that the key condition (KE) still holds.
- In addition, when $m \geq k$, as long as the complexity of the first step of the attack is less than the complexity of the second step, we may try to increase M, compute the lower possible d, and see if we get a better result (Cf. Section 7.6).

8   Application to Some Known Stream Cipher Constructions

8.1   Application to Modified LILI-128

Our attack can be applied to the second component of LILI-128 cipher [36]: we have an LFSR with $n = 89$ bits, and a Boolean function with $k = 10$ inputs. There is no memory bits ($m = 0$). In [15], a generic attack on LILI-128 is given, that requires $n^{5\omega}$ computations, (whatever is the Boolean function used). From our Theorem 5.1 we see that if in LILI we use simultaneously several Boolean functions, the complexity of the generic attack will substantially decrease. It will be $\binom{n}{d}^{\omega}$ with d given by Theorem 5.1. The resulting degree d quickly decreases with m:

m	1	2	3	5	7
d	5	4	3	2	1

Following closely [15], each of these attacks on the second component of LILI-128 can be transformed into an attack on the whole LILI-128 cipher in two possible ways. Either (A:) the complexity is multiplied by 2^{39} (one needs to guess the 39-bit state of LFSR in the clocking component), or (B:) the keystream requirements are multiplied by about 2^{39} (at each step the first component is clocked $2^{39} - 1$ times). See [15] for more details. This gives the following generic attack on modified LILI-128 with several outputs:

Table 1. Generic attacks on modified LILI-128 with m outputs

m	1		2		3		5		7	
M	1		1		1		1		1	
d	5		4		3		2		1	
keystream	2^{25}	2^{64}	2^{21}	2^{60}	2^{17}	2^{56}	2^{12}	2^{51}	$\mathbf{2^{6}}$	2^{45}
time(Step 1.)	2^{25}	2^{25}	2^{25}	2^{25}	2^{25}	2^{25}	2^{25}	2^{25}	2^{25}	2^{25}
time(Step 2.)	2^{107}	2^{68}	2^{95}	2^{56}	2^{83}	2^{44}	2^{69}	2^{30}	$\mathbf{2^{54}}$	2^{15}

We see that for ciphers that combine LFSR and Boolean functions, such as LILI-128, if we replace a Boolean function by a component that outputs a few bits at a time, the security will be dramatically reduced, and this for any component (worst case).

Note: There are attacks on LILI-128 itself, that are faster than the generic attack given here for $m = 1$, see [15, 16]. However for some of the modified versions of LILI-128 with many outputs, our attack will probably be the fastest general attack known on such ciphers.

8.2   Application to Modified E0

For the basic component of the stream cipher E0, we have $n = 128$, $k = 4$, $l = 4$, $m = 1$. The Krause-Armknecht theorem gives $d = 10$, see [2]. With our Theorem 5.1 we get the following results:

Table 2. Generic attacks on modified E0 with m outputs

m	1	2	3	4	5	6
M	5	3	2	3	1	1
d	10	5	3	2	2	1
keystream	2^{48}	2^{28}	2^{18}	2^{13}	2^{13}	2^{7}
time(Step 1.)	2^{64}	2^{42}	2^{30}	2^{30}	2^{19}	2^{19}
time(Step 2.)	2^{131}	2^{76}	2^{49}	2^{33}	2^{33}	2^{16}

We see that for ciphers that combine LFSRs and a combiner with 4 inputs, and 4 memory bits, such as E0, if one outputs several bits at a time (computed in an arbitrary way), the complexity of the attack and the keystream amount required dramatically decreases.

Note: Here we treat the worst case by a generic method, for E0 itself there are attacks faster than what we get for $m = 1$, see [2, 1, 16]. However most of modified versions of E0 with many outputs, our attack is probably the fastest attack known.

8.3   Application to Snow and Modified Versions of Snow

We consider both Snow and Snow 2.0. that have an LFSR with $n = 512$ bits that is connected to a stateful combiner that outputs $m = 32$ bits at a time. We obtain:

1. In Snow 1.0. we have $k = 64$, $l = 64$ and $m = 32$. With Theorem 7.4 we get $M = \lceil (l+1)/m \rceil = 3$ and $d = \lceil kM/2 \rceil = 96$ that can be lowered to $d = 54$ and still satisfies the requirements of the Theorem 5.1 (for reasons explained in Section 7.5).
2. Similarly, in Snow 2.0. we have $k = 96$, $l = 64$ and $m = 32$. With Theorem 7.4 we get $M = \lceil (l+1)/m \rceil = 3$ and $d = \lceil kM/2 \rceil = 144$ that can be lowered to $d = 92$.

These degrees are by far too large to hope for practical attacks on Snow.

Algebraic Attacks on Modified Snow

We will look how the complexity of the attack on Snow 1.0. and 2.0. when the number of output bits increases. This could arise if, in order to build a faster cipher, we add to Snow some **arbitrary** S-boxes or Boolean functions that derive some additional output bits, from the k inputs and the l memory bits of Snow combiner. Since the size of LFSR is 512 bits, an attack will be considered significant if it takes less than 2^{512}. (We study academic attacks on modified Snow, and do not claim to break the actual Snow in which the key is expanded from a shorter key of 128 or 256 bits.)

We see that when the number of outputs increases, the security of the cipher collapses. The complexity of the first step of the attack may be $< 2^{512}$ but

Table 3. Generic attacks on modified Snow ciphers with m outputs

Snow 1.0. $n = 512, l = 64, k = 64$							Snow 2.0. $n = 512, l = 64, k = 96$					
m	32	64	65	80	100	120	m	32	64	65	120	150
M	3	2	1	1	1	1	M	3	2	1	1	1
d	54	16	32	16	7	2	d	92	35	48	9	2
keystream	2^{245}	2^{99}	2^{169}	2^{99}	2^{51}	2^{17}	keystream	2^{344}	2^{352}	2^{226}	2^{62}	2^{17}
time(Step 1.)	2^{715}	2^{536}	2^{356}	2^{356}	2^{356}	2^{356}	time(Step 1.)	2^{985}	2^{715}	2^{446}	2^{446}	2^{446}
time(Step 2.)	2^{684}	2^{276}	2^{471}	2^{276}	2^{139}	$\mathbf{2^{45}}$	time(Step 2.)	2^{962}	2^{503}	2^{631}	2^{172}	$\mathbf{2^{45}}$

remains very high. However, one should not think that Snow with added outputs will be very secure: we only gave here the complexity of the generic method to find a useful equation. For a specific cipher, in many cases, there could be a much faster method that exploits the description of the cipher.

Note: Our attacks are very general. For the original cipher Snow 1.0. itself, much faster attacks are known, see [10, 7, 27].

8.4 Application to a Modified Turing Cipher

Turing is a stream cipher proposed in 2003 by Rose and Hawkes [34]. It is a new kind of stream cipher which outputs many bits at a time, and in which the combiner is key-dependent. We have $n = 17 * 32 = 544$, $m = 5 * 32 = 160$, $l = 0$ (no memory), $k = 9 * 32 = 288$. These values are very large and any attack faster than the exhaustive search of all possible states $2^n = 2^{544}$ should be considered as interesting.

We will study a modified version of Turing, in which the combiner is NOT key-dependent. Then with Theorem 7.4 we get $M = \lceil (l+1)/m \rceil = 1$ and $d = \lceil kM/2 \rceil = 160$ that can be lowered to $d = 37$ and still satisfies the requirements of the Theorem 5.1 (for reasons explained in Section 7.5). This degree $d = 37$ is still by far too large to give any hope for practical attacks on Turing. We get an attack on modified Turing with time(Step 1.) $= 2^{805}$ and time(Step 2.) $= 2^{534}$. The second step is faster then the exhaustive search which would be in 2^{544}. The first step can also probably be improved to be faster than 2^{544}.

9   Extension to Ciphers with Unknown or Key Dependent Combiners

The results of this paper can be also applied to ciphers in which the combiner is only partially known (or key dependent). For example, at some place inside the combiner we XOR the data with the secret key. Or, we use Boolean functions with some coefficients being unknown or key-dependent. Let l' be the total number of unknown bits in the combiner with parameters (k, l, m). Then we may just consider this cipher as a cipher in which the combiner is known with parameters

$(k, l + l', m)$: we just have l' additional memory bits that are not updated, they remain the same all the time. All the attacks described in the present paper will apply and when l' is not too big, or when $m \geq k$, they should be (relatively) efficient general attacks on stream ciphers.

10   Conclusion

In this paper we studied generic algebraic attacks on stream ciphers built with an LFSR and a combiner having a small number of memory bits. Our main result is that the complexity of algebraic attacks on stream ciphers will substantially decrease if the cipher outputs more bits at a time. We substantially extended and gave a much simpler proof of the important Theorem of [2]. Our new Theorem can be applied to substantially decrease the complexity of the best worse-case (generic) algebraic attack (whatever is the internal structure of the combiner component) for modified versions of four well known stream ciphers E0, LILI-128, Snow and Turing.

Acknowledgements. Many thanks to the reviewers of Crypto 2004, SAC 2004 and ICISC 2004 for careful reading and valuable comments.

References

1. Frederik Armknecht: *A Linearization Attack on the Bluetooth Key Stream Generator,* Available on http://eprint.iacr.org/2002/191/. 13 December 2002
2. Frederik Armknecht, Matthias Krause: *Algebraic Atacks on Combiners with Memory,* Crypto 2003, LNCS 2729, pp. 162-176, Springer.
3. Frederik Armknecht: *Improving Fast Algebraic Attacks,* FSE 2004, LNCS, 2004.
4. Ross Anderson: *Searching for the Optimum Correlation Attack,* FSE'94, LNCS 1008, Springer, pp 137-143.
5. Elad Barkan, Eli Biham, and Nathan Keller: *Instant Ciphertext-Only Cryptanalysis of GSM Encrypted Communication,* Crypto 2003, LNCS 2729, pp. 600-616.
6. Bluetooth CIG, Specification of the Bluetooth system, Version 1.1, February 22 2001, available from www.bluetooth.com.
7. Christophe De Canniere, *Guess and Determine Attack on SNOW,* Nessie public report, 12/11/2001, NES/DOC/KUL/WP5/011/a, available from www.cryptonessie.org.
8. Claude Carlet, Emmanuel Prouff: *On a new notion of nonlinearity relevant to multi-output pseudo-random generators,* SAC 2003, LNCS 3006, pp. 291-305.
9. Will Meier, Enes Pasalic and Claude Carlet: *Algebraic Attacks and Decomposition of Boolean Functions,* In Eurocrypt 2004, pp. 474-491, LNCS 3027, Springer, 2004.
10. Don Coppersmith, Shai Halevi and Charanjit Jutla, Cryptanalysis of stream ciphers with linear masking, Crypto 2002, LNCS 2442, Springer, 2002. Available at http://eprint.iacr.org/2002/020/
11. Paul Camion, Claude Carlet, Pascale Charpin and Nicolas Sendrier, *On Correlation-immune Functions,* Crypto'91, LNCS 576, Springer, pp. 86-100.

12. Don Coppersmith, Shmuel Winograd: *Matrix multiplication via arithmetic progressions*, J. Symbolic Computation (1990), 9, pp. 251-280.
13. Nicolas Courtois: *The security of Hidden Field Equations (HFE)*, Cryptographers' Track Rsa Conference 2001, LNCS 2020, Springer, pp. 266-281.
14. Nicolas Courtois: *Higher Order Correlation Attacks, XL algorithm and Cryptanalysis of Toyocrypt*, ICISC 2002, LNCS 2587, pp. 182-199, Springer.
15. Nicolas Courtois and Willi Meier: *Algebraic Attacks on Stream Ciphers with Linear Feedback*, Eurocrypt 2003, Warsaw, Poland, LNCS 2656, pp. 345-359, Springer. An extended version is available at http://www.minrank.org/toyolili.pdf
16. Nicolas Courtois: *Fast Algebraic Attacks on Stream Ciphers with Linear Feedback*, Crypto 2003, LNCS 2729, pp: 177-194, Springer.
17. Nicolas Courtois: *The Inverse S-box, Non-linear Polynomial Relations and Cryptanalysis of Block Ciphers*, in AES 4 Conference, LNCS, Springer.
18. Nicolas Courtois: *General Principles of Algebraic Attacks and New Design Criteria for Components of Symmetric Ciphers*, in AES 4 Conference, LNCS, Springer.
19. Nicolas Courtois and Josef Pieprzyk, *Cryptanalysis of Block Ciphers with Overdefined Systems of Equations*, Asiacrypt 2002, LNCS 2501, pp.267-287, Springer.
20. Patrik Ekdahl, Thomas Johansson, *SNOW - a new stream cipher*, Proceedings of First NESSIE Workshop, Heverlee, Belgium, 2000.
21. Patrik Ekdahl, Thomas Johansson, *A new version of the stream cipher SNOW*, in SAC 2002, LNCS 2595, Springer, pp. 47-61. Available from http://www.it.lth.se/cryptology/snow/
22. Jean-Charles Faugère: "A new efficient algorithm for computing Gröbner bases without reduction to zero (F5)" Workshop on Applications of Commutative Algebra, Catania, Italy, ACM Press, 2002.
23. Jovan Dj. Golic: *On the Security of Nonlinear Filter Generators*, FSE'96, LNCS 1039, Springer, pp. 173-188.
24. Jovan Dj. Golic: *Correlation Properties of a General Binary Combiner with Memory*. Journal of Cryptology vol. 9(2), pp. 111-126 (1996).
25. Jovan Dj. Golic, Vittorio Bagini, Guglielmo Morgari: *Linear Cryptanalysis of Bluetooth Stream Cipher*, Eurocrypt 2002, LNCS 2332, Springer, pp. 238-255.
26. Thomas Jakobsen: *Cryptanalysis of Block Ciphers with Probabilistic Non-Linear Relations of Low Degree*, Crypto 98, LNCS 1462, Springer, pp. 212-222, 1998.
27. Philip Hawkes, Gregory Rose: *Guess-and-determine attacks on SNOW*, in SAC 2002, LNCS 2595, Springer, pp. 37-46.
28. Philip Hawkes, Gregory Rose: *Rewriting Variables: the Complexity of Fast Algebraic Attacks on Stream Ciphers*, in Crypto 2004, LNCS 3152, pp. 390-406, Springer, 2004. Available from eprint.iacr.org/2004/081/.
29. Willi Meier and Othmar Staffelbach: *Fast correlation attacks on certain stream ciphers*, Journal of Cryptology, 1(3):159-176, 1989.
30. Willi Meier and Othmar Staffelbach: *Correlation Properties of Combiners with Memory in Stream Ciphers*, Journal of Cryptology 5(1): pp. 67-86 (1992).
31. Alfred J. Menezes, Paul C. van Oorschot, Scott A. Vanstone: *Handbook of Applied Cryptography*, Chapter 6, CRC Press.
32. Nessie Security Report v2.0. or Nessie deliverable D20, available from www.cryptonessie.org.
33. Jacques Patarin: *Cryptanalysis of the Matsumoto and Imai Public Key Scheme of Eurocrypt'88*, Crypto'95, Springer, LNCS 963, pp. 248-261, 1995.
34. Gregory G. Rose and Philip Hawkes: *Turing: a Fast Stream Cipher*, FSE 2003, LNCS, Springer.

35. Adi Shamir, Jacques Patarin, Nicolas Courtois, Alexander Klimov, *Efficient Algorithms for solving Overdefined Systems of Multivariate Polynomial Equations*, Eurocrypt'2000, LNCS 1807, Springer, pp. 392-407.
36. L. Simpson, E. Dawson, J. Golic and W. Millan: *LILI Keystream Generator*, SAC'2000, LNCS 2012, Springer, pp. 248-261,
37. Volker Strassen: *Gaussian Elimination is Not Optimal*, Numerische Mathematik, vol 13, pp 354-356, 1969.
38. Muxiang Zhang, Agnes Chan: *Maximum Correlation Analysis of Nonlinear S-boxes in Stream Ciphers*. In Crypto 2000, LNCS 1880, pp. 501-514, Springer 2000.

New Method for Bounding the Maximum Differential Probability for SPNs and ARIA

Hong-Su Cho[1], Soo Hak Sung[2], Daesung Kwon[3], Jung-Keun Lee[3],
Jung Hwan Song[4], and Jongin Lim[1]

[1] Graduate School of Information Security, Korea University,
1, 5-Ka, Anam-dong, Sungbuk-ku, Seoul 136-701, Korea
{karma3432,jilim}@korea.ac.kr
[2] Department of Computing information & mathematics, Paichai University,
426-6 Doma-dong, Seo-gu, Daejeon 302-735 Korea
sungsh@mail.pcu.ac.kr
[3] National Security Research Institute,
161 Gajeong-dong, Yuseong-gu, Daejeon 305-350, Korea
{ds_kwon, jklee}@etri.re.kr
[4] Department of Mathematics, Hanyang University,
17 Haengdang-dong, Seongdong-gu, Seoul 133-791, Korea
camp123@hanyang.ac.kr

Abstract. By considering the number of independent variables, we present a new method for finding an upper bound on the maximum differential probability (MDP) for $r(r \geq 2)$-round substitution-permutation networks (SPNs). It first finds an upper bound for 2-round SPNs and then uses a recursive technique for $r(r \geq 3)$-round SPNs. Our result extends and sharpens known results in that it is more effective for calculating MDP for $r(r \geq 3)$-round SPNs and applicable to all SPNs. By applying our method to ARIA, we get an estimated bound of 1.5×2^{-98} on MDP for 6-round ARIA.

Keywords: Cryptography, Differential cryptanalysis, Linear cryptanalysis, Substitution-permutation networks, Branch number, Independent variables, AES, ARIA.

1 Introduction

The substitution-permutation network (SPN) is one of the most widely used structures in block ciphers. The Advanced Encryption Standard (AES) [6] and ARIA [12] have the SPN structure. The security of SPNs against differential cryptanalysis (DC) [2] and linear cryptanalysis (LC) [13] depends on the maximum differential probability (MDP) and the maximum linear hull probability (MLHP), respectively.

Many researchers obtained upper bounds on the MDP and MLHP for SPNs (for 2-round SPNs, see Hong et al. [7], Kang et al. [8], and Chun et al. [3], for $r(r \geq 2)$-round SPNs, see Keliher et al. [9, 10] and Park et al. [15, 16]). Keliher

C. Park and S. Chee (Eds.): ICISC 2004, LNCS 3506, pp. 21–32, 2005.

et al. [9, 10] proposed an algorithm for finding an upper bound on the MLHP for SPNs. They applied the algorithm to AES, and obtained an upper bound of 2^{-92} when 9 or more rounds are approximated. The running time is 200,000 hours on a single Sun Ultra 5. Park et al. [15, 16] proposed a method for upper bounding on the MDP and MLHP for AES-like structures. By applying the method to AES, they obtained that the MDP and MLHP for 4-round AES are bounded by 1.144×2^{-111} and 1.075×2^{-106}, respectively. Although the method of Park et al. [15, 16] produces good upper bounds for AES-like structures, it does not apply to other SPNs (for example ARIA).

To find upper bounds on the MDP and MLHP for SPNs, one first finds upper bounds for 2-round and then use a recursive technique for 3 or more rounds. Thus it is very important to find *effective* upper bounds for 2-round SPNs. The *effective* upper bounds mean that they can be used to find good upper bounds for $r(r \geq 3)$-round SPNs. It seems that the upper bounds of Chun et al. [3] are the best known upper bounds on the MDP and MLHP for 2-round SPNs. Although the upper bounds of Chun et al. [3] were used to find good upper bound for 4-round AES by Park et al. [16], the upper bounds are not effective for general SPNs (see Theorem 6).

In this paper, we obtain a new method for finding upper bounds by considering the number of independent variables. We present only the method for finding an upper bound on the MDP because the method for finding an upper bound on the MLHP can be obtained similarly. Our results consist of three parts. Firstly, we improves the upper bounds of Chun et al. [3] on the MDP for 2-round SPNs by the factor of p^α where p is the MDP for s-boxes and α is a nonnegative integer depending on input/output differences, the number of independent variables. Secondly, we present an efficient recursive formula for the upper bound of the MDP for $r(r \geq 3)$-round SPNs using the number of independent variables. Finally, by applying our method to ARIA, we get an estimated bound of 1.5×2^{-98} on MDP for 6-round ARIA. It is expected that the full running time will be about 256 days on a single 2.6 GHz PC.

2   Preliminaries

Let $S : \mathbb{Z}_2^m \to \mathbb{Z}_2^m$ be a bijective mapping. For any given $a, b, \Gamma_a, \Gamma_b \in \mathbb{Z}_2^m$, the differential probability and linear (hull) probability for the S are defined as

$$DP^S(a, b) = \frac{\#\{x \in \mathbb{Z}_2^m | S(x) \oplus S(x \oplus a) = b\}}{2^m}$$

and

$$LP^S(\Gamma_a, \Gamma_b) = \left(\frac{\#\{x \in \mathbb{Z}_2^m | \Gamma_a \bullet x = \Gamma_b \bullet S(x)\}}{2^{m-1}} - 1 \right)^2,$$

respectively, where $x \bullet y$ denotes the parity (0 or 1) of bitwise product of x and y. The a and b are called the input difference and output difference, respectively, for the S. Also, the Γ_a and Γ_b are called the input mask and output mask, respectively, for the S.

One round of SPNs consists of three layers of key addition, substitution, and linear transformation, see Fig. 1. Substitution layer is made up of n small nonlinear substitutions referred to s-boxes. In $r(r \geq 2)$-round SPNs, the linear transformation of the last round is omitted because it has no cryptographic significance. Thus a 2-round SPN can be described as in Fig. 2.

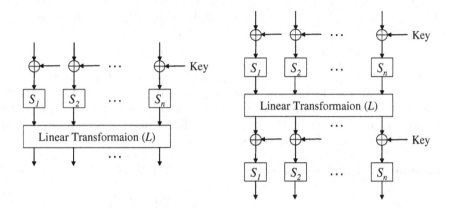

Fig. 1. One Round of SPN   **Fig. 2.** 2-Round SPN

The s-boxes and linear transformation (denoted by L) should be invertible in order to decipher. Therefore we assume that all s-boxes are bijective mappings defined on $\mathbb{Z}_2^m$. We also assume that the round keys are independent and uniformly distributed.

A differentially active s-box is defined as an s-box given a non-zero input (and output) difference, and a linearly active s-box is defined as an s-box given a non-zero output (and input) mask.

The MDP and MLHP for s-boxes are denoted by p and q, respectively, i.e.,

$$p = \max_{1 \leq i \leq n} \max_{a \neq 0, b} DP^{S_i}(a, b), \quad q = \max_{1 \leq i \leq n} \max_{\Gamma_a, \Gamma_b \neq 0} LP^{S_i}(\Gamma_a, \Gamma_b).$$

The linear transformation $L : (\mathbb{Z}_2^m)^n \rightarrow (\mathbb{Z}_2^m)^n$ can be represented by an $n \times n$ matrix $M = (m_{ij})_{n \times n}$, where $m_{ij} \in \mathbb{Z}_2^m$, i.e., $L(x) = Mx = (\sum_{j=1}^n m_{1j} \circ x_j, \sum_{j=1}^n m_{2j} \circ x_j, \cdots, \sum_{j=1}^n m_{nj} \circ x_j)$ where $x = (x_1, \cdots, x_n) \in (\mathbb{Z}_2^m)^n$ and $\circ$ is a multiplication in the finite field $\mathbb{Z}_2^m$.

The branch number of the linear transformation L on the DC is defined as

$$\beta_d = \min_{x \neq 0}\{wt(x) + wt(L(x))\},$$

where, for $x = (x_1, \cdots, x_n) \in (\mathbb{Z}_2^m)^n$, $wt(x) = \#\{i | 1 \leq i \leq n, x_i \neq 0\}$. When $x = (x_1, \cdots, x_n) \in (\mathbb{Z}_2)^n$, $wt(x)$ is equal to the Hamming weight of x.

If a is the input difference of L, then the output difference is $L(a)$ [4]. Thus the branch number β_d is also defined as the minimum number of differentially active s-boxes in the 2-round SPN [5]. Similarly, the branch number β_l of L on the LC is defined as the minimum number of linearly active s-boxes in the 2-round SPN [5]. The branch numbers β_d, β_l are used for finding upper bounds on MDP and MLHP for $r(r \geq 2)$-round SPNs, respectively.

The following are used to reduce the computation time of MDP, MLDP for $r(r \geq 2)$-round SPNs and are important ingredients of our new method. For $x = (x_1, \cdots, x_n)$, the *pattern* of x, γ_x, is defined as $\gamma_x = (\gamma_1, \cdots, \gamma_n) \in (\mathbb{Z}_2)^n$, where $\gamma_i = 0$ if $x_i = 0$, and $\gamma_i = 1$ if $x_i \neq 0$. For $\delta, \rho \in (\mathbb{Z}_2)^n$, we define $B_L(\delta, \rho)$ as

$$B_L(\delta, \rho) = \{x = (x_1, \cdots, x_n) | \gamma_x = \delta, \gamma_{L(x)} = \rho\}.$$

Since $wt(x) = wt(a)$ for $x \in B_L(\gamma_a, \gamma_b)$, $B_L(\gamma_a, \gamma_b)$ has $wt(a)$ variables when $B_L(\gamma_a, \gamma_b) \neq \emptyset$. Due to the condition $\gamma_{L(x)} = \rho$, there are some linear dependencies among these variables. The number $t(\delta, \rho)$ is defined as the size of the maximal set of variables in $B_L(\delta, \rho)$ without linear dependencies and is called the number of independent variables.

Example 1. *Let the linear transformation* $L : (\mathbb{Z}_2^m)^4 \to (\mathbb{Z}_2^m)^4$ *be defined by*

$$L(x) = L(x_1, x_2, x_3, x_4) = \begin{pmatrix} 0 & 1 & 1 & 1 \\ 1 & 0 & 1 & 1 \\ 1 & 1 & 0 & 1 \\ 1 & 1 & 1 & 0 \end{pmatrix} \begin{pmatrix} x_1 \\ x_2 \\ x_3 \\ x_4 \end{pmatrix}.$$

Then $B_L(\delta, \rho) = \emptyset$ *if* $wt(\delta) + wt(\rho) < \beta_d = 4$. *When* $\delta = \rho = (0, 0, 1, 1) = 3_x$, $B_L(\delta, \rho) = \{(0, 0, x_3, x_4) | x_3 = x_4, x_3 \neq 0\}$ *and so* $t(\delta, \rho)$ *is 1.*

3   New Upper Bound for r-Round SPNs

In this section, we give a new method for finding an upper bound on the MDP for r-round SPNs.

The r-round differential probability with input difference $a = (a_1, \cdots, a_n)$ and output difference $b = (b_1, \cdots, b_n)$ is denoted by $DP_r(a, b)$. In particular, $DP_1(a, b)$ denotes the differential probability for 1-round without the linear transformation. Therefore $DP_1(a, b) = \prod_{i=1}^{n} DP^{S_i}(a_i, b_i)$. The maximum r-round differential probability with input difference pattern δ and output difference pattern ρ is denoted by $dp_r(\delta, \rho)$. Namely,

$$dp_r(\delta, \rho) = \max_{\gamma_a = \delta, \gamma_b = \rho} DP_r(a, b).$$

We first find an upper bound for 2-round SPNs. The differential probability $DP_2(a, b)$ is given as

$$DP_2(a, b) = \sum_x \left(\prod_{i=1}^{n} DP^{S_i}(a_i, x_i) \right) \left(\prod_{j=1}^{n} DP^{S_j}(y_j, b_j) \right),$$

where $y = L(x), x = (x_1, \cdots, x_n)$, and $y = (y_1, \cdots, y_n)$. Since the s-boxes are bijective, $DP^{S_i}(u, v) = 0$ if $u = 0$ and $v \neq 0$, or $u \neq 0$ and $v = 0$. Therefore the $DP_2(a, b)$ can be written by

$$DP_2(a, b) = \sum_{x \in B_L(\gamma_a, \gamma_b)} \left(\prod_{i \in \chi(\gamma_a)} DP^{S_i}(a_i, x_i) \right) \left(\prod_{j \in \chi(\gamma_b)} DP^{S_j}(y_j, b_j) \right),$$

where, for $\delta = (\delta_1, \cdots, \delta_n) \in (\mathbb{Z}_2)^n, \chi(\delta) = \{i | 1 \leq i \leq n, \delta_i = 1\}$. To find an upper bound of $dp_2(\gamma_a, \gamma_b)$, we need the following lemma which is proved in Chun et al. [3].

Lemma 2. *Let* $\{x_i^{(j)}\}_{i=1}^n, 1 \leq j \leq m$, *be sequences of real numbers. Then the following inequality is satisfied.*

$$\sum_{i=1}^n |x_i^{(1)} \cdots x_i^{(m)}| \leq \max \left\{ \sum_{i=1}^n |x_i^{(1)}|^m, \cdots, \sum_{i=1}^n |x_i^{(m)}|^m \right\}.$$

Theorem 3. *Let* $L : (\mathbb{Z}_2^m)^n \rightarrow (\mathbb{Z}_2^m)^n$ *be the linear transformation used in the 2-round SPN. Then* $dp_2(\gamma_a, \gamma_b)$ *is bounded by* $p^{wt(a)+wt(b)-\beta_d-t+1}Q$, *where* t *is the number of independent variables in* $B_L(\gamma_a, \gamma_b)$, *and* Q *is defined by*

$$Q = \max \left\{ \max_{\substack{1 \leq i \leq n \\ 1 \leq u \leq 2^m - 1}} \sum_{j=1}^{2^m-1} \{DP^{S_i}(u, j)\}^{\beta_d}, \max_{\substack{1 \leq i \leq n \\ 1 \leq u \leq 2^m - 1}} \sum_{j=1}^{2^m-1} \{DP^{S_i}(j, u)\}^{\beta_d} \right\}.$$

Proof. Let $wt(a) = k$ and $wt(b) = l$. Without loss of generality, we assume that $a_1 \neq 0, \cdots, a_k \neq 0, a_{k+1} = 0, \cdots, a_n = 0, b_1 \neq 0, \cdots, b_l \neq 0, b_{l+1} = 0, \cdots, b_n = 0$. Then $B_L(\gamma_a, \gamma_b)$ is the set of solutions of the following system.

$$x_1 \neq 0, \cdots, x_k \neq 0, x_{k+1} = 0, \cdots, x_n = 0,$$
$$y_1 \neq 0, \cdots, y_l \neq 0, y_{l+1} = 0, \cdots, y_n = 0,$$

where $y = L(x)$. Then $DP_2(a, b)$ can be written by

$$DP_2(a, b) = \sum_{(x_1, \cdots, x_k, 0, \cdots, 0) \in B_L(\gamma_a, \gamma_b)} \left(\prod_{i=1}^k DP^{S_i}(a_i, x_i) \right) \left(\prod_{j=1}^l DP^{S_j}(y_j, b_j) \right).$$

For convenience of notation, we assume that $x_1, \cdots, x_t$ are independent variables in $B_L(\gamma_a, \gamma_b)$. Then the other $k - t$ variables are generated by these independent variables. That is, $x_{t+1}, \cdots, x_k$ in $B_L(\gamma_a, \gamma_b)$ are generated by $x_1, \cdots, x_t$. Also $y_1, \cdots, y_l$ are generated by $x_1, \cdots, x_t$. When $t - 1$ independent variables $x_1, \cdots, x_{t-1}$ are fixed, there exist many variables depending x_t. To specify such variables, we define U, V as

$$U = \{i | 1 \leq i \leq k, x_i \text{ depends on } x_t\},$$
$$V = \{j | 1 \leq j \leq l, y_j \text{ depends on } x_t\}.$$

Let $\alpha = \#U + \#V$. Note that $\alpha \geq \beta_d$ by the definition of the branch number. By defining $U^c = \{1, \cdots, k\} - U, V^c = \{1, \cdots, l\} - V$, we have that

$$\sum_{(x_1, \cdots, x_k, 0, \cdots, 0) \in B_L(\gamma_a, \gamma_b)} \left(\prod_{i=1}^{k} DP^{S_i}(a_i, x_i) \right) \left(\prod_{j=1}^{l} DP^{S_j}(y_j, b_j) \right)$$

$$\leq \sum_{x_1, \cdots, x_{t-1}} (\prod_{i \in U^c} DP^{S_i}(a_i, x_i)) (\prod_{j \in V^c} DP^{S_j}(y_j, b_j)) \Phi_t.$$

where $\Phi_t = \sum_{x_t} (\prod_{i \in U} DP^{S_i}(a_i, x_i))(\prod_{j \in V} DP^{S_j}(y_j, b_j))$. From Lemma 2, we obtain that

$$\sum_{x_t} (\prod_{i \in U} DP^{S_i}(a_i, x_i))(\prod_{j \in V} DP^{S_j}(y_j, b_j))$$

$$\leq \max \left\{ \max_{\substack{1 \leq i \leq n \\ 1 \leq u \leq 2^m - 1}} \sum_{j=1}^{2^m-1} \{DP^{S_i}(u, j)\}^{\alpha}, \max_{\substack{1 \leq i \leq n \\ 1 \leq u \leq 2^m - 1}} \sum_{j=1}^{2^m-1} \{DP^{S_i}(j, u)\}^{\alpha} \right\}$$

$$\leq p^{\alpha - \beta_d} Q.$$

Therefore

$$DP_2(a, b) \leq p^{\alpha - \beta_d} Q \sum_{x_1, \cdots, x_{t-1}} (\prod_{i \in U^c} DP^{S_i}(a_i, x_i))(\prod_{j \in V^c} DP^{S_j}(y_j, b_j))$$

$$\leq p^{\alpha - \beta_d} Q p^{\sharp U^c - (t-1) + \sharp V^c} \sum_{x_1, \cdots, x_{t-1}} (\prod_{i=1}^{t-1} DP^{S_i}(a_i, x_i))$$

$$= p^{k+l-\beta_d-t+1} Q.$$

Since the upper bound $p^{k+l-\beta_d-t+1} Q$ depends only on γ_a and γ_b, it is also an upper bound of $dp_2(\gamma_a, \gamma_b)$.   □

The following theorem gives an upper bound on the number of independent variables in $B_L(\gamma_a, \gamma_b)$.

Theorem 4. *Let $L : (\mathbb{Z}_2^m)^n \to (\mathbb{Z}_2^m)^n$ be a linear transformation with its associated matrix $A = (a_{ij})_{n \times n}$. Then the number t of independent variables in $B_L(\delta, \rho)$ is less than or equal to $wt(\delta) + wt(\rho) - \beta_d + 1$.*

Proof. Assume $B_L(\delta, \rho)$ is generated by $x_1, \cdots, x_t$. Consider the input difference $a \in B_L(\delta, \rho)$ which is generated only by x_1. Then $wt(a) \leq wt(\delta) - (t-1)$ and $wt(L(a)) \leq wt(\rho)$. Since $wt(a) + wt(L(a)) \geq \beta_d$, we have

$$\beta_d \leq wt(a) + wt(L(a)) \leq wt(\delta) - (t-1) + wt(\rho).$$

                                                                                            □

By using Theorems 3 and 4, we can obtain the following corollary which is proved by Chun et al. [3]. So our upper bound for the 2-round SPNs in Theorem 3 improves and generalizes the upper bound of Chun et al. [3].

Corollary 5. *Let $L : (\mathbb{Z}_2^m)^n \to (\mathbb{Z}_2^m)^n$ be the linear transformation used in the 2-round SPN. Then $dp_2(\gamma_a, \gamma_b)$ is bounded by Q.*

An upper bound of $dp_r(\gamma_a, \gamma_b)$ can be easily obtained by the following theorem. The maximum of upper bounds of $dp_r(\gamma_a, \gamma_b), 0 < \gamma_a, \gamma_b < 2^n$, is an upper bound on the MDP for the r-round.

Theorem 6. *Let $L : (\mathbb{Z}_2^m)^n \to (\mathbb{Z}_2^m)^n$ be the linear transformation used in the SPN. Then an upper bound of $dp_r(\gamma_a, \gamma_b)$ is obtained recursively as follows.*

$$dp_r(\gamma_a, \gamma_b) \leq \begin{cases} p^{wt(a)+wt(b)-\beta_d-t(\gamma_a,\gamma_b)+1}Q, & \text{if } r = 2, \\ \sum_{\delta \in (\mathbb{Z}_2)^n} dp_{r-1}(\gamma_a, \delta)p^{wt(b)-t(\delta,\gamma_b)}, & \text{if } r \geq 3, \end{cases}$$

where $t(\delta, \rho)$ is the number of independent variables in $B_L(\delta, \rho)$.

Proof. When $r = 2$, the result follows from Theorem 3. Now let $r \geq 3$. Then

$$DP_r(a, b) = \sum_x DP_{r-1}(a, x)DP_1(y, b)$$

$$= \sum_{\delta \in (\mathbb{Z}_2)^n} \sum_{x:\gamma_x=\delta} DP_{r-1}(a, x)DP_1(y, b)$$

$$\leq \sum_{\delta \in (\mathbb{Z}_2)^n} dp_{r-1}(\gamma_a, \delta) \sum_{x:\gamma_x=\delta} DP_1(y, b)$$

$$= \sum_{\delta \in (\mathbb{Z}_2)^n} dp_{r-1}(\gamma_a, \delta) \sum_{x \in B_L(\delta,\gamma_b)} DP_1(y, b)$$

where $y = L(x), x = (x_1, \cdots, x_n)$ and $y = (y_1, \cdots, y_n)$. Note that $L(B_L(\delta, \rho)) = \{y|\gamma_y = \rho, \gamma_{L^{-1}(y)} = \delta\}$. It is easy to see that the number of independent variables in $B_L(\delta, \rho)$ is equal to the number of independent variables in $L(B_L(\delta, \rho))$. Let $y_{j_1}, \cdots, y_{j_t}$ be the independent variables in $L(B_L(\delta, \gamma_b))$. Then we have that

$$\sum_{x \in B_L(\delta,\gamma_b)} DP_1(y, b) = \sum_{y \in L(B_L(\delta,\gamma_b))} DP_1(y, b)$$

$$= \sum_{y \in L(B_L(\delta,\gamma_b))} \prod_{i \in \chi(\gamma_b)} DP^{S_i}(y_i, b_i)$$

$$\leq p^{wt(b)-t} \sum_{y_{j_1}, \cdots, y_{j_t}} DP^{S_{j_1}}(y_{j_1}, b_{j_1}) \cdots DP^{S_{j_t}}(y_{j_t}, b_{j_t})$$

$$= p^{wt(b)-t}.$$

It follows that $DP_r(a, b) \leq \sum_{\delta \in (\mathbb{Z}_2)^n} dp_{r-1}(\gamma_a, \delta)p^{wt(b)-t(\delta,\gamma_b)}$. Since the upper bound of $DP_r(a, b)$ depends only on γ_a and γ_b, it is also an upper bound of $dp_r(\gamma_a, \gamma_b)$. $\quad\square$

4   Application of Our Upper Bound to ARIA

In this section, we compute an upper bound on the MDP for r-round ARIA. ARIA [12] is a SPN block cipher with parameters $n = 16, m = 8, \beta_d = 8, p = 2^{-6}$. The linear transformation L of ARIA is a 16×16 binary matrix. To describe it more simple, we denote column vector M^i by hexadecimal.

$$M^1 = 1ac6_x, \quad M^2 = 25c9_x, \quad M^3 = 4a39_x, \quad M^4 = 8536_x,$$
$$M^5 = a493_x, \quad M^6 = 5863_x, \quad M^7 = a16c_x, \quad M^8 = 529c_x,$$
$$M^9 = c925_x, \quad M^{10} = c61a_x, \quad M^{11} = 3685_x, \quad M^{12} = 394a_x,$$
$$M^{13} = 6358_x, \quad M^{14} = 93a4_x, \quad M^{15} = 9c52_x, \quad M^{16} = 6ca1_x.$$

ARIA has two s-boxes (S_1, S_2) and their inverses (S_1^{-1}, S_2^{-1}). The layer of substitution in the odd number of rounds is composed of 16 s-boxes as follows.

$$S_1, S_2, S_1^{-1}, S_2^{-1}, S_1, S_2, S_1^{-1}, S_2^{-1}, S_1, S_2, S_1^{-1}, S_2^{-1}, S_1, S_2, S_1^{-1}, S_2^{-1}.$$

Similarly, the layer of substitution in the even number of rounds is composed of 16 s-boxes as follows.

$$S_1^{-1}, S_2^{-1}, S_1, S_2, S_1^{-1}, S_2^{-1}, S_1, S_2, S_1^{-1}, S_2^{-1}, S_1, S_2, S_1^{-1}, S_2^{-1}, S_1, S_2.$$

If nonzero $u \in \mathbb{Z}_2^8$ is fixed, and v varies over $\mathbb{Z}_2^8$, then the distribution of differential probability values $DP^{S_i}(u, v)$ is independent of u and i ($1 \le i \le 16$), and is given in Table 1. Here ρ_i^1 is the differential probability value, and ϕ_i^1 is the number of occurrences of ρ_i^1. If nonzero $v \in \mathbb{Z}_2^8$ is fixed, and u varies over $\mathbb{Z}_2^8$, then the same distribution is obtained. From Table 1 and Theorem 3, we

Table 1. The distribution of differential probability values for ARIA s-box

ρ_i^1	2^{-6}	2^{-7}	0
ϕ_i^1	1	126	129

obtain that

$$Q = \sum_{j=1}^{255} \{DP^{S_1}(1, j)\}^8 = \frac{97,792}{(2^8)^8} \sim 1.49 \times 2^{-48}.$$

We now compute an upper bound of $dp_2(\gamma_a, \gamma_b)$. For example, let $\gamma_a = 3_x = (0, \cdots, 0, 1, 1)$, $\gamma_b = f0f3_x = (1, 1, 1, 1, 0, 0, 0, 0, 1, 1, 1, 1, 0, 0, 1, 1)$. Then it is easy to see that $B_L(\gamma_a, \gamma_b) = \{(0, \cdots, 0, u, u) | u = 1, \cdots, 255\}$, and so the number of independent variables in $B_L(\gamma_a, \gamma_b)$ is 1. Thus we have from Theorem 3 that the $dp_2(\gamma_a, \gamma_b)$ is bounded by

$$p^{wt(a)+wt(b)-\beta_d-t+1}Q \sim p^4 \times 1.49 \times 2^{-48} \sim 1.49 \times 2^{-72}.$$

Since the number of possible cases of (γ_a, γ_b) is $2^{16} \times 2^{16}$, by using a computer program, we can obtain all upper bounds of $dp_2(\gamma_a, \gamma_b), \gamma_a \in (\mathbb{Z}_2)^{16}, \gamma_b \in (\mathbb{Z}_2)^{16}$. The MDP for 2-round ARIA is bounded by $Q \sim 1.49 \times 2^{-48}$.

We next compute an upper bound of $dp_3(\gamma_a, \gamma_b)$. For example, let $\gamma_a = \gamma_b = 1_x = (0, \cdots, 0, 1)$. From Theorem 6, we obtain that

$$dp_3(\gamma_a, \gamma_b) \leq \sum_{\delta \in (\mathbb{Z}_2)^{16}} dp_2(\gamma_a, \delta) p^{wt(b)-t(\delta,\gamma_b)}.$$

In the above summation, it is enough to consider only δ satisfying $B_L(\gamma_a, \delta) \neq \emptyset$ and $B_L(\delta, \gamma_b) \neq \emptyset$. Such δ is only $6ca1_x$. The numbers of independent variables in $B_L(\gamma_a, 6ca1_x)$ and $B_L(6ca1_x, \gamma_b)$ are all 1. Thus we have from Theorem 6 that

$$dp_3(\gamma_a, \gamma_b) \leq dp_2(\gamma_a, 6ca1_x) p^{wt(b)-t(6ca1_x,\gamma_b)}$$
$$\leq p^{wt(a)+wt(6ca1_x)-\beta_d-t(\gamma_a,6ca1_x)+1} Q p^{wt(b)-t(6ca1_x,\gamma_b)}$$
$$= Q \sim 1.49 \times 2^{-48}.$$

Note that the differential probabilities for 3-round are smaller than or equal to the MDP for 2-round. Since $\max_{\gamma_a \neq 0, \gamma_b} dp_3(\gamma_a, \gamma_b) \leq Q$ and $dp_3(1_x, 1_x) \leq Q$, it follows that the MDP for 3-round ARIA is bounded by $Q \sim 1.49 \times 2^{-48}$.

When 4 or more rounds are approximated, we can not use Theorem 6 directly, because the computation time of the numbers of independent variables is very much. When r-round is approximated, for each (γ_a, γ_b), the numbers of independent variables are computed about $2^{16(r-2)+1}$ times. To decrease such times, we need the following theorem.

Theorem 7. *For any $\epsilon > 0$, the following inequality holds.*

$$dp_r(\gamma_a, \gamma_b) \leq \epsilon + \sum_{\delta \in (\mathbb{Z}_2)^n, dp_{r-1}(\gamma_a,\delta) > \epsilon} dp_{r-1}(\gamma_a, \delta) p^{wt(b)-t(\delta,\gamma_b)}.$$

Proof. Observe that

$$DP_r(a, b) = \sum_x DP_{r-1}(a, x) DP_1(y, b)$$
$$= \sum_{\delta \in (\mathbb{Z}_2)^n} \sum_{x:\gamma_x=\delta} DP_{r-1}(a, x) DP_1(y, b)$$
$$\leq \sum_{\delta \in (\mathbb{Z}_2)^n} dp_{r-1}(\gamma_a, \delta) \sum_{x \in B(\delta,\gamma_b)} DP_1(y, b)$$

and

$$\sum_{\delta \in (\mathbb{Z}_2)^n} dp_{r-1}(\gamma_a, \delta) \sum_{x \in B(\delta,\gamma_b)} DP_1(y, b)$$
$$= \sum_{\delta \in (\mathbb{Z}_2)^n, dp_{r-1}(\gamma_a,\delta) \leq \epsilon} dp_{r-1}(\gamma_a, \delta) \sum_{x \in B(\delta,\gamma_b)} DP_1(y, b)$$
$$+ \sum_{\delta \in (\mathbb{Z}_2)^n, dp_{r-1}(\gamma_a,\delta) > \epsilon} dp_{r-1}(\gamma_a, \delta) \sum_{x \in B(\delta,\gamma_b)} DP_1(y, b)$$
$$=: I + II.$$

I is bounded by

$$\epsilon \sum_{\delta \in (\mathbb{Z}_2)^n, dp_{r-1}(\gamma_a, \delta) \leq \epsilon} \sum_{x \in B(\delta, \gamma_b)} DP_1(y, b) \leq \epsilon \sum_x DP_1(y, b) = \epsilon,$$

where $y = L(x), x = (x_1, \cdots, x_n)$ and $y = (y_1, \cdots, y_n)$. From the proof of Theorem 6, $\sum_{x \in B(\delta, \gamma_b)} DP_1(y, b) \leq p^{wt(b) - t(\delta, \gamma_b)}$, which implies that

$$II \leq \sum_{\delta \in (\mathbb{Z}_2)^n, dp_{r-1}(\gamma_a, \delta) > \epsilon} dp_{r-1}(\gamma_a, \delta) p^{wt(b) - t(\delta, \gamma_b)}.$$

Therefore $dp_r(\gamma_a, \gamma_b)$ is bounded by

$$\epsilon + \sum_{\delta \in (\mathbb{Z}_2)^n, dp_{r-1}(\gamma_a, \delta) > \epsilon} dp_{r-1}(\gamma_a, \delta) p^{wt(b) - t(\delta, \gamma_b)}.$$

$\square$

To find upper bounds for 4, 5, 6-round ARIA, we use the following algorithm. Note that denominators (2^{6-i} in Step 3, 2 in Step 5) are selected by trial-and-error so that so they may not be optimal. In the algorithm, $udp_i(\gamma_a, \gamma_b)$ denotes an upper bound of $dp_i(\gamma_a, \gamma_b)$.

Algorithm 1. Find an upper bound on the MDP for r-round ARIA, where $4 \leq r \leq 6$.

Step 0. Select an integer T such that $2^{-T} \sim 2^{-128}$.
Step 1. $U_r \leftarrow 2^{-T}$
Step 2. For each (γ_a, γ_b), $udp_2(\gamma_a, \gamma_b) \leftarrow p^{wt(a) + wt(b) - \beta_d - t(\gamma_a, \gamma_b) + 1} Q$
Step 3. For i from 3 to $r - 1$,

$$udp_i(\gamma_a, \gamma_b) \leftarrow \frac{U_r}{2^{6-i}} + \sum_{\delta \in (\mathbb{Z}_2)^{16}, udp_{i-1}(\gamma_a, \delta) > \frac{U_r}{2^{6-i}}} udp_{i-1}(\gamma_a, \delta) p^{wt(b) - t(\delta, \gamma_b)}$$

Step 4. Compute

$$udp_r(\gamma_a, \gamma_b) \leftarrow U_r + \sum_{\delta \in (\mathbb{Z}_2)^{16}, udp_{r-1}(\gamma_a, \delta) > U_r} udp_{r-1}(\gamma_a, \delta) p^{wt(b) - t(\delta, \gamma_b)}$$

Step 5. If $\max\limits_{\gamma_a \neq 0, \gamma_b} udp_r(\gamma_a, \gamma_b) \leq U_r(1 + \frac{1}{2})$, then an upper bound is $\max\limits_{\gamma_a \neq 0, \gamma_b} udp_r$ (γ_a, γ_b) and STOP the program.
Else $T \leftarrow T - 1$, and go to Step 1.

For 4,5,6-round ARIA, Algorithm 1 gives upper bounds with running time as given in Table 2 on a single 2.6 GHz PC. The estimated value of the upper bound for 6-round ARIA is obtained after performing about 50% of computation.

On the other hand, Keliher [11] proposed an algorithm for finding an upper bound on the MLHP for AES. They obtained an upper bound of 1.778×2^{-107} when 8 or more rounds are approximated.

Table 2. Upper bounds on the MDP for r-round ARIA

Rounds	Upper bounds	Running Time
4	1.5×2^{-78}	4 days
5	1.5×2^{-84}	16 days
6	1.5×2^{-98}	256 days (estimated)

5   Conclusion

In this paper, we have given a new method for finding an upper bound on the MDP for r-round SPNs by using the number of independent variables. Our method consists of two steps. In the first step, we find an upper bound on the MDP for 2-round SPNs and in the second step, we use a recursive formula for finding an upper bound on the MDP for $r(r \geq 3)$-round SPNs. The result in the first step is better than known results in the sense that it can be used to find good upper bounds for $r(r \geq 3)$-round SPNs. The result in the second step is an efficient recursive formula for finding an upper bound on the MDP for $r(r \geq 3)$-round SPNs. Both results use the number of independent variables which is newly introduced in this paper. Finally, by applying our method to ARIA, we get an estimated bound of 1.5×2^{-98} on MDP for 6-round ARIA.

References

1. K. Aoki, T. Ichikawa, M. Kanda, M. Matsui, S. Moriai, J. Nakajima, T. Tokita, Camellia: A 128-bit block cipher suitable for multiple platforms - design and analysis, LNCS 2012, pp. 39-56. 2000.
2. E. Biham and A. Shamir, Differential cryptanalysis of DES-like cryptosystems, Advance in Cryptology-Crypto'90, LNCS Vol. 537, Springer-Verlag, pp. 2-21, 1991.
3. K. Chun, S. Kim, S. Lee, S.H. Sung, S. Yoon, Differential and linear cryptanalysis for 2-round SPNs, IPL 87, pp. 277-282, 2003.
4. J. Daemem, R. Govaerts, and J. Vandewlle, Correlation matrices, Fast Software Encryption-FSE'94, LNCS Vol. 1008, Springer-Verlag, pp. 275-285, 1995.
5. J. Daemen, L. Knudsen, and V. Rijmen, The block cipher SQUARE, Fast Software Encryption-FSE'97, LNCS Vol. 1267, Springer-Verlag, pp. 149-165, 1997.
6. FIPS 197, The Advanced Encryption Standard (AES), Federal Information Processing Standard, NIST, US Department of Commerce, Washington DC, 2001.
7. S. Hong, S Lee, J. Lim, J. Sung, and D. Cheon, Provable security against differential and linear cryptanalysis for the SPN structure, FSE 2000, LNCS Vol. 1978, Springer-Verlag, pp. 273-283, 2000.
8. J.-S. Kang, S. Hong, S. Lee, O. Yi, C. Park, and J. Lim, Practical and provable security against differential and linear cryptanalysis for substitution-permutation networks, ETRI J. 23, pp. 158-167, 2001.
9. L. Keliher, H. Meijer, and S. Tavares, New method for the upper bounding the maximum average linear hull probability for SPNs, Advances in Cryptology - Eurocrypt 2001, LNCS Vol. 2045, Springer-Verlag, pp. 420-436, 2001.

10. L. Keliher, H. Meijer, and S. Tavares, Improving the upper bound on the maximum average linear hull probability for Rijndael, Selected Areas in Cryptography - SAC 2001, LNCS Vol. 2259, Springer-Verlag, pp. 112-128, 2001.
11. L. Keliher, Refined Anlaysis of Bounds Related to Linear and Differential Cryptanalysis for the AES, Fourth Conference on the Advanced Encryption Standard (AES4).
12. D. Kwon, J. Kim, S. Park, S.H. Sung, Y. Sohn, J.H. Song, Y. Yeom, E-J. Yoon, S, Lee, J. Lee, S. Chee, D. Han, and J. Hong, New block cipher: ARIA, ICISC 2003, LNCS 2971, pp. 432-445, 2004.
13. M. Matsui, Linear cryptanalysis method for DES cipher, Advance in Cryptology-Eurocrypt'93, LNCS Vol. 765, Springer-Verlag, pp. 386-397, 1994.
14. NTT-Nippon Telegraph and Telephone Corporation, Specification of E2 - a 128 bit block cipher, AES proposal(available at http://info.isl.ntt.co.jp/e2/), 1998.
15. S. Park, S.H. Sung, S. Chee, E-J. Yoon, J. Lim, On the security of Rijndael-like structures against differential and linear cryptanalysis, Aciacrypt 2002, LNCS 2051, pp. 176-191, 2002.
16. S. Park, S.H. Sung, S. Lee, J. Lim, Improving the upper bound on the maximum differential and maximum linear hull probability for SPN Structures and AES, FSE 2003, LNCS 2887, pp. 247-260, 2003.

Dragon: A Fast Word Based Stream Cipher[*]

Kevin Chen[1], Matt Henricksen[1], William Millan[1], Joanne Fuller[1],
Leonie Simpson[1], Ed Dawson[1], HoonJae Lee[2], and SangJae Moon[3]

[1] Information Security Research Centre, Queensland University of Technology,
GPO Box 2434, Brisbane Qld 4001, Australia
{k.chen, m.henricksen, b.millan, j.fuller, lr.simpson,
e.dawson}@qut.edu.au
[2] School of Internet Engineering, Dongseo University,
San 69-1, Churye-2 Dong, SaSang-Ku, Pusan 617-716, Korea
hjlee@dongseo.ac.kr
[3] School of Electronic and Electrical Engineering, Kyungpook National University,
1370, Sankyuk-dong, Taegu 702-701, Korea
sjmoon@knu.ac.kr

Abstract. This paper presents Dragon, a new stream cipher constructed
using a single word based non-linear feedback shift register and a non-
linear filter function with memory. Dragon uses a variable length key and
initialisation vector of 128 or 256 bits, and produces 64 bits of keystream
per iteration. At the heart of Dragon are two highly optimised 8×32
s-boxes. Dragon uses simple operations on 32-bit words to provide a high
degree of efficiency in a wide variety of environments, making it highly
competitive when compared with other word based stream ciphers. The
components of Dragon are designed to resist all known attacks.

Keywords: word based stream cipher, nonlinear feedback shift register,
nonlinear filter.

1 Introduction

Traditionally stream cipher design has focussed on bit based linear feedback shift
registers (LFSRs), as these are well studied and produce sequences which satisfy
common statistical criteria. In these ciphers, non-linearity is introduced into the
keystream either by some type of non-linear combining function or filter function,
or by irregular clocking, or both. However, bit based LFSRs are notoriously slow
in software. Each iteration of the cipher's update function produces only one bit
of keystream. Sparse LFSR feedback functions may be exploited in an attack,
but increasing the number of feedback taps results in a decrease in efficiency.
Also, the security of some LFSR based stream ciphers is threatened by algebraic
attacks [6].

---

[*] This research was supported by Australian Research Grant No DP0450920 and South
Korean University IT Research Center Project for Mobile Network Security Technology Research Center.

Word based stream ciphers may provide a solution to the security-efficiency tradeoff. These produce many times the amount of keystream per iteration than do bit-based LFSRs, depending on the word size. The word size used for current word based stream cipher proposals range from 8-bit words for RC4 to 32-bit words for Turing [16]. Many of these ciphers are very fast in software, outperforming even fast block ciphers like the Advanced Encryption Standard [15]. Although it is easy to assess the speed of these word based stream ciphers, it is difficult to quantify their security precisely.

This paper presents Dragon, a new word based stream cipher designed with both security and efficiency in mind. Dragon uses a word based non-linear feedback shift register (NLFSR), in conjunction with a non-linear filter to produce keystream as 64-bit words. Dragon has a throughput of gigabits per second in both modern software and hardware, and requires little more than four kilobytes of memory, so is suitable for use in constrained environments. Not only is Dragon fast for keystream generation, it is also very efficient when rekeying. This makes Dragon especially suitable for applications that require frequent rekeying, such as mobile and wireless communications.

Dragon can be considered an evolution of the output feedback mode (OFB) of block ciphers. The modifications overcome a shortcoming of block ciphers in OFB mode: the output keystream is also the feedback to the internal state. We have analysed the security of Dragon using modern cryptanalytic techniques, and believe it is suitable for use as a secure cryptographic primitive. Collision attacks based on the birthday paradox can exploit this knowledge of the feedback. These attacks are prevented in the Dragon cipher by producing separate output and feedback words from the update function. Also, ciphers with small internal states are easily attacked by time/memory/data tradeoff attacks [3]. A minimum requirement to overcome these type of attacks is to have an internal state size at least twice the designed security. Time/memory/data tradeoff attacks are prevented in the Dragon cipher by having a large internal state. To increase the difficulty of guess and determine attacks [10], Dragon selects taps from the NLFSR according to a Full Positive Difference Set (FPDS).

Section 2 presents the specification of the cipher. Section 3 describes the design decisions behind the Dragon algorithm. Section 4 and 5 includes a security analysis of Dragon using modern cryptanalytic techniques. Section 6 discusses the performance of Dragon in software and hardware, and associated implementation issues.

2   Specification of Dragon

Dragon is a stream cipher constructed using a single word based NLFSR. Dragon has a large NLFSR of 1024 bits, an update function, denoted F, and a 64-bit memory, denoted M. Dragon-256 uses a secret master key of 256 bits, and a publicly known initialisation vector (IV), also of 256 bits to accommodate rekeying scenarios, while Dragon-128 uses 128-bit key and IV. The F function, which is called once per round, manipulates the internal state to generate 64 bits of pseudo-random keystream. The cipher's key setup converts the master key and initialisation vector for use in Dragon's large internal state.

Table 1. Dragon's F Function

Input = { a, b, c, d, e, f }
Pre-mixing Layer:
1. $b = b \oplus a;$ $d = d \oplus c;$ $f = f \oplus e;$
2. $c = c \boxplus b;$ $e = e \boxplus d;$ $a = a \boxplus f;$
S-box Layer:
3. $d = d \oplus G_1(a);$ $f = f \oplus G_2(c);$ $b = b \oplus G_3(e);$
4. $a = a \oplus H_1(b);$ $c = c \oplus H_2(d);$ $e = e \oplus H_3(f);$
Post-mixing Layer:
5. $d' = d \boxplus a;$ $f' = f \boxplus c;$ $b' = b \boxplus e;$
6. $c' = c \oplus b;$ $e' = e \oplus d;$ $a' = a \oplus f;$
Output = { a', b', c', d', e', f' }

2.1 Dragon's State Update Function (*F* Function)

The F function is used in both key setup and keystream generation. The F function is a reversible mapping of 192 bits (six 32-bit words) to 192 bits. It takes six 32-bit words as input and produces six 32-bit words as output. In Table 1 the input words are denoted a, b, c, d, e, f and the output words a', b', c', d', e', f'. The F function has six component functions denoted G_1, G_2, G_3, H_1, H_2 and H_3, as described below. The G and H functions provide algebraic completeness [11] and high non-linearity. A network of modular and binary additions are used for diffusion in the F function. It can be divided to three parts: pre-mixing, substitution, and post-mixing. Each step is designed to allow for parallelisation, giving Dragon its speed. The F function is shown in Table 1 where $\oplus$ denotes XOR and $\boxplus$ denotes addition modulo 2^{32}.

G and H Functions. The G and H functions are constructed from two 8×32-bit s-boxes, S_1 and S_2 to form virtual 32×32 s-boxes. The G functions contains three S_1s and one S_2, while the H functions have three S_2s and one S_1. S_1 and S_2 are included in Appendix B. The 32-bit input is broken into four bytes $(x = x_0 \| x_1 \| x_2 \| x_3$. Each byte is passed through an 8×32 s-box and the four 32-bit outputs combined using binary addition.

G and H functions are defined as

$$G_1(x) = S_1(x_0) \oplus S_1(x_1) \oplus S_1(x_2) \oplus S_2(x_3)$$

$$G_2(x) = S_1(x_0) \oplus S_1(x_1) \oplus S_2(x_2) \oplus S_1(x_3)$$

$$G_3(x) = S_1(x_0) \oplus S_2(x_1) \oplus S_1(x_2) \oplus S_1(x_3)$$

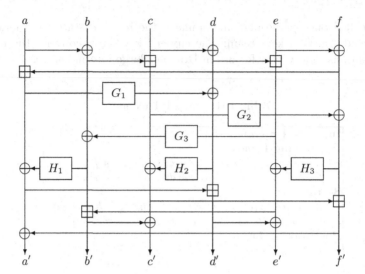

Fig. 1. F Function

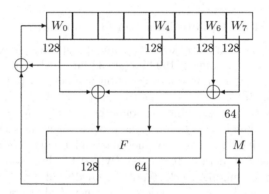

Fig. 2. Initialisation

$$H_1(x) = S_2(x_0) \oplus S_2(x_1) \oplus S_2(x_2) \oplus S_1(x_3)$$
$$H_2(x) = S_2(x_0) \oplus S_2(x_1) \oplus S_1(x_2) \oplus S_2(x_3)$$
$$H_3(x) = S_2(x_0) \oplus S_1(x_1) \oplus S_2(x_2) \oplus S_2(x_3)$$

2.2   Initialisation

Dragon can be used with two different key and initialisation vector lengths: 128-bit and 256-bit. We denote the 256-bit key and initialisation vector K and IV respectively. The 128-bit key and initialisation vector are denoted k and iv.

Dragon has a simple keying (and rekeying) strategy using the key and the publicly known initialisation vector. The 1024-bit internal state is divided into

Table 2. Dragon's Key Initialisation function

Input = { K, IV } (256-bit) Input = { k, iv } (128-bit)	
1.	$W_0 \parallel ... \parallel W_7 = K \parallel K \oplus IV \parallel \overline{K \oplus IV} \parallel IV$ (256-bit)
	$W_0 \parallel ... \parallel W_7 = k \parallel k' \oplus iv' \parallel iv \parallel k \oplus iv' \parallel k' \parallel k \oplus iv \parallel iv' \parallel k' \oplus iv$ (128-bit)
2.	$M = 0x0000447261676F6E$
Perform steps 3-8 16 times	
3.	$a \parallel b \parallel c \parallel d = (W_0 \oplus W_6 \oplus W_7)$
4.	$e \parallel f = M$
5.	$\{a', b', c', d', e', f'\} = F(a, b, c, d, e, f)$
6.	$W_0 = (a' \parallel b' \parallel c' \parallel d') \oplus W_4$
7.	$W_i = W_{i-1}$, for $i = 7$ down to 1 (shifting the state by one word)
8.	$M = e' \parallel f'$
Output = { $W_0 \parallel ... \parallel W_7$ }	

Table 3. Dragon's Keystream Generation Function

Input = { $B_0 \parallel ... \parallel B_{31}, M$ }	
1.	$(M_L \parallel M_R) = M$
2.	$a = B_0, b = B_9, c = B_{16}, d = B_{19}, e = B_{30} \oplus M_L, f = B_{31} \oplus M_R$
3.	$(a', b', c', d', e', f') = F(a, b, c, d, e, f)$
4.	$B_0 = b', B_1 = c'$
5.	$B_i = B_{i-2}, 2 \leq i \leq 31$
6.	$M = M + 1$
7.	$k = a' \parallel e'$
Output = { $k, B_0 \parallel ... \parallel B_{31}, M$ }	

eight 128-bit words, labelled W_0 to W_7. The internal state is initially filled by concatenating the key and the initialisation vector. The state initialisation process makes extensive use of the F function. The initialisation involves 16 iterations of the F function as shown in Table 2 where $\overline{x}$ denotes the complement of x and x' denotes the swapping of the upper half and the lower half of x. To protect against unknown future attacks, and against attacks that require large amounts of keystream, the cipher should be rekeyed at least once for every 2^{64} bits of keystream generated. The use of existing components for both initialisation and keystream generation simplifies analysis and increases implementation efficiency.

2.3 Keystream Generation

Dragon has a large NLFSR of 1024 bits divided into thirty two 32-bit words $B_i, 0 \leq i \leq 31$. During each round, six words from the internal state are used as inputs to the F function. The indices to these words are 0, 9, 16, 19, 30, 31, and form a Full Positive Difference Set (FPDS). Additionally, a 64-bit memory component, M, acts as a counter in keystream generation, with the initial value for keystream generation being the final value of M defined by the key initialisation process. Each round of the keystream generation results in the output of a 64-bit word k. Table 3 shows one round of keystream generation. Note the output of the process is a keystream word k, and an updated state B and memory M.

3   Design Principles of Dragon

3.1   Design of F Function

All the keystream words and feedback words are dependent on all input words, both at the bit level and word level. A single bit change in any of the six input words results in completely different keystream and feedback words.

3.2   Design of S-Boxes

Dragon uses two 8×32 S-boxes that have been designed heuristically to satisfy a range of important security related properties. They are used to create two nonlinear 32×32 mappings G and H. The simple construction (shown in Section 2.1) allows the non-linearity properties of the output bits of G and H to be calculated exactly from the known properties of the components of the underlying 8×32 S-boxes, S_1 and S_2. Both s-boxes were designed to have balanced component Boolean functions with:

- best known non-linearity of 116,
- optimum algebraic degree 6 or 7 according to Siegenthaler's tradeoff [18],
- low autocorrelation,
- distinct equivalence classes,
- all XOR pairs satisfying:
  - better than random non-linearity with 102 minimum,
  - almost balanced (the imbalance is not more than 16),
  - distinct equivalence classes,
  - same optimal degree as the components.

We adopt a standard notation (n, t, d, x, y) to describe Boolean function properties where n is the number of variables, t is the order of resiliency (where $t = 0$ indicates a balanced function), d is the algebraic degree, x is the non-linearity and y is the largest magnitude in the autocorrelation function. All the components of S_1 are $(8, 1, 6, 116, y)$ where $32 \leq y \leq 48$ which is considered sufficiently low. S_1 functions achieve the highest non-linearity possible for resilient functions. All the components of S_2 are $(8, 0, 7, 116, 24)$, where we note that the achieved autocorrelation of 24 is the lowest known for balanced functions of this size.

These s-boxes were created one output bit at a time using heuristic techniques. Existing methods [14] were adopted to generate the individual functions, then they were compared to the existing s-box functions to check the above-listed requirements for the XOR pairs. When the candidate function was acceptable it was appended to the s-box, else another function was tested. We found it was possible to generate 32 functions for each s-box while satisfying the stringent requirements outlined above.

Finally we remark that the output functions of resulting Dragon virtual s-boxes G and H have higher non-linearity (at 116) than other popular 32×32 cryptographic mappings, such as the SBOX/MIXCOL operation from AES [15] and Mugi[19], which use only 112 non-linearity functions. Also Dragon's s-boxes avoid the linear redundancy weakness that is intrinsic to finite field operation based s-boxes [9] which are used in the international standard ciphers AES [15], Camellia [2].

3.3   Design of Key Initialisation

The key setup and keystream generation of Dragon both use the F function, for ease of implementation and efficiency. However, the key setup of Dragon is deliberately designed to be different to keystream generation, so that the mapping of internal state to the feedback is different.

There are three differences between the key setup and the keystream generation: the use of the 64-bit memory M, the size of the feedback and the FPDS selection used.

From Section 3.1, F is a reversible mapping, and the design of the key setup network uses this property of F to produce a bijective process. For any unique pair of K and IV, the key setup procedure initialises the internal state and M to unique values. Note that M is used as memory in key setup, but as a counter in keystream generation.

The feedback of Dragon consists of four words of the F function outputs, totalling 128 bits in key setup (the feedback size is 64 bits in keystream generation). This means that F can mix K and IV effectively in a minimum number of rounds. A smaller number of rounds in key setup translate directly into high rekeying performance. This makes Dragon very competitive in practical applications that require frequent rekeying, such as mobile and wireless transmissions that usually use the frame number as the IV.

A different FPDS is chosen for the key setup because of the change in the size of the feedback. The taps from the internal state, $\{0,4,6,7\}$, form a FPDS both in the forward and reverse direction. This is designed to frustrate the cryptanalysis of key setup by guess and determine techniques.

Dragon-128 and Dragon-256 are designed to have very similar initialisation process so that the speed is identical. However, another important design consideration is the use 128-bit and 256-bit key and initialisation vector pairs. We ensured that no pair of 256-bit K and IV can initialise Dragon to the same state as any arbitrary pair of 128-bit k and iv. This avoids the cryptanalyst reducing the search space in a brute force attack from 256-bit to 128-bit.

4   Analysis of Cipher

4.1   Statistical Tests

Statistical tests provided by the CRYPT-X [8] package were performed on keystream produced by the Dragon cipher. The frequency, binary derivative, change point, subblock and runs tests were executed with 30 streams of Dragon output, each eight megabits in length. The sequence and linear complexity tests were executed for the 30 streams with two hundred kilobits each. Dragon passed all pertinent statistical tests.

4.2   Period Length

Given that Dragon has a 1024-bit internal state, the expected period of the internal state is 2^{512}, assuming the mapping is pseudo-random [4]. For cryptographic use, establishing the lower bound for the period of the output sequence is critical. Each round of Dragon is under the influence of a 64-bit counter, M. Since the counter M has a period of 2^{64}, the period of Dragon's internal state is lower bounded by 2^{64}. Taken together, the internal state and the counter M give Dragon an expected period of 2^{576}.

The amount of keystream produced by a unique pair of K and IV is limited to 2^{64} bits (in most applications the actual keystream would be much smaller) in the specification of Dragon. This is a small fraction of the lower bound of the period (and a very small fraction of the expected period), and therefore avoids the possibility of keystream collision attacks.

4.3   Weak Keys

Weak keys are those keys that bypass some operations of the cipher. That is, the operations have no effect in the calculation of the feedback or the output keystream.

Dragon is designed to avoid weak keys. The internal state is an NLFSR, therefore the all zero state is not a problem as Dragon is designed to avoid fixed points. While it is easy to bypass the pre-mixing phase in a single iteration of the F function by having repetitive inputs such as all zeros or all ones, it is only possible for the first of the 16 iterations of F in the key setup. Also, selected values are limited to the first four inputs of the F function, as the last two inputs take the value of M. The network of G and H functions ensure that the initial states which bypass the pre-mixing phase cannot bypass any other operations in F. We believe that the above design features provide a strong guarantee that there are no weak keys for Dragon.

5   Cryptanalysis of Dragon

5.1   Related Key and IV Attacks

The Dragon rekeying strategy is simple, and the use of initialisation vectors provides a way to reuse a master key without generating identical keystreams. The rekeying strategy prevents related key and IV attacks before even the first word of output is generated, by mixing each bit of the key into all words of the initial state over 16 rounds of the highly non-linear F function defined for the keystream generator module. This function has six 32-bit inputs and six 32-bit outputs. During rekeying, the F function is iterated 16 times, each time populating the leftmost side of the internal state with 128 bits comprising four outputs. After eight rounds, all of the initial keying material in the state has been replaced by unknown output from the F function.

Of the six inputs to the F function, four words are taken directly from the keyed internal state, while two are taken from a 64-bit memory M. The contents of this memory are initially known, since they are determined by a published constant. Also, the memory can not be manipulated by the attacker in the same way as the internal state, since it is not keyed. Two outputs from the F function feedback to the memory, making its value hard to determine after the first round. All output words of F are affected by the memory, increasing the difficulty that the attacker faces in controlling inputs to subsequent rounds.

Diffusion. One strategy in an attack is to minimise the number of words with a non-zero difference in the internal state. The aim of this strategy is controllability. The larger the number of non-zero words used as input to the non-linear function, the more complex the resulting output. The key schedule of Dragon is designed so that after 12 rounds, even an initial difference of single word difference is propagated to all words in the internal state (see Table 4). Since there are 16 rounds, this is an ample margin to ensure an attacker is unable to determine the state contents after rekeying. The speed of this diffusion is aided by the fact that the first word of the state is used as input to F function, and the output of the F function replaces the first word.

Table 4. Propagation of non-zero difference in internal state of the rekeying

1	0	ΔA	0	0	0	0	0	0
2	0	0	ΔA	0	0	0	0	0
3	0	0	0	ΔA	0	0	0	0
4	0	0	0	0	ΔA	0	0	0
5	ΔA	0	0	0	0	ΔA	0	0
6	ΔB	ΔA	0	0	0	0	ΔA	0
7	ΔC	ΔB	ΔA	0	0	0	0	ΔA
8	ΔD	ΔC	ΔB	ΔA	0	0	0	0
9	ΔE	ΔD	ΔC	ΔB	ΔA	0	0	0
10	ΔF	ΔE	ΔD	ΔC	ΔB	ΔA	0	0
11	ΔG	ΔF	ΔE	ΔD	ΔC	ΔB	ΔA	0
12	ΔH	ΔG	ΔF	ΔE	ΔD	ΔC	ΔB	ΔA

Even a single round of Dragon F function prevents high probability differentials due to its use of the G and H functions, and high diffusion. A single input difference is propagated to differences in each of the outputs. The F function consists of three layers: pre-mixing, confusion through s-box application, and post-mixing. Referring to the notation of section 2.1, only inputs a, b, c and d can be initially and indirectly controlled by an attacker, since e and f come from internal and inaccessible memory.

The attacker may wish to make use of the fact that b and d are mixed with only one other word in the pre-mixing phase, while a and c are mixed with two others. For the input $-(e \oplus f), b, -(b \oplus e \oplus f), -(b \oplus e \oplus f), e, e \oplus f)$ the pre-mixing stage produces the output $(0, b \oplus -(e \oplus f), 0, 0, e, e \oplus f)$. For difference input Δb, this produces the difference $(0, \Delta b, 0, 0, 0, 0)$ since e and f are at this stage constants. This bypasses the G row of s-boxes and activates a single s-box in the second row to produce the post-mixing input $(\Delta H_1(\Delta b), \Delta b, 0, 0, 0, 0)$. The post-mixing output is $(\Delta H_1(\Delta b), \Delta b, \Delta b, (\Delta H_1(\Delta b), 0, 0))$.

At this stage, all of the feedback words to the internal state are non-zero. However, the difference of the feedback to the internal state is still zero. This fact cannot be exploited by the attacker since the input differences to this round are not reproducible in later rounds, and thus the difference of the internal memory cannot be maintained. Consequently Dragon is not vulnerable to related key attacks that are more efficient than a brute force search of the 256-bit key.

5.2   Time-Memory Tradeoff Attacks

Time-Memory tradeoff attacks [3] rely on pre-computation to reduce the effort required for a key recovery attack on a keystream. The attack comprises two steps. The first, the preprocessing step, sees the attacker calculating a table of keys or internal states and corresponding keystream prefixes. The table is ordered upon the prefix. The second step involves observing keystreams and attempting to match each against a prefix in the table. If the match is successful, then with some likelihood the internal state is known by reading the opposing entry in the table.

The parameters in an attack are time (T), memory (M), and amount of data (D). Generally, $T \times M^2 \times D^2 = S^2$ where S is the state space of the cipher, and $D^2 \leq T$ [3]. The pre-computation time P is equal to $S \div D$.

Dragon has an internal state space of 1088 bits (including the 64-bit memory). Since the design strength of Dragon is 256 bits, the time-memory tradeoff attack is infeasible. For the brute-force equivalent attack with $T = 2^{256}$, data requirements are limited to 2^{64} bits, which imposes a lower bound on memory for the attack of 2^{896} bits.

5.3   Guess and Determine Attacks

The indices $\{0, 9, 16, 19, 30, 31\}$ of the state elements used in Dragon's update function form a full positive difference set. This is a design decision to prevent guess and determine attacks [10].

In keystream generation, guessing six inputs (192 bits) to F in a round allows an attacker to calculate the feedback words b' and c' and the keystream words a' and e', which can be used to discard most incorrect guesses. At this point the attacker has knowledge of the state words at indices $\{0, 1, 10, 17, 20\}$ and some information about the value of B_{31} and M. However, the FPDS selection of the internal state means that to obtain the next pair of keystream words, guessing a further five inputs (160 bits) is necessary. The attacker can attempt to jump ahead to a future keystream word pair, but again the FPDS means that the attacker needs to guess five inputs. This rapid increase in the number of possible guess pathways makes the attack infeasible. In addition, the interplay of B_{30}, B_{31} and M means there will be more than one set of values for these three elements for a unique pair of e and f, further complicating the cryptanalytic attempt by guess and determine.

The attacker is unable to reduce the complexity of a guess and determine attack by guessing individual state bytes, rather than whole words. The use of large s-boxes (G and H functions are effectively 32×32 s-boxes) means that guessing three of the four input bytes is insufficient to deduce any byte of the s-box output.

To calculate keystream words from two rounds of Dragon, the attacker is required to guess more than 256 bits of the internal state. This is worse than exhaustive key search, and makes guess and determine attacks on Dragon infeasible.

5.4   Distinguishing Attacks

If the output sequence of a stream cipher can be statistically distinguished from a random sequence, then the cipher is not strong enough for cryptographic applications. Dragon is designed with a large state and complex initialisation and update function. It has no linear masking, and therefore immune to this type of distinguishing attacks [5]. It is expected to have a very large period of 2^{582} (with a lower bound of 2^{64} because of the influence of the 64-bit counter) and it passes standard statistically tests for randomness. The amount of keystream output for an unique pair key and initialisation vector is limited to 2^{64} bits. We conjecture that it is impractical to collect an amount of output sufficient to distinguish Dragon keystream output from a random binary sequence.

5.5   Linear Approximations

From Lemma 15 of [17], the non-linearity of the sum of disjoint functions can be calculated as follows. Let $g(x_1, \ldots, x_s, y_1, \ldots, y_t) = f_1(x_1, \ldots, x_s) \oplus f_2(y_1, \ldots, y_t)$. Then the non-linearity of g satisfies $N_g \geq 2^{s+t-1} - \frac{1}{2}P_1 \cdot P_2$ where P_1 and P_2 are the maximum Walsh-Hadamard transform values of f_1 and f_2, respectively.

The G and H functions of Dragon are composed from two 8×32 s-boxes, S_1 and S_2. Both s-boxes have all outputs with non-linearity 116, therefore $P_{S_1} = P_{S_2} = 2^8 - 2 \cdot 116 = 24$. The non-linearity of the output bits of the G and H functions can then be calculated as $N_G = N_H \geq 2^{8+8+8+8-1} - \frac{1}{2} \cdot 24 \cdot 24 \cdot 24 \cdot 24 = 2^{31} - 165888$. The best affine approximation to the G or H function output bits has bias no greater than $\frac{2^{31} - 2^{31} + 165888}{2^{31}} = 2^{-14.66}$. At any given round, the keystream words of Dragon are the results of five G or H functions each, hence the best affine approximation to the Dragon F function output bits has bias no greater than $(2^{-14.66})^5 = 2^{-73.3}$.

Linear cryptanalysis requires equations relating the key bits to the internal state bits, and in turn the keystream bits, where the internal state variables can be cancelled. The complete mixing of Dragon's key setup avoids the divide and conquer approach, therefore all the internal state variables are needed in the linear equations. The output keystream will be dependent on all 1,024 bits of the initial internal state after 8 iterations of F. The bias of the best affine approximation over 8 iterations of F is no greater than $(2^{-73.3})^8 = 2^{-586.4}$. As the key size of Dragon is 256 bits, attack on Dragon using linear approximation has complexity greater than exhaustive key search.

5.6   Algebraic Attacks

Successful algebraic attacks on keystream generators [6] have so far been restricted mainly to LFSR based generators. The general attack model consists of the internal state S, the linear update function L and the output function f. Let S_0 denote the internal state at time $t = 0$, and $L^t(S_0)$ denote the internal state at time t. The attacker constructs a system of equations relating the internal state bits with the observed keystream bits, where $z_t = f(L^t(S_0))$ at time t. The attacker can set up a large number of equations just by merely collecting keystream bits, since the internal state at time t can easily be derived from the linear nature of LFSRs.

This model cannot be applied to Dragon since the update function is non-linear. Let the non-linear update function be N, then the equation becomes $z_t = f(N^t(S_0))$. Note that N has a poor linear approximation of $2^{-73.3}$ as shown in Section 5.5. The lack of the linear update function means the attacker can not simply calculate the internal state at time t to construct the system of equations.

When constructing the system of equations for Dragon, the degree of equations would grow exponentially. This is easy to see as any output of G or H is a degree 7 function of the inputs since S_2 has algebraic order 7. If we approximate $\boxplus$ with $\oplus$, we can then write equations of degree $7^2 = 49$ that maps the 192 input bits to the first 64 output keystream bits. However, the feedback is used immediately in the production of the next 64 bits of keystream, and results in equations of degree $7^4 = 2,401$. Note that at this point, the inputs consist of only 352 bits, and therefore the equations would be limited to degree 352. The degree of the equations would grow to the full 1024 bits of the internal state, after 8 iterations of the F function, or 512 bits of keystream produced.

Using the technique published in [7] to describe the 8×32 s-boxes of Dragon using quadratic equations results in 565 quadratic equations in 256 monomials for each s-box (identical to the analysis of CAST [1]). Again, let us approximate $\boxplus$ with $\oplus$, then after 8 iterations of F, the system of equations has degree 1,024 as well. This is to say, even if there existed some annihilators [13] that reduce Dragon's Boolean functions right down to quadratic, the degree of the overall equations would still grow to unmanageable sizes.

It is clear that the system of equations for Dragon will be very difficult to solve, if it is solvable at all. Furthermore, it will require far more effort than exhaustive key search since solving techniques all have complexities exponential in the degree of the equations. It is interesting to note that the above analysis approximates modular addition with XOR, and thus resulting in a weaker version of Dragon. With the modular addition in place, it will be even more difficult for algebraic attacks to succeed against Dragon (see similar example of the effect of modular addition in CAST [1]).

6   Implementation and Performance

Dragon is designed to be efficient in both software and hardware, in terms of throughput and a small implementation footprint. Its 32-bit word size is chosen to match that of the ubiquitous Intel Pentium family, since this leads to the best software efficiency on that platform. Note that the results presented in this Section apply to both Dragon-128 and Dragon-256.

6.1   Software

On an Intel Pentium 4, a naïve C implementation of Dragon produces one byte of keystream every 6.74 clock cycles, and 1,395 cycles per rekeying operation. On a 3.2GHz Pentium 4, the throughput of Dragon is 3.8Gbps. This is competitive with many of its peers, including SNOW 2 (5.5 cycles/byte), Turing (6.1 cycles/byte) and RC4 (7.1 cycles/byte).

Storage requirements include 2,048 bytes to store Dragon's two 8×32 s-boxes, 1,024 bits (128 bytes) for the internal state, and a further 8 bytes for the 64-bit counter. Including temporary variables and an object code size of 2,810 bytes, Dragon has memory requirements totalling 4,994 bytes. This is suitable for even very constrained environments.

A reference implementation of Dragon written in C can be obtained from http://www.isrc.qut.edu.au/resource/dragon/.

6.2   Hardware

The design of Dragon allows high degree of parallelisation in hardware. The operations on the six inputs of the F function can be divided into three groups, each operating on two inputs. The pre-mixing and the post-mixing are implemented using 32-bit modular adders. The G and H functions are implemented using look-up tables and XOR operations. The hardware complexity is about 6,524 gates and 196,672 bits of memory. On Samsung 0.13um ASIC running at 2.6GHz, the minimum delay is 2.774ns with a throughput of 23Gbps.

The speed in hardware can be improved by using m-parallel-structure proposed in [12]. This hardware implementation strategy applies to all shift registers, and achieves an m times increase in efficiency with m times increase in hardware complexity. On Altera FPGA/CPLD running at 16.67MHz, an implementation of Dragon achieves a throughput of 1.06Gbps with 16 times hardware complexity.

7 Conclusion

This paper presents Dragon, a new stream cipher constructed upon a word based non-linear feedback shift register. The key and initialisation vector are 128 bits for Dragon-128 and 256 bits for Dragon-256. Dragon is designed with both security and efficiency in mind. It has been shown that Dragon is secure against all known cryptanalytic attacks.

References

1. C. Adams. Designing Against the 'Overdefined System of Equations' Attack, May 2004. http://eprint.iacr.org/2004/110/.
2. K. Aoki, T. Ichikawa, M. Kanda, M. Matsui, S. Moriai, J. Nakajima, and T. Tokita. Camellia: A 128-Bit Block Cipher Suitable for Multiple Platforms - Design and Analysis. In D. Stinson and S. Tavares, editors, *Selected Areas in Cryptography: 7th Annual International Workshop, SAC 2000, Waterloo, Ontario, Canada, August 14-15, 2000. Proceedings*, volume 2012 of *Lecture Notes in Computer Science*, pages 39–56, Berlin Heidelberg, 2000. Springer-Verlag.
3. A. Biryukov and A. Shamir. Cryptanalytic Time/Memory/Data Tradeoffs for Stream Ciphers. In T. Okamoto, editor, *Advances in Cryptology - ASIACRYPT 2000, 6th International Conference on the Theory and Application of Cryptology and Information Security, Kyoto, Japan, December 3-7, 2000. Proceedings*, volume 1976 of *Lecture Notes in Computer Science*, pages 1–13, Berlin Heidelberg, 2000. Springer-Verlag.
4. W. Chambers. On Random Mappings and Random Permutations. In B. Preneel, editor, *Fast Software Encryption: Second International Workshop. Leuven, Belgium, 14-16 December 1994. Proceedings*, volume 1008 of *Lecture Notes in Computer Science*, pages 22–28, Berlin Heidelberg, 1995. Springer-Verlag.
5. D. Coppersmith, S. Halevi, and C. Jutla. Cryptanalysis of Stream Ciphers with Linear Masking. In M. Yung, editor, *Advances in Cryptology - CRYPTO 2002, 22nd Annual International Cryptology Conference, Santa Barbara, California, USA, August 18-22, 2002, Proceedings*, volume 2442 of *Lecture Notes in Computer Science*, pages 515–532, Berlin Heidelberg, 2002. Springer-Verlag.
6. N. Courtois. Higher Order Correlation Attacks, XL Algorithm and Cryptanalysis of Toyocrypt. In P. Lee and C. Lim, editors, *Information Security and Cryptology - ICISC 2002, 5th International Conference Seoul, Korea, November 28-29, 2002, Revised Papers*, volume 2587 of *Lecture Notes in Computer Science*, pages 182–199, Berlin Heidelberg, 2003. Springer-Verlag.
7. N. Courtois and J. Pieprzyk. Cryptanalysis of Block Ciphers with Overdefined Systems of Equations. In Y. Zheng, editor, *Advances in Cryptology - ASIACRYPT 2002, 8th International Conference on the Theory and Application of Cryptology and Information Security, Queenstown, New Zealand, December 1-5, 2002. Proceedings*, volume 2501 of *Lecture Notes in Computer Science*, pages 267–287, Berlin Heidelberg, 2002. Springer-Verlag.
8. E. Dawson, A. Clark, G. Gustafson, and L. May. *CRYPT-X'98 User Manual*, 1999.
9. J. Fuller and W. Millan. Linear Redundancy in S-Boxes. In T. Johansson, editor, *Fast Software Encryption, 10th International Workshop, FSE 2003, Lund, Sweden, February 24-26, 2003. Revised Papers*, volume 2887 of *Lecture Notes in Computer Science*, pages 74–86, Berlin Heidelberg, 2003. Springer-Verlag.

10. P. Hawkes and G. Rose. Guess-and-Determine Attacks on SNOW. In K. Nyberg and H. Heys, editors, *Selected Areas in Cryptography, 9th Annual International Workshop, SAC 2002, St. John's, Newfoundland, Canada, August 15-16, 2002. Revised Papers*, volume 2595 of *Lecture Notes in Computer Science*, pages 37–46, Berlin Heidelberg, 2003. Springer-Verlag.

11. J. Kam and G. Davida. Structured Design of Substitution-Permutation Encryption Networks. *IEEE Transactions on Computers*, 28(10):747–753, October 1979.

12. H. Lee and S. Moon. Parallel Stream Cipher for Secure High-Speed Communications. *Signal Processing*, 82(2):137–143, February 2002.

13. W. Meier, E. Pasalic, and C. Carlet. Algebraic Attacks and Decomposition of Boolean Functions. In C. Cachin and J. Camenisch, editors, *Advances in Cryptology - EUROCRYPT 2004, International Conference on the Theory and Applications of Cryptographic Techniques, Interlaken, Switzerland, May 2-6, 2004, Proceedings*, volume 3027 of *Lecture Notes in Computer Science*, pages 474–491, Berlin Heidelberg, 2004. Springer-Verlag.

14. W. Millan, J. Fuller, and E. Dawson. New Concepts in Evolutionary Search for Boolean Functions in Cryptology. In *The 2003 Congress on Evolutionary Computation, 2003. CEC '03.*, volume 3, pages 2157–2164. IEEE, 2003.

15. National Institute of Standards and Technology. Federal Information Processing Standards Publication 197, 2001.

16. G. Rose and P. Hawkes. Turing: A Fast Stream Cipher. In T. Johansson, editor, *Fast Software Encryption, 10th International Workshop, FSE 2003, Lund, Sweden, February 24-26, 2003. Revised Papers*, volume 2887 of *Lecture Notes in Computer Science*, pages 290–306, Berlin Heidelberg, 2003. Springer-Verlag.

17. J. Seberry, X. Zhang, and Y. Zheng. Nonlinearly Balanced Boolean Functions and Their Propagation Characteristics. In D. Stinson, editor, *Advances in Cryptology - CRYPTO '93, 13th Annual International Cryptology Conference, Santa Barbara, California, USA, August 22-26, 1993. Proceedings*, volume 773 of *Lecture Notes in Computer Science*, pages 49–60, Berlin Heidelberg, 1994. Springer-Verlag.

18. T. Siegenthaler. Correlation Immunity of Nonlinear Combining Functions for Cryptographic Applications. *IEEE Transactions on Information Theory*, 30(5):776–780, September 1984.

19. D. Watanabe, S. Furuya, H. Yoshida, K. Takaragi, and B. Preneel. A New Keystream Generator MUGI. In J. Daemen and V. Rijmen, editors, *Fast Software Encryption, 9th International Workshop, FSE 2002, Leuven, Belgium, February 4-6, 2002. Revised Papers*, volume 2365 of *Lecture Notes in Computer Science*, pages 179–194, Berlin Heidelberg, 2002. Springer-Verlag.

A   Test Vectors

128-BIT KEY AND IV

KEY:
00001111 22223333 44445555 66667777
IV:
00001111 22223333 44445555 66667777
KEYSTREAM:
99B3AA14 B63BD02F E14358A4 54950425 F4B0D3FD 8BA69178 E0392938 A718C165
2E3BEB1E 11613D58 9EABB9F5 43A1C51C 73C1F227 9D1CAEA8 5C55F539 BAFD3C59
ECAC88BD 17EB1C9D A28DD63E 9093C913 3032D918 3A9B33BC 2933A79D 75669827
20EF3004 C53B0253 7A1BE796 29F8D9A3 8DC1FD31 ED9D1100 B07DFFB1 AC75EB31

KEY:
00112233 44556677 8899AABB CCDDEEFF
IV:
00112233 44556677 8899AABB CCDDEEFF
KEYSTREAM:
98821506 0E87E695 EB7AEF36 313FF910 E6C7312F 30357424 4922043D 98146EE2
202D4D49 6C602ECC 937DD3F4 E39BE26C 849DB415 F04C540E 88588C7A A3C65A31
E2156229 1E86028B 3F5A21B9 4A94C135 B3A01527 747E6521 FFEE14F0 FA1FCC73
74C8B204 4009F57D 1D63007E F1D8D221 E429EBA8 60F56098 45891D74 716694B2

256-BIT KEY AND IV

KEY:
00001111 22223333 44445555 66667777 88889999 AAAABBBB CCCCDDDD EEEEFFFF
IV:
00001111 22223333 44445555 66667777 88889999 AAAABBBB CCCCDDDD EEEEFFFF
KEYSTREAM:
BC020767 DC48DAE3 14778D8C 927E8B32 E086C6CD E593C008 600C9D47 A488F622
3A2B94D6 B853D644 27E93362 ABB8BA21 751CAAF7 BD316595 2A37FC1E A3F12FE2
5C133BA7 4C15CE4B 3542FDF8 93DAA751 F5710256 49795D54 31914EBA 0DE2C2A7
8013D29B 56D4A028 3EB6F312 7644ECFE 38B9CA11 1924FBC9 4A0A30F2 AFFF5FE0

KEY:
00112233 44556677 8899AABB CCDDEEFF 00112233 44556677 8899AABB CCDDEEFF
IV:
00112233 44556677 8899AABB CCDDEEFF 00112233 44556677 8899AABB CCDDEEFF
KEYSTREAM:
8D3AB9BA 01DAA3EB 5CBD0F6D E3ECFCAB 619AF808 CF9C4A42 E2877766 6D2D7037
EE6F94AC 29D1EEE5 340DB047 8E91A679 480D8D88 2367CE2A 31C96AD4 49E70756
815EBEB2 290DBA7A 3CCB76A2 257BD122 2B0B7AED 917FAFFF 6B58B2B2 B05F24F6
E271A016 9E897BEF F5C22451 DA6F9E40 52B78BE5 6C97C1A5 C6F8E791 0F7B9C98

B   Dragon's S-Boxes

```
sbox1[256]={
0x393BCE6B,0x232BA00D,0x84E18ADA,0x84557BA7,0x56828948,0x166908F3,
0x414A3437,0x7BB44897,0x2315BE89,0x7A01F224,0x7056AA5D,0x121A3917,
0xE3F47FA2,0x1F99D0AD,0x9BAD518B,0x99B9E75F,0x8829A7ED,0x2C511CA9,
0x1D89BF75,0xF2F8CDD0,0x2DA2C498,0x48314C42,0x922D9AF6,0xAA6CE00C,
0xAC66E078,0x7D4CB0C0,0x5500C6E8,0x23E4576B,0x6B365D40,0xEE171139,
0x336BE860,0x5DBEEEFE,0x0E945776,0xD4D52CC4,0x0E9BB490,0x376EB6FD,
0x6D891655,0xD4078FEE,0xE07401E7,0xA1E4350C,0xABC78246,0x73409C02,
0x24704A1F,0x478ABB2C,0xA0849634,0x9E9E5FEB,0x77363D8D,0xD350BC21,
0x876E1BB5,0xC8F55C9D,0xD112F39F,0xDF1A0245,0x9711B3F0,0xA3534F64,
0x42FB629E,0x15EAD26A,0xD1CFA296,0x7B445FEE,0x88C28D4A,0xCA6A8992,
0xB40726AB,0x508C65BC,0xBE87B3B9,0x4A894942,0x9AEECC5B,0x6CA6F10B,
0x303F8934,0xD7A8693A,0x7C8A16E4,0xB8CF0AC9,0xAD14B784,0x819FF9F0,
0xF20DCDFA,0xB7CB7159,0x58F3199F,0x9855E43B,0x1DF6C2D6,0x46114185,
0xE46F5D0F,0xAAC70B5B,0x48590537,0x0FD77B28,0x67D16C70,0x75AE53F4,
0xF7BFECA1,0x6017B2D2,0xD8A0FA28,0xB8FC2E0D,0x80168E15,0x0D7DEC9D,
0xC5581F55,0xBE4A2783,0xD27012FE,0x53EA81CA,0xEBAA07D2,0x54F5D41D,
0xABB26FA6,0x41B9EAD9,0xA48174C7,0x1F3026F0,0xEFBADD8E,0x387E9014,
0x1505AB79,0xEADF0DF7,0x67755401,0xDA2EF962,0x41670B0E,0x0E8642F2,
0xCE486070,0xA47D3312,0x4D7343A7,0xECDA58D0,0x1F79D536,0xD362576B,
0x9D3A6023,0xC795A610,0xAE4DF639,0x60C0B14E,0xC6DD8E02,0xBDE93F4E,
0xB7C3B0FF,0x2BE6BCAD,0xE4B3FDFD,0x79897325,0x3038798B,0x08AE6353,
0x7D1D20EB,0x3B208D21,0xD0D6D104,0xC5244327,0x9893F59F,0xE976832A,
0xB1EB320B,0xA409D915,0x7EC6B543,0x66E54F98,0x5FF805DC,0x599B223F,
0xAD78B682,0x2CF5C6E8,0x4FC71D63,0x08F8FED1,0x81C3C49A,0xE4D0A778,
0xB5D369CC,0x2DA336BE,0x76BC87CB,0x957A1878,0xFA136FBA,0x8F3C0E7B,
0x7A1FF157,0x598324AE,0xFFBAAC22,0xD67DE9E6,0x3EB52897,0x4E07E855,
0x87CE73F5,0x8D046706,0xD42D18F2,0xE71B1727,0x38473B38,0xB37B24D5,
0x381C6AE1,0xE77D6589,0x6018CBFF,0x93CF3752,0x9B6EA235,0x504A50E8,
0x464EA180,0x86AFBE5E,0xCC2D6AB0,0xAB91707B,0x1DB4D579,0xF9FAFD24,
0x2B28CC54,0xCDCFD6B3,0x68A30978,0x43A6DFD7,0xC81DD98E,0xA6C2FD31,
0x0FD07543,0xAFB400CC,0x5AF11A03,0x2647A909,0x24791387,0x5CFB4802,
0x88CE4D29,0x353F5F5E,0x7038F851,0xF1F1C0AF,0x78EC6335,0xF2201AD1,
0xDF403561,0x4462DFC7,0xE22C5044,0x9C829EA3,0x43FD6EAE,0x7A42B3A7,
0x5BFAAAEC,0x3E046853,0x5789D266,0xE1219370,0xB2C420F8,0x3218BD4E,
0x84590D94,0xD51D3A8C,0xA3AB3D24,0x2A339E3D,0xFEE67A23,0xAF844391,
0x17465609,0xA99AD0A1,0x05CA597B,0x6024A656,0x0BF05203,0x8F559DDC,
0x894A1911,0x909F21B4,0x6A7B63CE,0xE28DD7E7,0x4178AA3D,0x4346A7AA,
0xA1845E4C,0x166735F4,0x639CA159,0x58940419,0x4E4F177A,0xD17959B2,
0x12AA6FFD,0x1D39A8BE,0x7667F5AC,0xED0CE165,0xF1658FD8,0x28B04E02,
0x1FA480CF,0xD3FB6FEF,0xED336CCB,0x9EE3CA39,0x9F224202,0x2D12D6E8,
0xFAAC50CE,0xFA1E98AE,0x61498532,0x03678CC0,0x9E85EFD7,0x3069CE1A,
0xF115D008,0x4553AA9F,0x3194BE09,0xB4A9367D,0x0A9DFEEC,0x7CA002D6,
0x8E53A875,0x965E8183,0x14D79DAC,0x0192B555};
```

```
sbox2[256]={
 0xA94BC384,0xF7A81CAE,0xAB84ECD4,0x00DEF340,0x8E2329B8,0x23AF3A22,
 0x23C241FA,0xAED8729E,0x2E59357F,0xC3ED78AB,0x687724BB,0x7663886F,
 0x1669AA35,0x5966EAC1,0xD574C543,0xDBC3F2FF,0x4DD44303,0xCD4F8D01,
 0x0CBF1D6F,0xA8169D59,0x87841E00,0x3C515AD4,0x708784D6,0x13EB675F,
 0x57592B96,0x07836744,0x3E721D90,0x26DAA84F,0x253A4E4D,0xE4FA37D5,
 0x9C0830E4,0xD7F20466,0xD41745BD,0x1275129B,0x33D0F724,0xE234C68A,
 0x4CA1F260,0x2BB0B2B6,0xBD543A87,0x4ABD3789,0x87A84A81,0x948104EB,
 0xA9AAC3EA,0xBAC5B4FE,0xD4479EB6,0xC4108568,0xE144693B,0x5760C117,
 0x48A9A1A6,0xA987B887,0xDF7C74E0,0xBC0682D7,0xEDB7705D,0x57BFFEAA,
 0x8A0BD4F1,0x1A98D448,0xEA4615C9,0x99E0CBD6,0x780E39A3,0xADBCD406,
 0x84DA1362,0x7A0E984B,0xBED853E6,0xD05D610B,0x9CAC6A28,0x1682ACDF,
 0x889F605F,0x9EE2FEBA,0xDB556C92,0x86818021,0x3CC5BEA1,0x75A934C6,
 0x95574478,0x31A92B9B,0xBFE3E92B,0xB28067AE,0xD862D848,0x0732A22D,
 0x840EF879,0x79FFA920,0x0124C8BB,0x26C75B69,0xC3DAAAC5,0x6E71F2E9,
 0x9FD4AFA6,0x474D0702,0x8B6AD73E,0xF5714E20,0xE608A352,0x2BF644F8,
 0x4DF9A8BC,0xB71EAD7E,0x6335F5FB,0x0A271CE3,0xD2B552BB,0x3834A0C3,
 0x341C5908,0x0674A87B,0x8C87C0F1,0xFF0842FC,0x48C46BDB,0x30826DF8,
 0x8B82CE8E,0x0235C905,0xDE4844C3,0x296DF078,0xEFAA6FEA,0x6CB98D67,
 0x6E959632,0xD5D3732F,0x68D95F19,0x43FC0148,0xF808C7B1,0xD45DBD5D,
 0x5DD1B83B,0x8BA824FD,0xC0449E98,0xB743CC56,0x41FADDAC,0x141E9B1C,
 0x8B937233,0x9B59DCA7,0xF1C871AD,0x6C678B4D,0x46617752,0xAAE49354,
 0xCABE8156,0x6D0AC54C,0x680CA74C,0x5CD82B3F,0xA1C72A59,0x336EFB54,
 0xD3B1A748,0xF4EB40D5,0x0ADB36CF,0x59FA1CE0,0x2C694FF9,0x5CE2F81A,
 0x469B9E34,0xCE74A493,0x08B55111,0xEDED517C,0x1695D6FE,0xE37C7EC7,
 0x57827B93,0x0E02A748,0x6E4A9C0F,0x4D840764,0x9DFFC45C,0x891D29D7,
 0xF9AD0D52,0x3F663F69,0xD00A91B9,0x615E2398,0xEDBBC423,0x09397968,
 0xE42D6B68,0x24C7EFB1,0x384D472C,0x3F0CE39F,0xD02E9787,0xC326F415,
 0x9E135320,0x150CB9E2,0xED94AFC7,0x236EAB0F,0x596807A0,0x0BD61C36,
 0xA29E8F57,0x0D8099A5,0x520200EA,0xD11FF96C,0x5FF47467,0x575C0B39,
 0x0FC89690,0xB1FBACE8,0x7A957D16,0xB54D9F76,0x21DC77FB,0x6DE85CF5,
 0xBFE7AEE9,0xC49571A9,0x7F1DE4DA,0x29E03484,0x786BA455,0xC26E2109,
 0x4A0215F4,0x44BFF99C,0x711A2414,0xFDE9CDD0,0xDCE15B77,0x66D37887,
 0xF006CB92,0x27429119,0xF37B9784,0x9BE182D9,0xF21B8C34,0x732CAD2D,
 0xAF8A6A60,0x33A5D3AF,0x633E2688,0x5EAB5FD1,0x23E6017A,0xAC27A7CF,
 0xF0FC5A0E,0xCC857A5D,0x20FB7B56,0x3241F4CD,0xE132B8F7,0x4BB37056,
 0xDA1D5F94,0x76E08321,0xE1936A9C,0x876C99C3,0x2B8A5877,0xEB6E3836,
 0x9ED8A201,0xB49B5122,0xB1199638,0xA0A4AF2B,0x15F50A42,0x775F3759,
 0x41291099,0xB6131D94,0x9A563075,0x224D1EB1,0x12BB0FA2,0xFF9BFC8C,
 0x58237F23,0x98EF2A15,0xD6BCCF8A,0xB340DC66,0x0D7743F0,0x13372812,
 0x6279F82B,0x4E45E519,0x98B4BE06,0x71375BAE,0x2173ED47,0x14148267,
 0xB7AB85B5,0xA875E314,0x1372F18D,0xFD105270,0xB83F161F,0x5C175260,
 0x44FFD49F,0xD428C4F6,0x2C2002FC,0xF2797BAF,0xA3B20A4E,0xB9BF1A89,
 0xE4ABA5E2,0xC912C58D,0x96516F9A,0x51561E77};
```

An Efficient and Verifiable Solution to the Millionaire Problem

Kun Peng[1], Colin Boyd[1], Ed Dawson[1], and Byoungcheon Lee[1,2]

[1] Information Security Research Centre,
IT Faculty, Queensland University of Technology
{k.peng, c.boyd, e.dawson}@qut.edu.au
http://www.isrc.qut.edu.au
[2] Joongbu University, Korea
sultan@joongbu.ac.kr

Abstract. A new solution to the millionaire problem is designed on the base of two new techniques: zero test and batch equation. Zero test is a technique used to test whether one or more ciphertext contains a zero without revealing other information. Batch equation is a technique used to test equality of multiple integers. Combination of these two techniques produces the only known solution to the millionaire problem that is correct, private, publicly verifiable and efficient at the same time.

Keywords: millionaire problem, efficiency, verifiability, zero test, batch equation.

1 Introduction

In the millionaire problem, two millionaires want to compare their richness without revealing their wealth. This problem can be formulated as a comparison of two ciphertexts without decrypting them. Many solutions to the millionaire problem [3, 6, 8, 9, 10, 13, 16, 15, 24, 19, 20, 4, 5, 31, 12, 2] have been proposed. However, none of them are both verifiable and efficient.

In this paper, a new solution to the millionaire problem is proposed. This new solution is based on two new techniques: zero test and batch equation. The zero test is an interactive multiparty protocol which takes as input one or more ciphertexts and outputs 0 if at least one ciphertext is an encryption of 0, and outputs 1 otherwise. Batch equation allows equality between multiple pairs to be checked simultaneously, using randomised inputs. A circuit to solve the millionaire problem is reduced to a zero test with the help of homomorphism of the employed encryption algorithm and batch equation. Then the zero test is performed by some participants (e.g. the two millionaires themselves) without revealing their wealth. This scheme is the only known correct, private, publicly verifiable and efficient solution to the millionaire problem.

The structure of the rest of this paper is as follows. In Section 2, the millionaire problem is introduced and drawbacks of the currently existing solutions are

C. Park and S. Chee (Eds.): ICISC 2004, LNCS 3506, pp. 51–66, 2005.
© Springer-Verlag Berlin Heidelberg 2005

pointed out. In Section 3, fundamental cryptographic primitives to be employed in this paper are recalled. In Section 4 and Section 5, an original cryptographic primitive — zero test — and a theorem about batch equation are proposed. In Section 6 and Section 7, a novel solution to the millionaire problem is presented and analysed. In Section 8, the paper is concluded.

2   The Millionaire Problem

In the millionaire problem, two millionaires want to compare who is richer without revealing their wealth. So they encrypt their wealth and the two ciphertexts should be compared. Some participants (often the two millionaires themselves) are employed to process the two ciphertexts and find out which contains a larger message. Any solution to the millionaire problem must be correct, private and verifiable according to the following standards.

- Correctness: If every participant is honest, the correct result is obtained.
- Privacy: After the computation, different entities' knowledge about the two messages is as follows:
  - a millionaire: his wealth, the result and what can be deduced from them;
  - others: at most the result and what can be deduced from it.
- Verifiability: Each participant can verify that the other participants are honest in their computation.

Fischlin [12] argued that the participants have no motivation to deviate from the protocol if they are input providers (millionaires). To support his claim, he gave an example, the "flighting ticket" problem and solved it as an application of the millionaire problem. However, verifiability is necessary to a solution to the millionaire problem in a general sense and needed in many of its applications like auctions. Even in the "flighting ticket" problem, motivation to deviate cannot be completely omitted and verifiability may still be needed.

Most current solutions to the millionaire problem are general-purpose and can deal with other applications than the millionaire problem, while some (like [12]) only deal with the millionaire problem. Two methods have been used to solve the millionaire problem. The first method is based on encrypted truth tables. Namely, a truth table of each logic gate in a circuit is encrypted and the rows in every table are shuffled, so that each gate can be evaluated with its inputs and output in ciphertext. The second method is based on logic homomorphism of encryption schemes. As special encryption algorithms are designed to be homomorphic in regard of the logical relation in the gates in the circuit, the evaluation can be realized by computing the ciphertexts of the inputs to the function without the help of any truth table.

The recent schemes employing the first method include [24], [20], [10], [19] and [5]. In [24], a circuit is generated by an authority AI and sent to another authority A, who uses it to process the ciphertext inputs. A hash function is employed in the truth tables to link their inputs to their outputs. Oblivious transfer is employed to submit the inputs to the function confidentially. Correctness of the circuit is guaranteed by a cut-and-choose mechanism. Correctness of the computation is guaranteed by one-wayness of the hash function and an assumption that AI and A do not conspire. As the oblivious transfer primitive employed in [24] is not verifiable, AI can modify the inputs to the circuit without being detected. This problem was fixed by Juels and Szydlo [20]. They design a primitive called verifiable 1-out-of-2 oblivious transfer, which is slightly less efficient than the 1-out-of-2 oblivious transfer in [24], but prevents AI from cheating alone. Other drawbacks of [24] are (1) the cut-and-choose mechanism to guarantee circuit correctness is highly inefficient in communication as a few circuits must be transported from AI to A; (2) correctness of the auction relies on trust that the two authorities do not collude and is not publicly verifiable[1]; (3) the oblivious transfer used for bid submission is not efficient (both in computation and communication).

In [10], [19] and [5], correctness of circuit and evaluation can be publicly verified with help of public-key cryptology. So the costly cut-and-choose mechanism is removed and correctness is not based on any trust. However, public-key cryptology is much less efficient in computation than the hash function computation in [24] and [20]. As the number of gates in a circuit is not small and construction, evaluation and the corresponding validity verification in each gate requires hundreds of exponentiations, an extremely high computational cost.

The recent schemes employing the second method include [31], [2] and [12]. The schemes in [31] and [12] limit the computation to a two-input-provider one-participant situation and employ a technique called non-interactive cryptocomputing, where one input provider encrypts his input and the other input provider acts as the participant to perform the computation on the encrypted input. The scheme in [2] is a multiparty version of [31]. In [31], NOT and OR gates are used to construct the circuits while Goldwasser-Micali encryption or ElGamal encryption are extended to be NOT and OR homomorphic to calculate NOT and OR logic in ciphertext. Extension of logic homomorphism is implemented by expanding and combining ciphertexts. This expansion and combination mechanism brings two problems. The first problem is that the distribution of the expression of the final result is dependent on the circuit (namely the input of the participant), which violates privacy. The second problem is that the length of ciphertexts increases quickly as the computations go on, which brings a heavy burden on computation and communication. The efficiency pressure is so great, that depth of the circuit (thus number and length of the inputs) is strictly limited in [31]. The scheme in [2] also has these two drawbacks. Non-interactive

---

[1] Although it is said in [24] that "A naive verification procedure is to require the auctioneer to publish the tables and garbled input values of the circuit (signed by the AI), and allow suspecting bidders to simulate its computation", this verification procedure violates the basic rule in [24] that AI alone cannot know the bids. So another verification method based on "signed 'translation' table" in [24] has to be employed. Soundness of this verification method is based on an assumption that AI does not reveal the 'translation' table to A.

cryptocomputing brings a third problem to [31]: lack of verifiability. Although Sander *et al* suggested the usage of a fault tolerance mechanism, it requires to run the non-interactive cryptocomputing protocol many times, so is impractical in efficiency. To overcome the first two problems in [31], Fischlin [12] proposed a non-interactive cryptocomputing protocol, which sacrifices generality in [31] and only deals with the millionaire problem. In [12], NOT, XOR and AND gates are used to construct the circuits while Goldwasser-Micali encryption (which is NOT and XOR homomorphic) is extended to be AND homomorphic. Ciphertext expansion in [12] is not continual and does not bring the influence of the circuit to the distribution of the expression of the final result. Although the first two problems are overcome, [12] lacks verifiability too. As it employs non-interactive cryptocomputing, there is no practical verification mechanism.

So far, there is not any correct, private, verifiable and efficient solution to the millionaire problem. A correct, private, verifiable and efficient solution to the millionaire problem will be designed in this paper. The new technique employs the second method, but in a novel way.

3   Fundamental Primitives

Two fundamental primitives to be used in this paper are introduced in this section.

3.1   Mix Network

A mix network [18, 29, 30] mixes a number of encrypted inputs to the same number of outputs, while the link between the inputs and the outputs is kept secret. A mix network is usually composed of some mixing servers, each of which re-encrypts (or decrypts) and permutes the inputs in turn. The following two properties are usually required.

1. Correctness: the plaintexts of outputs must be a permutation of the plaintexts of the inputs.
2. Privacy: the permutation between the inputs and the outputs is unknown.

Correctness of a mix network must be publicly verifiable. There are two methods to verify correctness of a mix network.

1. Global verification: after all the servers have finished their mixing and the outputs are decrypted, a final verification is performed on the outputs in plaintext.
2. Individual verification: immediately after each server's mixing, he has to prove that his mixing is correct and the proof is verified instantly.

Usually, mix network with global verification is more efficient, but global verification requires that some parties must know the plaintexts in the inputs and can only be performed after the outputs are decrypted.

3.2   Modified ElGamal Encryption

ElGamal encryption is modified slightly as follows to be additive homomorphic.

- Integers p and q are large primes, such that $p = 2q+1$. Integer g is a generator of the cyclic subgroup of order q in Z_p^*.
- The private key is an integer x in Z_q and the public key is $(p, g, y = g^x \bmod p)$.
- Encryption: a message m in Z_q is encrypted into $c = (a, b) = (g^r \bmod p, g^m y^r \bmod p)$ where r is randomly chosen from Z_q.
- Decryption: given a ciphertext $c = (a, b)$, firstly $d = b/a^x \bmod p$ is calculated, then a search is performed to find $m = \log_g d$.

As in this paper decryption is only employed to test whether a ciphertext contains a zero or not, the search becomes a comparison of d and 1 ($m = 0$ iff $d = 1$).

In the rest of this paper,

- unless specified all the computations are performed modulo p;
- when c_1 and c_2 are two modified ElGamal ciphertexts and $c_1 = (a_1, b_1)$, $c_2 = (a_2, b_2)$
  - $c_1 c_2 = (a_1 a_2, b_1 b_2)$;
  - $c_1/c_2 = (a_1 a_2^{-1}, b_1 b_2^{-1})$;
  - $c_1^\gamma = (a_1^\gamma, b_1^\gamma)$.

4   A Building Block—Zero Test

Zero test is a new technique to test whether one or more ciphertexts contain a zero without decrypting them. The employed encryption algorithm $E()$ must be semantically secure[2] and additive homomorphic: $E(m_1)E(m_2) = E(m_1 + m_2)$. The corresponding decryption function is $D()$, which must be a distributed decryption function. The modified ElGamal encryption in Section 3.2 is employed in this paper[3]. In this modified ElGamal encryption, an exponentiation with the message as its exponent is encrypted using normal ElGamal encryption, so it is additive homomorphic. As ElGamal encryption is based on DL problem, distributed key generation [11, 27, 14] without any trusted party can be implemented efficiently. Although usually a search of logarithm is needed in the decryption function of the modified ElGamal encryption, the costly search is not

---

[2] Roughly, an encryption algorithm is said to be semantically secure if given m_0, m_1 and $c = E(m_k)$ where $k = 0$ or 1, the difference between the probability that k can be correctly guessed and 0.5 is negligible. See [23–Page 306] for more formal definition.

[3] Paillier encryption [25] with distributed decryption could be used, but distributed generation of an encryption system based on factorization problem is highly inefficient.

necessary in this paper, where any decryption is only performed to test whether the encrypted message is a known certain value, say zero. So application of the modified ElGamal encryption in this paper is efficient.

4.1   Simple Zero Test

We start with a simple case: to test a single ciphertext. Given a ciphertext c, it is required to test whether $D(c) = 0$ without revealing $D(c)$. The test is denoted as simple zero test $ZT(c)$, which outputs 0 if $D(c) = 0$, 1 if $D(c) \neq 0$. This technique is similar to a equality test technique in [22]. However, a multiparty test is used here while 2 two-party test is used in [22]. Suppose the private key is shared by some authorities A_1, A_2, ..., A_m with a threshold secret sharing. The simple zero test is as follows.

1. Randomization
   Each A_l chooses a random integer r_l from $\{2, 3, \ldots, q - 1\}$ and calculates $c_l = c_{l-1}^{r_l}$ to randomise the ciphertext where $c_0 = c$. It is publicly verifiable that $c_l \neq 1$ ($a_l \neq 1$ is verified where $c_l = (a_l, b_l)$) and $c_l \neq c_{l-1}$, so it is ensured that $r_l > 1$. Each A_l proves c_l is an exponentiation of c_{l-1} for $l = 1, 2, \ldots, m$ using a proof of equality of logarithms [7]: $\log_{a_{l-1}} a_l = \log_{b_{l-1}} b_l$.
2. Decryption
   The authorities cooperate to calculate $d = D(c_m)$ and prove the correctness of the distributed decryption using Chaum-Pedersen proof of equality of logarithms in [7] (see [27, 28] for details of distributed ElGamal decryption and its correctness verification). The output of the zero test is then as follows.

$$ZT(c) = \begin{cases} 0 & \text{if } d = 0 \\ 1 & \text{if } d \neq 0 \end{cases} \tag{1}$$

Theorem 1. $ZT()$ *is correct (if $D(c) = 0$, then $ZT(c) = 0$) and sound (if $D(c) \neq 0$, then $ZT(c) = 1$).*

Proof: As the correctness proof of randomization (proof of equality of logarithms [7]) is sound (if A_l does not know r_l such that $c_l = c_{l-1}^{r_l}$, he can pass the verification with only a negligible probability), $c_m = c^{\prod_{l=1}^{m} r_l}$ with an overwhelmingly large probability (in regard to the length of the challenge in the proof of knowledge of logarithm in [32] or in the proof of equality of logarithms in [7]) if the randomization is verified to be valid.

As the proof of correctness of decryption (Chaum-Pedersen proof of equality of logarithms in [7]) is sound (incorrect decryption can pass the decryption verification with only a negligible probability), $ZT(c) = D(c_m)$ with an overwhelmingly large probability (in regard to the length of the challenge in the Chaum-Pedersen proof of equality of logarithms in [7]) if the decryption is verified to be valid. So $ZT(c) = D(c^{\prod_{l=1}^{m} r_l})$ with an overwhelmingly large probability.

As $E()$ is additive homomorphic, $D(c^{\prod_{l=1}^{m} r_l}) = D(c) \prod_{l=1}^{m} r_l$. So $ZT(c) = D(c) \prod_{l=1}^{m} r_l$ if the whole verification succeeds. So, if $D(c) = 0$, $ZT(c) = 0$, therefore $ZT()$ is correct. As it is publicly verifiable that $c_m \neq 1$, it is guaranteed $\prod_{l=1}^{m} r_l \neq 0 \bmod q$. So if $D(c) \neq 0$, then $ZT(c) \neq 0$. □

Theorem 2. *$ZT()$ is private. (No information about $D(c)$ is revealed except whether it is zero if at least one participant conceals his mixing and the number of dishonest authorities is not over the sharing threshold.)*

Proof: The correctness proof of randomization (proof of equality of logarithms in [7]) is special honest-verifier zero knowledge, so r_l is kept secret in the proof. Moreover, if the number of dishonest authorities is not over the threshold, none of $c_0, c_1, \ldots c_{m-1}$ can be decrypted. So the only revealed information about $D(c)$ is $D(c) \prod_{l=1}^{m} r_l$, from which the cooperation of all the authorities is necessary to deduce $D(c)$. Moreover, $D(c) \prod_{l=1}^{m} r_l$ is uniformly distributed in the message space of the encryption algorithm. So, if the number of dishonest authorities is not over the sharing threshold (thus no ciphertext but c_m can be decrypted and at least one authority A_l chooses r_l randomly), $D(c) \prod_{l=1}^{m} r_l$ is uniformly distributed and independent of $D(c)$. Therefore, if the number of dishonest authorities is not over the sharing threshold, no information about $D(c)$ is revealed except whether it is zero. □

Simple zero test is publicly verifiable as both randomization and distributed decryption are publicly verifiable.

4.2 Complex Zero Test

As $ZT()$ can only test whether a single ciphertext is an encryption of zero, it cannot work when there are more than one ciphertext to test (to be zero or not). In a complex zero test, it is required to test whether there is at least one encryption of zero in multiple ciphertexts without revealing any other information about the messages encrypted in the ciphertexts. Suppose ciphertexts c_i for $i = 1, 2, \ldots, n$ are the encrypted inputs to test. The complex zero test is denoted as $ZM(c_1, c_2, \ldots, c_n)$, which returns 0 iff there is at least one encryption of zero in c_i for $i = 1, 2, \ldots, n$. The test $ZM(c_1, c_2, \ldots, c_n)$ is implemented as follows where the decryption key is shared by authorities $A_1, A_2, \ldots, A_m$.

1. Mix network
   The authorities act as mixing servers and set up a mix network to mix c_i for $i = 1, 2, \ldots, n$ to ciphertexts c_i' for $i = 1, 2, \ldots, n$. The mix network must be correct and private. As it is not desired to decrypt the outputs c_i' for $i = 1, 2, \ldots, n$ in this paper, correctness of the mixing must be publicly verifiable without decrypting them. So mix networks with global verification [26, 17, 29] cannot be employed although they are very efficient. Among the mix networks employing individual verification, [18] and [30] are good choices here. Both of them are efficient and their correctness and privacy are strong enough for many applications including zero test. [30] is more efficient than [18], but achieves weaker privacy.[4]

---

[4] Shuffling in groups and batch verification of validity of shuffling are employed in [30]. The grouping operation leads to high efficiency, but weakens privacy a little.

2. Simple zero tests

The authorities then cooperate to perform $ZT(c_i')$ for $i = 1, 2, \ldots, n$ one by one until a zero is found in one simple zero test or all the n simple tests finish. The output of the zero test is then as follows.

$$ZM(c_1, c_2, \ldots, c_n) = \begin{cases} 0 & \text{if a zero is found in one simple zero test} \\ 1 & \text{if no zero is found after all the simple tests finish} \end{cases}$$

(2)

Theorem 3. $ZM()$ *is correct and sound ($ZM(c_1, c_2, \ldots, c_n) = 0$ iff there is at least one encryption of zero in c_i for $i = 1, 2, \ldots, n$).*

Proof: Equation (2) indicates that $ZM(c_1, c_2, \ldots, c_n) = 0$ iff $ZT(c_i') = 0$ for some i in $\{1, 2, \ldots, n\}$. As each $ZT()$ is correct and sound, $ZT(c_i') = 0$ iff c_i' encrypts a zero. So $ZM(c_1, c_2, \ldots, c_n) = 0$ iff c_i' encrypts a zero for some i in $\{1, 2, \ldots, n\}$.

As the employed mix network ([18] or [30]) is correct, $\{D(c_1'), D(c_2'), \ldots, D(c_n')\} = \{D(c_1), D(c_2), \ldots, D(c_n)\}$. So $ZM(c_1, c_2, \ldots, c_n)$ returns zero iff $D(c_i) = 0$ for some i in $\{1, 2, \ldots, n\}$. □

Theorem 4. $ZM()$ *is private. (No information about $D(c_i)$ for $i = 1, 2, \ldots, n$ is revealed except whether there is at least one zero among them if at least one authority conceals his mixing and the number of dishonest authorities is not over the sharing threshold.)*

Proof: As $ZT()$ is private, no information about $D(c_i')$ for $i = 1, 2, \ldots, n$ is revealed in $ZM()$ except whether at least one of them is zero and the index of the first zero among them (if there is at least one zero). As the employed mix ([18] or [30]) network is private, no link is known between $c_1, c_2, \ldots, c_n$ and $c_1', c_2', \ldots, c_n'$ if at least one authority conceals his mixing and the number of dishonest authorities is not over the sharing threshold. So no information about $D(c_i')$ for $i = 1, 2, \ldots, n$ is revealed in $ZM()$ except whether at least one of them is zero. □

Complex zero test is publicly verifiable as both the mix network and simple zero test are publicly verifiable.

5   Batch Equation

To apply zero test to the millionaire problem, the following theorem about batch equation is necessary. Batch equation is a technique to test equality of each pair of integers in multiple pairs. The idea is similar to the so called "batch verification" [1]. However, unlike batch verification, no zero-knowledge proof or verification is involved in batch equation.

Theorem 5. *When there exists $y_i \neq z_i \bmod q$ with any i in $\{1, 2, \ldots, n\}$, $\sum_{i=1}^{n} y_i t_i = \sum_{i=1}^{n} z_i t_i \bmod q$ with a probability no more than 2^{-T} if q is a prime, t_i is randomly chosen from $\{0, 1, 2, \ldots, 2^T - 1\}$ for $i = 1, 2, \ldots, n$ and $q \geq 2^T$.*

To prove Theorem 5, a lemma is proved first.

Lemma 1. *Suppose t_1, t_2, ..., t_{v-1}, t_{v+1}, t_{v+2}, ..., t_n are constant. If q is a prime, $y_v \neq z_v \bmod q$, $q \geq 2^T$ and $\sum_{i=1}^n y_i t_i = \sum_{i=1}^n z_i t_i \bmod q$, then there is only one possible T-bit solution for t_v.*

Proof: If this lemma is not correct, then the following two equations can be satisfied simultaneously where $y_v \neq z_v \bmod q$, $|t_v| = |\hat{t}_v| = T$ and $t_v \neq \hat{t}_v$.

$$\sum_{i=1}^n y_i t_i = \sum_{i=1}^n z_i t_i \bmod q \tag{3}$$

$$(\sum_{i=1}^{v-1} y_i t_i) + y_v \hat{t}_v + \sum_{i=v+1}^n y_i t_i = (\sum_{i=1}^{v-1} z_i t_i) + z_v \hat{t}_v + \sum_{i=v+1}^n z_i t_i \bmod q \tag{4}$$

Subtracting (4) from (3) yields

$$y_v(t_v - \hat{t}_v) = z_v(t_v - \hat{t}_v) \bmod q$$

So

$$(y_v - z_v)(t_v - \hat{t}_v) = 0 \bmod q$$

Note that $t_v - \hat{t}_v \neq 0 \bmod q$ because $q \geq 2^T$ and $|t_v| = |\hat{t}_v| = T$. As q is a prime, $y_v - z_v = 0 \bmod q$. A contradiction to the statement $y_v \neq z_v \bmod q$ is found. Therefore, the lemma is correct.   □

Proof of Theorem 5: Lemma 1 implies that among the $(2^T)^n$ possible combinations of $\{t_1, t_2, \ldots, t_n\}$ in $\{0, 1, 2, \ldots, 2^T - 1\}^n$, at most $(2^T)^{n-1}$ of them can satisfy $\sum_{i=1}^n y_i t_i = \sum_{i=1}^n z_i t_i \bmod q$ when $y_v \neq z_v \bmod q$. So if $y_v \neq z_v \bmod q$ and t_i are randomly chosen from $\{0, 1, 2, \ldots, 2^T - 1\}$ for $i = 1, 2, \ldots, n$, then $\sum_{i=1}^n y_i t_i = \sum_{i=1}^n z_i t_i \bmod q$ is satisfied with probability no more than 2^{-T}.   □

6   Solution to the Millionaire Problem

The millionaire problem is solved in a circuit to compare two ciphertexts to determine which one contains a larger message without decrypting them. The circuit is implemented through three levels of computation as shown in Statement (5), which is true iff $D(c_1) > D(c_2)$.

$$(D(c_{1,1}) = 1 \wedge D(c_{2,1}) = 0) \vee (D(c_{1,1}) = D(c_{2,1}) \wedge D(c_{1,2}) = 1 \wedge D(c_{2,2}) = 0)$$
$$\vee \ldots \vee (D(c_{1,1}) = D(c_{2,1}) \wedge D(c_{1,2}) = D(c_{2,2}) \tag{5}$$
$$\wedge \ldots \wedge D(c_{1,L-1}) = D(c_{2,L-1}) \wedge D(c_{1,L}) = 1 \wedge D(c_{2,L}) = 0)$$

At the innermost level, there are tests of bit equality and tests of bit difference, which can be implemented with the help of homomorphism of the employed encryption algorithm. At the middle level, there are computations of "AND" logic, which can be implemented with the help of batch equation and homomorphism of the employed encryption algorithm. At the outermost level, there are computations of "OR" logic, which can be implemented using zero test.

Suppose the two messages are encrypted bit by bit as $c_1 = (c_{1,1}, c_{1,2}, \ldots, c_{1,L})$ and $c_2 = (c_{2,1}, c_{2,2}, \ldots, c_{2,L})$ where the most significant bit is on the left. The solution is as follows.

1. The participants (e.g. the two millionaires) corporately and randomly choose t_i from $\{0, 1, 2, \ldots, 2^T - 1\}$ for $i = 1, 2, \ldots, n$. For example, one millionaire randomly chooses $t_{1,i}$ for $i = 1, 2, \ldots, n$ and publishes $H(t_{1,i})$ for $i = 1, 2, \ldots, n$ while the other randomly chooses $t_{2,i}$ for $i = 1, 2, \ldots, n$ and publishes $H(t_{2,i})$ for $i = 1, 2, \ldots, n$ where $H()$ is a one-way and collision-resistant hash function. Then they publish $t_{i,1}$, $t_{i,2}$ for $i = 1, 2, \ldots, n$ and calculate $t_i = t_{i,1} + t_{i,2} \bmod 2^T$ for $i = 1, 2, \ldots, n$.

2. The participants act as the authorities in $ZM()$ and perform

$$
ZM \ (\ c_{1,1}/(E(1)c_{2,1})\ ,\ (c_{1,1}/c_{2,1})^{t_1}(c_{1,2}/(E(1)c_{2,2}))^{t_2}, \ldots
$$

$$
(\prod_{i=1}^{L-1}(c_{1,i}/c_{2,i})^{t_i})\ \ (c_{1,L}/(E(1)c_{2,L}))^{t_L}\ ) \tag{6}
$$

where the modified ElGamal encryption in Section 3.2 is employed and $E(1) = (1, g)$. Then $D(c_1)$ is declared to be larger than $D(c_2)$ iff Statement (6)=0.

Theorem 6. *The solution to the millionaire problem through Statement (6) is a correct and sound with an overwhelmingly large probability $(D(c_1) > D(c_2)$ iff Statement (6)=0 with an overwhelmingly large probability).*

Proof: $D(c_1) > D(c_2)$ iff Statement (5) is true. According to additive homomorphism of the modified ElGamal encryption, Statement (5) is equivalent to

$$
D(c_{1,1}/(E(1)c_{2,1})) = 0 \ \lor\ (D(c_{1,1}/c_{2,1}) = 0 \land D(c_{1,2}/(E(1)c_{2,2})) = 0) \ \lor
$$
$$
\ldots \ \lor\ (D(c_{1,1}/c_{2,1}) = 0 \land D(c_{1,2}/c_{2,2}) = 0 \land \ldots \tag{7}
$$
$$
\land D(c_{1,L-1}/c_{2,L-1}) = 0 \land D(c_{1,L}/(E(1)c_{2,L})) = 0)
$$

According to Theorem 5, with an overwhelmingly large probability Statement (7) is equivalent to

$$
D(c_{1,1}/(E(1)c_{2,1})) = 0 \ \lor\ t_1 D(c_{1,1}/c_{2,1}) + t_2 D(c_{1,2}/(E(1)c_{2,2})) = 0 \ \lor
$$
$$
\ldots \ \lor\ t_1 D(c_{1,1}/c_{2,1}) + t_2 D(c_{1,2}/c_{2,2}) + \ldots \tag{8}
$$
$$
+ t_{L-1} D(c_{1,L-1}/c_{2,L-1}) + t_L D(c_{1,L}/(E(1)c_{2,L})) = 0
$$

According to additive homomorphism of the modified ElGamal encryption, Statement (8) is equivalent to

$$
(D(c_{1,1}/(E(1)c_{2,1})) = 0 \ \lor\ D((c_{1,1}/c_{2,1})^{t_1}(c_{1,2}/(E(1)c_{2,2}))^{t_2}) = 0 \tag{9}
$$
$$
\lor\ \ldots\ \lor\ D((\prod_{i=1}^{L-1}(c_{1,i}/c_{2,i})^{t_i})(c_{1,L}/(E(1)c_{2,L}))^{t_L}) = 0
$$

So $D(c_1) > D(c_2)$ iff one of the L clauses in Statement (9) is true. As $ZM()$ is correct and sound method to test whether there is any zero encrypted in some ciphertexts, Statement (9) can be evaluated through Statement (6). Therefore, $D(c_1) > D(c_2)$ iff Statement (6)=0 with an overwhelmingly large probability. □

Theorem 7. *The solution to the millionaire problem through Statement (6) is private.*

Proof: The computations in Statement (6) before the zero test are in ciphertext and involves no decryption, so are private. The computation in $ZM()$ is private as proved in Theorem 4. So the computations in Statement (6) are private. □

7   Analysis

Suppose the two millionaires act as the participants, the cost of the solution of the millionaire problem includes:

- $(L+2)(L-1)$ short exponentiations (T-bit exponent)
- $2L$ divisions;
- about $16L$ full-length exponentiations for the mix network in [18] or $2(2L + k(4k-2))$ full-length exponentiations for the mix network in [30] where k is a small parameter unrelated to L;
- average of $L/2$ simple zero tests: $2L$ full-length exponentiations, L proofs of equality of logarithms (costing $2L$ full-length exponentiations) and $L/2$ distributed decryptions and validity proof of decryptions (costing $3L$ full-length exponentiations).

A comparison between the new solution to the millionaire problem and solutions based on the existing solutions is provided in Table 1. The schemes in [10] and [2] are similar to [19] and [31] respectively, so are not analysed separately. In the schemes in Table 1, only [12] and our proposed scheme provided a concrete circuit to solve the millionaire problem. In [12], $L + L(L-1)$ two-input gates are needed, while in our scheme, $L + L(L-1)/2$ two-input gates are needed. In the other schemes, at least $7L$ two-input gates are needed as analysed in [21]. In [31], the number of gates should be larger than $7L$ as only "NOT" and "OR" gates are used. Even if only $7L$ two-input gates are employed in [31], ciphertext expansion bring an intolerable cost for encryption and communication. As each gate has only two inputs, those $7L$ gates must be in $\log_2 7L$ levels. So in [31] at least $8^{\log_2 7L} = 343L^3$ ciphertexts must be transmitted. It is pointed out in [12] that at least L^4 multiplications are needed in [31]. In [12], $L(L+1)/2$ "AND" gates are needed, so $\lambda L(L+3)/2$ encryptions (each costly 1.5 multiplication in average) and $\lambda L(L+1)/2$ multiplications are needed for "AND" gates. There are $L(L+1)/2$ multiplications for "XOR" and "NOT" computation in [12]. In this analysis, the number of multiplications are accounted in computation and transportation of integers with significant length (several hundred bits or longer) is accounted in regard of communication where K is the bit length of full-length exponent (e.g 1024 bits). One full-length exponentiations is regarded as $1.5K$ multiplications and computation of the product of n short exponentiations is regarded as $n + 0.5nT$ multiplications.

In the example in Table 1, $K = 1024$ and $L = 100$. Let t, the number of cuttings in [20] and λ, the parameter in [12] be 40. The parameter k in the mix network [30] is set to be 5, which is big enough to provide strong privacy.

The analysis in Table 1 indicates that both [12] and the proposed solution can efficiently solve the millionaire problem with short inputs. Compared to the proposed schemes, the scheme in [12] has a drawback: lack of verifiability. Is it possible to overcome the drawback and make the scheme in [12] verifiable? A naive method is to employ the shuffling-then-decryption technique from the proposed scheme to [12]. However, this method is infeasible. Firstly, non-interactive cryptocomputing implies that validity of the circuit is not verifiable. As the circuit is dependent on the participant's input, revealing the circuit violates privacy of the the participant's input. Even if non-interactive cryptocomputing is replaced by two-party computation in [12] and the circuit is independent on any input, verifiability still cannot be practically achieved in [12]. Distributed key generation for Goldwasser-Micali encryption (distributed generation of a secret factorization) is much more costly than distributed key generation for ElGamal encryption and distributing the Goldwasser-Micali private key between the two participants without any trusted party is highly inefficient. Moreover, before the encrypted result is output for decryption, the $L\lambda$ ciphertexts in it must be shuffled using a re-encryption mix, otherwise privacy is violated. Although it may be possible to design a verifiable re-encryption mix for Goldwasser-Micali ciphertexts[5], the large-scale shuffling ($L\lambda = 4000$ when $L = 100$ and $\lambda = 40$) and proof-verification of validity of the shuffling is too impractical. On the other hand, the proposed solution to the millionaire problem can be modified to non-interactive cryptocomputing to improve its efficiency. After the modification, distributed decryption and the costly proof operations can be omitted, so that the proposed scheme becomes much more efficient by sacrificing verifiability. Without verifiability, the proposed scheme becomes more efficient than the scheme in [12]. In summary, the new solution achieves the best trade-off between security and efficiency and provides the only general, correct, private, verifiable and efficient solution for the millionaire problem.

8   Conclusion

A new solution to the millionaire problem is designed to achieve correctness, privacy, verifiability and high efficiency, which have never been achieved simultaneously before. In the future, the possibility of using the techniques in this paper to compute other functions will be investigated.

---

[5] There is no such shuffling or mix at present.

Table 1. Comparison of solutions of the millionaire problem

	Correctness	Privacy	Verifiability	Computation		Communication	
				cost	example	cost	example
[24]	Yes	Yes	No	$\geq 10KLt$	≥ 40960000	$\geq 37Lt+2t$	≥ 148080
[20]	Yes	Yes	Yes	$\geq 10KLt$	≥ 40960000	$\geq 37Lt+2t$	≥ 148080
[19]	Yes	Yes	Yes	average $3110KL+6$	318464006	average $1626L+6$	162606
[5]	Yes	Yes	Yes	average $\geq 2693KL$	≥ 275763200	$\geq 1543L$	≥ 154300
[31]	Yes	No	No	L^4	≥ 100000000	$\geq 343L^3$	≥ 343000000
[2]	Yes	Yes	No	$(1.5\lambda L(L+3)+(\lambda+1)L(L+1))/2$	516050	$L(\lambda+2)$	4200
Proposed scheme using [18] mix	Yes	Yes	Yes	average $(L+2)(L-1)(1+0.5T)+25KL$	2772058	average $25L$	2500
Proposed scheme using [30] mix	Yes	Yes	Yes	average $(L+2)(L-1)(1+0.5T)+K(2L+2(5.5L+4k(k-2)))$	1666138	average $2L+2(5.5L+2k^2)$	1400

Acnowledgement

This paper is sponsored by DP 0345458 ARC DISCOVERY/ 2003-2005 and LX 0346868.

References

1. Riza Aditya, Kun Peng, Colin Boyd, and Ed Dawson. Batch verification for equality of discrete logarithms and threshold decryptions. In *Second conference of Applied Cryptography and Network Security, ACNS 04*, volume 3089 of *Lecture Notes in Computer Science*, pages 494–508, Berlin, 2004. Springer-Verlag.
2. D. Beaver. Minimal-latency secure function evaluation. In *EUROCRYPT '00, Bruges, Belgium, May 14-18, 2000, Proceeding*, volume 1807 of *Lecture Notes in Computer Science*, pages 335–350. Springer, 2000.
3. Michael Ben-Or, Shafi Goldwasser, Joe Killian, , and Avi Wigderson. Multi-prover interactive proofs: How to remove intractability assumptions. In *Proceedings of the Twentieth Annual ACM Symposium on Theory of Computing, STOC1988*, pages 113–131, 1988.
4. Christian Cachin. Efficient private bidding and auctions with an oblivious third party. In *the 6th ACM Conference on Computer and Communications Security*, 1999. Available at http://www.tml.hut.fi/~helger/crypto/link/protocols/auctions.html.
5. Christian Cachin and Jan Camenisch. Optimistic fair secure computation (extended abstract). In *CRYPTO '00*, pages 94–112, Berlin, 2000. Springer-Verlag. Lecture Notes in Computer Science 1880.
6. D. Chaum, I. B. Damgård, and J. van de Graaf. Multiparty computations ensuring privacy of each party's input and correctness of the result. In *CRYPTO '87*, pages 87–119, Berlin, 1987. Springer-Verlag. Lecture Notes in Computer Science Volume 293.
7. D. Chaum and T. P. Pedersen. Wallet databases with observers. In *CRYPTO '92*, pages 89–105, Berlin, 1992. Springer-Verlag. Lecture Notes in Computer Science Volume 740.
8. David Chaum, Claude Crepeau, and Ivan Damgård. Multiparty unconditionally secure protocols (extended abstract). In *Proceedings of the Twentieth Annual ACM Symposium on Theory of Computing, STOC1988*, pages 11–19, 1988.
9. Ronald Cramer, Ivan Damgård, Stefan Dziembowski, Martin Hirt, and Tal Rabin. Efficient multiparty computations secure against an adaptive adversary. In *EURO-CRYPT '99*, volume 1592 of *Lecture Notes in Computer Science*, pages 311–326. Springer, 1999.
10. Ronald Cramer, Ivan Damgård, and Jesper Buus Nielsen. Multiparty computation from threshold homomorphic encryption. In *EUROCRYPT '01, Innsbruck, Austria, May 6-10, 2001, Proceeding*, volume 2045 of *Lecture Notes in Computer Science*, pages 280–299. Springer, 2001.
11. P Feldman. A practical scheme for non-interactive verifiable secret sharing. In *28th Annual Symposium on Foundations of Computer Science*, pages 427–437, 1987.
12. Marc Fischlin. A cost-effective pay-per-multiplication comparison method for millionaires. In *Topics in Cryptology - CT-RSA 2001, The Cryptographer's Track at RSA Conference 2001, San Francisco, CA, USA, April 8-12, 2001, Proceedings*, volume 2020 of *Lecture Notes in Computer Science*, pages 457–472. Springer, 2001.

13. Matthew K. Franklin and Stuart Haber. Joint encryption and message-efficient secure computation. *Journal of Cryptology 9(4)*, pages 217–232, 1996.

14. R Gennaro, S Jarecki, H Krawczyk, and T Rabin. Secure distributed key generation for discrete-log based cryptosystems. In *EUROCRYPT '99*, pages 123–139, Berlin, 1999. Springer-Verlag. Lecture Notes in Computer Science Volume 1592.

15. Rosario Gennaro, Michael O. Rabin, and Tal Rabin. Simplified VSS and fast-track multiparty computations with applications to threshold cryptography. In *Proceedings of the seventeenth annual ACM symposium on Principles of distributed computing, PODC'98*, pages 101 – 111, 1987.

16. Oded Goldreich, Silvio Micali, and Avi Wigderson. How to play any mental game or a completeness theorem for protocols with honest majority. In *Proceedings of the Nineteenth Annual ACM Symposium on Theory of Computing, STOC 1987*, pages 218–229, 1987.

17. Philippe Golle, Sheng Zhong, Dan Boneh, Markus Jakobsson, and Ari Juels. Optimistic mixing for exit-polls. In *ASIACRYPT '02*, pages 451–465, Berlin, 2002. Springer-Verlag. Lecture Notes in Computer Science Volume 1592.

18. Jens Groth. A verifiable secret shuffle of homomorphic encryptions. In *Public Key Cryptography 2003*, pages 145–160, Berlin, 2003. Springer-Verlag. Lecture Notes in Computer Science Volume 2567.

19. M Jakobsson and A Juels. Mix and match: Secure function evaluation via ciphertexts. In *ASIACRYPT '00*, pages 143–161, Berlin, 2000. Springer-Verlag. Lecture Notes in Computer Science Volume 1976.

20. A. Juels and M. Szydlo. An two-server auction protocol. In *Proc. of Financial Cryptography*, pages 329–340, 2002.

21. Kaoru Kurosawa and Wakaha Ogata. Bit-slice auction circuit. In *7th European Symposium on Research in Computer Security, ESORICS2002*, volume 2502 of *Lecture Notes in Computer Science Volume 2339*, pages 24 – 38, Berlin, 2002. Springer-Verlag.

22. H. Lipmaa. Verifiable homomorphic oblivious transfer and private equality test. In *ASIACRYPT '03*, volume 2894 of *Lecture Notes in Computer Science*, pages 416–433, Berlin, 2003. Springer.

23. A. Menezes, P. van Oorschot, and S. Vanstone. *Handbook of Applied Cryptography*. CRC press, 1996.

24. Moni Naor, Benny Pinkas, and Reuben Sumner. Privacy perserving auctions and mechanism design. In *ACM Conference on Electronic Commerce 1999*, pages 129–139, 1999.

25. P Paillier. Public key cryptosystem based on composite degree residuosity classes. In *EUROCRYPT '99*, pages 223–238, Berlin, 1999. Springer-Verlag. Lecture Notes in Computer Science Volume 1592.

26. C. Park, K. Itoh, and K. Kurosawa. Efficient anonymous channel and all/nothing election scheme. In *EUROCRYPT '93*, pages 248–259, Berlin, 1993. Springer-Verlag. Lecture Notes in Computer Science Volume 765.

27. Torben P. Pedersen. A threshold cryptosystem without a trusted party. In *EUROCRYPT '91*, pages 522–526, Berlin, 1991. Springer-Verlag. Lecture Notes in Computer Science Volume 547.

28. Torben P. Pedersen. *Distributed Provers and Verifiable Secret Sharing Based on the Discrete Logarithm Problem*. PhD thesis, Computer Science Department, Aarhus University,Aarhus, Denmark, 1992.

66   K. Peng et al.

29. Kun Peng, Colin Boyd, Edward Dawson, and Kapali Viswanathan. Efficient implementation of relative bid privacy in sealed-bid auction. In *The 4th International Workshop on Information Security Applications, WISA 2003*, volume 2908 of *Lecture Notes in Computer Science*, pages 244–256, Berlin, 2003. Springer-Verlag.
30. Kun Peng, Colin Boyd, Edward Dawson, and Kapali Viswanathan. A correct, private and efficient mix network. In *2004 International Workshop on Practice and Theory in Public Key Cryptography*, pages 439–454, Berlin, 2004. Springer-Verlag.
31. Tomas Sander, Adam Young, and Moti Yung. Non-interactive cryptocomputing for NC^1. In *40th Annual Symposium on Foundations of Computer Science, New York, NY, USA, FOCS '99*, pages 554–567, 1999.
32. C Schnorr. Efficient signature generation by smart cards. *Journal of Cryptology, 4, 1991*, pages 161–174, 1991.

All in the XL Family: Theory and Practice

Bo-Yin Yang[1,*] and Jiun-Ming Chen[2]

[1] Department of Mathematics, Tamkang University, Tamsui, Taiwan
by@moscito.org
[2] Chinese Data Security, Inc., & National Taiwan U
jmchen@math.ntu.edu.tw

Abstract. The XL (EXTENDED LINEARIZATION) equation-solving algorithm belongs to the same extended family as the advanced Gröbner Bases methods $\mathbf{F_4}/\mathbf{F_5}$. XL and its relatives may be used as direct attacks against multivariate Public-Key Cryptosystems and as final stages for many "algebraic cryptanalysis" used today. We analyze the applicability and performance of XL and its relatives, particularly for generic systems of equations over medium-sized finite fields.

In examining the extended family of Gröbner Bases and XL from theoretical, empirical and practical viewpoints, we add to the general understanding of equation-solving. Moreover, we give rigorous conditions for the successful termination of XL, Gröbner Bases methods and relatives. Thus we have a better grasp of how such algebraic attacks should be applied. We also compute revised security estimates for multivariate cryptosystems. For example, the schemes SFLASH^{v2} and HFE Challenge 2 are shown to be unbroken by XL variants.

Keywords: algebraic analysis, finite field, Gröbner Bases, multivariate quadratics, multivariate cryptography, XL.

1 Introduction

Public Key Cryptography depends on the intractibility of "hard problems". Solving a system of quadratic equations over a finite field is one such (known to be NP-hard, [33]) problem. Further, often in a cryptographical primitive we find a polynomial system of equations to hold with good probability. This is called *algebraic cryptanalysis*, currently a very hot topic. Ergo, knowing how fast we can solve polynomial systems is important.

XL is an equation-solving method related to Gröbner Bases ([2,54]). It was proposed[1] by Courtois-Klimov-Patarin-Shamir ([20]). Claims of cryptanalysis involving XL-like system-solving have been made against many primitives: stream

---

[*] Supported by National Science Council of Taiwan under grant NSC 93-2115-M-032-008.

[1] XL is often regarded as a descendant of Kipnis-Shamir's relinearization ([37]), used in an algebraic attack on HFE, but we will discuss only XL-related methods from now on.

C. Park and S. Chee (Eds.): ICISC 2004, LNCS 3506, pp. 67–86, 2005.
© Springer-Verlag Berlin Heidelberg 2005

ciphers like Toyocrypt ([15]) and $E0$ (the Bluetooth protocol, [16]), block ciphers like Rijndael/AES and Serpent ([21]), and multivariate PKC's like HFE and SFLASH^{v2} ([17]).

XL does not operate on underdetermined systems, we must first take guesses to make it determined or over-determined. *Henceforth we concern ourselves with solving the system* $\ell_1(\mathbf{x}) = \ell_2(\mathbf{x}) = \cdots = \ell_m(\mathbf{x}) = 0$ *of* $m \geq n$ *(quadratic unless otherwise specified) equations in* n *variables* $\mathbf{x} = (x_1, x_2, \ldots, x_n)$ *over a field* $K = \mathrm{GF}(q)$.

We will study the time complexity of XL- and Gröbner-Bases-related algorithms. For generic systems, this depend primarily on the minimum degree of operation, which varies with m and n and other parameters. We hope to achieve the following:

– obtain exact and asymptotic time complexity of several XL-like methods; and hence:
– show some previous claims of cryptanalysis to be over-optimistic, and give updated security estimates for the primitives of SFLASH^{v2} and HFE challenge 2 (neither of which now decreasing below 2^{80}) by various methods;
– demonstrate that XL with the XL2 adjunct is a primitive version of $\mathbf{F_5}$.

2   The XL Algorithm

The "Basic XL" ([24] terms it "reduced XL") at degree D proceeds as follows:

1. "X" is for eXtend (or multiply). Generate equations $\mathcal{R}^{(D)} = \{\mathbf{x}^{\mathbf{b}}\ell_i(\mathbf{x}) = 0 : i = 1 \cdots m, |\mathbf{b}| \leq D - 2\}$. $|\mathbf{b}| = \sum_i b_i$ is the degree of monomial $\mathbf{x}^{\mathbf{b}} = x_1^{b_1} x_2^{b_2} \cdots x_n^{b_n}$.
2. "L" is for Linearize. Run an elimination on the equations $\mathcal{R}^{(D)}$, treating each monomial $\mathbf{x}^{\mathbf{b}}$ in the set $\mathcal{T} = \mathcal{T}^{(D)}$ of monomials of total degree $\leq D$ as a variable. The number of variables and equations are denoted T and R respectively. The number of independent equations (i.e., the rank of the system, denoted I) cannot exceed $T - 1$ if the original system has a solution. Indeed, if $I = T - 1$ we expect the algorithm to terminate with a unique solution. However, it is sufficient that the elimination results in an equation to solve for (say) x_1.
3. *If necessary, solve the univariate equation giving x_1, and repeat as needed.*

If solving M linear equations in N variables takes $E(N, M)$, then XL runs in time

$$C_{\mathrm{XL}} = E(T, R) = E\left(\binom{n+D}{D}, m\binom{n+D-2}{D-2}\right), \tag{1}$$

for larger fields because $R = m\binom{n+D-2}{D-2}$ and $T = \binom{n+D}{D}$. If we are dealing with small fields, then both T and R would be smaller. A reasonable terminating condition is then $I \geq T - \min(D, q - 1)$, as this final equation may have up to the $D + 1$ terms $1, x_1, \ldots, x_1^D$ (or up to x_1^{q-1} if $q \leq D$) instead of $T = I - 1$). Surprisingly (cf. Sec. 6.2) this may offer little practical improvement over $T - I = 1$.

3   The Family of XL Variants

When proposing XL ([20]) the authors noted that we need $m - n \geq 2$ for good performance. Which brings us the "FXL" method as the first of several XL variants.

3.1   FXL: Guessing as Aid to Equation-Solving

The "F" in FXL stands for "fix" ([20]). The attacker assigns random values to f variables, in effect guessing at them, hoping to decrease (cf. also XL', Sec. 3.3) the degree D needed for XL. After guessing, we run XL and test at the end if any solution found is valid. The complexity for f variables fixed at degree D is

$$C_{\text{FXL}} = q^f \left[C_0 + E \left(\binom{n-f+D}{D}, m\binom{n-f+D-2}{D-2} \right) \right], \tag{2}$$

where C_0 is a presumed small cost of collation. We will establish the worthiness of FXL by demonstrating its gains, and give some guidelines for its profitable application in Sec. 6.3. We note the fixing concept applies almost verbatim to the $\mathbf{F_4}$ and $\mathbf{F_5}$. I.e., we may also guess at a few variables before applying a Gröbner Bases method. We shall show that this can be a good idea in general.

3.2   XL2: Gaining Extra Equations via the T' Method

This was first proposed ([22]) as an addendum to XL over GF(2), to add useful equations. Let T' count the monomials that when multiplied by a given variable will still be in $\mathcal{T} = \mathcal{T}^{(D)}$. I.e. $T' = |\mathcal{T}_i'|$, where $\mathcal{T}_i' = \{\mathbf{x^b} : x_i \mathbf{x^b} \in \mathcal{T}\}$ for each i. Suppose I is not as large as $T - D$, but $C \equiv T' + I - T > 0$ (i.e. we have enough equations to eliminate all monomials not in $\mathcal{T}_i'$), then:

1. Eliminate from the system $\mathcal{R} = \mathcal{R}^{(D)}$ the monomials not in $\mathcal{T}_1'$ first. We are then left with relations $\mathcal{R}_1$ that gives each monomial in $\mathcal{T} \setminus \mathcal{T}_1'$ as a linear combination of those in $\mathcal{T}_1'$, plus C equations $\mathcal{R}_1'$ with only monomials in $\mathcal{T}_1'$.
2. Repeat for $\mathcal{T}_2'$ to get the equations $\mathcal{R}_2$ and $\mathcal{R}_2'$ (we should also have $|\mathcal{R}_2'| = C$).
3. For each $\ell \in \mathcal{R}_1'$, monomial in the equation $x_1 \ell = 0$ are either in $\mathcal{T}_2'$ or can be reduced (using $\mathcal{R}_2$) into $\mathcal{T}_2'$. Ditto each $x_2 \ell$ ($\ell \in \mathcal{R}_2'$) and we get $2C$ new equations.

XL2 is described as a sequence of Buchberger relations by [54]. It is important it is similar to the final stage (T'-method) of the related XSL (extended sparse linearization, [21]) method that purports to break block ciphers with *sparse quadratic structure*, including AES. We do not analyze XSL itself here. [22] claims that most of the $2C$ equations *"are likely"* to be linearly independent, and that XL2 can be repeated for an eventual solution. We seek to clarify the heuristics below.

3.3   XL': Searching as the Final Step

XL' ([22]) is XL except that we come down to a system in r variables and at least r equations, then end by brute-force search. The total time complexity for large q is

$$C_{XL'} \approx E\left(\binom{n+D}{D}, \, m\,\binom{n+D-2}{D-2}\right) + \frac{q^r D}{1-\frac{1}{q}}\binom{r+D}{D}. \tag{3}$$

The new terminations conditions are: instead of requiring $T - I \leq D$, we only require $T - I \leq \binom{r+D}{D} - r$. Note: It is usually 1-in-q for any polynomial to vanish on random inputs, and we must test degree-D polynomials with r variables and up to $\binom{r+D}{D}$ terms. We need a suitably small q^r and make some changes. This D is smaller than the D_0 for regular XL. We will check how much smaller in Sec. 7.

3.4   XLF: Using the Field Relations

[17] proposes to use the field relations $x^q = x$ to advantage when $q = 2^k$:

- Consider $(x_i^2), (x_i^4), \ldots, (x_i^{2^{k-1}})$ independent variables in K in addition to x_i.
- Equations are generated as in every other XL method, then each generated equation is raised to the second, fourth,... powers easily (since this is a linear operation) as equations in $(x_i^2), \ldots, (x_i^{2^{k-1}})$, for k times as many variables *and* equations.
- That all equivalent monomials are *ipso facto* equal become new equations, which may let the algorithm execute with a lower D (see Sec. 7).

3.5   XFL: Guessing with a Twist

Another variant proposed with the name "improved FXL" and later XFL ([17, 59]):

1. Choose f ("to fix") variables. Multiply the equations by all monomials up to degree $D - 2$ *in the other $n - f$ variables only*.
2. Order the monomials so that all monomials of exactly degree D with no "to-fix" factor comes first. Eliminate all such monomials from the top-degree block.
3. Substitute actual values for "to-fix" variables, then collate the terms and try to continue XL, re-ordering the monomials if needed, until we find at least one solution.

There are $\binom{n-f+D-1}{D}$ monomials of degree D with no "to-fix" variable, so $T' = \binom{n-f+D}{D} - \binom{n-f+D-1}{D} = \binom{n-f+D-1}{D-1}$ variables remain and the complexity is:

$$C_{XFL} = C_0'' + q^f\left[C_0' + E\left(\binom{n-f+D-1}{D-1}, \, m\binom{n-f+D-2}{D-2} - \binom{n-f+D-1}{D}\right)\right]. \tag{4}$$

C_0'' the cost of the initial elimination. What happens is that the max-degree block of the elimination need not repeat with the guessing. We shall see how this does later.

4 Gröbner Bases Algorithms F_4-F_5

Gröbner Bases have come a long way since the early days of Buchberger. The reader is referred to [6, 10, 11, 40] for general theory on the topic, although the speed estimates there can be considered superseded. The most advanced implementations are detailed in [29, 30, 31]. Summaries can also be found in [2, 54], here we only give a synopsis:

0. Initialize: The original are reduced according to some (usu. Degree Reverse Lexicographic) monomial order to a system in row-echelon form.
1. Multiply/Extend: Increase the maximal degree by 1. The resulting equations are multiplied by all monomials such that the product does not exceed the maximal degree. In F_5 the Frobenius selection criteria avoids redundant equations.
2. Linearize/Reduce: Run a Gaussian-like elimination to row-echelon form, such that every row/equation is only reduced against preceeding rows.
3. Repeat: If we do not yet have a Gröbner Basis, go to Step 1. We will find a Gröbner Bases as in $x_1 = f_1(x_2, x_3, \ldots, x_n)$, $x_2 = f_2(x_3, \ldots x_n)$, $\ldots$, maybe ending with $f_{k+1}(x_{k+1}, \ldots, x_n) = 0$ when the system variety has positive Krull dimension.

Please refer to the abovementioned articles for technical details. Lazard (cf. [40]) notes long ago that a Gröbner Basis for a set of equations ℓ_i may be found by a reduction on the extended version of the *Macaulay matrix* at some degree D. This matrix contains exactly the coefficients of the equations $\mathcal{R}^{(D)}$, and the reduction of this matrix is exactly XL. Hence [2] and [54] explains XL as a special case of Gröbner Algorithms.

5 Termination Conditions of XL and Gröbner Bases

How many independent equations do we get in the basic XL? Not all equations are independent: If we write $\ell_i(\mathbf{x}) = \sum_{j \leq k} a_{ijk} x_j x_k + \sum_j b_{ij} x_j + c_i$, then

$$
\begin{aligned}
[\ell_i \ell_{i'}] &= \sum_{j \leq k} a_{ijk} [x_j x_k \ell_{i'}] + \sum_j b_{ij} [x_j \ell_{i'}] + c_i [\ell_{i'}] \\
&= \sum_{j \leq k} a_{i'jk} [x_j x_k \ell_i] + \sum_j b_{i'j} [x_j \ell_i] + c_{i'} [\ell_i],
\end{aligned}
$$

where $[x_j \ell_i]$ denotes the equation $x_j \ell_i(\mathbf{x}) = 0$ in the XL system, etc., i.e., two ideals spanned by each pair of (ℓ_i, ℓ_j) intersect, hence there will be a corresponding dependency at every degree $D > 4$. We may compute the number of free equations assuming no other source of dependencies than the above:

Proposition 1 ([24, 58]). *If all dependencies result from* $\ell_i[\ell_{i'}] = \ell_{i'}[\ell_i]$ *then*

$$
T - I = [t^D] \left\{ (1-t)^{m-n-1} (1+t)^m \right\} = \sum_{j=0}^{\infty} (-1)^j \binom{m-n-1}{j} \binom{m}{D-j}, \tag{5}
$$

for all $D < \min(q, D_{reg})$. Here D_{reg} is the <u>degree of regularity</u> given by

$$D_{reg} := \min\{D : [t^D] \left((1-t)^{m-n-1}(1+t)^m\right) \leq 0\}, \qquad (6)$$

and $[t^k]\, p$ means "the coefficient of t^k in the expansion of p". E.g. $[x^2](1+x)^4 = 6$. This implies that the minimum D required for the reliable termination of XL is given by

$$D_0 := \min\{D : [t^D] \left((1-t)^{m-n-1}(1+t)^m\right) \leq D\}. \qquad (7)$$

Historical Remark: The [58] proof was faulty and did not prove D_0 to be a lower bound. T. Moh ([44]) states without proof a result similar to this one. C. Diem has the first and only derivation ([24]) showing D_0 to be a lower bound if the *Maximum Rank Conjecture* (originally due to Fröberg, [32]) is generally valid.

Corollary 2. *When there are no extraneous dependencies (i.e., Eq. 5 holds), then D_0 is: 2^n if $m = n$, m if $m - n = 1$, and $\lceil (m+1)/2 \rceil$ if $m - n = 2$.*

Proof. If $m - n = 0$, then $T - I = [t^D] \left((1+t)^m/(1-t)\right) = \sum_{j=0}^{D} \binom{m}{j}$, which increases rapidly. It stays constant after reaching 2^m at $D = m$, so $D_0 \geq \max(2^m, q)$.

If $m - n = 1$, then $T - I = [t^D] (1+t)^m = \binom{m}{D} > D$ up to and including $D = m - 1$ (whence $T - I = m > D$). Finally at $D = m$ we have $T - I = 1 < m$. The $D_0 = m - 1 = n$ of [17, 20] is due to a slightly different XL in [20].

If $m - n = 2$, then $T - I = [t^D] \left((1-t)(1+t)^m\right) = \binom{m}{D} - \binom{m}{D-1} = \binom{m}{D-1} \frac{m-2D+1}{D}$. Obviously $T - I \leq 0$ iff $D \geq (m+1)/2$, and $T - I > D$ early on. So $D_0 \leq \lceil (m+1)/2 \rceil$. When D is incremented by 1, $T - I$ increases by $\left(\binom{m}{D} - \binom{m}{D-1}\right) - \left(\binom{m}{D-1} - \binom{m}{D-2}\right) = \binom{m}{D+1} - 2\binom{m}{D} + \binom{m}{D-1}$, which starts out positive and when $D > \frac{1}{2}(m - \sqrt{m+2})$ turns negative (see below), so we only need to check the case of $D = \lceil (m+1)/2 \rceil - 1$ first, which is the last D before $T - I$ decreases down to or past 0. Combinatorics texts (e.g. [52]) tell us that $\binom{2D}{D} - \binom{2D}{D-1} = \frac{(2D)!}{D!(D+1)!}$ is the Catalan number c_D, which satisfy $c_0 = 1$, $c_n = \sum_{j=0}^{n-1} c_j c_{n-1-j}$. Since $c_2 = 2$, c_D for $D > 2$ will be the sum of D positive integers, not all of them 1, so $c_D > D$. Similarly for even m we have $\binom{2D+1}{D} - \binom{2D+1}{D-1} = \frac{(2D+2)!}{(D+1)!(D+2)!} = c_{D+1} > D$.

So the bounds are tight. Eqs. 5 and 7, says that as described in [20], as $m - n$ increases D_0 decreases, although formulas are more complicated for larger $m - n$ (Tab. 1).

Corollary 3. *Good approximations to D_0 for fixed small $f = m - n$ is given by Tab. 1.*

D_0 is not easily expressed as a function of m (or n) for larger $f = m - n$. We may approximately assume that $D_0 \approx D_{reg}$. D_0 is then the smallest D such that

$$T - I = [t^D] \left((1-t)^{f-1}(1+t)^m\right) = \sum_{j=0}^{f-1} (-1)^j \binom{f-1}{j}\binom{m}{D-j} < 0.$$

or, after dividing out $(m!)/[D!(m - D + f - 1)!]$, we get this inequality in D:

$$\sum_{j=0}^{f-1}(-1)^j \binom{f-1}{j} \frac{D!}{(D-j)!} \frac{(m-D+f-1)!}{(m-D+j)!} \le 0 \tag{8}$$

From Eq. 8 we can find D_{reg} explicitly (and D_0 approximately) for $f \le 10$ using lots of roots. We tabulate (cf. Table 1) the results for smaller $f = m - n$. This shows that the earlier estimate of $D_0 \approx \sqrt{n}$ ([20]) for small f is not very good. Indeed, [24] points out for any fixed f, $D_0/n \to 1/2$.

Table 1. Relationship between $f = m - n$ and minimal degree $D_0 = D_0(m)$

f	D_0 (as approximate function of m)	10	15	20	25	30	35
0	2^m (but only up to q, cf. Sec. 5)	2^{10}	2^{15}	2^{20}	2^{25}	2^{30}	2^{35}
1	m	10	15	20	25	30	35
2	$\lceil \frac{1}{2}(m+1) \rceil$	6	8	11	13	16	18
3	$\lceil \frac{1}{2}(m+2-\sqrt{m+2}) \rceil$	5	7	9	11	14	16
4	$\lceil \frac{1}{2}(m+3-\sqrt{3m+7}) \rceil$	4	6	8	10	12	14
5	$\left\lceil \frac{1}{2}(m+4-\sqrt{3m+8+\sqrt{6m^2+30m+40}}) \right\rceil$	3	5	7	9	11	13
6	$\left\lceil \frac{1}{2}(m+5-\sqrt{5m+15+\sqrt{10m^2+50m+76}}) \right\rceil$	3	4	6	8	10	12

The predictions of Eq. 5 is confirmed for random quadratics ℓ_i by simulations due to N. Courtois ([18]) up to very high dimensions and degrees, including all the parameters listed in [17] and earlier works. The public polynomials of several PKC's including SFLASH^{v2} also behaves like random polynomials at low degrees. This verifies our own simulations, which are not so extensive.

5.1   XL2, Gröbner Bases, and Their Relationship

Corresponding to Eq. 7 for Gröbner Bases algorithms such as $\mathbf{F_5}$ we have ([4], later [2]) is this result for semi-regular sequences of equations (i.e., no extra dependencies):

$$D_g := \min\{D : [t^D](1-t)^{m-n}(1+t)^m < 0\}. \tag{9}$$

Its resemblence to D_0 for XL means that some results for XL can extend directly to Gröbner Bases: We can think of this as corresponding to exactly one fewer variable, or we can think of the extra factor of $(1-t)$ to mean that the elimination is run on the highest-ranked monomials only. One variant method of XL does exactly that — the XL2 adjunct (Sec. 3.2), otherwise known as the T'-method. In [58,59], it was pointed out that one can run XL2 on all variables to achieve effectively going one degree higher. [59] comments that XL2 may not repeat even if it works once. It may be possible using the original, overly optimistic estimate $(T-I < T')$ as opposed to one that focuses on the top degree monomials; indeed, we *prove* below that it is not the case for large q.

Proposition 4. *XL2 (for large q) on all variables will run when $D \geq D_g$, and will then repeat until we find a solution or prove the system self-contradictory.*

Proof. We need to eliminate all top-degree monomials ([59]), and can think of regular XL being run on homogenized equations with one variable assigned to represent the constant 1, and we may apply Eq. 6, we get the first half of the statement.

We know that multiplying by $(1-t)^{-1}$ represent taking the sum of coefficients of a generating function up to some degree, i.e., $g_{m,f}(D) = \sum_{d=0}^{D} g_{m,f+1}(d)$ (here $f = m - n$). We know that the zeroes of $g_{m,f}(D) := [t^D] \left((1-t)^{f-1} (1+t)^m \right)$ and $g_{m,f+1}(D)$ alternate (see below) because their dominant terms are associated Hermite functions (which form an orthogonal sequence). So once $D \geq D_g$, $g_{m,f+1}(D)$ will not become positive until $g_{m,f}(D)$ becomes non-positive also.

We have shown that XL+XL2 and $\mathbf{F_5}$ operates at the same apparent degree D_g. Further ([30]) the *signature*, i.e. the underlying degree, of the polynomials in the matrix/system built by $\mathbf{F_5}$ is the same as the classical Buchberger algorithm, which is the same as classical XL (this remark was also made in [2]). This is scarcely surprising, given that Imai *et al* have shown that XL2 is equal to a sequence of Buchberger-like operations. Therefore, we can think of XL+XL2 as a less polished version of $\mathbf{F_5}$.

5.2   FXL and Asymptotic Estimates for D_0

With $m = n$ equations and variables in practice the attacker would most often run the variant FXL, i.e., guesses at f variables, then attempts to run XL on the remaining system. It is hence of particular interest to obtain asymptotic behavior when $m - n = f$ remains small compared to m or n. Eq. 7 is valid only when $D < q$, but for GF(2^8) we can cover all m up to about 500, which is large enough to bring in asymptotics. This requires first asymptotically estimating a coefficient then approximating a sign-change position in the following manner via Cauchy's integral formula ([35]),

$$g_{m,f}(D) := [t^D] \left((1-t)^{f-1}(1+t)^m \right) = \frac{1}{2\pi i} \oint z^{-D-1}(1-z)^{f-1}(1+z)^m dz.$$

Standard asymptotic analysis recipes (cf. [12, 35, 57]) can be applied to find ([3]) that

$$D_{reg} = \frac{m}{2} - (h_{f-1,1})\sqrt{\frac{m}{2}} + O(1) \sim \frac{m}{2} - \sqrt{fm}. \tag{10}$$

Here $f = o(m^{1/3})$ and $h_{k,1}$ is the largest zero of the Hermite polynomial $H_k(x)$, given by Szegö ([55]) as $\sqrt{2k+1} + O(k^{-1/6})$. And when we have $f \sim cm$

$$D_{reg} \sim \left(\frac{1}{2} - \sqrt{c} + \frac{c}{2} \right) m + O(m^{\frac{1}{3}}). \tag{11}$$

via the Coalescent Saddle Point method ([3, 12]). We note that Eqs. 10 and 11 are compatible which is necessary if the asymptotics are uniform.

One consequence of the above is that that an optimal f for FXL (cf. Sec. 6.3) exist, which also applies to $\mathbf{F_4}$-$\mathbf{F_5}$. Let us start with a medium-large $m = n$ (asymptotics come into play as low as in the teens), and start with the assumption that $D = m/2$, then we may compute $\lg T = 1.377m + O(\log m)$ via Stirling's formula:

$$T = \binom{n+D}{n} = \binom{3m/2}{m} = \frac{(3m/2)!}{m!(m/2)!} \sim \sqrt{\frac{3}{2\pi m}} \left(\frac{(3/2)^{3/2}}{(1/2)^{1/2}}\right)^m$$

When guessing at f variables, Eq. 10 means that n and D goes down respectively by f and roughly $\sqrt{fm}$. We find that T goes down by a factor of $\approx \left((3/2)^f \cdot 3^{\sqrt{fm}}\right)$, hence $\lg T \sim 1.377\,m - 1.585\sqrt{fm} + 0.585f$, and if we assume small f, $q = 2^8$, E to be degree $\omega = 2 + \varepsilon$ (a Lanczos solver, see below) and $R \sim T$ then FXL has

$$\lg C_{\text{FXL}} \sim 2 \lg T + f \lg q \sim (\lg q + 1.170)f - 3.170\sqrt{fm} + 2.755m. \tag{12}$$

If Eq. 12 holds for all f, then $\lg C_{\text{FXL}}$ will take a minimum of $\lg C_{\text{FXL}} \approx 2.63m$ at $f \sim 0.014m$, a significant gain. However, Eq. 12 is actually valid up for small f, actually to only $f = o(\sqrt[3]{m})$. We may still conclude that for FXL, there is some small $\epsilon > 0$ and δ such that we should take at least $f = \delta \cdot m^{1/3 - \epsilon}$ guessed variables, and we can say more since we have compatible asymptotics, for which we refer you to [59]. The result is that we should guess even more variables: For $q = 2^8$ and $\omega = 2$ the minimum occurs at around $c := f/m \sim 0.049$, when $\lg C_{\text{FXL}} \sim 2.4m$. Similarly when $\omega = \lg 7$ (Strassen blocking), the minimum $\lg C_{\text{FXL}} \sim 3.0m$ when $f \sim 0.096m$.

Even supposing that our numbers are slightly off, this shows that FXL is a better way to apply XL on non-small fields. As Gröber Bases methods theoretically and asymptotically resemble XL, the phenomena should be nearly the same. I.e., starting from $m = n$, one should guess at a very small percentage of the variables before starting the Gröbner Bases computations. Indeed, for $m = n$ and $\omega = 2.8$ (Strassen Blocking), we see that $\lg T = \lg \binom{2n}{n} \sim 2n$, hence $\lg C_{\mathbf{F_4}} \sim \lg C_{\mathbf{F_5}} \sim 5.6n$, as opposed to $4.2n$ for guessing at *one* variable, and $3.0n$ with the optimal guessing. For $\omega = 2$, the coefficients are 4, 3.0 and 2.4 respectively. This seems very natural, but has not been seen in print previously. Results for smaller q can be found in [4, 58, 59]. As C. Diem points out, a critical proof in [58] is inaccurate, its results are not always valid lower bounds. However [59] shows FXL (and likewise $\mathbf{FF_4}/\mathbf{FF_5}$ worthwhile for all values of (q, ω).

6   Pragmatic Issues in XL-Related Methods

We first mention some theoretical and practical aspects of XL-related method, particularly the parameters we shall use when estimating the complexity (security level).

6.1   On a Pragmatic Cost of Elimination

Naive cubic-time elimination ([8]) is inadequate for large matrices, and a cost estimate $T^{\lg 7}$ (where T counts the monomials) or even lower is cited in all XL articles ([15, 17, 20, 21, 22]). However, Strassen's ([53]) original $2^{\lg 7}$ algorithm does not reliably invert a known nonsingular square matrix. The XL situation is even more complex: The matrix (with $R > T$) is not square, and we want our elimination algorithm to (a) run despite the redundant rows (equations); (b) compute a useful basis for the kernel (e.g. reduced row-echelon form) if the matrix is not full-rank (i.e. $T - I = 1$). To pivot inerrantly around singular submatrices in $O(n^{\lg 7})$ is quite nontrivial ([7]). Similar caveats apply to adapting other sub-cubic matrix multiplications for equation-solving.

The best all-around result for dense matrices known to us is D. J. Bernstein's GGE (Generalized Gaussian Elimination, [9]) which computes the *quasi-inverse*, which can (method "S") solve a system of equations, and even (method "N") find a basis of the kernel of a matrix (a row-reduced echelon form)! *Assume M equations, N variables, and the time cost $\sim \alpha N^\omega$ to multiply two $N \times N$ matrices, then GGE uses time*

$$E_S(N, M) = \frac{2\alpha(1 + \gamma)}{(2^\omega - 2)} M^{\omega - 1} N + \frac{\alpha M^\omega}{(2^\omega - 1)}; \tag{13}$$

$$E_N(N, M) = \frac{2\alpha(2 + \gamma)}{(2^\omega - 2)} M^{\omega - 2} N^2 + \frac{4\alpha\gamma}{7} M^{\omega - 1} N + \frac{\alpha M^\omega}{(2^\omega - 1)}. \tag{14}$$

Here the coefficient $\gamma = (7\alpha)/(2^\omega - 4)$. We shall look at how to do better in Sec. 6.2.

6.2   A Need for Sparse-Matrix Algorithms

The systems generated by XL are obviously sparse. A respected textbook on sparse matrices ([27]) remarks that in not using a matrix algorithms more tailored for the situation *"you would just be pushing milliards of zeros around"*. Moving around gigabytes full of zeroes not only slows down the computation direcly, but increases the amount of memory required. With $n = 15$, $m = 20$, $q = 2^8$ and $D = 7$ (these are practical dimensions), we have $T = 170544$ monomials and $R = 310080$ equations. A full elimination will take about 50 GB ($\approx 2^{36}$ bytes). A normal procedure ([27]) is to find block structures with graph-coloring analysis. The elimination cost is then dominated by $E(N_0, M_0)$, the elimination cost for the largest block. *The XL equations are structured such that the largest block comprise the equations with the highest degrees (and this naturally happen in $\mathbf{F_5}$).* Still, if we know that there is at most one solution, then it must be better to use Lanczos, Conjugate Gradient (CG), or Wiedemann methods, each solving an $N \times N$ system $\mathsf{M}\mathbf{x} = \mathbf{b}$ using N multiplications of M to an $N \times 1$ vector. An $M \times N$ system ($M > N$) in Lanczos (or CG) is converted to $N \times N$ by solving $(\mathsf{M}^T \mathsf{D} \mathsf{M})\mathbf{x} = \mathsf{M}^T \mathsf{D}\mathbf{b}$ instead. Here D is invertible, diagonal, and suitably random. *For sparse systems with t terms per equation, the expected time cost drops to order $2 + \varepsilon$:*

$$E_L(M, N) \approx (c_0 + c_1 \lg N)\, tMN. \tag{15}$$

The log-factor is because accessing memory no longer take negligible time, and tags are $\propto \lg N$ long. Lanczos, CG and Wiedmann methods all have comparable speeds. Consensus seems to peg the Wiedemann algorithm as intrinsically slower but more reliable, and to get better results Lanczos methods must be randomized which adds to the cost (cf. [8, 28, 38, 56]). Warning: *Lanczos (or CG) is known to terminate sometimes incorrectly over a finite field. Wiedemann is not known to terminate always correctly for nonsquare matrices. Proper operating conditions are not fully understood.* However, we were informed ([13]) that such methods are usable and widely used in practice.

6.3   Practical Parameters for Assessing FXL (for Large q)

Since we ultimately want reasonable security estimates, we need concrete values for c_0 and c_1 in the Lanczos estimate E_L. What are reasonable estimates? *We will use $c_0 = 4$ and $c_1 = \frac{1}{4}$ in Eq. 15 to arrive at complexity estimates (for Lanczos-like sparse solvers) in field multiplications. Calibrating against our own test data, we should divide the number of multiplications by roughly 2^6 to get numbers in 3DES encryption blocks (comparable to but a little longer than AES blocks).*

Furthermore, if the dimensions become very large, then asymptotically we will eventually see R/T in the hundreds. However, we may generate fewer equations ([13, 18]), e.g., via a randomly picked set of equations (taking say 20% more equations than variables) and solve ten such random systems to ensure not missing a solution. Hence it makes much more sense to assume the equations to have roughly as many equations as there are variables, and we may assume R/T to be a constant on the order of "a few". Of course, for smaller dimensions, it may make more sense to run a more robust elimination. For Gröbner Bases methods, obviously $T = R$ and this is a smaller T because it is only the top degree portion, but this contribution dominates the number of monomials for large q and a typical case of XL/FXL anyway. With $\alpha = 7$, $\omega = \lg 7$ ([41]), we get

$$E_{\text{sparse}}(N, M) = (\frac{1}{4}\lg N + 4) \cdot MN \cdot (\text{ avg. \# of terms per equation}); \quad (16)$$

$$E_{\text{dense}}(N, M) = 51.33 M^{0.8} N^2 + 65.33 M^{1.8} N + 1.167 M^{2.8}.$$

These numbers are for processors with enough cache only. We hear that some IBM servers do have 100+GB of RAM and a mind-boggling 512MB cache per CPU, so we assume that processing power, memory size and bandwidth all pose no problems.

7   Practical Security Assessment of XL Variants

Infeasibility of the cryptanalysis against SFLASH^{v2} and HFE Challenge 2 as mentioned in [59] is given, using some results that we prove rigorously for semi-regular sequences.

7.1   Inefficiency of XL' and XLF for Small $m - n$ and Large q

Proposition 5. *The number of extra equations provided by XLF (Sec. 3.4) is given by*

$$\Delta T = k \binom{n + \lfloor D/2 \rfloor}{\lfloor D/2 \rfloor} - 1. \tag{17}$$

Proof. We need not track the redundant monomials explicitly as in [17]. These monomials are the degree $\leq D$ monomials in the (x_i)'s that can also be written as monomials of the (x_i^2)'s at a lower degree, copied k times. So we just count monomials in the (x_i^2)'s of total degree $\leq D/2$, which number $\binom{n+\lfloor D/2 \rfloor}{\lfloor D/2 \rfloor}$. The final -1 comes from the fact that the monomial 1 is counted as duplicated k times, once too many.

Corollary 6. *When $D < q$, XLF can be expected to work (most of the time) when*

$$[t^D] \left((1-t)^{m-n-1} (1+t)^m \right) - \binom{n+\lfloor D/2 \rfloor}{\lfloor D/2 \rfloor} < \lceil D/2 \rceil. \tag{18}$$

Note: This is likely only asymptotically correct (extra dependencies are possible). If the elimination ends with all odd powers of $(x_1^{2^j})$ left we can still solve for x_1.

Proposition 7. *The following holds about XL' and XLF for q large:*

XL': *When $m - n$ is 1 or 2, XL' operates if and only if $D > m - r$; if $m = n$, XL' will not run at $D = m + 1 - r$, but will at $D = m + 2 - r$ for r large enough (around $r > m/2$). When r is small, we need a much larger D, around $2^{m/r} (r!)^{1/r}$.*
XLF: *When $m - n \leq 2$, XLF need at least $D > n/2$ to operate.*

Proof. We can use the approximations $\binom{2k}{k} \approx 2^{2k}/\sqrt{\pi k}$ and $k! \approx \sqrt{2\pi k}(\frac{k}{e})^k$.

XL': $m = n$: From the description of XL' above and Eq. 5, we see that $\binom{r+D}{D} - r \geq \sum_{i=0}^{D} \binom{m}{i} > \binom{m+1}{D} + \binom{m+1}{D-2}$. Since $\binom{m}{k}$ is increasing in m, we need $r + D > m + 1$ which suffices for $r > m/2$. For small r, we need $\binom{r+D}{D} > 2^m$.

   $m = n + 1$: Need $\binom{r+D}{D} - \binom{m}{D} \geq r$; when $r + D = m + 1$, the left hand side becomes $\binom{m}{D-1} \geq m > r$, so it does the job, and no smaller r can do that.

   $m = n + 2$: Need $\binom{r+D}{D} = \binom{r+D+1}{D} - \binom{r+D}{D-1} > \binom{m}{D} - \binom{m}{D-1}$; again $r + D = m + 1$ will do, and barely, because nothing smaller works.
XLF: Let $m - n = 1$. As we can presume D small, we have gives $\binom{n+1}{D} < \binom{n+\lfloor D/2 \rfloor}{\lfloor D/2 \rfloor}$. At $D = n/2$ we have the left side $\approx \frac{2^{n+1}}{\sqrt{\pi n/2}}(1 - \frac{1}{n+2})$, and the right

$$\approx \frac{\sqrt{2\pi \cdot 5n/4} \cdot (\frac{5n}{4e})^{5n/4}}{\left(\sqrt{2\pi \cdot n/4} \cdot (\frac{n}{4e})^{n/4}\right) \left(\sqrt{2\pi n} \cdot (\frac{n}{e})^n\right)} = \sqrt{\frac{5}{2\pi n}} \left[\frac{(5/4)^{5/4}}{(1/4)^{1/4}} \right]^n.$$

We see $\left[\frac{(5/4)^{5/4}}{(1/4)^{1/4}}\right] \approx 1.87 < 2$ and $\frac{2}{\sqrt{\pi n/2}}(1 - \frac{1}{n+2}) > \sqrt{\frac{5}{2\pi n}}$. For $m = n$ we need a higher D (we can check that $D \approx 3m/4$ is needed).
Now consider $m - n = 2$. We want $\binom{n+2}{D} - \binom{n+2}{D-1} = \binom{n+1}{D} - \binom{n+1}{D-2} < \binom{n+\lfloor D/2 \rfloor}{\lfloor D/2 \rfloor}$, which we can verify to happen only when $D \leq n/2$. The LHS is about $1/D$ of what it was at $m - n = 1$, which is covered by the exponential factor $(1.87/2)^n$.

XL' (designed for GF(2)) work suboptimally on a larger field. XLF is hindered by the fact that the dependencies are multiplied along with the independent ones ([18]).

7.2   XFL Is Really a Space-Time Tradeoff

XFL of Sec. 3.5 appears to be an improvement, but there are important drawbacks. Essentially, for the initial elimination stage, the memory requirement is increased $(m - f)$-fold. More importantly, once the initial substitution is made, the resultant second-highest-degree block is no longer sparse and requires the equivalence of GGE (Sec. 6.1).
   There is no particular reason that XFL should fail. Indeed, it is better than XL' or XLF. However we believe that FXL works better due to the availability of Lanczos.

7.3   Reassessing XL'/XLF/XFL Versus SFLASH^{v2} and HFE Challenge 2

Proposition 8 ([58]). *If $2q > D \geq q$, and the system is semi-regular up to degree D, then $T - I = [t^D] \left((1-t)^{m-n-1} (1 - nt^q) (1 - t^2)^m\right)$. [Also similarly for $\mathbf{F_4}/\mathbf{F_5}$.]*

This is a yardstick we need for the complexity of some XL variants, and we look at how three XL variants apply to extant schemes SFLASH^{v2} and HFE Challenge 2.

Did XL Variants really break SFLASH^{v2} in 2^{80}? To recap, SFLASH^{v2} ([51]) is a NESSIE finalist. It is an instance of C^{*-} ([49], descendant of C^*, [43])) with $K = GF(2^7)$, $m = 26$ and $n = 37$, and reputed to be very fast, suitable for smart card implementations ([1]). Although the NESSIE writeup contained some extraneous private data that can be recovered ([34]), SFLASH^{v2} was previously considered safe ([47]). It is claimed ([17]) that after n is reduced to 26 by guessing at eleven variables, any of the variants XL', XLF, and XFL can provide a cryptanalysis within 2^{80} 3DES operations. None of the cryptanalysis attempts can function as given:

XL': [17] gives $D = 7$, $r = 5$. From Sec. 7.1, we can see that at $r = 5$, XL' should not work until $D = 92$. We actually ([18]) need $D = 93$. By trial and error, we get best result is around $r = 16$, which gives a complexity of $\sim 2^{118}$.

XFL: [17] gives $f = 4$, $D = 6$. Actually we see from Table 1 that $D_0 \geq 10$.
    With Strassen and Bernstein, we get 2^{104} multiplications (2^{98} 3DES blocks).
XLF: [17] gives $D = 10$. Using Sec. 7.1 and Prop. 8 we verify that XLF only
    works at $D = 18$ (complexity $\sim 2^{92}$ even with Lanczos).

Reports of the demise of SFLASH^{v2} is exaggerated and justifies the design
decisions of Patarin *et al.* This is significant as SFLASH^{v3} ([19]) is much slower
with bulkier keys, and has security concerns due to unlucky choice in dimensions
([18, 26], cf. also http://www.minrank.org/sflash/). The best cryptanalysis is
FF$_5$ if it works with Lanczos (complexity $\gtrsim 2^{81}$). Else the best try is likely FXL
(complexity $\sim 2^{85}$). If an attack works, it probably will be an algebraic attack
resembling [26].

Did XL Variants really break HFE challenge 2 in 2^{80}? HFE Challenge 2
is an HFE instance with $q = 2^4$ and $m = n = 32$. We believe that the parameters
as given in [17] does not lead to cryptanalysis under 2^{80}, after double-checking
against Sec. 7.1:

XL' [17] gives $m = n = 32$, $D = 10$, for which $T - I = 107594213$. We can do
    (cf. Prop. 7) XL' using $(D, r) = (15, 19)$ or $(14, 20)$, which is very sufficient.
XFL: $D = 7$ and $f = 2$ ([17]) won't function (since $T - I = 2459664$). We
    recommend $(f, D) = (12, 6)$ with complexity $\sim 2^{97}$ 3DES blocks.
XLF: [17] gives $D = 10$, which we can verify not to work (as above). We need
    (cf. Prop. 8) all the way to $D = 23$, with a complexity 2^{112} even for Lanczos.

8   Discussions and Conclusion

With all the results we have gathered, we may tabulate the complexity in various
schemes. Two points need explaining. F1F$_5$ and F2XL means **F$_5$** guessing at *one*
variable and FXL always guessing at two variables respectively. The asterisk
means that we are assuming Lanczos-class speed solvers to work with **F$_5$**, which
is not a given.

Table 2. Time Estimates (3DES blocks): Blocking ($\omega = 2.8$) v. Lanczos ($*$: may not
work)

Attack Method		FXL	F2XL	F$_5$	F1F$_5$	FF$_5$
$n = 26$ $q = 2^7$	B	2^{101}	2^{104}	2^{117}	2^{102}	2^{99}
(SFLASH^{v2})	L	2^{85}	2^{87}	2^{95}_*	2^{85}_*	2^{82}_*
$n = 32$, $q = 2^4$	B	2^{97}	2^{106}	2^{111}	2^{105}	2^{93}
(HFE Chal. #2)	L	2^{87}	2^{92}	2^{98}_*	2^{89}_*	2^{82}_*
$n = 20$, $q = 2^8$	B	2^{91}	2^{91}	2^{109}	2^{88}	2^{86}
	L	2^{78}	2^{78}	2^{84}_*	2^{77}_*	2^{74}_*
Asymptotic for	B	$2^{3.0n}$	$2^{3.86n}$	$2^{5.6n}$	$2^{3.86n}$	$2^{3.0n}$
big $m = n$, q	L	$2^{2.4n}$	$2^{2.75n}$	$2^{4.0n}_*$	$2^{2.75n}_*$	$2^{2.4n}_*$

8.1   Comments on the Relationship of Gröbner Bases to XL

Imai *et al* ([54]) shows XL to be variation of the $\mathbf{F_4}$-$\mathbf{F_5}$ algorithms. However, practical differences remain even if the theory of XL might be considered subsumed by Gröbner Bases. Gröbner Bases is a general and elegant mathematical theory that applies well to everything under the sun including symbolic computation. When $\mathbf{F_5}$ terminates , we should always obtain all solutions, including those in extension fields. Shamir *et al* proposed XL as a cryptanalytical tool, with one purpose: to find a known-or-conjectured-to-exist solution to a numerical set of equations. In FXL/Lanczos variant, this property is shown clearly: it is possible to find all solutions in K, but not in extensions of K.

Wiedemann and Lanczos algorithms are not suitable for computing reduced row-echelon forms; as a conclusion to XL, either works best with $T - I = 1$. In a Gaussian, we need not know $T - I$ beforehand and may come down to any number of monomials (between 1 and $\min(q-1, D)$) with no speed penalty and still terminate correctly; under Bernstein's GGE, we are penalized by the slower algorithm "N" (instead of "S", cf. Sec. 6.1–6.2); using Lanczos requires us to know $T - I$ in advance, and to run exactly that many different iterative sequences. [30, 31] seem to agree with the above assessment, and the critical step of $\mathbf{F_4}/\mathbf{F_5}$ appears to be an elimination on the top block.

The aversion to Gaussion or Generalized Gaussian elimination is also why we do not suggest XL2. We do not see how to link it reliably into a Lanczos-like sparse solver.

[4] claims that Gaussian-like elimination in $\mathbf{F_5}$ can achieve time close to Lanczos algorithms. At least, $\omega = 2$ is "plausible". We hasten to agree! It is quite plausible that one can adapt these advanced Gröbner Bases methods for Lanczos, or achieve $\omega = 2 + \varepsilon$ regardless. However, it is also plausible that one cannot, because according to [30], the elimination is severely restricted in the order of operations. We may also use guessing in $\mathbf{F_5}$, and the two methods behaves very similarly (as expected). But in this event, the two methods could be described as having largely converged. The entries that require running a Lanczos-like sparse solver with $\mathbf{F_5}$ or Fix-then- $\mathbf{F_5}$ (denoted "FF$_5$") is marked with an asterisk in Tab. 2. If effectively for $\mathbf{F_5}$ we will always have ω measurably greater than 2, then these estimates are invalid. In this case FXL will eventually dominate methods that always compute a Gröbner Basis.

There is one further situation where FXL might work better, which is when we cannot hold the entire matrix of the $\mathbf{F_5}$ in memory. In turn, we can run FXL without generating the whole Macaulay matrix. All the possibilities takes further study.

8.2   Some Remarks on the Termination of Basic XL

Moh in [44] pointed out that Basic XL should not work if the system of equations has a positive-dimensional solution at "at infinity". It is our aim to help to clarify this often-cited remark by Moh. We thank C. Diem for pointing out Prop. 9 to us.

As above, let $\ell_1, \ldots, \ell_m \in K[x_1, \ldots, x_n]$. Let $\ell_1^h, \ldots, \ell_m^h \in K[x_0, \ldots, x_n]$ denote the homogenizations of the ℓ_i. Let $D \in \mathbb{N}$, and let us assume (without loss of generality) that $\ell_1, \cdots, \ell_a$ have degree $\leq D$ and $\ell_{a+1}, \cdots, \ell_m^h \in K[x_0, \ldots, x_n]$ have degree $> D$. For example, for $i = 1, \ldots, a$, the ℓ_i might be quadratic and for $i = a+1, \ldots, m$, the ℓ_i might be the field equations which might have a much higher degree.

Let V_D be the projective algebraic set defined by the equations

$$\ell_1^h = 0, \ldots, \ell_a^h = 0.$$

We note that if the system $\ell_1 = \cdots = \ell_a = 0$ defines a 0-dimensional algebraic set and the "set at infinity" is non-empty, the dimension of V_D equals the dimension of the "set at infinity". Let T and I be as above. Our want to relate $T - I$ to the dimension of V_D.

Proposition 9. If $\dim(V_D) = r$, then $T - I \geq \binom{r+D}{r}$.

Proof. Let $\mathcal{J}$ be the homogeneous ideal defined by $\ell_1^h, \cdots, \ell_a^h$, and let $(K[x_0, \ldots, x_n]/\mathcal{J})_D$ be the D-th homogeneous part of the quotient ring $K[x_0, \ldots, x_n]/\mathcal{J}$.

Note that $\dim(V_D) = \dim(K[x_0, \ldots, x_n]/\mathcal{J}) - 1$ (since V_D is projective), and

$$T - I = \dim((K[x_0, \ldots, x_n]/\mathcal{J})_D)$$

(see [24–Section 4] for a derivation of this formula). We can go from K to any field extension L without changing the numbers T, I and $\dim(V_D)$. By going to a sufficiently large extension L/K, we can apply the Noetherian Normalization Theorem in the form of [39–Theorem 2.2]. We obtain that the ring $L[x_0, \ldots, x_n]/(\mathcal{J})$ contains a polynomial ring $L[y_0, \ldots, y_r]$ such that the images of the y_i are linear combinations of the x_i, hence

$$\dim_K((K[x_0, \ldots, x_n]/\mathcal{J})_D) = \dim_L((L[x_0, \ldots, x_n]/(\mathcal{J}))_D) \geq$$
$$\dim_L((L[y_0, \ldots, y_r])_D) = \binom{r+D}{r}$$

Note that the proposition implies in particular that if $r \geq 1$, then $T - I \geq D + 1$, and if $r \geq 2$, then $T - I \geq \frac{(D+2) \cdot (D+1)}{2}$.

If $T - I \leq D$, then XL will find a univariate polynomial, whereas if $T - I$ is greater than this number it will *usually* not find such a polynomial. The interpretation of the proposition above is thus that *usually* XL does not terminate if $\dim(V_D) > 0$. We would like to stress that V_D is the variety defined by $\ell_1^h = 0, \ldots, \ell_a^h = 0$, the equations $\ell_{a+1}^h = 0, \ldots, \ell_m^h = 0$ of higher degree are disregarded.

Let us now set V as the projective algebraic set defined by *all* equations $\ell_1^h = 0, \ldots, \ell_m^h = 0$. Then we have the following good news ([44–Lemma 2] is a corollary).

Corollary 10. If $V = \emptyset$ or $\dim(V) = 0$, then XL terminates for some D.

Proof. Let $\mathcal{J}$ be the homogeneous ideal defined by $\ell_1^h, \ldots, \ell_m^h$. We now have

$$T - I = \dim((K[x_0, \ldots, x_n]/\mathcal{J})_D)$$

for all D (where T and I are defined with respect to D).

Now, under our assumptions on V, there exists a $\tilde{D} > 0$ such that for $D \geq \tilde{D}$, $\dim(K[x_0, \ldots, x_n]/\mathcal{J}_D)$ is equal to the number of non-trivial solutions (counted with multiplicities) of the system $\ell_1^h = 0, \ldots, \ell_m^h = 0$. This is one of the statements of Hilbert theory. It is essentially proven in [36–I, §7]. It follows that *for some $D > 0$, one has $T - I \leq D$, and the algorithm finds a univariate polynomial.*

Cor. 10 does not apply to XSL (cf. 3.2) or any other method in which an entire ideal $\mathcal{I} = K[\ell_1, \ldots, \ell_m, p_1, p_2, \ldots, p_\kappa]$ is not used (the p_i's are extra polynomials added by the attacker). It also does not say what D is. Even if it always works, it may be slow.

When XSL ([21]) proposes to break AES (and Serpent). The more optimistic claims of cryptanlysis in 2^{87} or 2^{100} is based on XSL applied to the Murphy-Robshaw structural equations ([45, 46]) of AES. It is claimed ([17, 21]) that XSL can sidestep the objections of [44] because M-R equations are formed with techniques similar to Sec. 3.4, and the final ("T'-method") stage is similar to XL2. But these variant methods may work correctly or not, independently of Cor. 10.

8.3   A Conclusion

With all the analysis given here we hope to have done a reasonable job in covering various aspects of XL. In passing we may have rehabilitated the reputation of SFLASH^{v2} to some extent. In conclusion: XL is a simplified version of current Gröbner Bases algorithms. Some prior claims about XL variants were clearly too ambitious, and sometimes unrealistic claims were put forward. Yet, the invention of XL (and particularly FXL) is clearly an advance, justifying the insights of Courtois, Klimov, Patarin and Shamir. We hope we have evaluated the capabilities of XL algorithms in a more rational and pragmatical manner. Still, much remains to be done in the practical arena. One important item is to settle the question of the validity of XSL.

The results of this work along with [59] should go some ways to show that FXL is the best XL variant, and the principle extends to $\mathbf{F_4}/\mathbf{F_5}$. We hope that this study will lead to better equation-solving methods based on $\mathrm{FF_5}$ (or $\mathrm{FF_4}$). On the theoretical side, there are also a couple of things that can use a little further study. One is the identification of situations where $\mathbf{F_5}$ (and XL/FXL) will work substantially better or much worse than the [4] bound. Another is a correct way to implement sparse matrix arithmetic so that $\mathbf{F_5}$ can reliably run with a solver with Lanczos-like speeds. While the MAGMA project ([42]) has an implementation of $\mathbf{F_4}$ that is very well optimized, even faster than Faugère's own $\mathbf{F_5}$, it is not yet pushing the limits of what such a solver can do. This is an area that can still be exciting and practically useful.

Acknowledgements

The authors would like to thank Dr. Nicolas Courtois and Dr. Claus Diem for helpful discussions, and the first author would like to dedicate this to the 65th birthday and imminent retirement of his father, Prof. Wei-Zhe Yang of National Taiwan University.

References

1. M. Akkar, N. Courtois, R. Duteuil, and L. Goubin, *A Fast and Secure Implementation of SFLASH*, PKC 2003, LNCS 2567, pp. 267–278.
2. G. Ars and J.-C. Faugère, H. Imai, M. Kawazoe, and M. Sugita, *Comparison of XL and Gröbner Bases Algorithms over Finite Fields*, ASIACRYPT'04, LNCS 3329, pp. 338–353.
3. M. Bardet, *Étude des systèmes algébriques surdéterminés. Applications aux codes correcteurs et à la cryptographie.*, Ph.D. thesis, Université Paris 6, 2004.
4. M. Bardet, J.-C. Faugère, and B. Salvy, *Complexity of Gröbner Basis Computations for Regular Overdetermined Systems*, INRIA Rapport de Recherche No. 5049; a slightly modified preprint is accepted by the International Conference on Polynomial System Solving.
5. M. Bardet, J.-C. Faugère, B. Salvy, and B.-Y. Yang, *Asymptotic Complexity of Gröbner Basis Algorithms for Semi-regular Overdetermined Systems over Large Fields*, manuscript.
6. B. Barkee *et al*, *Why You Cannot Even Hope to Use Gröbner Bases in Public-Key Cryptography*, J. Symbolic Computations, 18 (1994), pp. 497–501.
7. J. R. Bunch and J. E. Hopcroft, *Triangular Factorizations and Inversion by Fast Matrix Multiplication*, Math. Computations, 24 (1974), pp. 231–236.
8. R. Burden and J. D. Faires, *Numerical Analysis, 7th ed.*, PWS-Kent Publ. Co., 2000.
9. D. Bernstein, *Matrix Inversion Made Difficult*, preprint, stated to be superseded by a yet unpublished version, available at http://cr.yp.to.
10. L. Caniglia, A. Galligo, and J. Heintz, *Some New Effectivity Bounds in Computational Geometry*, AAECC-6, 1988, LNCS 357, pp. 131–151.
11. L. Caniglia, A. Galligo, and J. Heintz, *Equations for the Projective Closure and Effective Nullstellensatz*, Discrete Applied Mathematics, 33 (1991), pp. 11–23.
12. C. Chester, B. Friedman, and F. Ursell, *An Extension of the Method of Steepest Descents,* Proc. Camb. Philo. Soc. 53 (1957) pp. 599–611.
13. D. Coppersmith, private communication.
14. N. Courtois, *The Security of Hidden Field Equations (HFE)*, CT-RSA 2001, LNCS 2020, pp. 266–281.
15. N. Courtois, *Higher-Order Correlation Attacks, XL Algorithm and Cryptanalysis of Toyocrypt*, ICISC '02, LNCS 2587, pp. 182–199.
16. N. Courtois, *Fast Algebraic Attacks on Stream Ciphers with Linear Feedback*, CRYPTO'03, LNCS 2729, pp. 177-194.
17. N. Courtois, *Algebraic Attacks over GF(2^k), Cryptanalysis of HFE Challenge 2 and SFLASH^{v2}*, PKC '04, LNCS 2947, pp. 201-217.
18. N. Courtois, private communication.

19. N. Courtois, L. Goubin, and J. Patarin, *SFLASH^{v3}, a Fast Asymmetric Signature Scheme*, preprint available at `http://eprint.iacr.org/2003/211`.
20. N. Courtois, A. Klimov, J. Patarin, and A. Shamir, *Efficient Algorithms for Solving Overdefined Systems of Multivariate Polynomial Equations*, EUROCRYPT 2000, LNCS 1807, pp. 392–407.
21. N. Courtois and J. Pieprzyk, *Cryptanalysis of Block Ciphers with Overdefined Systems of Equations*, ASIACRYPT 2002, LNCS 2501, pp. 267–287.
22. N. Courtois and J. Patarin, *About the XL Algorithm over GF(2)*, CT-RSA 2003, LNCS 2612, pp. 141–157.
23. J. Daemen and V. Rijmen, *The Design of Rijndael, AES - The Advanced Encryption Standard*. Springer-Verlag, 2002.
24. C. Diem, *The XL-algorithm and a Conjecture from Commutative Algebra*, ASIACRYPT'04, LNCS 3329, pp. 323–337 and private communication.
25. W. Diffie and M. Hellman, *New Directions in Cryptography*, IEEE Trans. Info. Theory, vol. IT-22, 6 (1972), pp. 644-654.
26. J. Ding and D. Schmidt, *Cryptanalysis of SFlash^{v3}*, `eprint.iacr.org/2004/103`.
27. I. S. Duff, A. M. Erismann, and J. K. Reid, *Direct Methods for Sparse Matrices*, published by Oxford Science Publications, 1986.
28. W. Eberly and E. Kaltofen, *On Randomized Lanczos Algorithms*, Proc. ISSAC '97, pp. 176–183, ACM Press 1997.
29. J.-C. Faugére, *A New Efficient Algorithm for Computing Gröbner Bases (F4)*, Journal of Pure and Applied Algebra, 139 (1999), pp. 61–88.
30. J.-C. Faugère, *A New Efficient Algorithm for Computing Gröbner Bases without Reduction to Zero (F5)*, Proceedings of ISSAC 2002, pp. 75-83, ACM Press 2002.
31. J.-C. Faugère and A. Joux, *Algebraic Cryptanalysis of Hidden Field Equations (HFE) Cryptosystems Using Gröbner Bases*, CRYPTO 2003, LNCS 2729, pp. 44-60.
32. R. Fröberg, An Inequality for Hilbert Series of Graded Algebras, Math. Scand. 56(1985) pp. 117-144.
33. M. Garey and D. Johnson, *Computers and Intractability, A Guide to the Theory of NP-completeness*, W. H. Freeman New York 1979.
34. W. Geiselmann, R. Steinwandt, and T. Beth, *Revealing 441 Key Bits of* SFLASH^{v2}, 3rd NESSIE Workshop, 2002.
35. H.-K. Hwang, *Asymptotic estimates of elementary probability distributions*, Studies in Applied Mathematics, 99:4 (1997), pp. 393-417.
36. R. Hartshorne, *Algebraic Geometry*, Springer-Verlag, 1977.
37. A. Kipnis and A. Shamir, *Cryptanalysis of the HFE Public Key Cryptosystem by Relinearization*, CRYPTO'99, LNCS 1666, pp. 19–30.
38. B. LaMacchia and A. Odlyzko, *Solving Large Sparse Linear Systems over Finite Fields*, CRYPTO'90, LNCS 537, pp. 109–133.
39. S. Lang, *Algebra* (3rd edition), Addison-Wesley, 1993.
40. D. Lazard, *Gröbner Bases, Gaussian Elimination and Resolution of Systems of Algebraic Equations*, EUROCAL '83, LNCS 162, pp. 146–156.
41. C. McGeoch, *"Veni, Divisi, Vici"*, Appearing in the "Computer Science Sampler" column of the Amer. Math. Monthly, May 1995.
42. The MAGMA project, University of Sydney, see `http://magma.maths.usyd.edu.au/users/allan/gb`
43. T. Matsumoto and H. Imai, *Public Quadratic Polynomial-Tuples for Efficient Signature-Verification and Message-Encryption*, EUROCRYPT'88, LNCS 330, pp. 419–453.
44. T. Moh, *On The Method of XL and Its Inefficiency Against TTM*, available at `http://eprint.iacr.org/2001/047`

45. S. Murphy and M. Robshaw, *Essential Algebraic Structures Within the AES*, CRYPTO 2002, LNCS 2442, pp. 1–16.
46. S. Murphy and M. Robshaw, *Comments on the Security of the AES and the XSL Technique*, preprint available from the authors http://www.isg.rhul.ac.uk/~sean/
47. *NESSIE Security Report, V2.0*, available at http://www.cryptonessie.org
48. J. Patarin, *Hidden Fields Equations (HFE) and Isomorphisms of Polynomials (IP): Two New Families of Asymmetric Algorithms*, EUROCRYPT'96, LNCS 1070, pp. 33–48.
49. J. Patarin, L. Goubin, and N. Courtois, C^*_{-+} *and HM: Variations Around Two Schemes of T. Matsumoto and H. Imai*, ASIACRYPT'98, LNCS 1514, pp. 35–49.
50. J. Patarin, N. Courtois, and L. Goubin, *QUARTZ, 128-Bit Long Digital Signatures*, CT-RSA 2001, LNCS 2020, pp. 282–297. Update at http://www.cryptonessie.org
51. J. Patarin, N. Courtois, and L. Goubin, *FLASH, a Fast Multivariate Signature Algorithm*, CT-RSA 2001, LNCS 2020, pp. 298–307. Update with SFLASH^{v2} available at http://www.cryptonessie.org
52. R. Stanley, *Enumerative Combinatorics*, vol. 1, second printing 1996; vol. 2 in 1999. Both published by Cambridge University Press, Cambridge.
53. V. Strassen, *Gaussian Elimination is not Optimal*, Numer. Math. 13 (1969) pp. 354-356.
54. M. Sugita, M. Kawazoe, and H. Imai, *Relation between XL algorithm and Groebner Bases Algorithms*, preprint, http://eprint.iacr.org/2004/112. Part of this merged with [2].
55. G. Szegö, *Orthogonal Polynomials, 4th ed.*, publ.: American Math. Society, Providence.
56. D. Wiedemann, *Solving Sparse Linear Equations over Finite Fields*, IEEE Transaction on Information Theory, v. IT-32 (1976), no. 1, pp. 54–62.
57. R. Wong, *Asymptotic Approximations of Integrals*, Academic Press, San Diego, 1989.
58. B.-Y. Yang and J.-M. Chen, *Theoretical Analysis of XL over Small Fields*, ACISP 2004, LNCS 3108, pp. 277-288.
59. B.-Y. Yang, J.-M. Chen, and N. Courtois, *On Asymptotic Security Estimates in XL and Gröbner Bases-Related Algebraic Cryptanalysis*, ICICS '04, LNCS 3269, pp. 401-413. Formerly titled *Exact and Asymptotic Behavior of XL-Related Methods*.

Efficient Broadcast Encryption
Using Multiple Interpolation Methods

Eun Sun Yoo, Nam-Su Jho, Jung Hee Cheon, and Myung-Hwan Kim*

ISaC and Department of Mathematical Sciences, Seoul National University,
Seoul 151-747, Korea
{eunsun, drake, jhcheon, mhkim}@math.snu.ac.kr

Abstract. We propose a new broadcast encryption scheme based on polynomial interpolations. Our scheme, obtained from the Naor-Pinkas scheme by partitioning the user set and interpolating multiple polynomials, turns out to be better in efficiency than the best known broadcast schemes like the Subset Difference and the Layered Subset Difference methods, which are tree based schemes. More precisely, when r users are revoked among n users, our method requires $O(\log(n/m))$ user keys and $O(\alpha r + m)$ transmission overhead in the worst case, where m is the number of partitions of the user set and can be chosen to optimize its efficiency, and α is a predetermined constant satisfying $1 < \alpha < 2$. So, our scheme is always better in the storage than the tree based schemes (whose storage overhead is $O(\log^2 n)$ or $O(\log^{3/2} n)$). In the transmission overhead, our scheme beats those schemes except for a very small r/n. The computation cost is worse than the other schemes but is reasonable for systems with moderate computing power. The security proof is given based on the computational Diffie-Hellman problem.

Keywords: Broadcast encryption, interpolation, partition.

1 Introduction

Broadcast Encryption is a cryptographic method to efficiently broadcast information to a large set of users so that only privileged users can decrypt it. We assume that each user is given a user-key, which is a set of secrets, from the center before starting the broadcast and the user key is never updated, that is, users are *stateless* receivers. In each session, a message is encrypted by a session key and the encrypted message is transmitted over an insecure channel with the encrypted value, called a *header*, of the session key, which can be decrypted only by the user-keys. When a user needs to be removed from the set of the privileged users, the center should be able to make a header to be decrypted only by the non-revoked users. Each session can have an independent set of revoked users. That is, a revoked user in one session may subscribe another

* This author was partially supported by KRF Research Fund(2004-070-100001).

C. Park and S. Chee (Eds.): ICISC 2004, LNCS 3506, pp. 87–103, 2005.
© Springer-Verlag Berlin Heidelberg 2005

session. The application covers pay TV, internet multicast of movies or news, and mobile games. The best known broadcast schemes are the *Subset Difference* method (SD) [11] and the *Layered Subset Difference* method (LSD) [8], which is a variant of SD adopting the notion of layers. There is also the *Stratified Subset Difference* method (SSD) [6], which was proposed at Crypto'04.

The polynomial interpolation method was first introduced by Berkovits [2] and improved by Naor and Pinkas [12]. In the improved scheme, called the NP scheme, a user-key is only one field element and the header size is very small. However, since this scheme is originally designed for multicast with traitor tracing rather than broadcast, it has the best performance for small size user sets (of several hundred users, say) and requires key refreshment from time to time. The scheme can also be used for stateless receivers without key refreshment. But in this case, the NP scheme has shortcomings. First, the system cannot revoke more than d users, where d is the degree of a polynomial predetermined in the setup stage. Second, its header size depends on d, not on the number r of revoked users. If we increase d large enough to cover the maximum number of revoked users, the header size increases along with d even if r is small. Third, the computation cost for each user increases in the square order of d, which is too large to be practical even for $d \approx 1000$.

In this paper, we adopt two simple ideas, partitioning the users and interpolating multiple polynomials, to overcome the weaknesses of the NP scheme mentioned above as a broadcast encryption for a large number of stateless receivers. Our scheme covers any number n of users and is more efficient than SD, LSD and SSD in the user-key size and in the header size. More precisely, our method requires $O(\log(n/m))$ key storage and the $O(\alpha r + m)$ header size in the worst case, where m, which can be chosen to optimize efficiency, is the number of partitions of the user set, and α is a predetermined constant satisfying $1 < \alpha < 2$. So, the user-key size is always smaller than the tree based schemes whose key storage is $O(\log^\beta n)$ with $1 \leq \beta \leq 2$. Our scheme satisfies, in fact, the *log-key restriction* (the storage size is bounded by $\log n$) [6] with much smaller transmission overhead than SSD. The header size is also smaller than those schemes except when r/n is very small. The computation cost is worse than those schemes but can be adjusted according to computing power of the user device.

Outline of this paper is as follows: After a brief preliminaries in Section 2, we propose the basic scheme interpolating multiple polynomials and the extended scheme partitioning the user set, respectively, in Sections 4 and 5. We analyze the performance of our scheme in Section 5, and compare our scheme with SD, LSD and SSD in Section 6. We discuss a security proof of our scheme in Section 7, and conclude in Section 8. We present a detailed proof in Appendix.

2   Preliminaries

We use the following parameters:

   n: the number of total users
   r: the number of revoked users

I_u: The identifier of a user u

K_u: the set of keys stored by a user u, i.e., the *user-key* of u

A *session* is one broadcast of data to all users and the *session key* k is the key used to encrypt the data in the session. In order to broadcast a message M, the center encrypts M using the session key k and sends the encrypted message together with a *header* to the users. That is, the center sends

$$\langle\ \langle header\rangle,\ E_k(M)\ \rangle$$

to the users, where $E_k(M)$ is a symmetric encryption of M by k. Then, a privileged user u can easily compute k from a pre-defined function F satisfying $F(K_u, \langle header\rangle) = k$. With this k, u can recover M by

$$D_k(E_k(M)) = M.$$

But any revoked user u should not be able to rend k from K_u and $\langle header\rangle$. Furthermore, there should be no efficient polynomial time algorithm O such that

$$O(K_1, K_2, \ldots, K_r, \langle header\rangle) = k,$$

where $K_a = K_{u_a}$ and u_a's are revoked users for $a = 1, 2, \cdots, r$. The fixed function F is pre-distributed before the system starts to operate, and is computable in polynomial time under user level computing power. We call the length of the header the *transmission overhead* or *message overhead*. The computing time of F is called the *computation overhead* or *computation cost*. In a broadcast encryption scheme, the user-key size, called the *storage overhead*, the transmission overhead, and the computation overhead are three most important parameters determining the efficiency of the scheme.

3   Basic Scheme

Let n be the number of total users. Given a system parameter α satisfying $1 < \alpha < 2$, define a sequence of positive integers d_i by the recurrence relation:

$$d_1 = 1 \quad \text{and} \quad d_{i+1} = \lfloor \alpha(d_i + 1) \rfloor,$$

where $\lfloor x \rfloor$ denotes the largest integer not exceeding x. Let w be the minimal integer such that $n \leq (\alpha+1)(d_w+1)+1$. Then we take random w polynomials $f_{d_1}, f_{d_2}, \ldots, f_{d_w}$, where f_{d_i} is a polynomial of degree d_i. Let r be the number of revoked users. If $d_{i-1} < r \leq d_i$ for some $1 \leq i \leq w$ (we make a convention that $d_0 = 0$ for convenience), then the polynomial of degree d_i is to be used for revocation. For the case when $r = 0$ or $d_w < r \leq n$, see below.

In the initialization step, the center chooses an elliptic curve E over $\mathbb{F}_p$ and selects a point P in E which is of order q, where p and q are 160 bit primes. (As a matter of fact, any abelian group over which the computational Diffie-Hallman problem is hard will do.) Then it chooses a system parameter α satisfying $1 < \alpha < 2$, and selects a non-secret identifier $I_u(= I) \in \mathbb{Z}_q$ to be given to each user u. Then, the center chooses random polynomials f_{d_i} of degree d_i corresponding to (d_i+1)-out-of-n threshold secret sharing schemes for $i = 1, 2, \ldots, w$, where $d_{i+1} = \lfloor \alpha(d_i + 1) \rfloor$ over $\mathbb{Z}_q$ for $i = 1, 2, \ldots, w - 1$. Each user u receives his/her user-key

$$K_u = \langle \, I, f_{d_1}(I), f_{d_2}(I), \ldots, f_{d_w}(I), \phi, \psi_u \, \rangle$$

and the system information E, P, p and q over a private channel from the center, where ϕ is the key which will be used when there is no revocation and ψ_u is the key which will be used when of $r > d_w$.

Let $u_1, u_2, \ldots, u_r$ be the revoked users, where $d_{i-1} < r \leq d_i$. For revocation, the center first learns the identifiers $I_a = I_{u_a}$ ($a = 1, 2, \ldots, r$) of the revoked users. For a random $s \in \mathbb{Z}_q$, it takes $k = f_{d_i}(0)sP$ as a new session key that should be unknown to the revoked users. The center chooses distinct $d_i - r$ random points, say $I_{r+1}, I_{r+2}, \ldots, I_{d_i} \in \mathbb{Z}_q$, which are distinct from the identifiers of all users, and broadcasts

$$\langle \, index, \ Q, \ (I_1, f_{d_i}(I_1)Q), \ \ldots, \ (I_r, f_{d_i}(I_r)Q),$$
$$(I_{r+1}, f_{d_i}(I_{r+1})Q), \ \ldots, \ (I_{d_i}, f_{d_i}(I_{d_i})Q) \, \rangle,$$

where $Q = sP$ and $index$ is the indicator of the polynomial used in the session. Each privileged user u can compute $f_{d_i}(I)Q$ by his/her own secret key $f_{d_i}(I)$ and therefore can recover the session key $f_{d_i}(0)Q$. More precisely, the user u can compute $f_{d_i}(0)Q$ from the values

$$(I_0, f_{d_i}(I_0)Q), \ (I_1, f_{d_i}(I_1)Q), \ \ldots, \ (I_r, f_{d_i}(I_r)Q),$$
$$(I_{r+1}, f_{d_i}(I_{r+1})Q), \ \ldots, \ (I_{d_i}, f_{d_i}(I_{d_i})Q),$$

where $I_0 = I$, as follows:

$$f_{d_i}(0)Q = \left(\sum_{a=0}^{d_i} \lambda_a f_{d_i}(I_a) \right) Q = \sum_{a=0}^{d_i} \lambda_a \left(f_{d_i}(I_a)Q \right),$$

where

$$\lambda_a = \prod_{b \neq a} \frac{I_b}{I_b - I_a}. \tag{1}$$

If there is no revoked user, i.e., $r = 0$, then the center encrypts the session key k by symmetric encryption scheme with the key ϕ and broadcasts

$$\langle \, index, E_\phi(k) \, \rangle.$$

So only one encrypted session key is required. If $r > d_w$, then the center encrypts the session key k with each non-revoked user's private key ψ_u and broadcasts

$$\langle \, index, E_{\psi_{u_1}}(k), E_{\psi_{u_2}}(k), \ldots, E_{\psi_{u_{n-r}}}(k) \, \rangle,$$

where $u_1, \ldots, u_{n-r}$ are non-revoked users. This requires $n - r$ encrypted session keys.

Performance. Each user stores $w + 2$ private keys, the identifier I and the system parameters E, P, p and q.

If $d_{i-1} < r \leq d_i$ for $1 \leq i \leq w$, then the transmitted data consists of an *index*, one base point, and d_i points in $E(\mathbb{F}_p)$. Since $d_i \leq \alpha(d_{i-1} + 1) \leq \alpha r$, it is at most $\alpha r + 1$ elements in $E(\mathbb{F}_p)$. If $r = 0$, then the header contains only one encryption. If $r > d_w$, then the header contains $n - r$ encryptions, which is also bounded by $\alpha r + 1$ since $n - r \leq n - (d_w + 1) \leq \alpha r + 1$. In any cases, the transmission overhead is at most $\alpha r + 1$ points ignoring the *index*.

The computation overhead is $d_i + 1(\leq \alpha r + 1)$ scalar multiplications in $E(\mathbb{F}_p)$ if $d_{i-1} < r \leq d_i$. If $r > d_w$, then it is $n - r(\leq \alpha r + 1)$ symmetric encryptions. Since one symmetric encryption is faster than a scalar multiplication in $E(\mathbb{F}_p)$, the computation overhead is bounded by $(\alpha r + 1)$ scalar multiplications in $E(\mathbb{F}_p)$. We can save the computation using a simultaneous scalar multiplication method, which will be discussed in Section 5.

Security. If $r = 0$ or $r > d_w$, then the scheme is secure since the session key is encrypted by a symmetric encryption algorithm using the key ϕ or private keys ψ_u's of non-revoked users. If $d_{i-1} < r \leq d_i$, the session key $f_{d_i}(0)Q$ is secure against the coalition of the r revoked users. This is because every revoked user has only d_i values and cannot gain any further information. A more detailed security proof is given in Appendix.

4   Extended Scheme

Although the basic scheme is quite efficient when the number n of users is small, the scheme is not usable in practice when n is very large because the polynomial degree grows too big and so does the computation overhead. This problem, however, can be resolved easily by partitioning the users into small partitions and applying the basic scheme to each partition. Despite the number of revoked users in each partition can vary dynamically, we can handle the dynamics properly with the polynomials chosen in the basic scheme.

In the initialization step, the center divides n users into m partitions of size D. Actually, D and m are adjusted by the system parameter α in order to make the computation overhead reasonable. (This will be discussed in Section 5.) Because each user should belong to one and only one partition, we can apply the basic scheme to each partition. The center generates random polynomials

$$f_{1d_i}, f_{2d_i}, \ldots, f_{md_i} \quad \text{for } i = 1, 2, \ldots, w,$$

where f_{jd_i} is a polynomial of degree d_i over $\mathbb{Z}_q$ assigned to the j-th partition. Then it provides each user u in the j-th partition, via a private channel, his/her user-key

$$K_u = \langle\ I,\ f_{jd_1}(I),\ f_{jd_2}(I),\ \ldots,\ f_{jd_w}(I), \phi_j, \psi_u\ \rangle,$$

where I is the identifier of u, j is the partition number of the partition containing u, and ϕ_j, ψ_u are the keys corresponding to ϕ, ψ_u, respectively, in the basic scheme.

For revocation, the center first learns the identifiers of r revoked users and the partitions they belong to. Let $u_{j1}, u_{j2}, \ldots, u_{jr_j}$ be the revoked users in the j-th partition, where

$$r = r_1 + r_2 + \cdots + r_m.$$

If $d_{i-1} < r_j \leq d_i$, then the polynomial f_{jd_i} will be used for revocation in the j-th partition. Let's denote this d_i(depending on r_j) by t_j for convenience in the following. The center then chooses $t_j - r_j$ random distinct points, say $I_{j,r_j+1}, I_{j,r_j+2}, \ldots, I_{jt_j}$, from $\mathbb{Z}_q$, which are distinct from the identifiers of the users in the j-th partition, and broadcasts the following:

$$\langle \ \ldots, \ par(j), \ ind(j), \ Q_j, \ (I_{j1}, f_{jt_j}(I_{j1})Q_j), \ldots, (I_{jr_j}, \ f_{jt_j}(I_{jr_j})Q_j),$$

$$(I_{j,r_j+1}, \ f_{jt_j}(I_{j,r_j+1})Q_j), \ \ldots, (I_{jt_j}, \ f_{jt_j}(I_{jt_j})Q_j), \ \ldots \ \rangle,$$

where $par(j)$ is the indicator of the j-th partition, $ind(j) = t_j$, $Q_j = s_j P$, s_j is a random number of the j-th partition and I_{ja} is the identifier of u_{ja} for $1 \leq a \leq r_j$. Each non-revoked user u in the j-th partition can find which part of the header is for his/her partition from the partition indicator $par(j)$, and then he/she can compute the session key $k = f_{jt_j}(0)Q_j$ as in the basic scheme. If $r_j = 0$, then the center broadcasts

$$\langle \ \ldots, par(j), ind(j), E_{\phi_j}(k), \ldots \ \rangle.$$

If $r_j > d_w$, then $D - r_j < \alpha r_j$. So the center sends following:

$$\langle \ \ldots, par(j), ind(j), E_{\psi_{u_1}}(k), E_{\psi_{u_2}}(k), \ldots, E_{\psi_{u_{D-r_j}}}(k), \ldots \ \rangle,$$

where $u_1, u_2, \ldots, u_{D-r_j}$ are non-revoked users and $E_{\psi_u}(k)$ is a symmetric encryption of the session key k by ψ_u.

In the above, if $f_{jt_j}(0)Q_j$'s are distinct, then we have to insert more information in the header for all privileged users in each partition to compute the session key k. To avoid this, we make all $f_{jt_j}(0)Q_j$'s be equal as follows:

Choose a random $s \in \mathbb{Z}_q$ and let

$$\hat{s} = f_{1t_1}(0)f_{2t_2}(0)\cdots f_{mt_m}(0)s.$$

For $\hat{s}$, define

$$s_j = \frac{\hat{s}}{f_{jt_j}(0)} \quad (\text{mod } q) \quad \text{for all } j = 1, 2, \ldots, m.$$

Then we have

$$k = f_{1t_1}(0)Q_1 = f_{2t_2}(0)Q_2 = \cdots = f_{mt_m}(0)Q_m. \tag{2}$$

5 Analysis

In the basic scheme, given α and w, the maximum possible value of n is $\lfloor (d_w + 1)(\alpha + 1) \rfloor + 1$. Since the performance of the basic scheme depends only on α and w, we take D to be $\lfloor (d_w + 1)(\alpha + 1) \rfloor + 1$ in the extended scheme.

Storage Overhead. The storage overhead for each user is $w + 2$ elements in $\mathbb{F}_p$ as in the basic scheme. We estimate w in terms of D. Recall that d_i's are determined from recurrence relations: $d_1 = 1$ and $d_{i+1} = \lfloor \alpha(d_i + 1) \rfloor$. Since $d_{i+1} > \alpha(d_i + 1) - 1$, we have

$$d_w + 1 > \alpha(d_{w-1} + 1) > \cdots > \alpha^{w-1}(d_1 + 1) = 2\alpha^{w-1}.$$

Since $D > (\alpha + 1)(d_{w-1} + 1)$, we obtain $D > 2(\alpha + 1)\alpha^{w-2} > 2\alpha^{w-1}$ and hence $w < \log D / \log \alpha$ (we assume that $1 < \alpha < 2$).

Transmission Overhead. Since the size of $par(j)$ and $ind(j)$ are negligible, the transmission overhead for the j-th partition is $\alpha r_j + 1$ elements of $E(\mathbb{F}_p)$ in the worst case. So the total message overhead is at most

$$\sum_{j=1}^{m}(\alpha r_j + 1) = \sum_{j=1}^{m} \alpha r_j + m = \alpha r + m.$$

Computation Overhead. Most computation of this scheme consists of scalar multiplications in $E(\mathbb{F}_p)$. To speed up this, one can use a simultaneous scalar multiplication method [10]. If one computes c scalar multiplications using Non-Adjacent Form (NAF), it takes $c \log q$ (doublings) + $(c/3) \log q$ (additions) on the average. Using the standard simultaneous scalar multiplication method, it takes $\log q$ (doublings) + $(c/3) \log q$ (additions) for c scalar multiplications, which amounts about $(3 + c)/4$ scalar multiplications.

The computation overhead for each user is bounded by $d_w + 1$ scalar multiplications, as in the basic scheme. Using the standard simultaneous multiplication method, it is reduced to $(d_w/4) + 1$ scalar multiplications. If the revoked users are assumed to be uniformly distributed over all partitions, however, there are r/m revoked users in each partition on the average. In this case, the computation overhead is about $(r/4m) + 1$ scalar multiplications.

Optimization. In the extended scheme, we first fix a constant α with $1 < \alpha < 2$ and an upper bound M of the computation overhead (the number of scalar multiplications in $E(\mathbb{F}_p)$). And then, in order to optimize the scheme's efficiency, we choose the other system parameters as follows:

- Compute d_i's, where $d_1 = 1$ and $d_{i+1} = \lfloor \alpha(d_i + 1) \rfloor$
- Find the maximum w such that $(d_w/4) + 1 \leq M$
- Compute $D = \lfloor (d_w + 1)(\alpha + 1) \rfloor + 1$ (with this D, we can optimize the upper bound of w with $(\log D - 1)/\log \alpha$
- Divide all users into $m = \lceil n/D \rceil$ partitions of size D

The extended scheme with these optimized parameters is summarized in Fig. 1.

Parameters
- n : the number of total users in the system
- p, q : 160-bit primes
- E : an elliptic curve over $\mathbb{Z}_p$
- P : an element of $E(\mathbb{F}_p)$ of order q
- M : an upper bound of the computation overhead

Setup (partitioning)
- Pick $\alpha(1 < \alpha < 2)$ and compute d_i's, where $d_1 = 1$ and $d_{i+1} = \lfloor \alpha(d_i + 1) \rfloor$
- Find the maximum w satisfying $d_w \le 4(M - 1)$
- Compute $D = \lfloor (d_w + 1)(\alpha + 1) \rfloor + 1$
- Divide all users into $m = \lceil n/D \rceil$ partitions of size D

Key Generation of the j-th partition for $j = 1, 2, \ldots, m$
- Assign the identifier I to each user u
- Make w random polynomials $f_{jd_1}, f_{jd_2}, \ldots, f_{jd_w}$ over $\mathbb{Z}_p$, where $\deg(f_{jd_i}) = d_i$
- Give the set $K_u = \{I, f_{jd_1}(I), \ldots, f_{jd_w}(I), \phi, \psi_u\}$ to each user u

Encryption and Decryption

o **Case 1** $1 \le r_j \le d_w$:
- Assume that there are r_j revoked users $u_{j1}, u_{j2}, \ldots, u_{jr_j}$ in the j-th partition
- For each j find d_i satisfying $d_{i-1} < r_j \le d_i$ (convention : $d_0 = 0$) and let $t_j = d_i$
- Choose $s \in \mathbb{Z}_q$ randomly and compute $\hat{s} = f_{1t_1}(0)f_{2t_2}(0) \cdots f_{mt_m}(0)s$
- For $\hat{s}$, define $s_j = \hat{s}/f_{jt_j}(0) \pmod{q}$ for each $j = 1, 2, \ldots, m$
- Compute $Q_j = s_j P$
- Select distinct I_{ja}'s which are different from the identifiers assigned to the users in the j-th partition for $a = r_j + 1, r_j + 2, \ldots, t_j$
- Broadcast the following message:

$$\langle \ldots, par(j), ind(j), Q_j, (I_{j1}, f_{jt_j}(I_{j1})Q_j), (I_{j2}, f_{jt_j}(I_{j2})Q_j), \ldots,$$
$$(I_{jr_j}, f_{jt_j}(I_{jr_j})Q_j), (I_{j,r_j+1}, f_{jt_j}(I_{j,r_j+1})Q_j), \ldots, (I_{jt_j}, f_{jt_j}(I_{jt_j})Q_j), \ldots \rangle$$

- Assume that u with identifier $I = I_0$ is a privileged user in the j-th partition
- From the message, find out t_j
- Compute $\lambda_{ja} = \prod_{b \ne a} \frac{I_{jb}}{I_{jb} - I_{ja}}$ and the session key k as follows:

$$k = f_{jt_j}(0)Q = \left(\sum_{a=0}^{t_j} \lambda_{ja} f_{jt_j}(I_{ja}) \right) Q = \sum_{a=0}^{t_j} \lambda_{ja} \left(f_{jt_j}(I_{ja})Q \right)$$

o **Case 2** $r_j = 0$:
- In this case, the center transmits $\langle \ldots, par(j), ind(j), E_{\phi_j}(k), \ldots \rangle$, where $E_{\phi_j}(k)$ is a symmetric encryption of the session key k by ϕ_j
- Then all users in the j-th partition decrypt by ϕ_j

o **Case 3** $r_j > d_w$:
- Let $u_1, u_2, \ldots, u_{D-r_j}$ be the non-revoked users. Then the center broadcasts

$$\langle \ldots, par(j), ind(j), E_{\psi_{u_1}}(k), E_{\psi_{u_2}}(k), \ldots, E_{\psi_{u_{D-r_j}}}(k), \ldots \rangle$$

where $E_{\psi_u}(k)$ is a symmetric encryption of the session key k by ψ_u
- Then each privileged user u in the j-th partition decrypts by ψ_u

Fig. 1. Extended Scheme

6   Comparison

Table 1 provides the values of d_i's for some α's such that $1 < \alpha < 2$. In the following, because the computation overhead depends on the degrees of polynomials, we set 40 as an upper bound of the degrees. Interpolating a polynomial of degree 40 requires about 11 scalar multiplications in $E(\mathbb{F}_p)$, which, we believe, is reasonable in practice. The table also provides the value of D, the size of one partition, which is determined from given α and d_w. With these values, the transmission overhead is less than or equals to $\alpha r + 1$ for any number r of the revoked users in one partition.

Table 1. The values of d_i's

α	d_i's														size of D
$3/2$	1	3	6	10	16	25	39								101
$4/3$	1	2	4	6	9	13	18	25	34						82
$\sqrt[4]{2}$	1	2	3	4	5	7	9	11	14	17	21	26	32	39	88

In Table 2, we compare our extended scheme with SD [11] and LSD [8], which are regarded as the best known broadcast encryption schemes. The threshold column shows that our scheme has the smaller message length than SD and LSD except when r/n is very small. For example, our scheme with $\alpha = \sqrt[4]{2}$ has smaller message length than SD when the number of revoked users is larger than 1.41 % of the total users. Fig. 2 depicts the transmission overhead (TO) of each scheme with respect to $100r/n(\%)$.

Table 2. Efficiency Comparison

Scheme	Key Storage	# of Computations	Message Length	Threshold(vs SD)
SD [11]	$log^2 n$	$\log n$ hashes	$2r - 1$	-
LSD [8]	$log^{3/2} n$	$\log n$ hashes	$4r$	-
Ours	$\lfloor \frac{(\log D - 1)}{\log \alpha} \rfloor + 2$	$(d_w/4) + 1$ mul. in E	$\alpha r + n/D$	-
$\alpha = 3/2$	9	10.75 mul. in E	$(3r/2) + (n/101)$	$r > 1.99n/100$
$\alpha = 4/3$	11	9.5 mul. in E	$(4r/3) + (n/82)$	$r > 1.83n/100$
$\alpha = \sqrt[4]{2}$	16	10.75 mul. in E	$\sqrt[4]{2}r + (n/88)$	$r > 1.41n/100$

From this comparison, we can see that our scheme always has the smaller key storage as well than SD and LSD. The computation overhead, however, is worse. But even so, our scheme is still practical for systems with some computing power. Furthermore, considering the fact that the transmission cost is much more expensive than the computation cost in practice, this is a desirable trade off. In addition, if the revoked users are assumed to be uniformly distributed in each partition, the computation overhead $O(d_w)$ is reduced to $(r/4m) + 1$, which is 2.26, 2.03, and 2.1 scalar multiplications in $E(\mathbb{F}_p)$ when $r \leq 0.05n$ and $\alpha = 3/2, 4/3, \sqrt[4]{2}$, respectively.

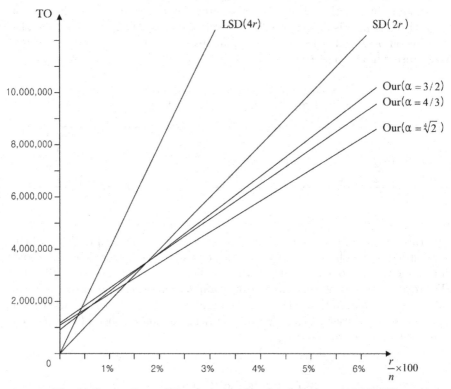

Fig. 2. Comparison of Transmission Overhead for n = 100,000,000

7   Security Proof

Naor and Pinkas [12] proved that their NP scheme is secure against coalitions of up to d revoked users under the computational Diffie-Hellman (CDH) assumption, where d is the degree of a polynomial predetermined. We generalize their proof to our basic scheme (with multiple polynomials) in the following lemma.

Lemma. *The basic scheme is secure against coalition of r revoked users under the CDH assumption for any r.*

Proof. See Appendix.   □

And then we prove the extended scheme (with multiple partitions) is also secure under the same assumption in the following theorem.

Theorem. *The extended scheme is secure against coalition of total r revoked users, where $r_1, r_2, \ldots, r_m$ are the numbers of revoked users in partitions*

$$G_1, G_2, \ldots, G_m,$$

respectively, such that

$$r = r_1 + r_2 + \cdots + r_m.$$

Proof. See Appendix.   □

Observe that if different $f_{jt_j}(0)Q_j$'s - called the *partition keys* - are used for different partitions in the extended scheme, the security proof is an immediate consequence of the lemma above. But this increases the transmission overhead by $O(m)$. In order to avoid this, the extended scheme makes all the partition keys be the same. The main part of the proof is, in fact, the part proving that the same partition key does not harm to the security of the scheme.

The authors expect that our proof is useful for security proofs in similar situations, that is, our security proof works for any broadcast encryption scheme with multiple partitions using the same partition key.

8   Conclusion

In this paper, we introduced two simple ideas, multiple interpolations and multiple partitions, to the NP scheme to obtain an efficient broadcast encryption scheme for a large number of stateless receivers. Our scheme fully satisfies the general requirements of broadcast encryption (even the log-key restriction, the notion of which was introduced in this year's Crypto) and the scheme's efficiency is comparable to the most efficient broadcast encryption schemes known. Our scheme, in fact, is better in the storage overhead and in the transmission overhead. It costs more computations but the computation overhead is reasonable for systems with moderate computing power. Our scheme is applicable to any abelian group over which the CDH is hard. Moreover, our scheme has another advantage: later entry of new users is very easy and cheap because we can simply add new partitions at anytime. We expect that this advantage is exploited in many applications in practice. We provide a detailed security proof in Appendix.

References

1. J. Anzai, N. Matsuzaki and T. Matsumoto, *A quick key distribution scheme with "Entity Revocation"*, Advances in Cryptology - Asiacrypt'99, Lecture Notes in Computer Science 1716, pp.333-347.

2. S. Berkovits, *How to Broadcast a secret*, Advances in Cryptology - Eurocrypt'91, Lecture Notes in Computer Science 547, pp.536-541.

3. G. Chick and S. Tavares, *Flexible access control with master keys*, Advances in Cryptology - Crypto'89, Lecture Notes in Computer Science, pp.316-322.

4. P. D'Aroco and D.R. Stinson, *Fault Tolerant and Distributed Broadcast Encrytion*, CT - RSA'03, Lecture Notes in Computer Science 2612, pp.263-280.

5. A. Fiat and M. Naor, *Broadcast Encryption*, Advances in Cryptology - Crypto'93, Lecture Notes in Computer Science 773, pp.480-491.

6. M.T. Goodrich, J.Z. Sun and R. Tamassia, *Efficient Tree-Based Revocation in Groups of Low-State Devices*, Advances in Cryptology - Crypto'04, Lecture Notes in Computer Science 3152, pp.511-527.

7. J. Garay, J. Staddon and A. Wool, *Long-Lived Broadcast Encryption*, Advances in Cryptology - Crypto'00, Lecture Notes in Computer Science 1880, pp.333-352.

8. D. Halevi and A. Shamir, *The LSD Broadcast Encryption Scheme*, Advances in Crytology - Crypto'02, Lecture Notes in Computer Science 2442, pp.47-60.

9. R. Kumar, S. Rajagopalan and A. Sahai, *Coding Constructions for blacklisting problems without Computational Assumptions*, Advances in Cryptology - Crypto'99, Lecture Notes in Computer Science 1666, pp.609-623.

10. B. Möller, *Algorithms for Multi-exponentiation*, Selected Areas in Cryptography - SAC'01. Lecture Notes in Computer Science 2259, pp.165-180.

11. D. Naor, M. Naor and J. Lotspiech, *Revocation and Tracing Schemes for Stateless Receivers*, Advances in Cryptology - Crypto'01, Lecture Notes in Computer Science 2139, pp.41-62.

12. M. Naor and B. Pinkas, *Efficient Trace and Revoke Schemes*, Financial Cryptography'00, Lecture Notes in Computer Science.

13. C.K. Wong, M. Gouda and S.S. Lam, *Secure Group Communication using Key Graphs*, ACM SIGGCOM'98 ACM.

14. M. Luby and J. Staddon, *Combinatorial Bounds for Broadcast Encryption*, Advances in Cryptology Eurocrypt'98, Lecture Notes in Computer Science 1403, pp.512-526.

15. A. Shamir, *How to Share a Secret*, Comm. ACM 22, pp.612-613.

Appendix: Security Proof

Naor and Pinkas [12] proved that the NP scheme is secure against coalitions of up to t revoked users under the computational Diffie-Hellman assumption.

Computational Diffie-Hellman Assumption: For a cyclic group G in which DLP is hard, let $g \in G$ be a generator of G. Then there is no efficient polynomial time algorithm that can compute g^{xy} from $\langle g, g^x, g^y \rangle$, where x and y are random integers in the interval $[1, |G|]$. In an elliptic curve, computational Diffie-Hellman assumption means that there is no efficient algorithm that can compute xyP from $\langle P, xP, yP \rangle$.

We will prove that our scheme is also secure under the same assumption.

Lemma. *The basic scheme is secure against coalition of r revoked users under CDH assumption for any r.*

Proof. Let r be the number of revoked users. When $r = 0$ or $r > d_w$, then the security of the scheme is guaranteed by the security of the symmetric encryption algorithm used. So we may assume that $r \in (d_{i-1}, d_i]$ for some $i = 1, 2, \ldots, w$, where $d_0 := 0$. We first consider the case of one time revocation. Since $r \in (d_{i-1}, d_i]$, the polynomial f_{d_i} of degree d_i is used for revocation. Points on other polynomials are useless to compute the session key since all polynomials are chosen randomly. Therefore, for revoked users, computing the session key is equivalent to finding polynomial of degree d_i with only d_i points, which is impossible by [15].

Next we consider the case of repeated revocations. Let r and r' be the number of revoked users in current session and previous session, respectively. If $r \in (d_{i-1}, d_i]$ but $r' \notin (d_{i-1}, d_i]$, then the scheme is secure because the current polynomial f_{d_i} is different from previous ones. We now assume that $r, r' \in (d_{i-1}, d_i]$. Then the same polynomial f_{d_i} of degree d_i is used in two different sessions, say the first two sessions. Assume also that a user u was not revoked in the first session but is revoked in the second session. Let the revoked users in the second session be $u_1 (= u), u_2, \ldots, u_r$. At least the following information are available to u with coalition of r revoked users:

$$s^{(1)}P, \ f_{d_i}(I_1)s^{(1)}P, \ \ldots, \ f_{d_i}(I_r)s^{(1)}P, \ f_{d_i}(I'_{r+1})s^{(1)}P, \ \ldots, f_{d_i}(I'_{d_i})s^{(1)}P,$$
$$f_{d_i}(0)s^{(1)}P, \ s^{(2)}P, \ f_{d_i}(I_1)s^{(2)}P, \ \ldots, \ f_{d_i}(I_r)s^{(2)}P, \ f_{d_i}(I_{r+1})s^{(2)}P, \ \ldots,$$
$$f_{d_i}(I_{d_i})s^{(2)}P, \ I_1, \ I_2, \ \ldots, \ I_r, \ f_{d_i}(I_1), \ f_{d_i}(I_2), \ \ldots, \ f_{d_i}(I_r),$$

where I_a is the identifier of u_a for each $a = 1, 2, \ldots, r$; I''_j's and I'_j's for $j = r + 1, \ldots, d_i$ are random and distinct from I_a's; and $s^{(b)}$ is a random number chosen in the b-th session for each $b = 1, 2$.

Although u doesn't know the value of the session key in session 2, u knows that the session key is of the form $f_{d_i}(0)s^{(2)}P$. In the following, we prove that u cannot find the value of $f_{d_i}(0)s^{(2)}P$. Suppose that the scheme is not secure, that is, there is an efficient algorithm O with input

$$Input = \langle\ s^{(1)}P,\ f_{d_i}(I_1)s^{(1)}P,\ \ldots,\ f_{d_i}(I_r)s^{(1)}P,\ f_{d_i}(I'_{r+1})s^{(1)}P,\ \ldots,$$
$$f_{d_i}(I'_{d_i})s^{(1)}P,\ f_{d_i}(0)s^{(1)}P,\ s^{(2)}P,\ f_{d_i}(I_1)s^{(2)}P,\ \ldots,$$
$$f_{d_i}(I_r)s^{(2)}P,\ f_{d_i}(I_{r+1})s^{(2)}P,\ \ldots, f_{d_i}(I_{d_i})s^{(2)}P,$$
$$I_1,\ I_2,\ \ldots,\ I_r,\ f_{d_i}(I_1),\ f_{d_i}(I_2),\ \ldots,\ f_{d_i}(I_r)\ \rangle,$$

which can compute $f_{d_i}(0)s^{(2)}P$. Let

$$O(\ Input\ )\ =\ f_{d_i}(0)s^{(2)}P$$

Then from the algorithm O, one can derive an efficient algorithm O' that can compute g^{xy} from arbitrarily given inputs g, g^x and g^y (see [12] for details). This, however, implies that u can solve the computational Diffie-Hellman problem. So the scheme is secure against coalition of r revoked users under CDH assumption.

□

Theorem. *The extended scheme is secure against coalition of total r revoked users, where $r_1, r_2, \ldots, r_m$ are the numbers of revoked users in partitions*

$$G_1, G_2, \ldots, G_m,$$

respectively, such that

$$r = r_1 + r_2 + \cdots + r_m.$$

Proof. We first show that the session key cannot be recovered from given secret shares for the revoked users. Since polynomials in this scheme are all chosen randomly, secret shares (points on polynomials) in one partition are useless in guessing the polynomials in the other partitions. So, the security of the whole scheme depends on the security of the session key of each partition. For all $j = 1, 2, \ldots, m$, we define t_j as d_i satisfying $d_{i-1} < r_j \le d_i$. Since the case of $r_j = 0$ or $r_j > d_w$ can be proved trivially as in the previous lemma, we may assume that $0 < r_j \le d_w$.

We now prove that using the same partition key for all partitions is also secure. In the extended scheme, non-revoked users in all partitions can compute the same partition key

$$k = f_{1t_1}(0)s_1P = f_{2t_2}(0)s_2P = \cdots = f_{mt_m}(0)s_mP$$

as in (2). This holds for all sessions. This, however, does not cause any weakness. In order to show this, we first consider the case of one time revocation with two partitions. In this case, the coalition of $r = r_1 + r_2$ revoked users know the values of s_1P, s_2P, the secret shares of r users, and the fact that $f_{1t_1}(0)s_1P = f_{2t_2}(0)s_2P$. Then,

$$f_{1t_1}(0)s_1P = \lambda_u f_{1t_1}(I_u)s_1P + \sum_{a=1}^{t_1} \lambda_{1a}f_{1t_1}(I_{1a})s_1P,$$

$$f_{2t_2}(0)s_2P = \lambda_v f_{2t_2}(I_v)s_2P + \sum_{a=1}^{t_2} \lambda_{2a} f_{2t_2}(I_{2a})s_2P,$$

where u and v are non-revoked users chosen randomly from the first and the second partitions, respectively, and λ_u and λ_v are the constants obtained from the formula (1). So the attackers, the r revoked users, can make the following equation:

$$A + \alpha = B + \beta,$$

where

$$A = \lambda_u f_{1t_1}(I_u)s_1P, \quad B = \lambda_v f_{2t_2}(I_v)s_2P,$$

$$\alpha = \sum_{a=1}^{t_1} \lambda_{1a} f_{1t_1}(I_{1a})s_1P, \quad \beta = \sum_{a=1}^{t_2} \lambda_{2a} f_{2t_2}(I_{2a})s_2P.$$

Here, α and β are known to the attackers, but A and B are not. Although there are many pairs of (A,B) satisfying the above equation, it is impossible for the attackers to find the right values of A and B, which render the correct session key. When there are three partitions, the attackers may obtain

$$A + \alpha = B + \beta = C + \gamma.$$

But it is still impossible for them to find the correct session key. It is obvious from mathematical induction that increasing the number of partitions does not do any harm to the security of the scheme.

Next, let's consider the case of repeated revocations. Assume that there are two partitions. We may assume that for each $j = 1,2$, the degree t_j of the polynomial corresponding to the number $r_j^{(b)}$ of the revoked users in the j-th partition dose not change in the b-th session for $b = 1, 2$. Assume further that in the first session, some of the r revoked users in the second session were not revoked so that they know the first session key. After receiving the header of the second session, those r revoked users know

$$s_1^{(1)}P, \quad s_2^{(1)}P, \quad f_{1t_1}(0)s_1^{(1)}P = f_{2t_2}(0)s_2^{(1)}P$$

and the private keys of their own, where $s_j^{(b)}$ denotes a random number used in the j-th partition in the b-th session. To break the scheme, they must obtain the value of

$$f_{1t_1}(0)s_1^{(2)}P = f_{2t_2}(0)s_2^{(2)}P.$$

As above, the attackers can set up the following equations:

$$A + \alpha = B + \beta \tag{3}$$

$$\tau A + \alpha' = \tau B + \beta',$$

where $\tau \in \mathbb{Z}_q$ satisfying $\tau s_1^{(1)}P = s_1^{(2)}P, \tau s_2^{(1)}P = s_2^{(2)}P$ and

$$\alpha' = \sum_{a=1}^{t_1} \lambda_{1a}^{(2)} f_{1t_1}(I_{1a}^{(2)})s_1^{(2)}P, \quad \beta' = \sum_{a=1}^{t_2} \lambda_{2a}^{(2)} f_{2t_2}(I_{2a}^{(2)})s_2^{(2)}P,$$

where $I_{ja}^{(b)}$ is the identifier of a user $u_{ja}^{(b)}$, a revoked user in the j-th partition in the b-th session for $1 \leq a \leq r_j^{(b)}$. Since α, α', β and β' satisfy

$$\tau A + \tau \alpha = \tau B + \tau \beta \quad \text{and} \quad (\tau \alpha) - (\alpha') = (\tau \beta) - (\beta'),$$

we obtain

$$\beta' - \alpha' = \tau(\beta - \alpha). \tag{4}$$

Suppose that there is an efficient algorithm O that can compute τA, which renders the second session key. In other words, let

$$O(A, \alpha, B, \beta, \alpha', \beta') = \tau A,$$

where $A, B, \alpha, \beta, \alpha', \beta'$ satisfy the conditions (3) and (4). Then, one can derive an efficient algorithm O' that can compute xyP with the input data P, xP, yP as follows:

$$O'(P, xP, yP) = O(\ xP,\ P,\ (x - \xi + 1)P,\ \xi P,\ \eta P,\ (y(\xi - 1) + \eta)P\ ),$$

where ξ and η are random. From the conditions (3) and (4), one can easily obtain that $\tau = y$ and $O'(P, xP, yP) = xyP$.

Since this contradicts to the CDH assumption, we may conclude that there is no efficient algorithm that can compute the second session key. For three partitions, we can prove using the algorithm O defined by

$$O(A, \alpha, B, \beta, C, \gamma, \alpha', \beta', \gamma') = \tau A.$$

where $A, B, C, \alpha, \beta, \gamma, \alpha', \beta', \gamma'$ satisfy the conditions

$$A + \alpha = B + \beta = C + \gamma$$

$$\tau(\beta - \alpha) = \beta' - \alpha'\ ,\ \tau(\gamma - \alpha) = \gamma' - \alpha'.$$

By the same argument, we can derive from O an efficient algorithm O' that can solve the CDH, which proves the security of the scheme. So, the security of the extended scheme with two sessions follows from mathematical induction on the number of partitions.

Assume that the extended scheme with m partitions is secure in the first $\ell(\geq 2)$ sessions for any m. We now suppose that if the attackers know the first ℓ session keys, then session $(\ell + 1)$ is not secure. Under this supposition, we prove that the second session is not secure if the attackers know the first session key, which is a contradiction, as follows:

From the first session, the attackers can set up the equation

$$A_1 + \alpha_1 = A_2 + \alpha_2 = \cdots = A_m + \alpha_m \tag{5}$$

and they know that the second session key is

$$\tau A_1 + \alpha_1' = \tau A_2 + \alpha_2' = \cdots = \tau A_m + \alpha_m'$$

for some τ. From (5) the attackers can compute

$$\tau^{(b)}A_1, \ \tau^{(b)}A_2, \ \ldots, \ \tau^{(b)}A_m, \ \alpha_1^{(b)}, \ \alpha_2^{(b)}, \ \ldots, \ \alpha_m^{(b)}$$

satisfying

$$\alpha_1^{(b)} = \tau^{(b)}\alpha_1 + t^{(b)}, \ \alpha_2^{(b)} = \tau^{(b)}\alpha_2 + t^{(b)}, \ \ldots, \ \alpha_m^{(b)} = \tau^{(b)}\alpha_m + t^{(b)}$$

from randomly chosen distinct $\tau^{(b)}$ and $t^{(b)}$. Then it is easy to check that

$$\tau^{(b)}A_1 + \alpha_1^{(b)} = \tau^{(b)}A_2 + \alpha_2^{(b)} = \cdots = \tau^{(b)}A_m + \alpha_m^{(b)} \tag{6}$$

and

$$\alpha_2^{(b)} - \alpha_1^{(b)} = \tau^{(b)}(\alpha_2 - \alpha_1), \ldots, \ \alpha_m^{(b)} - \alpha_1^{(b)} = \tau^{(b)}(\alpha_m - \alpha_1)$$

for $b = 1, 2, \ldots, \ell - 1$. With these $\ell - 1$ equations in (6) together with the equation (5), the attackers can find the second session key (regarded as the $(\ell + 1)$-st session) by the supposition.

Therefore, we may conclude that the extended scheme with any number of partitions is secure in any number of repeated sessions. □

We expect that our security proof may be applied to systems using partitions. In particular, when a broadcast encryption, which is efficient for a small set of users, is applied to many disjoint such sets, our security proof may help reducing the transmission overhead further securely.

On Private Scalar Product Computation for Privacy-Preserving Data Mining

Bart Goethals[1], Sven Laur[2], Helger Lipmaa[2], and Taneli Mielikäinen[1]

[1] HIIT Basic Research Unit,
Department of Computer Science,
University of Helsinki, Finland
{goethals, tmielika}@cs.helsinki.fi
[2] Laboratory for Theoretical Computer Science,
Department of Computer Science and Engineering,
Helsinki University of Technology, Finland
{slaur, helger}@tcs.hut.fi

Abstract. In mining and integrating data from multiple sources, there are many privacy and security issues. In several different contexts, the security of the full privacy-preserving data mining protocol depends on the security of the underlying private scalar product protocol. We show that two of the private scalar product protocols, one of which was proposed in a leading data mining conference, are insecure. We then describe a provably private scalar product protocol that is based on homomorphic encryption and improve its efficiency so that it can also be used on massive datasets.

Keywords: Privacy-preserving data mining, private scalar product protocol, vertically partitioned frequent pattern mining.

1 Introduction

Within the context of privacy-preserving data mining, several private (shared) scalar product protocols [DA01b, DA01a, DZ02, VC02] have been proposed. The goal is that one of the participants obtains the scalar product of the private vectors of all parties. Additionally, it is often required that no information about the private vectors, except what can be deduced from the scalar product, will be revealed during the protocol. Moreover, since data mining applications work with a huge amount of data, it is desirable that the scalar product protocol is also very efficient. A secure scalar product protocol has various applications in privacy-preserving data mining, starting with privacy-preserving frequent pattern mining on vertically distributed database [VC02] and ending with privacy-preserving cooperative statistical analysis [DA01a].

To give an idea of how such a protocol can be used, let us look at the protocol by Vaidya and Clifton for computing frequent itemsets from vertically partitioned transaction database [VC02]. A transaction database is a multi-set of subsets (*transactions*) of some finite set (of *items*). A transaction database

C. Park and S. Chee (Eds.): ICISC 2004, LNCS 3506, pp. 104–120, 2005.

can be seen also as a binary matrix where each row corresponds to a transaction, each column corresponds to an item, and there is one in the entry (i, j) if and only if the transaction i contains the item j. An itemset is a subset of items. The frequency of an itemset in a transaction database is the fraction of transactions containing the itemset as their subset. (The support of an itemset is its frequency multiplied by the number of transactions in the database.) The σ-frequent itemsets (i.e., the frequent itemsets with minimum frequency threshold σ) in a transaction database are the itemsets with frequency at least σ. Thus, mining the σ-frequent itemsets is equivalent to finding all subsets of columns of the binary matrix where at least a σ-fraction of rows have only ones in those columns. In a frequent itemset mining protocol for a vertically partitioned transaction database one party, Alice, has the projection of the database onto some items and another party, Bob, has the projection of database onto the rest of the items. The frequent itemset mining protocol of Vaidya and Clifton is based on the property that an itemset can be frequent only if all of its subsets are frequent. The candidate itemsets are generated and tested level-wise as in the APRIORI algorithm [AMS+96].

If an itemset contains items of only one party, then the party can compute the frequency privately and share it with the other parties without any additional privacy problems. The main challenge occurs when the support of a candidate itemset containing items from both parties needs to be computed. In that case, each party first computes which of the transactions contain the itemset within their own part of the database. This kind of information can be conveniently represented as binary vectors in which the ith entry represents whether or not the itemset is contained in the ith transaction. The number of transactions containing the itemset in the combined transaction database amounts to the scalar product between the corresponding binary vectors of Alice and Bob. A protocol, given by Vaidya and Clifton [VC02], attempts to compute the scalar product in a secure manner, by computing the scalar product on scrambled versions of the binary vectors, such that in the end of the protocol, both parties obtain the joint support without ever seeing each others vector. Their protocol reveals the supports of some infrequent itemsets, as not all candidate itemsets are frequent; this can be avoided by combining private shared scalar product protocols and Yao's circuits for frequency testing.

In this paper, we show that the private scalar product protocol of Vaidya and Clifton [VC02] is not private. Additionally, we are able to break another private (shared) scalar product protocol which was recently proposed by Du and Atallah [DA01a]. Our attacks against the Vaidya-Clifton and Du-Atallah protocols work in the simplest cryptographic model: namely, they enable one of the two parties to retrieve the private input of another party with probability, very close to 1, after the two parties have executed the corresponding protocol once.

While the attacks do not work for all possible private vectors of Alice and Bob, they show that before applying the Vaidya-Clifton and Du-Atallah protocols, one must carefully analyse whether it is safe to apply these protocols in any concrete case. Moreover, the provided attacks can be readily generalised to work for a much larger fraction of private vectors in a more complex model where attack's success probability does not have to be 1 (but just large enough for practical purposes, say 0.001) and/or when Alice and Bob re-execute the corresponding scalar product protocols from [DA01a, VC02] with similar private vectors. (Scalar product protocol from [DA01b] was recently analysed in [LL04].)

As a positive result, we describe a cryptographic protocol for computing scalar product. We prove that the new scalar product protocol is private—in a strong cryptographic sense—under standard cryptographic assumptions. More specifically, no probabilistic polynomial time algorithm substituting Alice (resp., Bob) can obtain a non-negligible amount of information about Bob's (resp., Alice's) private input, except what can be deduced from the private input and private output of Alice (resp., Bob). This means, in particular, that this protocol can be used a polynomial number of times (in the security parameter) with *any* private vectors of Alice and Bob in *any* context. In practice, the latter means "an arbitrary number of times". Finally, we show that by using some optimisation tricks, the proposed protocol can be made very efficient: we show how to separately optimise for Alice's and Bob's computation, and for the communication of the new protocol. In particular, the communication-optimal version is more communication-efficient than either of the Vaidya-Clifton or the Du-Atallah protocols.

Road-map. In Section 2, we describe the necessary cryptographic preliminaries. In Section 3, we analyse some previous private scalar product protocols. In Section 4, we propose a new scalar product protocol, prove its security and propose some important optimisations. We finish with conclusions and acknowledgements.

2   Cryptographic Preliminaries

Secure Multi-party and Two-Party Computation. To guarantee that a protocol is secure in as many applications as possible, one should use the secure multi-party and two-party techniques [Gol04]. Briefly, a two-party protocol between Alice and Bob is *secure* when privacy and correctness are guaranteed for both Alice and Bob. It is said that a protocol *protects privacy*, when the information that is leaked by the distributed computation is limited to the information that can be learned from the designated output of the computation [Pin02].

There are several different security models where one can prove the security of a protocol in. The simplest setting is the *semi-honest* model, where it is assumed that both Alice and Bob follow the protocol, but they are also curious: that is, they store all exchanged data and try to deduce information from it. In the *malicious model*, no assumption is made about the behaviour of Alice and Bob, and it is required that the privacy of one party is preserved even in the case of an arbitrary behaviour of the second party. Most of the papers on privacy-preserving data mining provide only security in the semi-honest model. Such a protocol can be made secure in the malicious model when accompanied with zero-knowledge proofs that both parties follow the protocol. However, such proofs are usually too inefficient to be used in data mining applications.

Homomorphic public-key cryptosystems. A public-key cryptosystem Π is a triple (Gen, Enc, Dec) of probabilistic polynomial-time algorithms for key-generation, encryption and decryption. The security of a public-key cryptosystem is determined by a security parameter k. For a fixed k, it should take more than polynomial in k operations to break the cryptosystem. Together with increased security, larger k means also larger keys and ciphertexts. The key generation algorithm generates, on input $1^k = 1 \ldots 1$ (k ones) a valid pair (sk, pk) of private and public keys that corresponds to the security parameter k. For a fixed key pair (sk, pk), let $P(\mathsf{sk})$ denote the plaintext space of Π. The encryption algorithm Enc takes as an input a plaintext $m \in P(\mathsf{sk})$, a random value r and a public key pk and outputs the corresponding ciphertext $\mathsf{Enc}_{\mathsf{pk}}(m; r)$. The decryption algorithm Dec takes as an input a ciphertext c and a private key sk (corresponding to the public key pk) and outputs a plaintext $\mathsf{Dec}_{\mathsf{sk}}(c)$. It is required that $\mathsf{Dec}_{\mathsf{sk}}(\mathsf{Enc}_{\mathsf{pk}}(m; r)) = m$ for any $m \in P(\mathsf{sk})$, pk and r.

A public-key cryptosystem is *semantically secure* (IND-CPA secure) when a probabilistic polynomial-time adversary cannot distinguish between random encryptions of two elements, chosen by herself. We denote the encryption of a message m by $\mathsf{Enc}_{\mathsf{pk}}(m; r)$, where pk is the corresponding public key and r is the used random string. A public-key cryptosystem is *homomorphic* when $\mathsf{Enc}_{\mathsf{pk}}(m_1; r_1) \cdot \mathsf{Enc}_{\mathsf{pk}}(m_2; r_2) = \mathsf{Enc}_{\mathsf{pk}}(m_1 + m_2; r_1 \cdot r_2)$, where $+$ is a group operation and $\cdot$ is a groupoid operation. This means that a party can add encrypted plaintexts by doing simple computations with ciphertexts, *without* having the secret key. Usually, $P(\mathsf{sk}) = \mathbb{Z}_m$ for some large m. One of the most efficient currently known semantically secure homomorphic cryptosystems was proposed by Paillier cryptosystem [Pai99] and then improved by Damgård and Jurik [DJ01]. In Paillier's case, $P(\mathsf{sk}) = \mathbb{Z}_m$ with $m \geq 2^{1024}$. One can effectively assume that m is as large as say 2^{4096}, when using the Damgård-Jurik cryptosystem [DJ01]. We will assume that k is the bit length of the plaintexts, thus $k \geq 1024$.

Oblivious transfer. In an $\binom{n}{1}$-oblivious transfer protocol, Bob has a database $(\mathcal{D}_1, \ldots, \mathcal{D}_n)$ and Alice has an index $i \in [n]$. The goal is for Alice to retrieve the element $\mathcal{D}_i$ without revealing her index i to Bob, and Bob does not want Alice to get to know anything about the other elements in his database apart from the element she asks for. Recently, Lipmaa [Lip04] proposed an asymptotically efficient $\binom{n}{1}$-oblivious transfer protocol with communication $\Theta(\log^2 n)k$.

3   Cryptanalysis of Proposed Private SP Protocols

Before cryptanalysing some of the previously proposed private scalar product and private shared scalar product protocols, we must define what does it mean to attack one. Next, we will give a somewhat intuitive definition. For simplicity, we will require that all arithmetic is done in $\mathbb{Z}_m$ for some m.

We call a protocol between Alice and Bob a *scalar product* (SP) protocol when Bob obtains, on Alice's private input $\boldsymbol{x} = (x_1, \ldots, x_N) \in \mathbb{Z}_m^N$ and on Bob's private input $\boldsymbol{y} = (y_1, \ldots, y_N) \in \mathbb{Z}_m^N$, the scalar product $\boldsymbol{x} \cdot \boldsymbol{y} = \sum_{i=1}^{N} x_i y_i$. A protocol is a *shared scalar product (SSP) protocol* when Alice receives a uniformly distributed random value $s_A \in \mathbb{Z}_m$ and Bob receives a dependent uniformly distributed random value $s_B \in \mathbb{Z}_m$, such that $s_A + s_B \equiv \boldsymbol{x} \cdot \boldsymbol{y} \pmod{m}$. A scalar product protocol is *private* when after executing the protocol, Bob obtains no more knowledge than $\boldsymbol{x} \cdot \boldsymbol{y}$ and Alice obtains no new knowledge at all. In particular, Alice gets to know nothing new about Bob's vector and Bob gets to know nothing about Alice's vector that is not implied by $\boldsymbol{x}$ and $\boldsymbol{x} \cdot \boldsymbol{y}$. A *private shared scalar product protocol* is defined analogously.

Recently, several researchers from the data mining community have proposed private SSP and SP protocols [DA01b, DA01a, DZ02, VC02], that were primarily meant to be used in the context of privacy-preserving data mining. Most of the proposed solutions try to achieve information-theoretical security—that is, without relying on any computational assumption—by using additive or linear noise to mask the values. In almost all such solutions, one can construct a system of linear equations based on the specification of the protocol, and solve it for the secret values. We will next demonstrate that explicitly in the case of the protocols from [DA01a, VC02].

3.1   Vaidya-Clifton Private Scalar Product Protocol

First, we analyse the Vaidya-Clifton private SP protocol [VC02], depicted by Protocol 1. For the sake of simplicity, we assume that the database size is $N = \ell n$, where n is a block size and ℓ is the number of blocks. We represent each N-dimensional vector $\boldsymbol{z}$ either as $\boldsymbol{z} = (z_1, \ldots, z_N)$ or $\boldsymbol{z} = (\boldsymbol{z}[1], \ldots, \boldsymbol{z}[\boldsymbol{\ell}])$, where $\boldsymbol{z}[i] = (z_{(i-1)n+1}, \ldots, z_{in})$. We denote the n-dimensional vectors $(1, \ldots, 1)$ and $(0, \ldots, 0)$ by $\boldsymbol{1}$ and $\boldsymbol{0}$.

Protocol 1 is a slight modification of the original Vaidya-Clifton protocol. Namely, in the original protocol all scalars belong to $\mathbb{R}$, while in Protocol 1 they belong to $\mathbb{Z}_m$ with $m > N$. Our modifications make the protocol more applicable and also more secure for the next reasons. First, as computers can use only limited precision, there will be stability and correctness problems when computing over real numbers. Second, adding random noise r from $\mathbb{R}$ to value x from $\mathbb{R}$ does not perfectly hide x since it is impossible to choose r uniformly at random from $\mathbb{R}$, or even from $\mathbb{N}$. Therefore, cryptanalysis of the original Vaidya-Clifton protocol is simpler and attacks against it are more dangerous when we consider their protocol as working in $\mathbb{R}$.

In the following, we explicitly assume that m is prime. Proposed attacks also work with composite m, but then one would have to tackle many insubstantial yet technical details. We will also establish some additional notation. First, for any $\mathcal{I} = \{i_1, \ldots, i_j\} \subseteq [N]$ with $|\mathcal{I}| = j$, any vector $\boldsymbol{x}$ and any matrix M, let $\boldsymbol{x}_\mathcal{I} = (x_{i_1}, \ldots, x_{i_j})$ and $M_\mathcal{I}$ denote the sub-matrix of M that consists of the

PRIVATE INPUT OF ALICE: $\boldsymbol{x} \in \{0,1\}^N$
PRIVATE INPUT OF BOB: $\boldsymbol{y} \in \{0,1\}^N$
PRIVATE OUTPUT OF BOB: Scalar product $\boldsymbol{x} \cdot \boldsymbol{y} \mod m$.

1. Alice and Bob jointly do:
     Generate a random invertible $N \times N$ matrix C.
2. Alice does:
     Generate a random vector $\boldsymbol{p} \in \mathbb{Z}_m^N$.
     Send $\boldsymbol{u} \leftarrow \boldsymbol{x} + C\boldsymbol{p}$ to Bob.
3. Bob does:
     Generate ℓ random values $s_1, \ldots, s_\ell \in \mathbb{Z}_m$.
     Send $\boldsymbol{v} \leftarrow C^{\mathrm{T}}\boldsymbol{y} + \boldsymbol{r}$, where $\boldsymbol{r}[i] \leftarrow s_i \boldsymbol{1}$, to Alice.
4. Alice does:
     Set $t_0 := \boldsymbol{v} \cdot \boldsymbol{p}$.
     For $i \in \{1, \ldots, \ell\}$, set $t_i := \sum_{j=1}^n p[i]_j$.
     Send $(t_0, t_1, \ldots, t_\ell)$ to Bob.
5. Bob does:
     Return $\boldsymbol{u} \cdot \boldsymbol{y} - t_0 + \sum_{i=1}^\ell s_i t_i$.

Protocol 1. Vaidya-Clifton private shared scalar product protocol. (All computations are done modulo a public m.)

rows $\mathcal{I} = \{i_1, \ldots, i_j\}$. Second, C is invertible and known to both Alice and Bob. Therefore, define $\boldsymbol{a_i} := (C^{\mathrm{T}})^{-1}\boldsymbol{e_i} \mod m$, where $\boldsymbol{e_i}[j] = 1$ if $i = j$ and $\boldsymbol{e_i}[j] = 0$, otherwise. Define $\boldsymbol{\omega} := (C^{\mathrm{T}})^{-1}\boldsymbol{v}$. Then $(C^{\mathrm{T}})^{-1}\boldsymbol{r} \equiv (C^{\mathrm{T}})^{-1}(s_1 \boldsymbol{1}, \ldots, s_\ell \boldsymbol{1}) \equiv \sum_{i=1}^\ell s_i \boldsymbol{a_i} \pmod{m}$, $\boldsymbol{\omega} \equiv \boldsymbol{y} + \sum_{i=1}^\ell s_i \boldsymbol{a_i} \pmod{m}$ and $t_i \equiv \boldsymbol{e_i} \cdot \boldsymbol{p} \equiv \boldsymbol{a_i} \cdot C\boldsymbol{p} \pmod{m}$ for $i \geq 1$.

First, we show that if the vector $\boldsymbol{y}$ has a low support then Alice is guaranteed to learn half coefficients y_i—and with a high probability the whole vector $\boldsymbol{y}$—after just executing Protocol 1 once.

Lemma 1. *As previously, let* $\mathrm{supp}(\boldsymbol{y}) := |\{y : y_i \neq 0\}|$ *be the support of* $\boldsymbol{y}$. *Assume that* $N \geq (2\,\mathrm{supp}(\boldsymbol{y}) + 1)\ell$. *After just executing Protocol 1 once, a semihonest Alice obtains at least half of the coefficients of* $\boldsymbol{y}$, *with probability 1, by solving* $2\,\mathrm{supp}(\boldsymbol{y}) + 1$ *systems of linear equations in* ℓ *variables.*

Proof. Let M be the matrix with column vectors $\boldsymbol{a_1}, \ldots, \boldsymbol{a_\ell}$. Let $\boldsymbol{s} = (s_1, \ldots, s_\ell)$. The attack is based on the observation that the equality $M\boldsymbol{s} \equiv \boldsymbol{\omega} - \boldsymbol{y} \pmod{m}$ gives Alice a system of N linear equations in ℓ unknowns s_j. The values v_i and vectors $\boldsymbol{a_1}, \ldots, \boldsymbol{a_\ell}$ are known to Alice; the values $y_i \in \{0,1\}$ are unknown. Alice partitions the set $[N]$ iteratively into $\geq N/\ell$ (non-empty) parts $\mathcal{I}_k$ as follows: Denote $\mathcal{J}_k := [N] \setminus \bigcup_{i<k} \mathcal{I}_i$. Alice chooses an $\mathcal{I}_k \subseteq \mathcal{J}_k$, such that the matrix $M_{\mathcal{I}_k}$ has the maximal possible rank with respect to $\mathcal{J}_k$ and $\mathcal{I}_k$ is minimal unless the rank of $M_{\mathcal{J}_k}$ is zero. In particular, $M_{\mathcal{J}_k} = D_k M_{\mathcal{I}_k}$ for some matrix D_k. If rank of $M_{\mathcal{J}_k}$ is zero then Alice chooses a random index from $\mathcal{J}_k$. Note that $M_{\mathcal{J}_k} = D_k M_{\mathcal{I}_k}$ still holds for an appropriate zero matrix D_k.

Now, there are at least $N/\ell \geq 2\,\mathrm{supp}(\boldsymbol{y})+1$ parts $\mathcal{I}_k$. For a majority of indices k (we say that such indices k are "good"), $\boldsymbol{y}_{\mathcal{I}_k}$ is a zero vector. Therefore, in the majority of the cases, Alice obtains the correct values $\boldsymbol{s}_{\mathcal{I}_k}$ by solving the equation $M_{\mathcal{I}_k}\boldsymbol{s} = \boldsymbol{\omega}_{\mathcal{I}_k}$. Since $M_{\mathcal{J}_k}\boldsymbol{s} = D_k M_{\mathcal{I}_k}\boldsymbol{s}$, the value of $\boldsymbol{y}_{\mathcal{J}_k}$ is uniquely determined by $\boldsymbol{s}_{\mathcal{I}_k}$. Moreover, the smallest "good" $k = k_0$ satisfies $k_0 \leq \mathrm{supp}(\boldsymbol{y}) + 1$. The solution $\boldsymbol{s}$ of $M_{\mathcal{I}_{k_0}}\boldsymbol{s} = (\boldsymbol{\omega})_{\mathcal{I}_{k_0}}$ is consistent with the solutions that correspond to other "good" k's, that is, $M_{\mathcal{I}_k} \cdot \boldsymbol{s}_{\mathcal{I}_{k_0}} = \boldsymbol{\omega}_{\mathcal{I}_k}$ for all "good" indices $k > k_0$. Therefore, Alice can find all "good" indices k by majority voting. She also obtains all coordinates of $\boldsymbol{y}_{\mathcal{J}_{k_0}}$. $\qquad\square$

If $|\mathcal{I}_{k_0}| = \ell$ then all coordinates of $\boldsymbol{y}$ are revealed, otherwise coefficients are revealed for all sets $|\mathcal{I}_k| \leq |\mathcal{I}_{k_0}|$, as any solution to $M_{\mathcal{I}_{k_0}}\boldsymbol{s} = \boldsymbol{\omega}_{\mathcal{I}_{k_0}}$ uniquely determines $\boldsymbol{y}_{\mathcal{J}_{k_0}} = \boldsymbol{\omega}_{\mathcal{J}_{k_0}} - D_{k_0}\boldsymbol{\omega}_{\mathcal{I}_{k_0}}$. The next result shows that $\boldsymbol{y}$ is revealed almost certainly.

Lemma 2. *Let $\mathcal{I}_k$ be defined as in the proof of the previous lemma. Then* $\Pr\left[\,|\mathcal{I}_k| = |\mathcal{I}_{k+1}|\,\right] = \prod_{i=0}^{d-1}\left(1 - m^{-|\mathcal{J}_k|+i}\right)$. *Thus, the probability that all coefficients are revealed is approximately* $(1 - m^{-N/2})^{\mathrm{supp}(\boldsymbol{y})\ell} \approx 1 - \mathrm{supp}(\boldsymbol{y})\ell m^{-N/2}$.

Proof. Consider all possible vector assignments of $\boldsymbol{a}_1, \ldots, \boldsymbol{a}_\ell$ that are consistent with the choice of $\mathcal{I}_1, \ldots, \mathcal{I}_k$; that is, such assignments, for which $M_{\mathcal{J}_k} = D'_k M_{\mathcal{I}_k}$ for some D'_k. The latter is equivalent to the assumption that rows of $M_{\mathcal{J}_k}$ are randomly sampled from a vector space of dimension $|\mathcal{I}_k|$. By a standard result [vLW92–p. 303], the probability that $\mathrm{rank}(M_{\mathcal{J}_k}) = |\mathcal{I}_k|$ is equal to $\prod_{i=0}^{|\mathcal{I}_k|-1}(1 - m^{-|\mathcal{J}_k|+i})$. Hence, the first claim is proven. Now, $\boldsymbol{y}$ is completely determined if $|\mathcal{I}_{\mathrm{supp}(\boldsymbol{y})+1}| = \ell$. As $|\mathcal{I}_1| = \ell$ by the protocol construction and for $k < \mathrm{supp}(\boldsymbol{y})$, $|\mathcal{J}_{\mathrm{supp}(\boldsymbol{y})}| > N/2$, the second claim follows from a straightforward calculation. $\qquad\square$

If we give more power to Alice, she will be able to do much better. Assume that Protocol 1 is run twice with the same input vector $\boldsymbol{y}$; let $\boldsymbol{a}_1, \ldots, \boldsymbol{a}_\ell$ and $\boldsymbol{a}'_1, \ldots, \boldsymbol{a}'_\ell$ be vectors, computed from the random matrices C and C' as previously. Then, $\boldsymbol{\omega} - \boldsymbol{\omega}' = \sum_{i=1}^{\ell} s_i \boldsymbol{a}_i - \sum_{i=1}^{\ell} s'_i \boldsymbol{a}'_i$. With high probability, this determines $\boldsymbol{s}$ and $\boldsymbol{s}'$ uniquely. To avoid similar attacks, Bob must never run Protocol 1 twice with the same input $\boldsymbol{y}$ but different matrices C. The next lemma shows that also Alice must never run Protocol 1 twice with the same input $\boldsymbol{x}$ but different matrices C.

Lemma 3. *If Protocol 1 is re-executed $k > N/\ell$ times with the same $\boldsymbol{x}$, Bob obtains $\boldsymbol{x}$ with probability higher than $\prod_{i=0}^{N-1}(1 - m^{-k\ell+i})$.*

Proof. Each execution of Protocol 1 provides ℓ linear equations $\boldsymbol{a}_i \cdot \boldsymbol{u} = \boldsymbol{a}_i \cdot \boldsymbol{x} + \boldsymbol{a}_i \cdot C\boldsymbol{p} = \boldsymbol{a}_i \cdot \boldsymbol{x} + t_i$ for $i \in \{1, \ldots, \ell\}$. As $\boldsymbol{a}_1, \ldots, \boldsymbol{a}_\ell$ are chosen randomly, similar argumentation as in Lemma 2 gives the probability estimate. $\qquad\square$

Finally, we get another efficient attack when we consider itemsets with almost the same support. For example, assume that Alice knows that $\mathrm{supp}(\boldsymbol{y} - \boldsymbol{y}') <$

$N/(4\ell) - 1/2$. Then, by using Lemma 1, Alice can determine s and s' from the equation $\omega - \omega' = y - y' + \sum_{i=1}^{\ell} s_i a_i - \sum_{i=1}^{\ell} s_i' a_i'$; therefore, she obtains y and y'. This attack works with any choice of C. The condition $\mathrm{supp}(y - y') \ll N$ is not so rare in the context of frequent itemset mining. Moreover, several optimisations of APRIORI are devised to exploit such shortcuts. To analyse the applicability of low support attacks, we need additional notations. Let $\mathrm{supp}(I)$ denote the support of the itemset I and y_I the corresponding vector, i.e. $y_{I,k} = 1$ iff the kth row contains items I. We say that I is a closed frequent itemset, iff $\mathrm{supp}(I)$ is over frequency threshold and for any proper superset $J \supsetneq I$, $\mathrm{supp}(I) > \mathrm{supp}(J)$. Now, if the frequent itemset I is not closed, then the APRIORI algorithm discovers $J \supset I$ such that $\mathrm{supp}(y_I - y_J) = 0$ and Alice can apply the attack. The ratio ρ between frequent and frequent closed sets describes the average number of vectors revealed by a single closed set. Empirical results [PHM00] on standard data mining benchmarks indicate that ρ can range from 2 to 100 depending on the frequency threshold, when the database contains some highly correlated items.

The analysis can be extended further by using notion of frequent δ-free sets. A itemset I is δ-free if and only if for any proper subset J of I, $\mathrm{supp}(y_J - y_I) > \delta$. In other words, an itemset I is not δ-free if and only if there is $J \subsetneq I$ with $\mathrm{supp}(y_J - y_I) \leq \delta$. Again, empirical results [BBR03, BB00] on standard data mining benchmarks show that the number of frequent δ-free sets with $\delta \in [0, 20]$ is several magnitudes smaller than the number of frequent sets, when database contain highly correlated items. To conclude, low support differences are quite common for many practical data sets and thus the Vaidya-Clifton scalar product protocol is insecure for frequent itemset mining.

Remark on [VC02–Section 5.2]. In [VC02–Section 5.2], Vaidya and Clifton note that the fact that x_i and y_i belong to $\{0, 1\}$ can create a disclosure risk. They propose two solutions. The first consists of "*cleverly*" selecting the matrix C so that it is not evident which of the values of x_i and y_i are 1's. Lemma 1 states that such a "*clever*" choice is impossible in general since at least a half of y's coordinates is revealed for every matrix C. Besides, the solution is not fully spelled out and no security proofs are given. Another solution from [VC02–Section 5.2] is said to increase the security of Bob but decrease the security of Alice, but again, no security proofs are given. Thus, it is difficult to estimate the exact security of the proposed solutions. It seems that neither of these mentioned solutions is secure against our attacks.

Communication and computation of Vaidya-Clifton protocol. Alice and Bob must both know C, thus the communication of the Vaidya-Clifton protocol is approximately $N^2 \log m$ bits. In the version of the scalar product protocol where no privacy is guaranteed, Alice just sends her vector (N bits) to Bob, who returns the scalar product ($\lceil \log_2 N \rceil$ bits). Define the communication overhead of a private scalar protocol P to be equal to $C(P)/N$, where $C(P)$ is the number of bits communicated in the protocol P. Thus, the communication overhead of

PRIVATE INPUTS: Vectors $x \in \{0,1\}^N$ and $y \in \{0,1\}^N$.
PRIVATE OUTPUTS: Shares $s_A + s_B \equiv x \cdot y \mod m$.

1. Alice does:
   Generate random $v_1, \ldots, v_{d-1} \leftarrow \mathbb{Z}_m^N$.
   Set $v_d := x - \sum_{i=1}^{d-1} v_i$ and $s_A := 0$.
2. For $i = 1$ to d do
   (a) Alice does:
       Generate random $\ell_i \in \{1, \ldots, p\}$.
       Set $h_{i\ell_i} := v_i$.
       For $j \in \{1, \ldots, \ell_i - 1, \ell_i + 1, \ldots, p\}$: Generate random $h_{ij} \in \mathbb{Z}_m^n$.
       Send $(h_{i1}, \ldots, h_{ip})$ to Bob.
   (b) Bob does:
       Generate random $r_i \in \mathbb{Z}_m$.
       For $j \in \{1, \ldots, p\}$: Set $z_{ij} := h_{ij} \cdot y + r_i$.
   (c) Alice does:
       Use $\binom{p}{1}$-oblivious transfer to retrieve $z_{i\ell_i}$ from $(z_{i1}, \ldots, z_{ip})$.
       Set $s_A := s_A + z_{i\ell_i}$.
3. Alice outputs s_A, Bob outputs $s_B = -\sum_{i=1}^{d} r_i$.

Protocol 2. Du-Atallah private SSP protocol. Here, $m > N$ is a public modulus

the Vaidya-Clifton private SP protocol is Nm. Computation is dominated by $\Theta(N^2)$ multiplications and additions in $\mathbb{Z}_m$. The new scalar product protocol, that we will propose in this paper, is both more secure and more efficient.

3.2   Du-Atallah Private Scalar Product Protocol

Du and Atallah proposed another private SSP protocol [DA01a], depicted by Protocol 2. We show that also this protocol cannot handle binary vectors with low support.

Since Protocol 2 chooses the values r_i randomly, s_A is a random value and therefore Alice does not learn anything about y. To learn x, Bob must guess correctly the values ℓ_i for all i. Since the probability of a random guess is p^{-d}, Du and Atallah argue that this protocol is secure when $p^d > 2^{80}$. Bob can do much better, however.

Lemma 4. *Assume $N \geq (2\operatorname{supp}(x) + 1)pd$. Then, with probability 1, Bob finds at least $N/2$ coordinates of x by solving $\operatorname{supp}(x) + 1$ systems of linear equations, each having dimension $pd - 1$. With high probability $\approx (1 - m^{-N/2})^{\operatorname{supp}(x)(pd-1)} \approx 1 - \operatorname{supp}(x)(pd-1)m^{-N/2}$, Bob obtains the whole vector x.*

Proof. Bob knows that $\sum_{i=1}^{d} h_{ij_i} = x$ for some values j_i. Equivalently,

$$\sum_{i=1}^{d} \sum_{j=1}^{p} c_{ij} h_{ij} = x \ ,$$

where $c_{ij} = 1$ if $j = j_i$ and $c_{ij} = 0$, otherwise. Exactly as Alice did in the proof of Lemma 1, Bob iteratively partitions $[N]$ into subsets $\mathcal{I}_k$ with maximal possible rank. Hence, a solution to $\sum_{i,j} c_{ij}(h_{ij})_{\mathcal{I}_{k_0}} = \mathbf{0}$ uniquely determines $x_{\mathcal{I}_k} = \sum_{i,j} c_{ij}(h_{ij})_{\mathcal{I}_k}$ for $k > k_0$. On the other hand, Bob creates at least $2\operatorname{supp}(x) + 1$ partitions $\mathcal{I}_k$. Thus, there exists a $k \leq \operatorname{supp}(x) + 1$, such that $x_{\mathcal{I}_k} = \mathbf{0}$. As in the proof of Lemma 1, we can determine the first "good" $k_0 \leq \operatorname{supp}(x) + 1$ by using majority voting.

To reduce the amount of computations, Bob can ignore all sets $|\mathcal{I}_k| = pd$. For any "good" k, $|\mathcal{I}_k| \leq pd - 1$, as $x_{\mathcal{I}_k} = \mathbf{0}$ and the homogeneous system $\sum_{i,j} c_{ij}(h_{ij})_{\mathcal{I}_k} = \mathbf{0}$ has a nontrivial solution.

The proof of the second claim is similar to the proof of Lemma 2, since it is sufficient that $pd - 1$ random vectors are linearly independent, and $|\mathcal{I}_1| \geq pd - 1$ by construction.   □

This protocol has another serious weakness, since with high probability slightly more than pd coordinates of x allow to determine correct c_{ij} and thus also reveal other coordinates. Therefore, a leakage of pd database entries, can reveal the whole vector (database) and thus pd must be large, say more than 200. On the other hand, this protocol is very inefficient when pd is large.

Communication and computation complexity. Assume $p^d > 2^{80}$. Then the communication of the Du-Atallah private SSP protocol is $dpN + dt_p$, where t_p is the communication complexity of the $\binom{p}{1}$-oblivious transfer protocol. This is minimal when d is maximised, i.e., when $p = 2$. Taking the efficient $\binom{p}{1}$-oblivious transfer protocol from [AIR01], one has $t_2 = 3k$, where $k \approx 1024$ is the security parameter. Then the communication is $2dN + 3dk$ bits for $d \geq 80$ and $k \geq 1024$. Taking $d = 80$ and $k = 1024$, we get communication $160N + 245760$ bits. However, Lemma 4 indicates that for the security of the Du-Atallah protocol, one should pick p and d such that pd is quite large. For example, picking $p = 2^{11}$ and $d = 8$ might result in an acceptable security level, but then the communication of the protocol will be $2^{14} \cdot N + dt_p$ bits.

4   Cryptographic Private SSP Protocol

In this section we describe a private SSP protocol (Protocol 3) that is based on homomorphic encryption. Note that a private SP protocol can be obtained from it by defining $s_B \leftarrow 0$.

Theorem 1. *Assume that $\Pi = (\mathsf{Gen}, \mathsf{Enc}, \mathsf{Dec})$ is a semantically secure homomorphic public-key cryptosystem with $P(\mathsf{sk}) = \mathbb{Z}_m$ for some large m. Set $\mu := \lfloor \sqrt{m/N} \rfloor$. Protocol 3 is a secure SSP protocol in the semi-honest model, assuming that $x, y \in \mathbb{Z}_\mu^N$. Alice's privacy is guaranteed when Bob is a probabilistic polynomial-time machine. Bob's privacy is information-theoretical.*

Proof. Clearly, the protocol is correct if the participants are honest. Since the cryptosystem is semantically secure, Bob only sees N random ciphertexts, for

PRIVATE INPUTS: Private vectors $x, y \in \mathbb{Z}_\mu^N$.
PRIVATE OUTPUTS: Shares $s_A + s_B \equiv x \cdot y \mod m$

1. Setup phase. Alice does:
      Generate a private and public key pair $(\mathsf{sk}, \mathsf{pk})$.
      Send pk to Bob.
2. Alice does for $i \in \{1, \ldots, N\}$:
      Generate a random new string r_i.
      Send $c_i = \mathsf{Enc}_{\mathsf{pk}}(x_i; r_i)$ to Bob.
3. Bob does:
      Set $w \leftarrow \prod_{i=1}^N c_i^{y_i}$.
      Generate a random plaintext s_B and a random nonce r'.
      Send $w' = w \cdot \mathsf{Enc}_{\mathsf{pk}}(-s_B; r')$ to Alice.
4. Alice does: Compute $s_A = \mathsf{Dec}_{\mathsf{sk}}(w') = x \cdot y - s_B$.

Protocol 3. Private homomorphic SSP protocol

which he cannot guess the plaintexts. In particular, this holds even when Bob has given two candidate vectors x_1 and x_2 to Alice and Alice has randomly chosen one of them, $x := x_b$. Even after a polynomial number of protocol executions with Alice's input, Bob will gain only an insignificant amount of information about x_b that will not help him in guessing the value of b. (This roughly corresponds to the standard notion of semantic security.) On the other hand, Alice only sees a random encryption of $s_A = x \cdot y - s_B$, where s_B is random. But Alice has the key anyways, so she can decrypt this message. Thus, Alice obtains no information at all.   □

(Note that if $m > 2^{1024}$ and $N \approx 2^{16}$ then $\mu \geq 2^{504}$.) In Appendix A, we describe an extension of this protocol to more than two parties.

Practical considerations. Note that when Alice and Bob need to execute this protocol several times, they can reuse public and private keys and thus the setup phase can be executed only once. Public key cryptography is computationally demanding. To estimate the computational cost of the new scalar product protocol, we must count encryptions, decryptions and multiplications of ciphertexts. Bob must perform N exponentiations and 1 encryption. Alice has to perform N encryptions and 1 decryption.

In the specifically interesting case when $x_i, y_i \in \{0, 1\}$ (e.g., when x and y correspond to characteristic functions of two sets X and Y; then $x \cdot y = |X \cap Y|$), this protocol can be further optimised. Namely, Alice can pre-compute and then store a large table of random encryptions of 0's and 1's. Then every "encryption" just corresponds of fetching a new element from the correct table; this can be done very quickly. Bob has to perform 1 encryption and $\mathsf{supp}(y)$ multiplications, since the exponents y_i are all Boolean. (When $y_i = 0$ then $c_i^{y_i} = 1$ and otherwise $c_i^{y_i} = c_i$.)

The current hardware allows to do approximately 10^5 multiplications per seconds and thus the computational complexity of both Alice and Bob is tolerable. A similar analysis applies for Protocol 4. Here, Alice and Bob must pre-compute N encryptions. Hence, we can conclude that the computational complexity is not a serious downside of the proposed protocols. Similar, although not as efficient, optimisation tricks can also be used to speed up Protocol 3 when x and y are not binary.

Estimated communication complexity. The only serious drawback of the new protocols is the communication overhead: since Alice sends N ciphertexts c_i, the overhead is k'/μ, where k' is just the size of each ciphertext in bits. When using any of the currently known most efficient semantically secure homomorphic cryptosystems (e.g., the one from [Pai99]), $k' \approx 2048$. For $x, y \in \mathbb{Z}_{m'}$ with very small m'—say, $m' \le 13$, this compares non-favourably with the overhead of the (insecure) Du-Atallah protocol which has the overhead of approximately 160 times with $d = 80$ and $k = 1024$. For a large m', the described protocol is already more communication-efficient than the Du-Atallah protocol.

Comparison with Freedman-Nissim-Pinkas protocol. Recently, Freedman, Nissim and Pinkas proposed a related cryptographically secure protocol for computing the set intersection cardinality [FNP04], a task that is equivalent to privately computing the scalar product of two binary vectors. In the non-shared case, the Freedman-Nissim-Pinkas protocol is more efficient than the new one, but then the participants also learn the values $\text{supp}(x)$ and $\text{supp}(y)$. However, recall that in the data mining applications it is preferable that both parties will get only shares $s_A + s_B = x \cdot y \mod m$ of the scalar product, otherwise frequency of some infrequent sets is revealed. Moreover, sometimes only a list of frequent sets without frequencies is required.

Freedman, Nissim and Pinkas proposed also a solution for shared version, but their protocol requires a secure circuit evaluation. Briefly, secure evaluation means that first Alice and Bob obtain $\text{supp}(x)$ different shares

$$
s_i + t_i = \left\{ \begin{array}{l} 0, \text{ if } x_i = 1, y_i = 1 \\ r_i, \text{ if } x_i = 1, y_i = 0 \end{array} \right\} \mod m
$$

where $r_i \in \mathbb{Z}_m$ is a random value and m is (say) a 1024-bit number. To securely compute $x \cdot y$ by secure circuit evaluation, one therefore needs to execute oblivious transfer for each $1024 \cdot \text{supp}(x)$ input bit pairs (s_i, t_i). Since a $\binom{2}{1}$-oblivious transfer protocol requires sending at least three encryptions, the communication overhead of the Freedman-Nissim-Pinkas protocol is lower than the communication overhead of Protocol 3 only if $\text{supp}(x) \le N/(3 \cdot 1024)$, i.e., if the candidate set is very infrequent.

Reducing communication overhead. We shall now discuss how to reduce the overhead if it is known that x and y are Boolean. Again, similar optimisation techniques can be used when $x, y \in \mathbb{Z}_{\mu'}$ for some $2 < \mu' \ll \mu$. In the following we assume that the plaintext space of the cryptosystem Π is a residue ring $\mathbb{Z}_m$ such that $\log m \geq 1024$. This is the case for all widely known homomorphic cryptosystems. When we assume that $x_i, y_i \in \{0, 1\}$, every ciphertext c_i in Protocol 3 only transfers a single bit x_i, which results in communication overhead.

The next technique for packing several bits into one plaintext is fairly standard in cryptography (it has been used at least in the context of electronic voting [CGS97, DJ01], electronic auctions [LAN02] and oblivious transfer [Lip04]). To pack k entries into a single message—recall that the plaintext length is k bits—, we fix a radix $B > N$, such that $B^k < m$, and work implicitly with B-ary numbers. Let $[v_k, \ldots, v_2, v_1] = v_1 + v_2 B + \cdots + v_k B^{k-1}$. Our method works only in the case when Alice and Bob do batch computation of scalar products, more precisely, when Alice and Bob need to compute $\boldsymbol{x_i} \cdot \boldsymbol{y}$ for several vectors $\boldsymbol{x_i}$, $i \in \{1, \ldots, k\}$, owned by Alice. (This is exactly what happens in the context of frequent itemset mining.)

The new batch scalar product protocol looks exactly like Protocol 3, except that Alice computes c_i as

$$
\begin{aligned}
c_i &= \mathsf{Enc}_{\mathsf{pk}}([x_{ki}, \ldots, x_{2i}, x_{1i}]; r_i) \\
&= \mathsf{Enc}_{\mathsf{pk}}(1; 0)^{x_{1i}} \mathsf{Enc}_{\mathsf{pk}}(B; 0)^{x_{2i}} \cdots \cdot \mathsf{Enc}_{\mathsf{pk}}(B^{k-1}; 0)^{x_{ik}} \mathsf{Enc}_{\mathsf{pk}}(0; r_i) \ .
\end{aligned}
$$

It takes at most k multiplications to compute c_i. Again, the encryptions $\mathsf{Enc}_{\mathsf{pk}}(B^j; 0)$ can be computed in the setup phase. Hence, during the first step, Alice's computation is N encryptions and $O(kN)$ multiplications.

At the second step of the protocol, Bob computes

$$
\begin{aligned}
w &= \prod_{i=1}^{N} \mathsf{Enc}_{\mathsf{pk}}(y_i[x_{ki}, \ldots, x_{2i}, x_{1i}]; r_i) \mathsf{Enc}_{\mathsf{pk}}(-s_B, r') \\
&= \mathsf{Enc}_{\mathsf{pk}}([\boldsymbol{x_k} \cdot \boldsymbol{y}, \ldots, \boldsymbol{x_1} \cdot \boldsymbol{y}] - s_B; r'') \ .
\end{aligned}
$$

Hence, if Bob reveals s_B, Alice can restore *all* scalar products $\boldsymbol{x_j} \cdot \boldsymbol{y}$. Sometimes it is also needed that Alice be able only to compute $\boldsymbol{x_j} \cdot \boldsymbol{y}$ for $j \in I$, where I is a proper subset of $\{1, \ldots, k\}$. One can do this efficiently by using standard cryptographic techniques.

Therefore, when using the Paillier cryptosystem, the resulting protocol for privately computing the scalar product of two binary vectors has almost optimal communication overhead of $\lceil \log N \rceil$ times. (When using the Damgård-Jurik cryptosystem, the communication overhead might even be smaller.) This should be compared to the 160 times overhead of the insecure Du-Atallah protocol.

Related work. After the preproceedings version of the current paper was published, we were acknowledged by Rebecca Wright of some previous work. In particular, in [WY04], Wright and Yang proposed essentially the same SSP protocol as Protocol 3, optimised for the case of binary data as in our paper. However, they did not consider the batch SSP case. (See [WY04] and the references therein.)

Security in malicious model. Protocol 3 can be made secure in the malicious model by letting Alice to prove in zero-knowledge, for every i, that c_i encrypts a value from $\mathbb{Z}_\mu$. This can be done efficiently in the random oracle (or common reference string) model [Lip03]. An alternative is to use conditional disclosure of secrets [AIR01] modified recently to the setting of Paillier's cryptosystem in [LL05]. Both methods guarantee that at the end of a protocol run, Alice is no better of than mounting the next *probing attack*: Alice creates a suitable valid input vector $\boldsymbol{x}'$, executes the protocol with Bob, and obtains $\boldsymbol{x}' \cdot \boldsymbol{y}$. If $\boldsymbol{x}'$ is suitably chosen (e.g., $x_i' = 1$ and $x_j' = 0$ for $j \neq i$), this may result in privacy leakage. However, this probing attack is unavoidable, no matter what private scalar product protocol is used instead of Protocol 3. The only way to tackle this attack is to let Alice to prove that her input $\boldsymbol{x}$ is "correctly" computed, whatever "correctly" means in the concrete application (e.g., in frequent itemset mining on vertically distributed databases). While such a functionality can be added to Protocol 3, it is not a part of the definition of a "private scalar product" protocol, but highly application-dependent (and thus should be left to be specified on a higher level), and very often, highly costly.

5   Conclusions

The secure computation of a scalar product is an important task within many data mining algorithms that require the preservation of privacy. Recently, several protocols have been proposed to solve this task. We have shown, however, that they are insecure. Moreover, we presented a private scalar product protocol based on standard cryptographic techniques and proved that it is secure. Furthermore, we described several optimisations in order to make it very efficient in practice.

Acknowledgements

We would like to thank Benny Pinkas for useful comments. The second and the third author were partially supported by the Estonian Information Technology Foundation, the Finnish Defence Forces Institute for Technological Research and by the Finnish Academy of Sciences.

References

[AIR01]   William Aiello, Yuval Ishai, and Omer Reingold. Priced Oblivious Transfer: How to Sell Digital Goods. In Birgit Pfitzmann, editor, *Advances in Cryptology — EUROCRYPT 2001*, volume 2045 of *Lecture Notes in Computer Science*, pages 119–135, Innsbruck, Austria, 6–10 May 2001. Springer-Verlag.

[AMS+96] Rakesh Agrawal, Heikki Mannila, Ramakrishnan Srikant, Hannu Toivo-
nen, and A. Inkeri Verkamo. Fast Discovery of Association Rules. In
Usama M. Fayyad, Gregory Piatetsky-Shapiro, Padhraic Smyth, and Ra-
masamy Uthurusamy, editors, *Advances in Knowledge Discovery and Data
Mining*, pages 307–328. AAAI/MIT Press, 1996.

[BB00]   Jean-Fran?is Boulicaut and Artur Bykowski. Frequent Closures as a Concise Representation for Binary Data Mining. In *PAKDD 2000*, volume 1805
of *Lecture Notes in Computer Science*, pages 62–73. Springer, 2000.

[BBR03]   Jean-Fran?is Boulicaut, Artur Bykowski, and Christophe Rigotti. Free-
Sets: A Condensed Representation of Boolean Data for the Approximation
of Frequency Queries. *Data Mining and Knowledge Discovery*, 7(1):5–22,
2003.

[CGS97]   Ronald Cramer, Rosario Gennaro, and Berry Schoenmakers. A Secure and
Optimally Efficient Multi-Authority Election Scheme. In Walter Fumy, editor, *Advances in Cryptology — EUROCRYPT '97*, volume 1233 of *Lecture
Notes in Computer Science*, pages 103–118, Konstanz, Germany, 11–15 May
1997. Springer-Verlag.

[DA01a]   Wenliang Du and Mikhail J. Atallah. Privacy-Preserving Statistical Analysis. In *Proceedings of the 17th Annual Computer Security Applications
Conference*, pages 102–110, New Orleans, Louisiana, USA, December 10–
14 2001.

[DA01b]   Wenliang Du and Mikhail J. Atallah. *Protocols for Secure Remote Database
Access with Approximate Matching*, volume 2 of *Advances in Information Security*, page 192.   Kluwer Academic Publishers, Boston, 2001.
http://www.wkap.nl/prod/b/0-7923-7399-5.

[DJ01]   Ivan Damgård and Mads Jurik. A Generalisation, a Simplification and
Some Applications of Paillier's Probabilistic Public-Key System.   In
Kwangjo Kim, editor, *Public Key Cryptography 2001*, volume 1992 of *Lecture Notes in Computer Science*, pages 119–136, Cheju Island, Korea, 13–
15 February 2001. Springer-Verlag.

[DZ02]   Wenliang Du and Zhijun Zhan. A Practical Approach to Solve Secure
Multi-party Computation Problems. In Carla Marceau and Simon Foley,
editors, *Proceedings of New Security Paradigms Workshop*, pages 127–135,
Virginia Beach, virginia, USA, September 23–26 2002. ACM Press.

[FNP04]   Michael J. Freedman, Kobbi Nissim, and Benny Pinkas. Efficient Private
Matching and Set Intersection. In Christian Cachin and Jan Camenisch,
editors, *Advances in Cryptology — EUROCRYPT 2004*, volume 3027 of
Lecture Notes in Computer Science, pages 1–19, Interlaken, Switzerland,
2–6 May 2004. Springer-Verlag.

[Gol04]   Oded Goldreich. *Foundations of Cryptography: Basic Applications*. Cambridge University Press, 2004.

[LAN02]   Helger Lipmaa, N. Asokan, and Valtteri Niemi. Secure Vickrey Auctions
without Threshold Trust. In Matt Blaze, editor, *Financial Cryptography —
Sixth International Conference*, volume 2357 of *Lecture Notes in Computer
Science*, pages 87–101, Southhampton Beach, Bermuda, 11–14 March 2002.
Springer-Verlag.

[Lip03]   Helger Lipmaa.   On Diophantine Complexity and Statistical Zero-
Knowledge Arguments. In Chi Sung Laih, editor, *Advances on Cryptology
— ASIACRYPT 2003*, volume 2894 of *Lecture Notes in Computer Science*,
pages 398–415, Taipei, Taiwan, 30 November–4 December 2003. Springer-
Verlag.

[Lip04]   Helger Lipmaa. An Oblivious Transfer Protocol with Log-Squared Total Communication. Technical Report 2004/063, International Association for Cryptologic Research, February 25 2004.

[LL04]   Sven Laur and Helger Lipmaa. On Private Similarity Search Protocols. In Sanna Liimatainen and Teemupekka Virtanen, editors, *Proceedings of the Ninth Nordic Workshop on Secure IT Systems (NordSec 2004)*, pages 73–77, Espoo, Finland, November 4–5, 2004.

[LL05]   Sven Laur and Helger Lipmaa. Additive Conditional Disclosure of Secrets. Manuscript, January 2005.

[Pai99]   Pascal Paillier. Public-Key Cryptosystems Based on Composite Degree Residuosity Classes. In Jacques Stern, editor, *Advances in Cryptology — EUROCRYPT '99*, volume 1592 of *Lecture Notes in Computer Science*, pages 223–238, Prague, Czech Republic, 2–6 May 1999. Springer-Verlag.

[PHM00]   J. Pei, J. Han, and R. Mao. CLOSET: An efficient algorithm for mining frequent closed itemsets. In *2000 ACM SIGMOD Workshop on Research Issues in Data Mining and Knowledge Discovery*, 2000.

[Pin02]   Benny Pinkas. Cryptographic Techniques for Privacy-Preserving Data Mining. *KDD Explorations*, 4(2):12–19, 2002.

[VC02]   Jaideep Vaidya and Chris Clifton. Privacy Preserving Association Rule Mining in Vertically Partitioned Data. In *Proceedings of The 8th ACM SIGKDD International Conference on Knowledge Discovery and Data Mining*, pages 639–644, Edmonton, Alberta, Canada, July 23–26 2002. ACM.

[vLW92]   Jacobus H. van Lint and Richard M. Wilson. *A Cource in Combinatorics*. Cambridge University Press, 1992.

[WY04]   Rebecca N. Wright and Zhiqiang Yang. Privacy-Preserving Bayesian Network Structure Computation on Distributed Heterogeneous Data. In *Proceedings of The Tenth ACM SIGKDD International Conference on Knowledge Discovery and Data Mining*, pages 713–718, Seattle, Washington, USA, August 22–25 2004. ACM.

A   Private Generalised Scalar Product Protocol

Next, we propose a secure generalised scalar product protocol (Protocol 4) for

$$\langle \boldsymbol{x_1}, \boldsymbol{x_2}, \dots, \boldsymbol{x_k} \rangle = \sum_{i=1}^{N} x_{1i} \cdots x_{ki} \ .$$

For the sake of simplicity, we consider only the three-party case but the protocol can be easily generalised. Again, Alice has a private key; Bob and Carol know only the corresponding public key. The security of the generalised scalar product protocol depends on Alice. Namely, when Alice colludes with other parties then privacy can be compromised. For example, colluding Alice and Carol can reveal y_i, unless $x_i = 0$, since $\mathsf{Dec_{sk}}(d_i) = x_i y_i$. Thus, we get the following result.

Theorem 2. *Assume that $\Pi = (\mathsf{Gen}, \mathsf{Enc}, \mathsf{Dec})$ is a semantically secure homomorphic public-key cryptosystem with $P(\mathsf{sk}) = \mathbb{Z}_m$ for some large m. Protocol 4 is a secure generalised scalar product protocol. In particular, it is secure against all possible coalitions provided that Alice does collude with other parties.*

The proof is a simple generalisation of the previous proof. Bob must re-randomise c_i's as $d_i = c_i \cdot \mathsf{Enc}_{\mathsf{pk}}(0; r_i')$, since otherwise the values of y_i's can be detected only by comparing the ciphertext that he receives from Alice with the one he sends to Carol. The sharing step 4 allows combine the outcome with other cryptographic protocols.

PRIVATE INPUTS: Private vectors $x, y, z \in \mathbb{Z}_\mu^N$.
PRIVATE OUTPUTS: Shares $s_A + s_B + s_C \equiv \langle x, y, z \rangle \mod m$

1. Alice does:
    Generate a key-pair $(\mathsf{sk}, \mathsf{pk})$.
    Send the public key pk to Bob and Carol.
2. Alice does for $i \in \{1, \ldots, N\}$:
    Send $c_i = \mathsf{Enc}_{\mathsf{pk}}(x_i; r_i)$ to Bob.
3. Bob does for $i \in \{1, \ldots, N\}$:
    Set $d_i = c_i^{y_i} \mathsf{Enc}_{\mathsf{pk}}(0; r_i')$.
    Send d_i to Carol.
4. Carol does:
    Set $w \leftarrow \prod_{i=1}^{N} c_i^{z_i}$.
    Generate a random plaintext s_C and a random nonce r'.
    Send $w' \leftarrow w \cdot \mathsf{Enc}_{\mathsf{pk}}(-s_C; r')$ to Bob.
5. Bob does:
    Generate a random plaintext s_B and a random nonce r''.
    Send $w'' \leftarrow w' \cdot \mathsf{Enc}_{\mathsf{pk}}(-s_B; r'')$ to Alice.
6. Alice computes $s_A \leftarrow \mathsf{Dec}_{\mathsf{sk}}(w'') = x \cdot y - s_B - s_C$.

Protocol 4. Private generalised homomorphic SSP protocol

The assumption that Alice does not collude with other parties is quite strong. When we modify the protocol so that $(\mathsf{sk}, \mathsf{pk})$ is generated jointly by Alice, Bob and Carol and that on the step 4, they do threshold decryption of w, we get a private SP protocol with the next security result:

Theorem 3. *Assume $\Pi = (\mathsf{Gen}, \mathsf{Enc}, \mathsf{Dec})$ is a semantically secure homomorphic threshold public-key cryptosystem. Then Protocol 4, generalised to κ parties, is secure against coalitions by $< \kappa/2$ parties.*

Separable Implicit Certificate Revocation*

Dae Hyun Yum[1,2] and Pil Joong Lee[1,3]

[1] Information Security Lab., EEE, Postech, Korea
[2] Department of Computer Science, New York University, USA
[3] KT Research Center, KT Corp., Korea
{dhyum, pjl}@postech.ac.kr

Abstract. The popular certificate revocation systems such as CRL and OCSP have a common drawback that they are explicit certificate revocation; the sender must obtain the revocation status information of the receiver's certificate, before sending an encrypted message. Recently, an implicit certificate revocation system called 'certificate-based encryption' was introduced. In this model, a receiver needs both his private key and an up-to-date certificate from the CA (Certification Authority) to decrypt a ciphertext, while senders need not be concerned about the certificate revocation problem. Hence, the certificate-based encryption system has the advantage of light infrastructure requirement. However, the certificate-based encryption system has an important drawback that it is inseparable; only the CA can handle the certificate revocation problem and the load cannot be distributed among multiple trusted authorities. In this paper, we propose a separable implicit certificate revocation system called 'status certificate-based encryption,' in which the authenticity of a public key is guaranteed by a (long-lived) certificate and the certificate revocation problem is resolved by a (short-lived) status certificate. We present a secure construction based on bilinear mappings as well as definitional works.

Keywords: PKI, certificate revocation, certificate-based encryption.

1   Introduction

PKI AND THE CERTIFICATE REVOCATION PROBLEM.   The main idea of PKI (Public Key Infrastructure) is a digital certificate, which is a digitally signed statement that binds a user and his public key. When a certificate is issued, its validity is limited by a pre-defined expiration time. However, instances occur where a certificate must be invalidated prior to its expiration time, and CRL (Certificate Revocation List) is the most common mechanism for determining whether a certificate is revoked or not [11]. CRL is a signed list of revoked certificates that is periodically issued by the CA (Certification Authority). The size

* This research was supported by University IT Research Center Project and the Brain Korea 21 Project.

C. Park and S. Chee (Eds.): ICISC 2004, LNCS 3506, pp. 121–136, 2005.

of the CRL can grow arbitrarily large, which causes unnecessary consumption of storage and bandwidth that cannot be tolerated in certain environments. Another popular mechanism that provides the revocation status of a certificate is OCSP (Online Certificate Status Protocol), where a client generates an OCSP request and an on-line server replies to the client with an OCSP response [12]. OCSP provides a real-time status and is appropriate for applications where timeliness is of high priority. A major drawback of OCSP is that the server must be on-line and involved in every transaction. Also, the OCSP server can be the main target of DoS (Denial-of-Service) attacks. Even though there are other certificate revocation systems [14, 13, 15], previous systems including CRL and OCSP have a common inherent drawback that they are explicit revocation; when Alice wants to send an encrypted message to Bob, she must obtain fresh information on Bob's certificate status in advance. Distributing large amounts of fresh status information on certificates requires infrastructure, and the apparent need for this infrastructure is often cited as a reason against the deployment and management of public key cryptosystems.

CERTIFICATE-BASED ENCRYPTION.   Recently, the notion of CBE (Certificate-Based Encryption) was introduced to construct an efficient PKI requiring light infrastructure [10]. In this model, a certificate acts as a partial decryption key as well as a traditional public key certificate. When Bob wants to decrypt a ciphertext sent from Alice, he needs both an up-to-date certificate from the CA and his private key. Hence, senders in a CBE system are not required to obtain fresh status information on the receiver's certificate and the up-to-date certificate (i.e., the revocation information) can be 'pushed' only to the owner of the public key by the CA. As with public key cryptography, each user in CBE generates his own public key/private key pair and requests a certificate from the CA. Then, the CA uses identity-based cryptography [3, 19] to generate the certificate. CBE retains the desirable properties of public key cryptography (no key escrow) and identity-based cryptography (implicit certification), while it is not subjected to the private key escrow problem inherent in identity-based cryptography.

Even though the CBE system has better characteristics than previous certificate revocation systems, it is questionable whether CBE systems are widely adopted in practice. This question stems from the fact that CBE systems are inseparable; only the CA can manage the certificate revocation problem and the burden cannot be distributed among other trusted parties. Therefore, a CBE system becomes inefficient when the CA has a large number of clients and performs frequent certificate updates. By contrast, the load distribution among multiple trusted parties can be accomplished in other certificate revocation systems; for example, the OCSP server can be (1) the CA who issued the certificate, (2) a Trusted Responder whose public key is trusted by the requester, or (3) a CA Designated Responder (Authorized Responder) who holds a specially marked certificate issued directly by the CA, indicating that the responder may issue OCSP responses for that CA [12].

OUR CONTRIBUTION. When the CA in CBE issues a certificate for the time period t, the CA guarantees the following two facts: (1) the binding between a user and his public key, and (2) the validity of the binding for the time period t. However, we find that the authenticity of a public key (i.e., the binding between a user and his public key) does not need to be checked at every time period. Instead, we propose the notion of SCBE (Status Certificate-Based Encryption), in which (1) the authenticity of a public key is guaranteed by a (long-lived) certificate and (2) the certificate revocation status information is carried by a (short-lived) status certificate. In an SCBE system, each user generates his own public key/private key pair and the CA issues a certificate. When Alice wants to send an encrypted message to Bob, Alice obtains Bob's public key from Bob's certificate with no need of checking the revocation status of the certificate. To decrypt Alice's message, Bob needs a decryption key which is computed from his private key and an up-to-date status certificate. Note that a status certificate can be issued by a dedicated trusted third party, say SCA (Status Certification Authority), which is different from the CA. Therefore, the CA guarantees the binding between a user and his public key, and the SCA guarantees the validity of the binding for the time period t. The advantageous properties of CBE (no key escrow and implicit revocation) are maintained in SCBE while the CA's burden of certificate revocation can be distributed, i.e., SCBE is a separable implicit certificate revocation system.

2   Certificate-Based Encryption

2.1   The Model of CBE

A CBE scheme [10] is specified by a 6-tuple of poly-time algorithms. In lieu of an original definition, we present a slightly modified version, i.e., a CBE scheme with only one security parameter instead of two. While eliminating the second security parameter from the input of CB_Set_User_Key, we add a parameter list *params* to the input of CB_Set_User_Key. Since *params* is derived from a security parameter k, this change makes no meaningful difference for all practical purposes. In addition, the definition is not based on either identity-based encryption or public key encryption. Thus, the definition is independent and more general.

Definition 1. *A certificate-based encryption scheme is a 6-tuple of poly-time algorithms* (CB_Gen, CB_Set_User_Key, CB_Upd^{CA}, CB_Upd^{User}, CB_Enc, CB_Dec) *such that:*

- CB_Gen, *the master key and parameter generation algorithm, is a probabilistic algorithm that takes as input a security parameter 1^k and (optionally) the total number of time periods N. It returns a master key $CBSK^*$ and a parameter list params.*
- CB_Set_User_Key, *the user key generation algorithm, is a probabilistic algorithm that takes as input a parameter list params, a user identity id, and*

(optionally) the total number of time periods N. It returns the user id's private key $CBSK_{id}$ and public key $CBPK_{id}$.

- CB_Upd^{CA}, *the CA update algorithm, is a deterministic algorithm that takes as input a parameter list params, a user identity id, a time period t, the user id's public key $CBPK_{id}$, and the CA's master key $CBSK^*$. It returns the user id's certificate $Cert_{(id,t)}$ for the time period t.*

- CB_Upd^{User}, *the user update algorithm, is a deterministic algorithm that takes as input a parameter list params, a time period t, a certificate $Cert_{(id,t)}$, and (optionally) $CBPD_{(id,t-1)}$. It returns a partial decryption key $CBPD_{(id,t)}$.*

- CB_Enc, *the encryption algorithm, is a probabilistic algorithm that takes as input a message M, a user identity id, a time period t, a parameter list params, and the user id's public key $CBPK_{id}$. CB_Enc$_{params}^{CBPK_{id}}(M, id, t)$ returns a ciphertext C.*

- CB_Dec, *the decryption algorithm, is a deterministic algorithm that takes as input a parameter list params, the private key $CBSK_{id}$, a ciphertext C, and the partial decryption key $CBPD_{(id,t)}$. CB_Dec$_{params}^{CBSK_{id}}(C, CBPD_{(id,t)})$ returns a message M or the special symbol $\perp$.*

We require that CB_Dec$_{params}^{CBSK_{id}}$(CB_Enc$_{params}^{CBPK_{id}}(M, id, t), CBPD_{(id,t)}) = M$ for all message M.

In a CBE scheme, CB_Gen and CB_Upd^{CA} are performed by the CA, and CB_Set_User_Key and CB_Upd^{User} are executed by a user. Since CB_Set_User_Key is executed by a user, the key escrow of a user's private key is not inherent in a CBE scheme. At the beginning of a time period t, a certificate $Cert_{(id,t)}$ is given to a user id by the CA. Then, the user computes a partial decryption key $CBPD_{(id,t)}$ based on $Cert_{(id,t)}$ and (possibly) $CBPD_{(id,t-1)}$. This partial decryption key $CBPD_{(id,t)}$ together with the private key $CBSK_{id}$ is used in the decryption algorithm. CB_Enc is carried out by a sender unconcerned about certificate revocation status.

For security analysis, we give the adversary access to three types of oracles. The first is a certification oracle CB_Upd$(\cdot, \cdot, \cdot, \cdot)$ that returns $Cert_{(id,t)}$ on input a user identity id, a time period t, and the user id's public key/private key pair. The second is a decryption oracle CB_Dec$(\cdot, \cdot, \cdot, \cdot, \cdot)$ that returns CB_Dec$_{params}^{CBSK_{id}}(C,$ $CBPD_{(id,t)})$ on input $(id, t, C, CBPK_{id}, CBSK_{id})$. The third is a left-or-right encryption oracle [2] CB_Enc$(\cdot, \cdot, \cdot, \cdot, LR(\cdot, \cdot, b))$ that given a user id, a time period t, the user id's public key/private key pair, and equal length messages M_0, M_1 returns a challenge ciphertext $C_{ch} =$ CB_Enc$_{params}^{CBPK_{id}}(M_b, id, t)$.

The security of a CBE scheme is directed against two different types of adversaries. The Type I adversary A_I has no access to the master key, but may make certification queries and decryption queries. The Type II adversary A_{II} equipped with the master key models an eavesdropping CA. For the Type II adversary A_{II}, $(params, CBSK^*)$ replaces the user id's public key/private key pair in the input of oracle queries. For detailed restrictions on the two types of adversaries and security notions, refer to [10].

Definition 2. *Let Π_{CB} be a certificate-based encryption scheme. For any adversary A, we may define the following:*

$$\mathsf{Succ}_{A,\Pi_{CB}}(k) = Pr[b' = b : (CBSK^*, params) \leftarrow \mathsf{CB\_Gen}(1^k); \ b \leftarrow \{0,1\};$$
$$b' \leftarrow A^{\mathsf{O}(\cdot),\mathsf{CB\_Dec}(\cdot,\cdot,\cdot,\cdot,\cdot),\mathsf{CB\_Enc}(\cdot,\cdot,\cdot,\cdot,LR(\cdot,\cdot,b))}(params, h)]$$

where $\mathsf{O}(\cdot) = \mathsf{CB\_Upd}(\cdot,\cdot,\cdot,\cdot)$, $h = \perp$ for A_I, and $\mathsf{O}(\cdot) = \perp$, $h = CBSK^*$ for A_{II}. The adversary may query oracles adaptively, except that it can make exactly one query to the left-or-right encryption oracle. A must follow the adversarial constraints given in [10]. Π_{CB} is said to be secure against chosen-ciphertext attacks if for any probabilistic polynomial time (PPT) adversary A, the advantage $\mathsf{Adv}_{A,\Pi_{CB}}(k) = 2 \times |\mathsf{Succ}_{A,\Pi_{CB}}(k) - 1/2|$ is negligible.

2.2   Remarks on CBE

THE RELIEF OF THE CA'S BURDEN.   When the CA in CBE issues an up-to-date certificate for the time period t, the CA guarantees both (1) the binding between a user and his public key and (2) the validity of the binding for the time period t. If certificate revocation information is to be refreshed day by day, the CA should update certificates everyday. While it is desirable that the revocation information is frequently refreshed, the authenticity of the public key need not be checked by the CA at each time period. Since the CA's private key is required to guarantee the condition (1), the CA's burden of certificate revocation management cannot be distributed in CBE.

To relieve the CA's burden of certificate revocation management, we deal with the two confirmations individually and introduce the notion of SCBE in which (1) a (long-lived) certificate by the CA guarantees the authenticity of a public key and (2) a (short-lived) status certificate by the SCA guarantees the validity of the certificate for a specific time period. If certificate revocation information is to be refreshed day by day, the CA issues a long-lived (say one year) certificate and the SCA issues status certificates everyday. Since the binding is guaranteed by the CA's long-lived certificate, the SCA only checks the validity of the binding; therefore the generation of a status certificate in SCBE is easier than that of a certificate in CBE.

KEY-INSULATED SECURITY.   As portable devices are spreading widely, the threat of exposure of private keys has increased considerably. Cryptographic computations are often performed on a relatively insecure device that cannot be trusted to maintain secrecy of the private key. To mitigate the damage of private key exposure, key-insulated security [7, 8] can be adopted. In this model, the decryption key stored on the insecure device is refreshed at discrete time periods via interaction with a physically-secure but computationally-limited device which stores a master private key. In a (T, N)-key-insulated scheme, an adversary who compromises the insecure device and obtains decryption keys for up to T time periods of his choice is unable to violate the security of the cryptosystem for any of the remaining $N - T$ periods.

In a CBE system, the decryption algorithm CB_Dec takes as input the private key $CBSK_{id}$ as well as the partial decryption key $CBPD_{(id,t)}$. When a user travels, he has to carry the private key $CBSK_{id}$ together with the partial decryption key $CBPD_{(id,t)}$ on his portable device. If the device is compromised, the private key $CBSK_{id}$ is revealed. Since the adversary can eavesdrop on the user id's certificate sent by the CA, the user id's security for the entire time period collapses in case of the exposure of the private key $CBSK_{id}$.

To minimize the damage caused by the exposure of secret information in a user's device, we remove the private key from the input of decryption algorithm in the SCBE system. Instead, the user update algorithm computes a decryption key for a time period t from the user's private key and a status certificate. Hence, the user need not store his private key on a portable device. When a decryption key for a time period t is exposed, the security for other time periods can be maintained.

3   Status Certificate-Based Encryption

3.1   The Model of SCBE

An SCBE scheme is specified by an 8-tuple of poly-time algorithms.

Definition 3. *A status certificate-based encryption scheme is an 8-tuple of poly-time algorithms* (SCB_Gen^{CA}, SCB_Gen^{SCA}, SCB_Set_User_Key, SCB_Init^{CA}, SCB_Upd^{SCA}, SCB_Upd^{User}, SCB_Enc, SCB_Dec) *such that:*

- SCB_Gen^{CA}, *the CA's master key and parameter generation algorithm, is a probabilistic algorithm that takes as input a security parameter 1^k and (optionally) the total number of time periods N. It returns a master key $SCBSK^{CA}$ and a parameter list $params^{CA}$.*
- SCB_Gen^{SCA}, *the SCA's status certification key generation algorithm, is a probabilistic algorithm that takes as input a parameter list $params^{CA}$. It returns a status certification key $SCBSK^{SCA}$ and an SCA parameter $params^{SCA}$.*
- SCB_Set_User_Key, *the user key generation algorithm, is a probabilistic algorithm that takes as input a parameter list $params^{CA}$, a user identity id, and (optionally) the total number of time periods N. It returns the user id's private key $SCBSK_{id}$ and public key $SCBPK_{id}$.*
- SCB_Init^{CA}, *the certificate issuance algorithm, is a deterministic algorithm that takes as input a parameter list $params^{CA}$, a user identity id, the user id's public key $SCBPK_{id}$, and the CA's master key $SCBSK^{CA}$. It returns the user id's certificate $Cert_{id}$.*
- SCB_Upd^{SCA}, *the SCA update algorithm, is a deterministic algorithm that takes as input a parameter list $params^{CA}$, the user id's certificate $Cert_{id}$, a time period t, and the status certification key $SCBSK^{SCA}$. It returns the user id's status certificate $StatusCert_{(id,t)}$ for the time period t.*

- SCB_Upd^{User}, *the user update algorithm, is a deterministic algorithm that takes as input a parameter list params^{CA}, a time period t, a status certificate StatusCert$_{(id,t)}$, the user id's public key/private key pair, and (optionally) SCBDK$_{(id,t-1)}$. It returns a decryption key SCBDK$_{(id,t)}$ for the time period t.*
- SCB_Enc,[1] *the encryption algorithm, is a probabilistic algorithm that takes as input a message M, a user identity id, a time period t, a parameter list params^{CA}, an SCA parameter params^{SCA}, and the user id's public key SCBPK$_{id}$. SCB_Enc$_{params^{CA},params^{SCA}}^{SCBPK_{id}}(M,id,t)$ returns a ciphertext C.*
- SCB_Dec, *the decryption algorithm, is a deterministic algorithm that takes as input a parameter list params^{CA}, a ciphertext C, and the decryption key SCBDK$_{(id,t)}$. SCB_Dec$_{params^{CA}}(C,SCBDK_{(id,t)})$ returns a message M or the special symbol $\perp$.*

When $SCBDK_{(id,t)}$ is derived from the private key and $StatusCert_{(id,t)}$ correctly, SCB_Dec$_{params^{CA}}$(SCB_Enc$_{params^{CA},params^{SCA}}^{SCBPK_{id}}(M,id,t),SCBDK_{(id,t)})=$ M holds for all message M.

In an SCBE scheme, SCB_Gen^{CA} and SCB_Init^{CA} are performed by the CA while SCB_Gen^{SCA} and SCB_Upd^{SCA} are performed by the SCA. Since SCB_Gen^{SCA} takes as input a parameter list $params^{CA}$, it is assumed that the CA generates the system parameter and multiple SCA's can be employed under a specific CA. After each user generates his own public key/private key pair using the SCB_Set_User_Key algorithm, the CA issues a (long-lived) certificate $Cert_{id}$ to guarantee the binding between a user and his public key. At the beginning of every time period, a (short-lived) status certificate $StatusCert_{(id,t)}$ is given to a user id by the SCA. Since SCB_Upd^{SCA} takes as input the user id's certificate $Cert_{id}$, the authenticity of the public key is guaranteed by the certificate and the SCA has only to check the revocation status of the public key in order to issue a status certificate. After receiving $StatusCert_{(id,t)}$ from the SCA, the user computes a decryption key $SCBDK_{(id,t)}$ for the time period t from his private key and the status certificate. When Alice wants to send an encrypted message to Bob, Alice obtains Bob's public key from a certificate without checking the revocation status. To decrypt Alice's message, Bob needs his decryption key $SCBDK_{(id,t)}$ for the time period t. Note that the private key $SCBSK_{id}$ is not required by the decryption algorithm SCB_Dec. If an SCBE system is constructed carefully, exposure of a decryption key $SCBDK_{(id,t)}$ does not compromise security for other time periods.

For security analysis, we give the adversary access to three types of oracles. The first is a status certification oracle SCB_Upd($\cdot,\cdot,\cdot,\cdot$) that returns $StatusCert_{(id,t)}$ on input a user identity id, a time period t, and the user id's public key/private key pair. The second is a decryption oracle SCB_Dec($\cdot,\cdot,\cdot,\cdot,\cdot$) that returns SCB_Dec$_{params^{CA}}(C,SCBDK_{(id,t)})$ on input $(id, t, C, SCBPK_{id},$

---

[1] The public key $SCBPK_{id}$ can be obtained from $Cert_{id}$ as an ordinary PKI. In addition, $Cert_{id}$ can include an SCA's parameter, especially when multiple SCA's exist.

$SCBSK_{id}$). The third is a left-or-right encryption oracle SCB_Enc($\cdot, \cdot, \cdot, LR(\cdot, \cdot, b)$) that given a user id, a time period t, the user id's public key, and two equal length messages M_0, M_1 returns a challenge ciphertext C_{ch} that is the encryption of M_b.

The security of an SCBE scheme is directed against three different types of adversaries: an eavesdropping CA, an uncertified user and a dishonest SCA. In traditional PKI, the CA is always assumed not to issue new certificates binding arbitrary public keys and a user, and especially not for those where the CA knows the corresponding private key [1]. If the CA forges certificates, then the CA can be identified as having misbehaved through the existence of two valid certificates for the same user. The CA who replaces a user's public key will be implicated in the event of a dispute. Hence, we can consider an uncertified user and an eavesdropping CA in a unified way. The Type I adversary who models an eavesdropping CA (and an uncertified user) has no access to the status certification key, but may make status certification queries and decryption queries. The Type II adversary A_{II} who models a dishonest SCA is equipped with the status certification key. For the Type II adversary A_{II}, the user id's public key and private key are replaced by $(params^{SCA}, SCBSK^{SCA})$ in the input of oracle queries.

Definition 4. *Let Π_{SCB} be a status certificate-based encryption scheme. For any adversary A, we may define the following:*

$$\mathsf{Succ}_{A,\Pi_{SCB}}(k)$$
$$= Pr[b' = b : (SCBSK^{CA}, params^{CA}) \leftarrow \mathsf{SCB\_Gen}^{CA}(1^k);$$
$$(SCBSK^{SCA}, params^{SCA}) \leftarrow \mathsf{SCB\_Gen}^{SCA}(params^{CA}); b \leftarrow \{0,1\};$$
$$b' \leftarrow A^{O(\cdot),\mathsf{SCB\_Dec}(\cdot,\cdot,\cdot,\cdot),\mathsf{SCB\_Enc}(\cdot,\cdot,\cdot,LR(\cdot,\cdot,b))}(params^{CA}, params^{SCA}, h)]$$

where $O(\cdot) = \mathsf{SCB\_Upd}(\cdot, \cdot, \cdot, \cdot)$, $h = SCBSK^{CA}$ *for* A_I, *and* $O(\cdot) = \perp$, $h = SCBSK^{SCA}$ *for* A_{II}. *The adversary may query oracles adaptively, except that it can make exactly one query to the left-or-right encryption oracle. A must follow the adversarial constraints given in [10]. Π_{SCB} is said to be secure against chosen-ciphertext attacks if the advantage* $\mathsf{Adv}_{A,\Pi_{SCB}}(k) = 2 \times |\mathsf{Succ}_{A,\Pi_{SCB}}(k) - 1/2|$ *is negligible for any PPT adversary A.*

To define key-insulated security, we give the adversary additional access to a key exposure oracle SCB_Exp($\cdot, \cdot$) that returns a decryption key $SCBDK_{(id,t)}$ on input a user identity id and a time period t.

Definition 5. *Let Π_{SCB} be a secure status certificate-based encryption scheme and A be an adversary who submits at most T requests to the key exposure oracle and never queries to the key exposure oracle the same (id, t) as the input of left-or-right encryption oracle. Π_{SCB} is said to be (T, N)-key-insulated, if the advantage* $\mathsf{Adv}_{A,\Pi_{SCB}}(k)$ *in Definition 4 is negligible for any PPT adversary A with additional access to* SCB_Exp($\cdot, \cdot$).

3.2   An SCBE Scheme Based on Bilinear Mappings

We provide a concrete example of a secure SCBE scheme $\Psi_{SCB} = ($SCB_Gen^{CA}, SCB_Gen^{SCA}, SCB_Set_User_Key, SCB_Init^{CA}, SCB_Upd^{SCA}, SCB_Upd^{User}, SCB_Enc, SCB_Dec) based on bilinear mappings that are implemented using Weil and Tate pairings on elliptic curves in practice. While the encryption algorithm of Full-CBE in [10] requires two executions of pairing computation, SCB_Enc of Ψ_{SCB} needs only one execution of pairing computation.

SCB_Gen$^{CA}(1^k)$

- Run a BDH parameter generator $\mathcal{G}$ [3] on input a security parameter 1^k, to generate groups G_1, G_2 of some prime order q and an admissible bilinear mapping $\hat{e} : G_1 \times G_1 \rightarrow G_2$.
- Choose a generator $P \in G_1$ and four cryptographic hash functions $H_1 : \{0,1\}^* \rightarrow G_1$, $H_2 : G_2 \rightarrow \{0,1\}^n$, $H_3 : \{0,1\}^n \times \{0,1\}^n \rightarrow \mathbf{Z}_q^*$, $H_4 : \{0,1\}^n \rightarrow \{0,1\}^n$ for some n.
- Choose a master key $s_{CA} \in \mathbf{Z}_q^*$ and sets $Q_{CA} = s_{CA}P$.
- Return the CA's master key $SCBSK^{CA} = s_{CA}$ and the parameter list $params^{CA} = (G_1, G_2, \hat{e}, P, Q_{CA}, H_1, H_2, H_3, H_4, q, n)$.

SCB_Gen$^{SCA}(params^{CA})$

- Choose a status certification key $s_{SCA} \in \mathbf{Z}_q^*$ and sets $Q_{SCA} = s_{SCA}P$.
- Return the SCA's status certification key $SCBSK^{SCA} = s_{SCA}$ and the SCA parameter $params^{SCA} = Q_{SCA}$.

SCB_Set_User_Key$(params^{CA}, id)$

- Select a private key $s_{id} \in \mathbf{Z}_q^*$ and sets $Q_{id} = s_{id}P$.
- Return the user id's private key $SCBSK_{id} = s_{id}$ and public key $SCBPK_{id} = Q_{id}$.

SCB_Init$^{CA}(params^{CA}, id, Q_{id}, s_{CA})$

- Verify the user id's information. If the information is invalid, abort.
- Compute $Cert_{id} = s_{CA}P_{id}$, where $P_{id} = H_1(id, Q_{id}, params^{CA}, id\text{-}info)$ and $id\text{-}info$ contains additional information about the user id.
- Return the user id's certificate[2] $Cert_{id} = s_{CA}P_{id}$.

SCB_Upd$^{SCA}(params^{CA}, Cert_{id}, t, s_{SCA})$

- Check the revocation status of the user id's certificate $Cert_{id}$ for the time period t. If $Cert_{id}$ is already revoked, abort.
- Compute $StatusCert_{(id,t)} = s_{SCA}P_{(id,t)}$, where $P_{(id,t)} = H_1(id, t, Q_{id}, params^{CA}, id\text{-}info)$.
- Return the user id's status certificate $StatusCert_{(id,t)} = s_{SCA}P_{(id,t)}$.

---

[2] For simplicity, we misuse the terminology since $Cert_{id}$ is a signature value and the public key cannot be extracted from $Cert_{id}$. In practice, $Cert_{id}$ should be a complete certificate, such as X.509 certificate [11].

SCB_Upd^{User}($params^{CA}, t, s_{SCA}P_{(id,t)}, Q_{id}, s_{id}$)

- Compute $SCBDK_{(id,t)} = s_{id}P_{(id,t)} + s_{SCA}P_{(id,t)}$.
- Return the decryption key $SCBDK_{(id,t)}$.

SCB_Enc($M, id, t, params^{CA}, Q_{SCA}, Q_{id}$)

- Choose a random $\sigma \in \{0,1\}^n$ and set $r = H_3(\sigma, M)$.
- Compute the ciphertext
  $C = [U, V, W] = [rP, \sigma \oplus H_2(g^r), M \oplus H_4(\sigma)]$, where $g = \hat{e}(Q_{SCA}+Q_{id}, P_{(id,t)})$.
- Return the ciphertext C.

SCB_Dec($params^{CA}, C, SCBDK_{(id,t)}$)

- Compute $\sigma = V \oplus H_2(\hat{e}(U, SCBDK_{(id,t)}))$.
- Compute $M = W \oplus H_4(\sigma)$ and set $r = H_3(\sigma, M)$.
- Test whether $U = rP$ or not.
- If the test succeeds, return M and otherwise, return $\perp$.

The certificate $Cert_{id}$ issued by the CA is a BLS signature [4]. If $(P, Q_{CA}, P_{id}, Cert_{id})$ is a valid co-Diffie-Hellman tuple, the CA's signature is accepted. From $Cert_{id}$, the sender obtains Q_{id} that is required by SCB_Enc.

If $C = [U, V, W]$ is the encryption of M sent to a user id for the time period t, the decryption is the inverse of the encryption.

$$V \oplus H_2(\hat{e}(U, SCBDK_{(id,t)}))$$
$$= V \oplus H_2(\hat{e}(rP, s_{id}P_{(id,t)} + s_{SCA}P_{(id,t)}))$$
$$= V \oplus H_2(\hat{e}((s_{id} + s_{SCA})P, P_{(id,t)})^r)$$
$$= V \oplus H_2(\hat{e}(Q_{id} + Q_{SCA}, P_{(id,t)})^r)$$
$$= \sigma \oplus H_2(g^r) \oplus H_2(\hat{e}(Q_{SCA} + Q_{id}, P_{(id,t)})^r)$$
$$= \sigma$$

Fujisaki-Okamoto transformation [9] was applied to the construction of Ψ_{SCB} and the security of Ψ_{SCB} can be proved in the random oracle model [5] by reduction to the security of FullIdent in [3].

Theorem 1. Ψ_{SCB} *is a secure SCBE scheme in the random oracle model under the Bilinear Diffie-Hellman assumption.*

Proof. (Sketch) We assume that readers are familiar with FullIdent and the security model of identity-based encryption. Otherwise, refer to Appendix A or [3].

Let A_I be a Type I attacker who can break Ψ_{SCB}. Suppose that A_I has advantage ϵ and runs in time t. We show how to construct from A_I an adversary A' against FullIdent = (ID_Gen, ID_Ext, ID_Enc, ID_Dec). At the beginning, the FullIdent adversary A' is given by the FullIdent challenger a parameter list $params = (G_1, G_2, \hat{e}, P, Q_{PKG}, H_1, H_2, H_3, H_4, q, n)$ and three oracles: key exposure oracle $\mathsf{ID\_Exp}_{params}^{IDSK^*}(\cdot)$, decryption oracle $\mathsf{ID\_Dec}_{params}^{IDSK^*}(\cdot, \cdot)$, and left-or-right encryption oracle $\mathsf{ID\_Enc}_{params}(\cdot, LR(\cdot, \cdot, b))$, where $IDSK^* = s$. To run A_I, A' simulates the $\mathsf{SCB\_Gen}^{\mathsf{CA}}(1^k)$ by supplying A_I with $params^{CA} = (G_1, G_2, \hat{e}, P, yP, H_1, H_2, H_3, H_4, q, n)$ where y is randomly chosen by A'. Then, all users register their public keys to the CA and A' can simulate $\mathsf{SCB\_Init}^{\mathsf{CA}}$ perfectly by using y. A' also gives $s_{CA} = y$ to A_I. Note that yP is independent of Q_{PKG}. Let q_c be the number of certificate issuance requests.

First, we consider the case for the adversary A_I attacking other users' security. A' chooses a random index $j \in \{1, \cdots, q_c\}$. Denote id_j as the user identity of j-th certificate issuance query. To simulate $\mathsf{SCB\_Gen}^{\mathsf{SCA}}(1^k)$, A' chooses a random $x \in \mathbf{Z}_q^*$ and sets $Q_{SCA} = xP$. Additionally, A' sets $Q_{id_j} = Q_{PKG} - Q_{SCA}$. Now, A' responds to A_I's oracle queries as follows.[3]

- Status certification oracle $\mathsf{SCB\_Upd}(\cdot, \cdot, \cdot, \cdot)$ queries: Suppose that A_I asks with an input (id, t, Q_{id}, s_{id}).
  1. If $id = id_j$, A' checks the validity of the query. If $Q_{id_j} = s_{id}P$, A' obtains $s = s_{id} + x$ and wins the game. Otherwise, A' aborts.
  2. If $id \neq id_j$, A' checks the validity of the query and computes the status certificate $StatusCert_{(id,t)} = xP_{(id,t)}$, where $P_{(id,t)} = H_1(id, t, Q_{id}, params^{CA}, id\text{-}info)$. A' sends $StatusCert_{(id,t)}$ to A_I.
- Decryption oracle $\mathsf{SCB\_Dec}(\cdot, \cdot, \cdot, \cdot, \cdot)$ queries: Suppose that A_I asks with an input $(id, t, C, Q_{id}, s_{id})$.
  1. If $id = id_j$, A' sends (id_α, C) to the $\mathsf{FullIdent}$ decryption oracle $\mathsf{ID\_Dec}_{params}^{IDSK^*}(\cdot, \cdot)$, where id_α is $(id, t, Q_{id}, params^{CA}, id\text{-}info)$. Since $Q_{id_j} + Q_{SCA} = Q_{PKG}$, $\mathsf{FullIdent}$ decryption oracle can decrypt this query correctly. A' sends the output of $\mathsf{ID\_Dec}_{params}^{IDSK^*}(\cdot, \cdot)$ to A_I.
  2. If $id \neq id_j$, A' checks the validity of the query and computes the decryption key $SCBDK_{(id,t)} = s_{id}P_{(id,t)} + xP_{(id,t)}$, where $P_{(id,t)} = H_1(id, t, Q_{id}, params^{CA}, id\text{-}info)$. A' returns $M = \mathsf{SCB\_Dec}(params^{CA}, C, SCBDK_{(id,t)})$.
- Left-or-right encryption oracle $\mathsf{SCB\_Enc}(\cdot, \cdot, \cdot, \cdot, LR(\cdot, \cdot, b))$ queries: Suppose that A_I asks with an input $(id, t, Q_{id}, M_0, M_1)$.
  1. If $id = id_j$, A' checks the validity of the query and sends (id_α, M_0, M_1) to the $\mathsf{FullIdent}$ left-or-right encryption oracle, where id_α is $(id, t, Q_{id}, params^{CA}, id\text{-}info)$. Let C_{ch} be the output of the $\mathsf{FullIdent}$ left-or-right encryption oracle. C_{ch} is also a valid output of the SCBE left-or-right encryption oracle from the relation of $Q_{PKG} = Q_{id_j} + Q_{SCA}$.
  2. If $id \neq id_j$, A' aborts.

When the Type I adversary A_I outputs b', A' outputs the same b' to the $\mathsf{FullIdent}$ challenger. If A' does not abort during the simulation, the A_I's view is identical to its view in the real attack. Since the index j is chosen randomly, the probability

---

[3] If the input is invalid, A' answers arbitrarily.

that A' does not abort during the simulation is $1/q_c$. Hence, the advantage of A' satisfies $\mathsf{Adv}_{A',\mathsf{FullIdent}}(k) = 2 \times |\Pr(b' = b) - 1/2| \geq \epsilon/q_c$ and A' runs in time $O(time(t))$.

Second, we consider the case for the adversary A_I trying to decrypt his own ciphertext without valid status certificate. Let Q_{A_I} be the public key of A_I. To simulate $\mathsf{SCB\_Gen}^{\mathsf{SCA}}(1^k)$, A' sets $Q_{SCA} = Q_{PKG} - Q_{A_I}$. Since A' does not know $\log_P Q_{SCA}$, A' maintains the list $H_1^{list} = \{(id, t, z)\}$ and simulates the random oracle H_1 by answering $P_{(id,t)} = zP$, which is a typical trick in the random oracle model. Then, A' can simulate three Ψ_{SCB} oracles. The other parts of the simulation are similar to the first case.

Let A_{II} be a Type II attacker who can break Ψ_{SCB}. Suppose that A_{II} has advantage ϵ and runs in time t. We construct from A_{II} an adversary A'' against FullIdent. At the beginning, A'' is given by a FullIdent challenger a parameter list $params = (G_1, G_2, \hat{e}, P, Q_{PKG}, H_1, H_2, H_3, H_4, q, n)$ and three oracles. To run A_{II}, A'' simulates the $\mathsf{SCB\_Gen}^{\mathsf{CA}}(1^k)$ by supplying A_{II} with $params^{CA} = (G_1, G_2, \hat{e}, P, yP, H_1, H_2, H_3, H_4, q, n)$ where y is randomly chosen by A''. Then, all users register their public keys to the CA and A'' can simulate $\mathsf{SCB\_Init}^{\mathsf{CA}}$ by using y. Let q_c be the number of certificate issuance requests. A'' chooses a random index $j \in \{1, \cdots, q_c\}$. Denote id_j as the user identity of j-th certificate issuance query. To simulate $\mathsf{SCB\_Gen}^{\mathsf{SCA}}(1^k)$, A'' chooses a random $x \in \mathbf{Z}_q^*$ and sets $Q_{SCA} = xP$. Additionally, A'' sets $Q_{id_j} = Q_{PKG} - Q_{SCA}$. Now, A'' gives x to A_{II} since A_{II} has access to the status certification key. This is practically equivalent to the case in which A_{II} chooses the status certification key. The remaining part of the proof is A''''s simulation of the decryption oracle and the left-or-right oracle. Here, one important point is that A_{II} does not have access to the users' private keys. If A_{II} has the users' private keys with the status certification key, no security can be guaranteed by definition. Therefore, A_{II} does not know users' private keys and A'' sets these values. Then, the simulation of the decryption oracle and the left-or-right encryption oracle is exactly the same as the first case of the Type I attacker A', since A'' has access to s_{id} for each id. After the simulation, the advantage of A'' satisfies $\mathsf{Adv}_{A'',\mathsf{FullIdent}}(k) \geq \epsilon/q_c$ and the running time of A'' is $O(time(t))$.

$$\text{Q.E.D. } \square$$

To study the damage of decryption key exposure, we consider a dishonest SCA who obtains a decryption key $SCBDK_{(id,t_a)} = s_{id}P_{(id,t_a)} + s_{SCA}P_{(id,t_a)}$. The dishonest SCA can get $s_{id}P_{(id,t_a)}$ from $SCBDK_{(id,t_a)}$, since he can compute the user id's status certificate $StatusCert_{(id,t_a)} = s_{SCA}P_{(id,t_a)}$. However, he cannot compute $s_{id}P_{(id,t_b)}$ for $a \neq b$, since $s_{id}P_{(id,t_a)}$ is a BLS signature and the BLS signature is unforgeable under an adaptive chosen message attack [4]. To prove the key-insulated security of Ψ_{SCB}, we convert an adversary attacking Ψ_{SCB} into the attacker against FullIdent. We leave the details to readers.

Theorem 2. Ψ_{SCB} is $(N-1, N)$-key-insulated.

3.3   Applications

COMPATIBILITY WITH LEGACY CA SYSTEMS. The CA's signature of Ψ_{SCB} is a BLS signature. However, any signature scheme including RSA [18] and DSA [16] can be used by the CA. As $Cert_{id}$ is used only for guaranteeing the authenticity of a public key, the change of the CA's signing algorithm does not affect other parts of Ψ_{SCB}.[4] Therefore, the legacy CA systems can be used in Ψ_{SCB} with little modification.

NEW BUSINESS MODEL. The new part of Ψ_{SCB} is the SCA. The separation of a certificate and a status certificate introduces different business models. For example, the department of motor vehicles (DMV) issues a driver's license guaranteeing the authenticity of a citizen and the assurance company makes a profit from assuring the current validity of the driver's license. This role of the SCA is similar to that of the key compromise agent (or the suicide bureau) in the Rivest's system [17].

MULTIPLE SCA'S.   When multiple SCA's are implemented, Ψ_{SCB} can be extended in diverse ways. For example, the ciphertext generation can be modified as follows, when a sender wants the validity of the receiver's certificate to be assured by both SCA_i and SCA_j:

$$C = [rP, \, \sigma \oplus H_2(g^r), \, M \oplus H_4(\sigma)],$$
$$\text{where } g = \hat{e}(Q_{SCA_i} + Q_{SCA_j} + Q_{id}, \, P_{(id,t)}).$$

To decrypt this ciphertext, the receiver has to obtain status certificates from both SCA_i and SCA_j. The required number of the pairing computation does not increase and this extension holds for any number of SCA's. Multiple SCA's are very useful for inter-domain applications and are also used for creating subdomains. Note that multiple SCA's can be naturally applied to access control and a similar technique was used in [6, 20].

Acknowledgement

The authors would like to thank the anonymous ICISC 2004 referees for their helpful comments.

References

1. S. S. Al-Riyami and K. G. Paterson, "Certificateless public key cryptography," ASIACRYPT 2003, LNCS 2894, pp. 452-473, 2003.
2. M. Bellare, A. Desai, D. Jokipii, and P. Rogaway, "A concrete security treatment of symmetric encryption: analysis of the DES modes of operation," FOCS 1997, IEEE, 1997.

---

[4] In this case, $params^{CA}$ contains conceptually two different parameters; one for the CA's algorithm and the other for the SCA's algorithm.

3. D. Boneh and M. Franklin, "Identity based encryption from the Weil pairing," CRYPTO 2001, LNCS 2139, pp. 213-229, 2001.
4. D. Boneh, B. Lynn, and H. Shacham, "Short signatures from the Weil pairing," ASIACRYPT 2001, LNCS 2248, pp. 514-532, 2001.
5. M. Bellare and P. Rogaway, "Random oracles are practical: a paradigm for designing efficient protocols," 1st ACM Conf. on Computer and Communications Security, pp. 62-73, 1993.
6. L. Chen, K. Harrison, D. Soldera, and N.P. Smart, "Applications of multiple trust authorities in pairing based cryptosystems," InfraSec 2002, LNCS 2437, pp. 260-275, 2002.
7. Y. Dodis, J. Katz, S. Xu, and M. Yung, "Key-insulated public key cryptosystems," EUROCRYPT 2002, LNCS 2332, pp. 65-82, 2002.
8. Y. Dodis, J. Katz, S. Xu, and M. Yung, "Strong key-insulated signature schemes," PKC 2003, LNCS 2567, pp. 130-144, 2003.
9. E. Fujisaki and T. Okamoto, "Secure integration of asymmetric and symmetric encryption schemes," CRYPTO 1999, LNCS 1666, pp. 537-554, 1999.
10. C. Gentry, "Certificate-based encryption and the certificate revocation problem," EUROCRYPT 2003, LNCS 2656, pp. 272-293, 2003.
11. R. Housley, W. Polk, W. Ford, and D. Solo, "Internet X.509 public key infrastructure certificate and certificate revocation list (CRL) profile," RFC 3280, IETF, 2002.
12. M. Myers, R. Ankney, A. Malpani, S. Galperin, and C. Adams, "X.509 Internet public key infrastructure online certificate status protocol - OCSP," RFC 2560, IETF, 1999.
13. P. McDaniel and S. Jamin, "Windowed certificate revocation," IEEE Infocom 2000, pp. 1406-1414, 2000.
14. S. Micali, "NOVOMODO: scalable certificate validation and simplified PKI management," 1st Annual PKI Research Workshop Proceedings, pp. 15-26, 2002.
15. M. Naor and K. Nissim, "Certificate revocation and certificate update," 7th USENIX Security Symposium, pp. 217-228, 1998.
16. NIST, "Digital siganture standard," FIPS PUB 186-2, 2000.
17. R. Rivest, "Can we eliminate certificate revocation lists?," Financial Cryptography 1998, LNCS 1465, pp. 178-183, 1998.
18. R. Rivest, A. Shamir, and L. Adleman, "A method for obtaining digital signature and public key cryptosystem," Comm. of the ACM, 21(2), pp. 120-126, 1978.
19. A. Shamir, "Identity-based cryptosystems and signature Schemes," CRYPTO 1984, LNCS 196, pp. 47-53, 1984.
20. N.P. Smart, "Access Control Using Pairing Based Cryptography," CT-RSA 2003, LNCS 2612, pp. 111-121, 2003.

Appendix A. Identity-Based Encryption

After reviewing the definition and security model of identity-based encryption, we present FullIdent of [3].

Definition 6. *An identity-based encryption scheme is a 4-tuple of poly-time algorithms* (ID_Gen, ID_Ext, ID_Enc, ID_Dec) *such that:*

- ID_Gen, *the master key and parameter generation algorithm, is a probabilistic algorithm that takes as input a security parameter* 1^k. *It returns a master key* $IDSK^*$ *and a parameter list params.*
- ID_Ext, *the decryption key issuance algorithm, is a deterministic algorithm that takes as input a user identity id, a parameter list params, and a master key* $IDSK^*$. *It returns the user id's decryption key* $IDSK_{id}$.
- ID_Enc, *the encryption algorithm, is a probabilistic algorithm that takes as input a message M, a user identity id, and a parameter list params.* ID_Enc$_{params}(M, id)$ *returns a ciphertext C.*
- ID_Dec, *the decryption algorithm, is a deterministic algorithm that takes as input a parameter list params, the decryption key* $IDSK_{id}$, *and a ciphertext C.* ID_Dec$_{params}^{IDSK_{id}}(C)$ *returns a message M or the special symbol* $\perp$.

We require that for all message M, ID_Dec$_{params}^{IDSK_{id}}$(ID_Enc$_{params}(M, id)$) $= M$.

In an identity-based encryption scheme, ID_Gen and ID_Ext are performed by a private key generator (PKG). A decryption key $IDSK_{id}$ is given to a user id by the PKG through a secure channel. Note that the key escrow of the user's private key is inherent in an identity-based encryption scheme.

For security analysis, we define a key exposure oracle ID_Exp$_{params}^{IDSK^*}(\cdot)$ that returns a decryption key $IDSK_{id}$ on input id. We also give the adversary access to a decryption oracle ID_Dec$_{params}^{IDSK^*}(\cdot, \cdot)$ that returns ID_Dec$_{params}^{IDSK_{id}}(C)$ on input (id, C). Finally, the adversary can access a left-or-right encryption oracle ID_Enc$_{params}(\cdot, LR(\cdot, \cdot, b))$ that given a user identity id_{ch} and equal length messages M_0, M_1 returns a challenge ciphertext $C_{ch} =$ ID_Enc$_{params}(M_b, id_{ch})$.

The security goal of an identity-based encryption scheme is chosen ciphertext security. This means that any PPT adversary A should have a negligible advantage of distinguishing the encryptions of two messages of his choice given access to the key exposure oracle ID_Exp$_{params}^{IDSK^*}(\cdot)$, the decryption oracle ID_Dec$_{params}^{IDSK^*}(\cdot, \cdot)$, and the left-or-right encryption oracle ID_Enc$_{params}(\cdot, LR(\cdot, \cdot, b))$. The key exposure oracle models the ability of the adversary to compromise any user of his choice, except the target user.

Definition 7. *Let* Π_{ID} *be an identity-based encryption scheme. For any adversary A, we may define the following:*

Succ$_{A,\Pi_{ID}}(k) =$
$$Pr[b' = b : (IDSK^*, params) \leftarrow \text{ID\_Gen}(1^k); b \leftarrow \{0, 1\};$$
$$b' \leftarrow A^{\text{ID\_Exp}_{params}^{IDSK^*}(\cdot), \text{ID\_Dec}_{params}^{IDSK^*}(\cdot, \cdot), \text{ID\_Enc}_{params}(\cdot, LR(\cdot, \cdot, b))}(params)]$$

where the adversary may query oracles adaptively subject to the restriction that it can make exactly one query to the left-or-right encryption oracle. Let id_{ch} *be the user identity of the query to the left-or-right encryption oracle and* C_{ch} *be the challenge ciphertext returned by the left-or-right encryption oracle. We say that A succeeds if* $b' = b$, id_{ch} *was never submitted to the key exposure oracle, and* (id_{ch}, C_{ch}) *was never submitted to the decryption oracle after* C_{ch} *was returned by the left-or-right encryption oracle.* Π_{ID} *is said to be secure against chosen-ciphertext attacks if for any PPT A, the advantage* $\mathsf{Adv}_{A,\Pi_{ID}}(k) = 2 \times |\mathsf{Succ}_{A,\Pi_{ID}}(k) - 1/2|$ *is negligible.*

FullIdent is an identity-based encryption scheme secure against chosen-ciphertext attacks in the random oracle model under the Bilinear Diffie-Hellman assumption. The description of FullIdent is as follows:

ID_Gen(1^k)

- Run a BDH parameter generator $\mathcal{G}$ on input a security parameter 1^k, to generate groups G_1, G_2 of some prime order q and an admissible bilinear mapping $\hat{e} : G_1 \times G_1 \to G_2$.
- Choose a generator $P \in G_1$ and four cryptographic hash functions $H_1 : \{0,1\}^* \to G_1$, $H_2 : G_2 \to \{0,1\}^n$, $H_3 : \{0,1\}^n \times \{0,1\}^n \to \mathbf{Z}_q^*$, $H_4 : \{0,1\}^n \to \{0,1\}^n$ for some n.
- Choose a master key $s \in \mathbf{Z}_q^*$ and sets $Q_{PKG} = sP$.
- Return the CA's master key $IDSK^* = s$ and the parameter list $params = (G_1, G_2, \hat{e}, P, Q_{CA}, H_1, H_2, H_3, H_4, q, n)$.

ID_Ext($id, params, IDSK^*$)

- Compute $P_{id} = H_1(id)$ and $IDSK_{id} = IDSK^* P_{id} = sP_{id}$.
- Return the user id's decryption key $IDSK_{id}$.

ID_Enc($M, id, params$)

- Choose a random $\sigma \in \{0,1\}^n$ and set $r = H_3(\sigma, M)$.
- Compute the ciphertext
  $C = [U, V, W] = [rP, \sigma \oplus H_2(g^r), M \oplus H_4(\sigma)]$, where $g = \hat{e}(Q_{PKG}, P_{id})$.
- Return the ciphertext C.

ID_Dec($params, IDSK_{id}, C$)

- Compute $\sigma = V \oplus H_2(\hat{e}(U, IDSK_{id}))$.
- Compute $M = W \oplus H_4(\sigma)$ and set $r = H_3(\sigma, M)$.
- Test whether $U = rP$ or not.
- If the test succeeds, return M and otherwise, return $\perp$.

Fractional Windows Revisited:
Improved Signed-Digit Representations for Efficient Exponentiation

Bodo Möller[*]

University of California, Berkeley
bmoeller@acm.org

Abstract. This paper extends results concerning efficient exponentiation in groups where inversion is easy (e.g. in elliptic curve cryptography). It examines the right-to-left and left-to-right signed fractional window (RL-SFW and LR-SFW) techniques and shows that both RL-SFW and LR-SFW representations have minimal weight among all signed-digit representations with digit set $\{\pm 1, \pm 3, \ldots, \pm m, 0\}$. (Fractional windows generalize earlier sliding-window techniques, providing more flexibility for exponentiation algorithms in order to make best use of the memory that is available for storing intermediate results.) Then it considers the length of representations: LR-SFW representations are an improvement over RL-SFW representations in that they tend to be shorter; further length improvements are possible by post-processing the representations.

Keywords: Efficient implementations, elliptic curve cryptography.

1   Introduction

Many public-key cryptosystems involve *exponentiation* in some finite group, and often (e.g. for elliptic curve cryptography) group elements are represented such that the inversion operation is almost immediate. It is well known that *signed-digit representations* of integers e are useful to perform exponentiations g^e in such groups. A particular signed-digit representation is the right-to-left signed fractional window (RL-SFW) representation introduced in [14]. Fractional windows provide more flexibility for exponentiation algorithms than the previously known sliding-window representations; the purpose of this flexibility is to make best use of the memory that is available for storing intermediate results. The present paper also considers the left-to-right signed fractional window (LR-SFW) representation (cf. [19] and [9]); it complements [14] by proving minimality of weight for both RL-SFW and LR-SFW representations. A general motive for preferring the left-to-right variant is that it generates the digits in the order in which they are usually needed for exponentiation. We also examine the length of

---

[*] Supported by a DAAD (German Academic Exchange Service) Postdoc fellowship.

C. Park and S. Chee (Eds.): ICISC 2004, LNCS 3506, pp. 137–153, 2005.

representations and see that the LR-SFW method actually provides an improvement in this respect, and that further improvements are possible. Finally, we observe that no finite-state machine can always generate a minimal-length representation among the representations of minimal weight employing a prescribed set of signed digits.

To motivate and explain the goals of this paper, let us first look at typical exponentiation techniques in some detail. Given integers $b_\ell, \ldots, b_0$ as *digits*, we write $(b_\ell \ldots b_0)_2$ for $\sum_{0 \le i \le \ell} b_i \cdot 2^i$. For $m \ge 1$ odd, let

$$B_m = \{\pm 1, \pm 3, \ldots, \pm m\}$$

be the set of odd integers with absolute values up to m. We call $b_\ell \ldots b_0$ a B_m-*representation* of e if $b_\ell, \ldots, b_0 \in B_m \cup \{0\}$ and $e = (b_\ell \ldots b_0)_2$. For example, for any m, $100\bar{1}$ is a B_m-representation of 7, where we use the convention that $\bar{b}$ denotes digit value $-b$. If we assume that $e \ne 0$ and that ℓ is chosen minimal (so that $b_\ell \ne 0$), the power g^e can be computed by the two-stage algorithm shown in Fig. 1. This algorithm processes the digits b_i from the most significant down to

---

{ **LR-exponentiation:** compute g^e where $e = (b_\ell \ldots b_0)_2$, $b_\ell \ne 0$ }

{ **Precomputation stage** }

$A \leftarrow g^2$
$G_1 \leftarrow g$
for $b = 3$ **to** m **step** 2 **do**
$\quad G_b \leftarrow G_{b-2} \cdot A \quad \{ = g^b \}$

{ **Evaluation stage** }

$A \leftarrow \left(b_\ell > 0 \,?\, G_{b_\ell} : G_{|b_\ell|}^{-1} \right)$
for $i = \ell - 1$ **down to** 0 **do**
$\quad A \leftarrow A^2$
$\quad$ **if** $b_i \ne 0$ **then**
$\quad\quad A \leftarrow A \cdot \left(b_i > 0 \,?\, G_{b_i} : G_{|b_i|}^{-1} \right)$
return A

Fig. 1. LR-exponentiation

the least significant one, i.e. *left to right* assuming big-endian notation; we speak of *LR-exponentiation*. Let $\mathcal{H}(b_\ell \ldots b_0)$ denote the *weight* (generalized Hamming weight) of the given B_m-representation, i.e. the number of non-zero digits. The LR-exponentiation algorithm in Fig. 1 performs the following numbers of group operations (where we distinguish between squarings and general multiplications since they will typically have different computational cost, but neglect inversions as these are assumed to be almost immediate):

- In the *precomputation stage*, one squaring and $\dfrac{m-1}{2}$ general multiplications;

- in the *evaluation stage*, ℓ squarings and $\mathcal{H}(b_\ell \ldots b_0) - 1$ general multiplications.[1]

Increasing parameter m makes additional digit values available, typically allowing for lower-weight representations at the cost of an increased precomputation stage effort. Parameter m also determines the amount of memory needed for storing the precomputed values $G_1, \ldots, G_m$, so implementations may have to take into account some upper limit on m.

For given m and e, there is a lower limit on ℓ (i.e. there is a lower limit on the *length* $\ell + 1$ of B_m-representations of e). There is no upper limit on ℓ (since e.g. $(1\bar{1}\bar{1} \ldots \bar{1})_2 = 1$), but low-weight B_m-representations never need be more than one digit longer than the binary representation: the bounds

$$\lfloor \log_2 |e| \rfloor - \lfloor \log_2 m \rfloor \leq \ell \leq \lceil \log_2 |e| \rceil$$

hold for the following well-known representations and for the newer representations that will be discussed afterwards.

- Let $w \geq 1$ be an integer parameter and $W = w + 1$. The *width-W nonadjacent form (W-NAF)* of e is a specific B_{2^w-1}-representation such that for $|e| \to \infty$, on average

$$\frac{\mathcal{H}(b_\ell \ldots b_0)}{\lceil \log_2 |e| \rceil} \approx \frac{1}{W + 1}$$

  assuming that e consists of random bits. As the W-NAF is a signed-digit equivalent of the well-known sliding-window technique for unsigned digits (cf. [4]), we also call it a *right-to-left signed window representation* with *window width W*.

  (The origin of the 2-NAF is "property M" in [18]; the generalization to arbitrary $W \geq 2$ was alluded to in [20] and described independently in [12], in [22] as an improvement of a technique from [21], and in [2].)

- Now consider an arbitrary odd $m \geq 1$ and let

$$w_m = \lfloor \log_2 m \rfloor + 1,$$
$$W_m = w_m + 1,$$
$$\Delta_m = \frac{2^{w_m} - 1 - m}{2^{w_m - 1}}$$

  (so that $0 \leq \Delta_m < 1$). Generalizing right-to-left signed window representations, there is a *right-to-left signed fractional window (RL-SFW) representation* of e, the *m-RL-SFW* representation (details follow in Section 2). This is a B_m-representation such that for $|e| \to \infty$, on average

$$\frac{\mathcal{H}(b_\ell \ldots b_0)}{\lceil \log_2 |e| \rceil} \approx \frac{1}{W_m - \Delta_m + 1}$$

---

[1] For $\ell \geq 1$, an immediate optimization to the algorithm as written is to skip the first evaluation stage assignment and squaring if $b_\ell = 1$ (just keep g^2 in A from the precomputation stage) or $b_\ell = -1$ (just invert A to obtain g^{-2}).

assuming that e consists of random bits. If m is of the form $2^w - 1$ (so that $\Delta_m = 0$), the m-RL-SFW representation is the same as the W_m-NAF; otherwise the *effective window width* $W_m - \Delta_m$ is a fraction between w_m and W_m:

m	1	3	5	7	9	11	13	15	17	19	$\cdots$
$W_m - \Delta_m$	2	3	$3\frac{1}{2}$	4	$4\frac{1}{4}$	$4\frac{2}{4}$	$4\frac{3}{4}$	5	$5\frac{1}{8}$	$5\frac{2}{8}$	$\cdots$

(The RL-SFW representation was introduced in [14].)

The (finite-state) algorithms to obtain these representations given the binary representation of e read the least significant bit first and output the least significant signed digit first, proceeding towards the most significant input bit and output digit. This means they work *right to left* assuming big-endian notation; thus we speak of *RL-transformations*.

The use of an RL-transformation with LR-exponentiation means that usually the B_m-representation would be computed and stored before the actual exponentiation begins. This is unfortunate if memory is scarce. (Alternatively, the RL-transformation could be used repeatedly to determine the signed digits in the order in which they are needed, but this would mean an increased computational cost.) It is possible to perform *RL-exponentiation* instead so that the signed digits are used in the order in which they are generated (using algorithms from [23] and [11–exercise 4.6.3-9] as summarized in [14]); however, there are drawbacks:

– The group operation count to perform an RL-exponentiation is slightly higher than for an LR-exponentiation, given the same B_m-representation.
– The technique of employing mixed coordinates [3] for elliptic curves requires LR-exponentiation. (This technique uses additional precomputation effort to convert the elements G_b into a representation that accelerates evaluation stage operations $A \cdot G_b$ or $A \cdot G_{|b|}^{-1}$.)
– The technique of interleaved exponentiation [13] for efficiently computing power products $\prod_{1 \leq j \leq k} g_j^{e_j}$ applies to LR-exponentiation only.

Hence, left-to-right analogues of the above low-weight representations are called for.

A left-to-right analogue of the 2-NAF was described in [8], and recently, general left-to-right analogues of the signed window representation have appeared in [16], [1], and [17]. The latter two publications use an inherently identical LR-transformation, but describe it differently; see also [7–Section 6]. Also recently, proofs have appeared that the right-to-left signed window representation and its left-to-right variants are optimal in the sense of always achieving minimal weight ([15], [16], [1]): that is, given any e and w, no B_{2^w-1} representation $b'_{\ell'} \ldots b'_0$ of e can have lower weight than the $(w+1)$-NAF or its left-to-right analogues.

We generalize and extend these results by examining the right-to-left signed fractional window (RL-SFW) technique from [14] as well as its left-to-right variant (LR-SFW) implied by the approach of [17] and [7–Section 6]. (For *unsigned* windows, right-to-left and left-to-right variants are equally simple: unsigned fractional windows, originally only presented for an RL-transformation in [14], have an immediate left-to-right analogue [5]. Signed-digit representations are trickier since they involve carries, but the approach of [17] and [7] makes it straightforward to come up with a left-to-right analogue of the RL-transformation from [14]; cf. [19] and [9].) We give minimality proofs for the weight of both RL-SFW and LR-SFW representations, and we examine the length of representations to study efficiency improvements beyond weight minimization. We always assume that e is positive: the case $e = 0$ is trivial; for negative e, apply the technique to $-e$ and replace all resulting digits by their negatives.

Section 2 looks at the RL-SFW representation and proves that it has minimal weight. Then Section 3 develops the LR-SFW representation and shows that it too has minimal weight. Section 4 points out that the left-to-right method is advantageous in that it tends to achieve slightly shorter lengths than the right-to-left method (and never yields a greater length), with details in Appendix A. It also considers how modified representations can further reduce the length in some cases; however, Appendix B observes that no finite-state transformation algorithm can always achieve minimal length among the representations of minimal weight.

2   Right-to-Left Signed Fractional Windows (m-RL-SFW)

The right-to-left signed fractional window (RL-SFW) representation was introduced (plainly as the "signed fractional window representation") in [14]. Here we describe the technique in a way that encompasses the non-fractional case as well (i.e. the right-to-left signed window representation, often called the W-NAF). Given any odd $m \geq 1$, let $w_m = \lfloor \log_2 m \rfloor + 1$ and $W_m = w_m + 1$; then we have $2^{w_m - 1} \leq m < 2^{w_m}$. The m-RL-SFW representation of any positive integer e is the B_m-representation $b_\ell \ldots b_0$ obtained as follows.

First we define a mapping $digit_m \colon \{0, 1, \ldots, 2^{W_m} - 1\} \to B_m \cup \{0\}$.

- If x is even, let $digit_m(x) = 0$;
- otherwise if $0 < x \leq m$, let $digit_m(x) = x$;
- otherwise if $m < x < 2^{W_m} - m$, let $digit_m(x) = x - 2^{w_m}$;
- otherwise (i.e. $2^{W_m} - m \leq x < 2^{W_m}$), let $digit_m(x) = x - 2^{W_m}$.

Observe that if x is odd, then $x - digit_m(x) \in \{0, 2^{w_m}, 2^{W_m}\}$. We extend the mapping to

$$digit_m \colon \mathbb{Z} \to B_m \cup \{0\}$$

by defining $digit_m(x) = digit_m(x \bmod 2^{W_m})$; it follows that $2^{w_m} \mid x - digit_m(x)$ for any odd $x \in \mathbb{Z}$. The RL-transformation algorithm in Fig. 2 on input any B_m-representation $b'_{\ell'} \ldots b'_0$ of a positive integer e (e.g. the binary representation) generates a B_m-representation $b_\ell \ldots b_0$ such that

$$b_i = digit_m \left(\frac{e - \sum_{0 \leq j < i} b_j \cdot 2^j}{2^i} \right).$$

{ **RL-transformation:** determine the
   right-to-left signed fractional window (m-RL-SFW) **representation**
   of $(b'_{\ell'} \ldots b'_0)_2$ }

$i \leftarrow 0$
$D \leftarrow (b'_{w_m} \ldots b'_0)_2$
while $D \neq 0 \ \vee \ i + w_m < \ell'$ **do**
$\quad d \leftarrow digit_m(D)$
$\quad b_i \leftarrow d$
$\quad i \leftarrow i + 1$
$\quad D \leftarrow b'_{i+w_m} \cdot 2^{w_m} + \dfrac{D-d}{2}$
$\ell \leftarrow i - 1$
return $(b_\ell, \ldots, b_0)$

Fig. 2. RL-transformation for fractional windows

The algorithm assumes that $b'_i = 0$ for $i > \ell'$. It is easy to verify that the algorithm in Fig. 2 will always terminate with $b_\ell \ldots b_0$ as given above, which implies $e = (b_\ell \ldots b_0)_2$. (Note that

$$(b'_{\ell'} \ldots b'_{i+W_m} \underbrace{0 \ldots 0}_{w_m \text{ zeros}} d\, b_{i-1} \ldots b_0)_2 = e$$

holds as a loop invariant.)

To see that the average weight for $|e| \to \infty$ with e composed of random bits satisfies

$$\frac{\mathcal{H}(b_\ell \ldots b_0)}{\lceil \log_2 |e| \rceil} \approx \frac{1}{W_m - \dfrac{2^{w_m}-1-m}{2^{w_m-1}} + 1} = \frac{1}{W_m + \dfrac{1+m}{2^{w_m-1}} - 1}$$

as claimed in Section 1, assume that the RL-transformation algorithm has to process an endless sequence of independently uniform random bits b'_i. Whenever the loop generates a non-zero digit b_i, the current value of D has its least significant bit set and is an otherwise random W_m-bit integer. Thus from the definition of $digit_m$ it is clear that with probability $p = \dfrac{1+m}{2^{w_m}}$ we have $2^{W_m} \mid D - digit_m(D)$, and with probability $1-p$ we have $D - digit_m(D) = 2^{w_m}$. In the latter case, the next non-zero output digit will follow after exactly $W_m - 2$ intermediate zeros; in the former case, the next non-zero output digit will follow after W_m intermediate zeros on average. This means that the total average a for the number of intermediate zeros is

$$a = pW_m + (1-p)(W_m - 2) = W_m + 2p - 2 = W_m + \frac{1+m}{2^{w_m-1}} - 2,$$

and thus the density $\dfrac{1}{a+1}$ of non-zero digits is as claimed above.

2.1   Minimality of Weight

To prove that the m-RL-SFW representation has minimal weight among all B_m-representations of any integer e, we show that

$$\mathcal{H}(b_\ell \ldots b_0) \leq \mathcal{H}(b'_{\ell'} \ldots b'_0)$$

always holds if the transformation algorithm is applied to any B_m-representation $b'_{\ell'} \ldots b'_0$ to obtain the corresponding RL-SFW representation $b_\ell \ldots b_0$. For the analysis, we look at a variant of the algorithm from Fig. 2, shown in Fig. 3. This variant is easily seen to generate results that are identical except for leading zeros. The algorithm as written assumes that all variables b_i are initially zero.

{ **RL-transformation** (variant): determine the
right-to-left signed fractional window (m-RL-SFW) **representation**
of $(b'_{\ell'} \ldots b'_0)_2$ }

$l \leftarrow \ell'$
$(b_l, \ldots, b_0) \leftarrow (b'_{\ell'}, \ldots, b'_0)$
$i \leftarrow 0$
while $i \leq l$ **do**
$\quad \{\, b_\ell, \ldots, b_{i+1}, b_{i-1}, \ldots, b_0 \in B_m \cup \{0\} \ \wedge \ |b_i| \leq 2m \,\}$
$\quad$ **if** b_i is even **then**
$\quad\quad b_{i+1} \leftarrow b_{i+1} + b_i/2$
$\quad\quad b_i \leftarrow 0$
$\quad\quad \{\, b_\ell, \ldots, b_{i+2}, b_i, \ldots, b_0 \in B_m \cup \{0\} \ \wedge \ |b_{i+1}| \leq 2m \,\}$
$\quad\quad i \leftarrow i + 1$
$\quad$ **else**
$\quad\quad D \leftarrow (b_{i+w_m} \ldots b_i)_2$
$\quad\quad d \leftarrow digit_m(D)$
$\quad\quad (b_{i+w_m}, \ldots, b_i) \leftarrow \left(\dfrac{D-d}{2^{w_m}}, \ \underbrace{0, \ldots, 0}_{w_m - 1 \text{ zeros}}, \ d \right)$
$\quad\quad \{\, b_\ell, \ldots, b_{i+w_m+1}, b_{i+w_m-1}, \ldots, b_0 \in B_m \cup \{0\} \ \wedge \ |b_{i+w_m}| \leq 2m \,\}$
$\quad\quad i \leftarrow i + w_m$
$\quad$ **if** $i > l \ \wedge \ b_i \neq 0$ **then**
$\quad\quad l \leftarrow i$
$\ell \leftarrow i - 1$
return $(b_\ell, \ldots, b_0)$

Fig. 3. RL-transformation (variant) for fractional windows

While the input and output consist only of digits from $B_m \cup \{0\}$, the variable b_i at the beginning of the loop body may contain other values; we call this digit the current *carry digit*. We can verify as a loop invariant that at the beginning of the loop body digits b_h other than the carry digit ($h \neq i$) will always satisfy $b_h \in B_m \cup \{0\}$ (thus $|b_h| \leq m$) while the carry digit will always satisfy $|b_i| \leq 2m$. This clearly holds for $i = 0$. If for any i it holds at the beginning of the loop body, then it will also hold at the end of the loop body:

- If at the beginning of the loop body b_i is even, then it follows from $|b_{i+1}| \leq m$ and $|b_i| \leq 2m$ that

$$|b_{i+1} + b_i/2| \leq 2m.$$

- If at the beginning of the loop body b_i is odd, then

$$|D| \leq m \cdot 2^{w_m} + \ldots + m \cdot 2^1 + 2m = m \cdot 2^{w_m+1}$$

and thus

$$|D - d| \leq m \cdot 2^{w_m+1} + m.$$

Now because of $2^{w_m} \mid D - d$ and $m < 2^{w_m}$, it follows that

$$|D - d| \leq m \cdot 2^{w_m+1}$$

and thus

$$\left| \frac{D - d}{2^{w_m}} \right| \leq 2m.$$

So in both cases, the absolute value of the subsequent carry digit indeed cannot exceed $2m$. It is clear that no other digit will be set to values not in $B_m \cup \{0\}$.

Now we consider the value

$$\widetilde{H} = \mathcal{H}(b_l \ldots b_0) + \#\{h; \; |b_h| > m + 1\}$$

as observed at the beginning and at the end of the loop body (remember that b_i is the only digit among $b_l \ldots b_0$ that is not necessarily an element of $B_m \cup \{0\}$, so $\widetilde{H}$ exceeds the weight of $b_l \ldots b_0$ at most by one). Given the loop invariant, we can show that $\widetilde{H}$ will never increase within the loop body. It is clear that the loop body will not change any of the digits and thus not $\widetilde{H}$ if b_i is zero. If b_i is non-zero and even, following the algorithm it is easy to see that the changes done to b_i and b_{i+1} cannot increase $\widetilde{H}$. For b_i odd, at the beginning of the loop body define

$$H = \mathcal{H}(b_{i+w_m} \ldots b_i) + \#\{h; \; |b_h| > m + 1 \land i + w_m \geq h \geq i\};$$

now we can distinguish between the following cases:

- $H = 1$. This implies $b_i = D = d$, so the loop body will not change any of the digits and thus not $\widetilde{H}$.
- $H = 2$. If $|b_i| \leq m$, then there are initially exactly two non-zero digits among $b_{i+w_m} \ldots b_i$, both of absolute value at most m, and thus we have $|D| \leq (2^{w_m} + 1) \cdot m$. If $|b_i| > m$, then b_i is the only non-zero digit among $b_{i+w_m} \ldots b_i$, which implies $D = b_i$ and thus $|D| \leq 2m$. In both cases, for $d = digit_m(D)$ it follows that

$$\left| \frac{D-d}{2^{w_m}} \right| \leq \frac{(2^{w_m}+2) \cdot m}{2^{w_m}},$$

and since $2^{w_m} \mid D - d$ and $m < 2^{w_m}$,

$$\left| \frac{D-d}{2^{w_m}} \right| \leq \left\lfloor \frac{(2^{w_m}+2) \cdot m}{2^{w_m}} \right\rfloor = m + \left\lfloor \frac{2m}{2^{w_m}} \right\rfloor = m+1.$$

Thus when the loop body overwrites the digits $b_{i+w_m} \ldots b_i$ with new values, the new carry digit will be of absolute value at most $m+1$; and since at most two of the new values will be non-zero, $\widetilde{H}$ cannot increase.

- $H \geq 3$. The digits $b_{i+w_m} \ldots b_i$ are overwritten with new values out of which at most two are non-zero, and at most one is of absolute value larger than $m+1$; so these new digit values will contribute at most 3 to $\widetilde{H}$. This means that $\widetilde{H}$ cannot increase.

Initially, $\widetilde{H}$ is the input weight $\mathcal{H}(b'_{\ell'} \ldots b'_0)$; in the end, it is the output weight $\mathcal{H}(b_\ell \ldots b_0)$. $\widetilde{H}$ never increases, so no B_m-representation can have lower weight than the RL-SFW representation generated by the transformation algorithm.

3   Left-to-Right Signed Fractional Windows (m-LR-SFW)

To arrive at a left-to-right version, we use an approach from [17] and (building on [6] and [1]) from [7–Section 6]. This provides another way to view the RL-SFW method, and it yields an LR-SFW method (which was new at time of writing, but meanwhile has independently been described in [19] and [9].)

Given the binary representation $\beta_\lambda \ldots \beta_0$ of any positive integer e, first let $\ell' = \lambda + 1$ and $b'_i = \beta_{i-1} - \beta_i$ for $i = \ell', \ldots, 0$ (where $\beta_{-1} = 0$). Since

$$(b'_{\ell'} \ldots b'_0)_2 = (\beta_\lambda \ldots \beta_0\, 0)_2 - (\beta_\lambda \ldots \beta_0)_2 = 2e - e = e,$$

this gives us a new B_1-representation $b'_{\ell'} \ldots b'_0$ of e. Observe that this representation can be obtained from the binary representation just as easily in left-to-right as in right-to-left direction. It is clear from the construction of this new representation that every digit $b'_i = 1$ indicates that β_{i-1} is the left-most digit in a sequence of successive ones in the binary representation ($\beta_i = 0$, $\beta_{i-1} = 1$), and that every digit $b'_i = \bar{1}$ indicates that β_i is the right-most digit in such a sequence of ones ($\beta_i = 1$, $\beta_{i-1} = 0$). Thus, there must be an even number of non-zero digits in $b'_{\ell'} \ldots b'_0$, and these show the following structure:

- The left-most non-zero digit is a 1.
- Skipping any zeros, the neighbors of a 1 digit will always have value $\bar{1}$ and the neighbors of a $\bar{1}$ digit will always have value 1.
- The right-most non-zero digit is a $\bar{1}$.

Because of this structure, we call $b'_{\ell'} \ldots b'_0$ a *sign-alternating* B_1-representation. (This representation has previously been called "reversed binary representation" [10–exercise 4.1-27], "alternating greedy expansion" [6], and "mutual opposite form" [17].) If we write $*$ for any digit of value either 1 or $\bar{1}$, we get a simplified form from which the B_1-representation $b'_{\ell'} \ldots b'_0$ can unequivocally be reconstructed due to its structure. We call $*$ and 0 *compressed digits*. The compressed-digit form of any subsequence of digits $b'_i \ldots b'_h$ allows reconstructing the digits except for possible sign reversal (i.e. reconstruction would yield a sequence of digits that is identical to either $b'_i \ldots b'_h$ or $\overline{b'_i} \ldots \overline{b'_h}$).

The approach from [17] and [7] to obtain related RL- and LR-transformations is to apply a sliding-window technique to the sign-alternating B_1-representation: this can be done right to left, giving the well-known right-to-left signed window representation (W-NAF); or left to right, giving a left-to-right signed window representation. Generalizing this technique, we describe how a similar approach can be used with fractional windows. As before, let $w_m = \lfloor \log_2 m \rfloor + 1$ and $W_m = w_m + 1$ given an odd integer $m \geq 1$.

The sliding-window technique scans the representation $b'_{\ell'} \ldots b'_0$ in one direction, either right to left *(RL-scanning)* or left to right *(LR-scanning)*, starting a new window whenever a non-zero digit is encountered. Observe that for a window $\boxed{b'_{i+w} \ldots b'_i}$ of any width $w + 1$ in a sign-alternating B_1-representation, the window value $(b'_{i+w} \ldots b'_i)_2$ will have absolute value at most 2^w, or $2^w - 1$ after dividing out powers of two. (In a B_1-representation that is not necessarily sign-alternating, the maximal absolute window value would be $2^{w+1} - 1$.) To accommodate fractional windows, the width of the current window is set to W_m if this is admissible, or w_m otherwise (or less than w_m when less than w_m digits are left for scanning). Here a window width is considered admissible if the window value is either some digit in B_m or even. (Window widths smaller than W_m are always admissible.) In any case, the window value will be the product of a power of two and a digit from B_m. Thus, each of the windows requires just one of the digits from B_m, appropriately positioned, to achieve the proper window value.

As an illustration of the transformations, we consider an example for $m = 5$. We have $w_m = 3$ and $W_m = 4$. Now windows of the form $\boxed{*\,0\,0\,*}$ and $\boxed{*\,0\,*\,*}$ are not admissible while windows of the form $\boxed{*\,*\,0\,*}$ and $\boxed{*\,*\,*\,*}$ are admissible (because $(100\bar{1})_2 = (10\bar{1}1)_2 = 7 > m$ but $(1\bar{1}01)_2 = (1\bar{1}1\bar{1})_2 = 5 \leq m$, and similarly for the corresponding negative cases). Let $e = 22369 = (101011101100001)_2$; the sign-alternating B_1-representation of e obtained by the rule $b'_i = \beta_{i-1} - \beta_i$ is $1\bar{1}1\bar{1}100\bar{1}10\bar{1}0001\bar{1}$. RL-scanning does not encounter any inadmissible width-W_m windows; it yields the window constellation

$$\boxed{1}\,\boxed{\bar{1}\,1\,\bar{1}\,1}\,0\,0\,\boxed{\bar{1}\,1\,0\,\bar{1}}\,0\,\boxed{0\,0\,1\,\bar{1}},$$

resulting in the B_5-representation

$$\boxed{1}\,\boxed{0\,0\,0\,5}\,0\,0\,\boxed{0\,0\,0\,\bar{5}}\,0\,\boxed{0\,0\,0\,1}.$$

LR-scanning has to use width w_m $(= 3)$ in one instance to avoid the inadmissible window $\boxed{1\,0\,0\,\bar{1}}$; it yields the window constellation

$$\boxed{1\,\bar{1}\,1\,\bar{1}}\;\boxed{1\,0\,0}\;\boxed{\bar{1}\,1\,0\,\bar{1}}\;0\,0\,0\;\boxed{1\,\bar{1}},$$

resulting in the B_5-representation

$$\boxed{5}\;\boxed{1\,0\,0}\;\boxed{0\,0\,0\,\bar{5}}\;0\,0\,0\;\boxed{0\,1}.$$

It is easy to see that this procedure with RL-scanning will always determine the same B_m-representation as the algorithms shown in Section 2 (ignoring any leading zeros); that is, this is just another way to view the RL-SFW technique. With LR-scanning, this is a new technique, giving us a *left-to-right signed fractional window (LR-SFW)* representation of e, the *m-LR-SFW* representation.

So far we have seen how to obtain LR-SFW representations using an intermediate sign-alternating B_1-representation (following [17] with appropriate changes for fractional windows). This intermediate step is helpful for describing and analyzing the method, but it is not necessary for implementation. Instead, the algorithm in Fig. 4 (following [1] with appropriate changes for fractional windows) can be used to obtain the m-LR-SFW representation $b_\ell \ldots b_0$ of a positive integer directly from the binary representation $\beta_\lambda \ldots \beta_0$. The algorithm as written assumes that $\beta_{\lambda+1} = 0$ and $\beta_i = 0$ for $i < 0$; also, all variables b_i are initially zero. In comments, we use b_i' as defined above ($b_i' = \beta_{i-1} - \beta_i$) to show that this algorithm expresses exactly the LR-transformation that we have introduced in terms of LR-scanning; to verify the correspondence, observe that for $i \geq h$

$$
\begin{aligned}
(b_i' \ldots b_h')_2 &= (\beta_{i-1} \ldots \beta_{h-1})_2 \;-\; (\beta_i \ldots \beta_h)_2 \\
&= (\beta_{i-1} \ldots \beta_h)_2 \cdot 2 \;+\; \beta_{h-1} \;-\; \beta_i \cdot 2^{i-h} \;-\; (\beta_{i-1} \ldots \beta_h)_2 \\
&= -\beta_i \cdot 2^{i-h} \;+\; (\beta_{i-1} \ldots \beta_h)_2 \;+\; \beta_{h-1} \\
&= (\overline{\beta_i}\beta_{i-1} \ldots \beta_h)_2 \;+\; \beta_{h-1}.
\end{aligned}
$$

3.1   Minimality of Weight

As discussed in Section 1, there are general advantages of LR-transformations over RL-transformations. A natural question to consider is whether despite of these advantages, the LR-SFW representation might be worse for exponentiation than the RL-SFW representation. To address this issue, here we show that for given m and e, the weight of the m-LR-SFW representation of e is the same as the weight of the m-RL-SFW representation. Thus, by the result from Section 2.1, the weight is minimal among all B_m-representations of e. (Later in Section 4 we will see that the LR-SFW representation actually has advantages beyond the general advantages of LR-transformations.)

Let ℓ' be a positive integer and

$$S = \left\{ s \in \{0, *\}^{\ell'} \mid * \text{ occurs an even number of times in } s \right\}$$

{ **LR-transformation:** determine the
left-to-right signed fractional window (m-LR-SFW) **representation**
of $(\beta_\lambda \dots \beta_0)_2$ on binary input }

$i \leftarrow \lambda + 1$
$\ell \leftarrow 0$

while $i \geq 0$ **do**
    if $\beta_i = \beta_{i-1}$ **then**   { $b_i' = 0$ }
        $i \leftarrow i - 1$
    else   { $b_i' \neq 0$ }
        $W \leftarrow W_m$
        $d \leftarrow (\overline{\beta_i}\beta_{i-1} \dots \beta_{i-W+1})_2 + \beta_{i-W}$
        if d is odd $\wedge |d| > m$ **then**
            $W \leftarrow w_m$
            $d \leftarrow (\overline{\beta_i}\beta_{i-1} \dots \beta_{i-W+1})_2 + \beta_{i-W}$
        { $d = (b_i' \dots b_{i-W+1}')_2$ }

        $next\_i \leftarrow i - W$
        $i \leftarrow next\_i + 1$
        while d is even **do**
            $i \leftarrow i + 1; \; d \leftarrow d/2$
        { d is odd, $|d| \leq m$, $d = (b_{next\_i+W-1}' \dots b_i')_2$ }

        $b_i \leftarrow d$
        if $i > \ell$ **then**
            $\ell \leftarrow i$
        $i \leftarrow next\_i$
return $(b_\ell, \dots, b_0)$

Fig. 4. LR-transformation for fractional windows

the set of all compressed-digit forms of length ℓ'. We can examine the scanning process in terms of compressed-digit forms. For $s \in S$, let $\#LR_m(s)$ denote the number of windows that LR-scanning yields; i.e., $\#LR_m(s) = \mathcal{H}(b_\ell \dots b_0)$ where $b_\ell \dots b_0$ is the m-LR-SFW representation of the integer determined by the compressed-digit form s. Similarly, let $\#RL_m(s)$ denote the number of windows that RL-scanning yields. LR-scanning and RL-scanning are mostly symmetric, except for admissibility of window width W_m (for example, $(b_{i+w_m}' \dots b_i')_2$ may be a digit in B_m when $(b_i' \dots b_{i+w_m}')_2$ is odd bot not in B_m). However, there is some symmetry that does respect admissibility: When a window is started and there are W_m compressed digits to look at (the first one of which in scanning direction is necessarily a $*$), there are 2^{w_m} possibilities what these W_m compressed digits might look like; and both for LR-scanning and for RL-scanning, window width W_m is admissible for exactly $m + 1$ of these possibilities and not admissible for the remaining $2^{w_m} - m - 1$ possibilities. Thus, there is a bijection

$$\alpha: S \twoheadrightarrow S$$

such that the window structure (i.e., the positioning and width of windows, not taking into account the actual compressed digits in the windows) obtained by LR-scanning of any $s \in S$ is the exact mirror image of the window structure obtained by RL-scanning of $\alpha(s)$. This implies $\#LR_m(s) = \#RL_m(\alpha(s))$.

Now assume that there was some specific $s_0 \in S$ such that $\#LR_m(s_0) \neq \#RL_m(s_0)$. By minimality of weight of the RL-SFW representation (Section 2.1), this would imply

$$\#LR_m(s_0) > \#RL_m(s_0),$$

i.e. $\#RL_m(\alpha(s_0)) > \#RL_m(s_0)$. Since α is a bijection, we have

$$\sum_{s \in S} \#RL_m(\alpha(s)) = \sum_{s \in S} \#RL_m(s),$$

so furthermore it would follow that there is some $s_1 \in S$ such that $\#RL_m(\alpha(s_1)) < \#RL_m(s_1)$. But this would mean

$$\#LR_m(s_1) < \#RL_m(s_1),$$

contradicting the minimality of weight of the RL-SFW representation. Thus no s_0 can exist for which $\#LR_m(s_0)$ differs from $\#RL_m(s_0)$.

4   Length

From Sections 2.1 and 3.1, we know that both the m-RL-SFW representation and the m-LR-SFW representation are optimal in the sense of having minimal weight among all B_m-representations. The efficiency of exponentiation given a B_m-representation $b_\ell \ldots b_0$ with $b_\ell \neq 0$ depends not just on the weight $\mathcal{H}(b_\ell \ldots b_0)$, but also on ℓ (see Section 1). Thus we are interested in representations that provide not only low weight, but also a short length $\ell + 1$. Sometimes these goals are in conflict: for example, for B_7-representations of $255 = (1000000\bar{1})_2 = (70071)_2$, minimal weight and minimal length exclude each other; one or the other representation might provide better efficiency for LR-exponentiation depending on the relative speed of squarings and general multiplications in the group. We prioritize weight over length and consider only ways to reduce the length that do not increase the weight.

A first observation is that the LR-SFW representation can never be longer than the RL-SFW counterpart: Consider the scanning process on sign-alternating B_1-representations as described in Section 3, which yields a B_m-representation when each window value is expressed through a single non-zero digit from B_m. The maximal index ℓ of such a B_m-representation is the index of the right-most non-zero digit within the left-most window over the sign-alternating B_1-representation. For RL-scanning, the left-most (final) window will cover some number of non-zero digits of the sign-alternating B_1-representation, including its most significant digit. All of these non-zero digits, and possibly more, would also be covered by the left-most (first) window obtained by LR-scanning. Thus, the maximal index ℓ cannot increase for LR-scanning compared with RL-scanning.

The example in Section 3 has already shown us that the m-LR-SFW representation is indeed shorter than the m-RL-SFW representation in some cases. In fact, the m-LR-SFW representation is advantageous for every m. For example, for $m = 1$, the average length saving for m-LR-SFW representations of long random integers e compared with m-RL-SFW representations is $1/6$; for $m = 3$, it is $1/2$; for $m = 7$, it is $37/40$. (Consider $m = 1$. If the left-most window over the sign-alternating B_1-representation is $\boxed{1\,0}$ or $\boxed{1}$, the 1-LR-SFW or 1-RL-SFW representation will be one digit longer than it is for the window $\boxed{1\,\bar{1}}$. The latter case happens with probability $1/2$ for LR-scanning, but only with probability about $1/3$ for RL-scanning. See Appendix A for more details.)

The m-LR-SFW representation does not guarantee minimal length among all B_m-representations of minimal weight. Substitution rules resulting in a modified (right-to-left) signed fractional window representation have been given in [14–Section 5.1], and these can similarly be applied to the left-to-right case to obtain shorter representations in some cases. Additional rules not mentioned in [14] are possible. For example, if $3 \leq m \leq 13$, the m-LR-SFW and m-RL-SFW representations both write $e = 15$ as $(100\bar{1})_2$; this can be improved into $(303)_2$, or even into $(55)_2$ if $5 \leq m$.

No LR-transformation implemented by a finite-state machine can always ensure minimal length among all minimal-weight B_m-representations; see Appendix B for examples. With the m-LR-SFW representation and its generally shorter length, there is less of a need to use modified representations to decrease the length than with the m-RL-SFW representation, in particular if m is relatively large. However, if program space permits, implementations can include a table of optimized substitution rules for certain prefixes that can be encountered in m-LR-SFW representations (such as $100\bar{1} \mapsto 31$ and $1000\bar{1} \mapsto 303$ for $m = 3$). While no such table could be complete for arbitrary lengths, this can help improve the average efficiency at least by a small margin.

Acknowledgement

Thanks to the anonymous reviewers for making me aware of [10–exercise 4.1-27], [6], and [7].

References

1. AVANZI, R. M. A note on the sliding window integer recoding and its left-to-right analogue. In *Selected Areas in Cryptography – SAC 2004*, Lecture Notes in Computer Science. To appear.

2. BLAKE, I. F., SEROUSSI, G., AND SMART, N. P. *Elliptic Curves in Cryptography*, vol. 265 of *London Mathematical Society Lecture Note Series*. Cambridge University Press, 1999.

3. COHEN, H., ONO, T., AND MIYAJI, A. Efficient elliptic curve exponentiation using mixed coordinates. In *Advances in Cryptology – ASIACRYPT '98* (1998), K. Ohta and D. Pei, Eds., vol. 1514 of *Lecture Notes in Computer Science*, pp. 51–65.

4. GORDON, D. M. A survey of fast exponentiation methods. *Journal of Algorithms* 27 (1998), 129–146.

5. GORIAC, I., AND IFTENE, S. Personal communication, 2003.

6. GRABNER, P. J., HEUBERGER, C., PRODINGER, H., AND THUSWALDNER, J. M. Analysis of linear combination algorithms in cryptography. Preprint, 2003. Available from http://www.opt.math.tu-graz.ac.at/~cheub/publications/.

7. HEUBERGER, C., KATTI, R., PRODINGER, H., AND RUAN, X. The alternating greedy expansion and applications to left-to-right algorithms in cryptography. Preprint, 2004. Available from http://www.opt.math.tu-graz.ac.at/~cheub/publications/.

8. JOYE, M., AND YEN, S.-M. Optimal left-to-right binary signed-digit recoding. *IEEE Transactions on Computers 49* (2000), 740–748.

9. KHABBAZIAN, M., AND GULLIVER, T. A. A new minimal average weight representation for left-to-right point multiplication methods. Cryptology ePrint Archive Report 2004/266, 2004. Available from http://eprint.iacr.org/.

10. KNUTH, D. E. *The Art of Computer Programming – Vol. 2: Seminumerical Algorithms*. Addison-Wesley, 1969.

11. KNUTH, D. E. *The Art of Computer Programming – Vol. 2: Seminumerical Algorithms (3rd ed.)*. Addison-Wesley, 1998.

12. MIYAJI, A., ONO, T., AND COHEN, H. Efficient elliptic curve exponentiation. In *International Conference on Information and Communications Security – ICICS '97* (1997), Y. Han, T. Okamoto, and S. Qing, Eds., vol. 1334 of *Lecture Notes in Computer Science*, pp. 282–290.

13. MÖLLER, B. Algorithms for multi-exponentiation. In *Selected Areas in Cryptography – SAC 2001* (2001), S. Vaudenay and A. M. Youssef, Eds., vol. 2259 of *Lecture Notes in Computer Science*, pp. 165–180.

14. MÖLLER, B. Improved techniques for fast exponentiation. In *Information Security and Cryptology – ICISC 2002* (2003), P. J. Lee and C. H. Lim, Eds., vol. 2587 of *Lecture Notes in Computer Science*, pp. 298–312.

15. MUIR, J. A., AND STINSON, D. R. Minimality and other properties of the width-w nonadjacent form. *Mathematics of Computation*. To appear; preprint available from http://www.cacr.math.uwaterloo.ca/tech_reports.html.

16. MUIR, J. A., AND STINSON, D. R. New minimal weight representations for left-to-right window methods. In *CT-RSA 2005*, Lecture Notes in Computer Science. To appear; preprint available from http://www.cacr.math.uwaterloo.ca/tech_reports.html.

17. OKEYA, K., SCHMIDT-SAMOA, K., SPAHN, C., AND TAKAGI, T. Signed binary representations revisited. In *Advances in Cryptology – CRYPTO 2004* (2004), M. Franklin, Ed., vol. 3152 of *Lecture Notes in Computer Science*, pp. 123–139.

18. REITWIESNER, G. W. Binary arithmetic. *Advances in Computers 1* (1960), 231–308.

19. SCHMIDT-SAMOA, K., SEMAY, O., AND TAKAGI, T. Analysis of some efficient window methods and their application to elliptic curve cryptosystems. Technical Report TI-3/04, 2004. Available from http://www.informatik.tu-darmstadt.de/ftp/pub/TI/TR/.

20. SCHROEPPEL, R., ORMAN, H., O'MALLEY, S., AND SPATSCHECK, O. Fast key exchange with elliptic curve systems. In *Advances in Cryptology – CRYPTO '95* (1995), D. Coppersmith, Ed., vol. 963 of *Lecture Notes in Computer Science*, pp. 43–56.

21. SOLINAS, J. A. An improved algorithm for arithmetic on a family of elliptic curves. In *Advances in Cryptology – CRYPTO '97* (1997), B. S. Kaliski, Jr., Ed., vol. 1294 of *Lecture Notes in Computer Science*, pp. 357–371.

22. SOLINAS, J. A. Efficient arithmetic on Koblitz curves. *Designs, Codes and Cryptography 19* (2000), 195–249.

23. YAO, A. C.-C. On the evaluation of powers. *SIAM Journal on Computing 5* (1976), 100–103.

A   LR-SFW versus RL-SFW: Length Comparison

We want to compare the expected lengths of left-to-right and right-to-left signed fractional window representations of long random integers by looking at LR-scanning and RL-scanning as described in Section 3. Assume an ℓ'-bit integer e is given (so that $b'_{\ell'} = 1$ in the sign-alternating B_1-representation). By the probabilities below, in the m-LR-SFW representation the expected maximum index ℓ is $\ell' - 1/2$ for $m = 1$, $\ell' - 5/4$ for $m = 3$, and $\ell' - 17/8$ for $m = 7$; in the m-RL-SFW representation it is about $\ell' - 1/3$ for $m = 1$, about $\ell' - 3/4$ for $m = 3$, and about $\ell' - 6/5$ for $m = 7$.

Consider $m = 2^w - 1$ such that windows over the sign-alternating form have width $w + 1$. For LR-scanning over random compressed-digit forms of sufficient length, the left-most window is

- $\boxed{* 0 \ldots 0}$ with probability 2^{-w};
- $\boxed{* * 0 \ldots 0}$ also with probability 2^{-w};
- of the form $\boxed{* ? * 0 \ldots 0}$ with probability 2^{1-w};
- ...;
- of the form $\boxed{* ? \ldots ? *}$ with probability 2^{-1}.

As the resulting B_m-representations will be successively shorter (the maximal index ℓ of the B_m-representation is the index of the right-most non-zero digit within the left-most window over the compressed-digit form), this is in favor of generating short B_m-representations. No $m \geq 2^w - 1$ will have average lengths longer than this.

For RL-scanning over long random compressed-digit forms, however, the left-most window is

- $\boxed{*}$ (width 1) with probability about $2/(w + 2)$;
- $\boxed{* *}$ (width 2) with probability about $1/(w + 2)$;
- of the form $\boxed{* ? *}$ (width 3) with probability about $1/(w + 2)$;
- ...;
- of the form $\boxed{* ? \ldots ? *}$ (width $w + 1$) with probability about $1/(w + 2)$.

Again the resulting B_m-representations will be successively shorter, but here the probabilities are in favor of generating long B_m-representations. No $m \leq 2^w - 1$ will have average lengths shorter than this. (To derive these estimates, assume that RL-scanning is applied to a long sequence of independently uniformly random compressed digits. Any given $*$ in a width-w window well within such a sequence, after earlier windows have provided plenty of randomization of window positions, will be at the right-most position of its window with probability about $2/(w+2)$, and at any other position of its window with probability about $1/(w+2)$ each. This is seen by counting how many of the 2^w possibilities for the content of a width-$(w+1)$ window in RL-scanning have a $*$ in the respective position: the right-most position sees twice the proportion of $*$'s as each other position. Since the start of a window does not depend on what follows, the left-most $*$ in a [finite but long] compressed-digit form is similar except that the window abruptly ends there.)

B   Limitations of Length Reduction

Let $\langle m \rangle$ be a shorthand notation for a digit string consisting of $w_m - 1$ zeros concatenated with the digit m, and $\langle \bar{m} \rangle$ for w_m zeros (e.g. $\langle 5 \rangle = 005$, $\langle \bar{5} \rangle = 000$). Now consider the integers with B_m-representations either of the form

$$m1\, 0\langle m \rangle\, 0\langle m \rangle \ldots 0\langle m \rangle\, 0\langle m \rangle\, \langle m \rangle$$

or of the form

$$m1\, 0\langle m \rangle\, 0\langle m \rangle \ldots 0\langle m \rangle\, 0\langle m \rangle\, \langle \bar{m} \rangle.$$

The m-LR-SFW representation for an integer of the first form is longer than the above representation and has lower weight; this is a weight minimization that would not be possible without the length increase. For example, for $m = 1$,

$$(11\,01\,01 \ldots 01\,01\,1)_2$$

is rewritten with lower weight as

$$(100\,\bar{1}0\,\bar{1}0 \ldots \bar{1}0\,\bar{1}0\,\bar{1})_2.$$

The m-LR-SFW representation for an integer of the second form is longer as well, but the weight remains unchanged; the original representation as given above already has minimal weight. For example,

$$(11\,01\,01 \ldots 01\,01\,0)_2$$

has no B_1-representation of lower weight. There is no bound on the number of digits that one might have to examine to distinguish between such cases, so no finite-state machine could always generate the shortest representation among those of minimal weight.

Improvement on Ha-Moon Randomized Exponentiation Algorithm*

Sung-Ming Yen[1], Chien-Ning Chen[1], SangJae Moon[2], and JaeCheol Ha[3]

[1] Laboratory of Cryptography and Information Security (LCIS),
Dept of Computer Science and Information Engineering,
National Central University, Chung-Li, Taiwan 320, R.O.C
{yensm, ning}@csie.ncu.edu.tw
http://www.csie.ncu.edu.tw/~yensm/
[2] School of Electronic and Electrical Engineering,
Kyungpook National University, Taegu, Korea 702-701
sjmoon@knu.ac.kr
[3] Dept of Computer and Information,
Korea Nazarene University, Choong Nam, Korea 330-718
jcha@kornu.ac.kr

Abstract. Randomized recoding on the exponent of an exponentiation computation into a signed-digit representation has been a well known countermeasure against some side-channel attacks. However, this category of countermeasures can only be applicable to those cryptosystems with fixed parameters on the base integer when evaluating exponentiation or to some classes of cryptosystems such that the inversion is performed only once or can be computed very efficiently. This paper considers the development of novel inversion-free exponentiation algorithms which bijectively map the signed-digit exponent into non-negative digits. These signed-digit based exponentiation algorithms are therefore applicable to cryptosystems with varying base integers, e.g., the RSA cryptosystem. We also propose a left-to-right version of the Ha-Moon recoding and integrate the recoding with the proposed non-inversion technique. The integrated algorithm leads to a more secure countermeasure of implementing exponentiation against side-channel attacks.

Keywords: Differential power analysis (DPA), Modular exponentiation, Physical cryptanalysis, Randomized recoding, Side-channel attack, Signed-digit recoding, Simple power analysis (SPA).

1 Introduction

Exponentiation algorithms play an important role in most public key cryptosystems, but most exponentiation algorithms when implemented on the smart IC cards are vulnerable to physical cryptanalysis such as side-channel attacks (including power analysis attacks [8] and timing attacks [7]) and fault-based

---

* This work was supported by University IT Research Center Project.

C. Park and S. Chee (Eds.): ICISC 2004, LNCS 3506, pp. 154–167, 2005.

attacks [1]. Randomized exponentiation algorithms were recently considered effective software-based countermeasures against most side-channel attacks and have been widely discussed in the past few years. Randomized algorithms were considered effective to be resistant to most side-channel attacks since the computational process is non-deterministic and the measured side-channel information does not correlate directly to the embedded secret within the device.

For the context of developing countermeasures for computing scalar multiplication on elliptic curve or modular exponentiation, Coron proposed three possible countermeasures [2], i.e., randomization of the private exponent, blinding on the base, and randomized projective coordinates. The first two countermeasures mentioned above are suitable for most cryptosystems based on computation of modular exponentiation, while the third countermeasure is only suitable for elliptic curve cryptosystems. Oswald and Aigner [12] proposed a countermeasure by exploiting randomization on addition and subtraction chain. Recently, Ha and Moon [5] proposed a countermeasure based on randomized recoding on the private exponent prior to performing the exponentiation computation. Both of Oswald-Aigner and Ha-Moon's countermeasures employ the technique of signed-digit representation (say with a digit set consisting of both positive and negative integers, e.g., the digit set $\{0, 1, -1\}$), and require the operation of inversion (or the operation of division). Unfortunately, the computational overhead of obtaining a multiplicative inversion in $\mathbf{Z}_n$ is basically approaching to an ordinary modular exponentiation. Therefore, existing countermeasures based on signed-digit recoding are only applicable to systems with fixed base or systems in which inversion computation is very efficient. This fact of course limits the applicability of all existing countermeasures of this category.

The main contribution of this paper is to propose a novel technique that removes the above mentioned limitation of calculating inversion. The inversion is no longer required even though the private exponent is signed-digit recoded, because the signed-digit exponent can be re-recoded into a new one with a positive digit set. Some configurations of the proposed recoding procedure are bijective maps except the first few bits of the exponent. Because of the one-to-one mapping, these "bijective" recoding procedures will not reduce the security of the original countermeasure.

A useful trick to convert the Ha-Moon method into a left-to-right version is also proposed. Recall that the original Ha-Moon method recodes the exponent in the right-to-left approach (i.e., LSB-to-MSB), but the proposed exponentiation algorithms (in fact, also in many existing exponentiation algorithms reported in the literature) scan the private exponent from MSB towards LSB. The left-to-right recoding is more suitable for the left-to-right exponentiation computation. In addition, the proposed non-inversion algorithm and the left-to-right recoding can employ the windowing technique. When the window size is two, it always takes one modular multiplication and two modular squaring operations for processing every two bits of the exponent. Therefore, a computational performance of $1.5n$ modular operations (including both modular multiplication and modular squaring) is achieved for an n-bit exponent.

The rest of this paper is organized as follows. In Sect. 2, some necessary notations are defined and preliminary of signed-digit based exponentiation algorithm is briefly reviewed. The original Ha-Moon method and the weakness of it are also described in this section. Improvements on the Ha-Moon method are proposed in Sect. 3. Section 4 provides randomized recoding schemes motivated from the Ha-Moon method. The proposed schemes are integrated solutions obtained by combining the left-to-right version of the Ha-Moon method and the proposed non-inversion technique. Security analysis of the proposed schemes is given in Sect. 5. Finally, Sect. 6 concludes this paper.

2   Notations and Preliminaries

Necessary notations are defined in the following.

Suppose K is the n-bit private exponent and $(k_{n-1} k_{n-2} \cdots k_1 k_0)_2$ denotes K's binary representation. Let $D = (d_n d_{n-1} \cdots d_1 d_0)_{\text{SD2}}$ be one of K's possible signed-digit representation. Notice that signed-digit representation is not unique for usual cases. In addition, K_i denotes the bit string of $(n-1)$th to ith bits of K, and D_i denotes the signed-digit bit string of nth to ith bits of D. The subscripts $()_2$ and $()_{\text{SD2}}$ indicate that the numbers in the bracket are binary (0 or 1) and signed-digit binary (-1, 0, or 1) represented, respectively. The following summarize the above definitions.

$$K = (k_{n-1} k_{n-2} \cdots k_1 k_0)_2 \quad K_i = (k_{n-1} \cdots k_i)_2$$
$$D = (d_n d_{n-1} \cdots d_1 d_0)_{\text{SD2}} \quad D_i = (d_n \cdots d_i)_{\text{SD2}}$$

2.1   Signed-Digit Based Exponentiation Algorithms

Randomized recoding on the private exponent by signed-digit representation is a straightforward approach to randomize an exponentiation computation. The above approach was considered as a countermeasure against most side-channel attacks because of its non-deterministic computational process. The basic algorithm for computing exponentiation with signed-digit represented exponent can be in a similar approach as in the conventional exponentiation except that an inverse element has to be prepared prior to the main computation.

The algorithm in Fig. 1 computes the exponentiation with exponent D in signed-digit representation. Within the algorithm, a division operation is carried out when encountering a negative bit in D. By another approach, this division can be replaced by multiplying the inverse of the base g. However, the computational complexity of finding g^{-1} in $\mathbf{Z}_n$ is roughly of the same order of computing a modular exponentiation, and this makes the algorithm be inefficient[1], espe-

---

[1] Of course, extended Euclidian-like algorithms may slightly speed up the computation, but those auxiliary algorithms themselves are somewhat complicated and need some additional temporary storages. This brings storage overhead for small device implementation, e.g., IC card.

INPUT: $g, D = (d_n \cdots d_0)_{\text{SD2}}$ where $d_n = 1$	
OUTPUT: g^D	
01	Precomputation: g^{-1}
02	$R = 1$
03	for i from n downto 0 do
04	$R = R^2$
05	if $d_i \neq 0$ then $R = R \times g^{d_i}$
06	return R

Fig. 1. Exponentiation with signed-digit represented exponent

cially for the cases that the base g is not a constant value. Pre-computation of the inverse g^{-1} is the primary disadvantage of the above randomized algorithm. This category of countermeasures only benefits those systems whose base integer is a constant or in systems computation of the inverse is very efficient.

2.2   Randomized Exponentiation Proposed by Ha and Moon

Ha and Moon proposed a randomized elliptic curve scalar multiplication algorithm in CHES 2002 [5]. It randomly recodes the original binary scalar (similar to the exponent in exponentiation computations) into a signed-digit scalar from LSB towards MSB. In the Ha-Moon method, the original scalar $K = (k_{n-1} \cdots k_0)_2$ ($k_{n-1} = 1$ since K is assumed to be an integer of n bits represented in binary form), is randomly recoded into a signed-digit scalar $D = (d_n \cdots d_0)_{\text{SD2}}$ by the following equations

$$d_i + 2c_{i+1} = c_i + k_i \tag{1}$$
$$(k_{i-1} k_{i-2} \cdots k_1 k_0)_2 = (c_i d_{i-1} d_{i-2} \cdots d_1 d_0)_{\text{SD2}}. \tag{2}$$

In the above equations, c_i is the carry bit ($c_i \in [0, 1]$, $c_0 = 0$, and $c_i = 0 \ \forall i \geq n+1$) and $c_i 2^i$ is the difference between the values of $(k_{i-1} \cdots k_0)_2$ and $(d_{i-1} \cdots d_0)_{\text{SD2}}$. The Ha-Moon method recodes the exponent from LSB towards MSB, i.e., it solves d_i and c_{i+1} from c_i and k_i in Eq (1). Table 1 lists the solutions of Eq (1) with various c_i and k_i. When there are two possible solutions in Eq (1), one of them will be selected randomly depending on the result of $k_{i+1} \oplus r_i$ where r_i is a random bit.

Table 1. The Ha-Moon randomized recoding method

c_i	k_i	$k_{i+1} \oplus r_i$	(c_{i+1}, d_i)
0	0	×	$(0, 0)$
0	1	0	$(0, 1)$
0	1	1	$(1, -1)$
1	0	0	$(0, 1)$
1	0	1	$(1, -1)$
1	1	×	$(1, 0)$

Proposition 1. *The Ha-Moon method generates all possible signed-digit recoded scalars of the original scalar. The output (the probability distribution) of the Ha-Moon method is identical to the output of the following steps:*

1. *For the n-bit scalar K, select an n-bit random number $R = (r_{n-1} \cdots r_0)_2$.*
2. *Adding a random number R to the scalar K makes a temporary result T.*
3. *Subtract the random number R from the temporary result T by the "signed-digit subtraction" and get the output D where the signed-digit subtraction means a subtraction with no borrow, e.g., $(100)_2 - (010)_2 = (1\ -1\ 0)_{SD2}$.*

2.3   Disadvantages of the Ha-Moon Method

There are several disadvantages and weaknesses in the Ha-Moon method. The main disadvantage is the requirement of the inverse operation. This is also the case in most randomized exponentiation algorithms based on the signed-digit representation. Since the computation of finding the multiplicative inverse is inefficient in $\mathbf{Z}_n$, only the elliptic curve cryptosystem is considered in the Ha-Moon method.

In addition, the Ha-Moon method recodes the scalar from LSB towards MSB, however the scalar multiplication algorithm performs inconsistently by scanning the bits of the recoded scalar from MSB towards LSB. The disadvantage is twofold: it disables simultaneous computation for recoding and scalar multiplication and it also needs additional memory to store the recoded scalar.

2.4   Security Weaknesses of the Ha-Moon Method

Several attacks against the Ha-Moon method were proposed and can be divided into two categories. One focuses on the Addition-Doubling (AD) sequence, and the other one focuses on the distribution of the intermediate results.

The scalar is recoded into a signed-digit representation in the Ha-Moon method. When the recoded scalar is employed in the scalar multiplication algorithm (similar to the algorithm in Fig. 1), a sequence of addition and doubling operations will be carried out. Because both of 1 and −1 in the recoded scalar cause an addition operation, the attacker cannot directly retrieve the private scalar even if he can distinguish addition and doubling operations.

However, the AD sequence still reveals partial information about the private scalar. The private scalar can be derived from several (less than ten in most cases) AD sequences corresponding to it. Several SPA attacks based on this weakness were proposed [6, 10, 11]. The "doubling and addition always" algorithm illustrated in Ha and Moon's paper can prevent those attacks because the AD sequence is independent of the private scalar. However, those dummy operations unfortunately may benefit another potential and practical physical attack, i.e., the safe-error attack [13]. The attacker can identify those dummy operations because faults induced on those dummy operations will not change the final result.

Another weakness of the Ha-Moon method is the monotone distribution of the intermediate results. When executing from right to left, the intermediate result

$(d_i \cdots d_0)_{\text{SD2}} \times P$ always equals to $(k_i \cdots k_0)_{\text{SD2}} \times P$ or $((k_i \cdots k_0)_{\text{SD2}} - 2^{i+1}) \times P$. Similarly, the intermediate result $D_i \times P$ always equals to $K_i \times P$ or $(K_i + 1) \times P$ when executing from left to right. This monotone distribution results in some attacks [3, 4]. It is not easy to completely prevent from attacks based on this weakness, but it is relatively easy to increase the complexity of those attacks by enlarging the set of all possible intermediate results. For example, considering the ZEMD (zero-exponentiation multiple data) attack [9] on the randomized exponentiation, half of the samples are meaningful and the others can be treated as noise when the size of the possible distribution is two. The rate of efficient samples can be cut down by enlarging the size of the possible distribution.

3   Improvements on the Ha-Moon Method

In this section, three improvements are proposed. The first one is the left-to-right recoding that allows the recoding algorithm and the left-to-right exponentiation algorithm to execute simultaneously. The second is the non-inversion technique that removes the requirement of inversion by transforming the signed-digit binary exponent into another form with a positive digit set. The last one is the windowing method which improves the performance.

3.1   Left-to-Right Recoding

The original Ha-Moon method recodes the exponent (or scalar in ECC) from right to left. In the following, the right-to-left recoding will be transformed into a left-to-right version by modifying Eq (1) and Eq (2) into the following forms

$$c_i - d_i = 2c_{i+1} - k_i$$
$$(k_{n-1} k_{n-2} \cdots k_{i+1} k_i)_2 + c_i = (d_n d_{n-1} \cdots d_{i+1} d_i)_{\text{SD2}}$$

where the carry bit c_{n+1} is initialized to be $c_{n+1} = 0$. The recoded d_i and carry bit c_i are solved from k_i and c_{i+1}, i.e., the exponent is recoded from MSB towards LSB. The solutions of c_{i+1} and k_i are listed in Table 2 in which if there are two possible solutions, then one of them will be selected randomly.

In the Ha-Moon method, the carry bit always ends with $c_{n+1} = 0$, but in the left-to-right recoding, the carry bit may end with $c_0 = 0$ or 1. When $c_0 = 1$

Table 2. Left-to-right randomized recoding

c_{i+1}	k_i	(c_i, d_i)
0	0	$(0, 0)$ or $(1, 1)$
0	1	$(0, 1)$
1	0	$(1, -1)$
1	1	$(0, -1)$ or $(1, 0)$

occurs, the recoded exponent is greater than the original exponent, and a division (multiplication by the inverse of the base) is required when implementation the exponentiation algorithm. This can be avoided by replacing the original K by $K' = K - 1$, but an extra multiplication is required when $c_0 = 0$.

Similar to the Ha-Moon method, the proposed recoding generates all possible signed-digit exponents equaling to the original exponent K, but it also generates some signed-digit exponents equaling to $K + 1$.

3.2   Non-inversion Technique

Basic idea of the proposed non-inversion technique is to reserve a small value b_i (called the *pre-borrow*) from the higher priority digit to the lower priority digit to prevent the possibility of inversion (division in $\mathbf{Z}_n$) occurred at the lower priority digit when executing the exponentiation. When an inversion occurs (i.e., $d_i = -1$ in D), it can be removed or replaced by a multiplication. For example, if $d_i = -1$ and a pre-borrow of "one" from its higher priority digit (i.e., $b_{i+1} = -1$), then the expected value at digit i becomes $d_i - b_{i+1} \times 2 = 1$ and we have the re-recoded $d'_i = 1$.

One of the main consideration when developing the above management on recoding d_i into d'_i (to be a non-negative value) and deciding b_i as soon as given the previous b_{i+1} and d_i is that the value of d'_i should be limited within a pre-determined range. The size of this mentioned range determines the storage cost of a look-up table required in the exponentiation algorithm. Necessary parameters and mathematical relationship within the above idea are described in the following.

$$D_i = 2D_{i+1} + d_i$$
$$D'_i = D_i + b_i = 2(D_{i+1} + b_{i+1}) + (b_i - 2b_{i+1} + d_i)$$
$$= 2D'_{i+1} + d'_i$$
$$d'_i - b_i = d_i - 2b_{i+1} \quad (\text{or } d'_i = d_i - 2b_{i+1} + b_i) \tag{3}$$

In the proposed technique, $g^{D_i} = (g^{D_{i+1}})^2 \times g^{d_i}$ (the temporary result when computing from MSB to d_i) will not be computed directly. Instead, $g^{D'_i} = (g^{D'_{i+1}})^2 \times g^{d'_i}$ is computed. The difference (or more precisely the quotient) between g^{D_i} and $g^{D'_i}$ is g^{-b_i} at any iteration i. It is interesting to notice that there will be no division occurred in this exponentiation algorithm since the re-recoded d'_i is always a non-negative integer.

Based on our previous derivation, the original digit d_i, the re-recoded digit d'_i, and the two consecutive pre-borrow b_{i+1}, and b_i should satisfy Eq (3). The pre-borrow and the re-recoded digit are calculated from MSB towards LSB by a function $(b_i, d'_i) = \mathcal{F}(b_{i+1}, d_i)$. The exact definition of $\mathcal{F}()$ depends on implementations, and some definitions will be provided in Sect. 3.3. In some implementations, there might be more than one possible pair (b_i, d'_i) satisfying Eq (3), and the algorithm can randomly select one of them as the output of function $\mathcal{F}()$. The size of the pre-computation table (look-up table) depends on the size of set $\{d'_i\}$.

3.3   Suggested Parameters for the Non-inversion Technique

Two definitions of $\mathcal{F}(\ )$ are provided. The first one is a simplified version with the fixed pre-borrow $b_i = -2$, and the second has two possible pre-borrows $b_i = -2$ or -3. In order to avoid the division operation, d_i' is non-negative in both of the proposed parameters.

In the first proposed parameter, the re-recoded d_i' equals to $d_i + 2$ for all $0 \le i \le n - 3$, because the pre-borrow is fixed to -2 and $d_i' = d_i - 2b_{i+1} + b_i$. The re-recoding procedure subtracts two from the first three bit $(d_n\, d_{n-1}\, d_{n-2})_{\text{SD2}}$ and replaces other d_i by $d_i' = d_i + 2$. The definition of $\mathcal{F}(\ )$ is described below.

$$(d_n'\, d_{n-1}'\, d_{n-2}') = (d_n d_{n-1} d_{n-2})_{\text{SD2}} - 2,\ b_{n-2} = -2,$$
$$(b_i, d_i') = \mathcal{F}(b_{i+1} = -2, d_i) = (-2, d_i + 2)\ \ \forall 0 \le i \le n - 3.$$

It is easy to verify that $(d_n\, d_{n-1}\, d_{n-2})_{\text{SD2}}$ is greater than two because the exponent D is greater than 2^{n-1}.

When the non-inversion technique with the first proposed parameter is integrated with the original Ha-Moon method, there exists a bijective mapping between the re-recoded exponent and the original signed-digit exponent except the first three bits $(d_n'\, d_{n-1}'\, d_{n-2}')$. Thus, the re-recoded exponent will keep the original properties because of the bijective mapping.

In the second proposed parameter, the pre-borrow b_i equals to -2 or -3. Table 3 provides a suggested definition of the function $\mathcal{F}(\ )$ to generate the pair (b_i, d_i') from the given values b_{i+1} and d_i. There are four distinct values of d_i', and this design requires three pre-computation values g^2, g^3, and g^4, where g is the base integer of the exponentiation.

Table 3. Proposed function $\mathcal{F}(\ )$ for $b_i = -2$ or -3

b_{i+1}	d_i	$(b_i, d_i') = \mathcal{F}(b_{i+1}, d_i)$
-1	0	$(-2, 0)$
-1	1	$(-2, 1)$ or $(-3, 0)$
-2	-1	$(-2, 1)$
-2	0	$(-2, 2)$ or $(-3, 1)$
-2	1	$(-2, 3)$ or $(-3, 2)$
-3	-1	$(-2, 3)$ or $(-3, 2)$
-3	0	$(-2, 4)$ or $(-3, 3)$
-3	1	$(-3, 4)$

The proposed exponentiation algorithm is given in Fig. 2 where the variable b is initially set to $-(d_n\, d_{n-1})_{\text{SD2}}$. Notice that $(d_n\, d_{n-1})_{\text{SD2}}$ is always non-negative because the exponent D is positive. Since the temporary result R after executing the for loop equals to $g^{D_0'} = g^{D_0} \times g^{b_0}$, therefore the final result g^D should be $g^{D_0} = g^{D_0'} \times g^{-b_0}$ (where $D_0 = D$). Recall that b_0 is a non-positive integer, so multiplication with g^{-b_0} is not a division operation.

In Table 3, the conditions of $(b_{i+1}, d_i) = (-1, 0)$ and $(-1, 1)$ are called the *initial transient state*. Due to the fact of $2^{n-1} \leq D \leq 2^n - 1$, it implies that $1 \leq (d_n d_{n-1})_{\text{SD2}} \leq 2$ and $2 \leq (d_n d_{n-1} d_{n-2})_{\text{SD2}} \leq 4$. There are only six possibilities of such $(d_n d_{n-1} d_{n-2})_{\text{SD2}}$, i.e., $(0\,1\,0)$, $(1\,-\!1\,0)$, $(0\,1\,1)$, $(1\,0\,-\!1)$, $(1\,-\!1\,1)$, and $(1\,0\,0)$. Given the above result, it can be proven easily that after processing the first 3 bits in D, this proposed algorithm will fall into the usual state (say, all cases exclude the initial transient state). The reason is that in the usual state, values of the pre-borrow will never be "-1" again.

INPUT: g, $D = (d_n \cdots d_0)_{\text{SD2}}$ where $2^{n-1} \leq D \leq 2^n - 1$
OUTPUT: g^D

---

01 Precomputation: all values of $g^{d_i'}$
02 $R = 1$
03 $b = -(d_n d_{n-1})_{\text{SD2}}$
04 for i from $n - 2$ downto 0 do
05      $(b, d') = \mathcal{F}(b, d_i)$
06      $R = R^2$
07      $R = R \times g^{d'}$
08 $R = R \times g^{-b}$
09 return R

Fig. 2. Exponentiation without division

If there are two possible output pairs (b_i, d_i') for some given pairs (b_{i+1}, d_i), these two pairs can be randomly selected each time. So, this may lead to a randomized algorithm and provides further security in the context of considering side-channel attacks. Totally, the algorithm requires n squarings and n multiplications for a given n-bit exponent.

3.4   The Windowing Technique

The proposed non-inversion technique can incorporate with the windowing method to improve the performance. Necessary mathematical relationships with window size of two are provided in the following.

$$D_i = 4D_{i+2} + (d_{i+1} d_i)_{\text{SD2}}$$
$$D_i' = D_i + b_i = 4(D_{i+2} + b_{i+2}) + (b_i - 4b_{i+2} + (d_{i+1} d_i)_{\text{SD2}})$$
$$= 4D_{i+2}' + d_i'$$
$$d_i' - b_i = (d_{i+1} d_i)_{\text{SD2}} - 4b_{i+2} \quad \text{(or } d_i' = (d_{i+1} d_i)_{\text{SD2}} - 4b_{i+2} + b_i) \tag{4}$$
$$g^{D_i'} = (g^{D_{i+2}'})^4 \times g^{d_i'}$$

This algorithm requires two squarings and one multiplication (if $d_i' \neq 0$) for every two bits of the exponent. Given the previous pre-borrow b_{i+2} and the two bits $(d_{i+1}, d_i)_{\text{SD2}}$ in present window block, the new pre-borrow b_i and the re-recoded digit d_i' should follow the rule provided in Eq (4).

Moderately increasing the window size may improve the performance, but this also increases the storage requirement for the look-up table. It is therefore a trade-off between computational performance and storage cost. For most practical applications (say with reasonable storage cost and also acceptable performance), in this paper we only concentrate on the non-windowing version (or windowing version with window size of one) and the windowing version with window size of two.

4   Randomized Exponentiation Without Inversion

In this section, the proposed non-inversion technique is integrated with the left-to-right version of the Ha-Moon method by combining the carry and the pre-borrow. By integrating these two algorithms, we obtain

$$d_i' - b_i = -2b_{i+1} + d_i$$
$$= -2b_{i+1} + (c_i - 2c_{i+1} + k_i)$$
$$d_i' - s_i = d_i' - (b_i + c_i) = k_i - 2(b_{i+1} + c_{i+1}) = k_i - 2s_{i+1},$$

where s_i is the sum of the carry c_i and the pre-borrow b_i.

The originally proposed non-inversion technique requires the randomly recoded signed-digit exponent as its input parameter, but in the above integrated version, the randomized exponent recoding and the non-inversion technique perform simultaneously. Table 4 provides one recoding rule with $s_i = -2$ or -3. During executing the algorithm from MSB towards LSB, the algorithm randomly selects (s_i, d_i') pair from the two possible pairs within Table 4.

Table 4. Proposed parameters for integrated algorithm

s_{i+1}	k_i	(s_i, d_i')
-2	0	$(-3, 1)$ or $(-2, 2)$
-2	1	$(-3, 2)$ or $(-2, 3)$
-3	0	$(-3, 3)$ or $(-2, 4)$
-3	1	$(-3, 4)$ or $(-2, 5)$

This integrated algorithm can also incorporate with the windowing method to improve the performance. Details of the algorithm with 2-bit window are given in Table 5 and Fig. 3. In Fig. 3, the variable s is initialized to $s = -(k_{n-1} k_{n-2})_2$ in Step 03 and is updated by a random number between -1 and -3 in Step 06. After each iteration, the temporary result $R[0]$ equals to $g^{(k_{n-1}\cdots k_i)_2} \times g^s$. Therefore, after executing the for loop, $R[0]$ equals to $g^K \times g^s$ and the expected result g^K can be obtained by $R[0] \times g^{-s}$. Notice again that the multiplication with g^{-s} is not a division since s is non-positive.

Table 5. Proposed parameters for integrated algorithm with 2-bit window

s_{i+2}	$(k_{i+1} k_i)_2$	(s_i, d_i')
-1	0	$(-3, 1)$, $(-2, 2)$, or $(-1, 3)$
-1	1	$(-3, 2)$, $(-2, 3)$, or $(-1, 4)$
-1	2	$(-3, 3)$, $(-2, 4)$, or $(-1, 5)$
-1	3	$(-3, 4)$, $(-2, 5)$, or $(-1, 6)$
-2	0	$(-3, 5)$, $(-2, 6)$, or $(-1, 7)$
-2	1	$(-3, 6)$, $(-2, 7)$, or $(-1, 8)$
-2	2	$(-3, 7)$, $(-2, 8)$, or $(-1, 9)$
-2	3	$(-3, 8)$, $(-2, 9)$, or $(-1, 10)$
-3	0	$(-3, 9)$, $(-2, 10)$, or $(-1, 11)$
-3	1	$(-3, 10)$, $(-2, 11)$, or $(-1, 12)$
-3	2	$(-3, 11)$, $(-2, 12)$, or $(-1, 13)$
-3	3	$(-3, 12)$, $(-2, 13)$, or $(-1, 14)$

INPUT: $g, K = (k_{n-1}, \cdots, k_0)_2$
  where n is even and $(k_{n-1}k_{n-2})_2 = (01)_2, (10)_2,$ or $(11)_2$
OUTPUT: g^K

```
01  R[0] = 1;  R[1] = g
02  Precomputation: R[2] = g², ···, R[14] = g¹⁴
03  s = -(kₙ₋₁kₙ₋₂)₂
04  for i from n - 4 downto 0 step -2 do
05    d = -4s
06    s = RandomInteger(-1, -3)
07    R[0] = R[0]⁴
08    R[0] = R[0] × R[d + s + (kᵢ₊₁kᵢ)₂]
09  R[0] = R[0] × R[-s]
10  output R[0]
```

Fig. 3. Randomized exponentiation without division

This algorithm requires thirteen pre-computation values $g^2 \sim g^{14}$ and have three intermediate states, $s = -1, -2,$ or -3. Since d_i' is a positive integer for all i, this algorithm always has two modular squarings and one modular multiplication in each iteration.

5   Security Analysis

5.1   SPA by Distinguishability Between Squaring and Multiplication

Several analysis on randomized scalar multiplication over elliptic curve (or probably also applicable to exponentiation) have been reported based on the distinguishability between doubling and addition over elliptic curve (or squaring and multiplication if some specific methods are employed to speed up the computation of squaring) [6, 10, 11]. If the attacker can distinguish doubling and addition (squaring and multiplication) by analyzing a given power consumption trace, then he might be able to deduce the secret key from the collected doubling and addition (squaring and multiplication) sequence.

The proposed algorithm in Sect. 4 is highly regular and makes the attacker be impossible to deduce the private exponent from the collected (if possible) squaring and multiplication sequence. The reason is obvious because there will be always two (one in the non-windowing version) squarings followed by a multiplication in each iteration whatever the private exponent will be. Therefore, the algorithm in Sect. 4 is immune to SPA by distinguishing the squaring and multiplication. In addition, there is no dummy operation within the proposed algorithm. It will not benefit the safe-error attack [13].

5.2 DPA on Intermediate Results

Differential power analysis (DPA) requires the target implementation under attack be a deterministic algorithm, e.g., the private exponent of RSA is uniquely represented but the base integers can be different for each computation. The proposed algorithm in Fig. 3 generates s randomly (and accordingly also for d) in each iteration. This randomization result makes the original DPA infeasible.

However, there are still some weaknesses, because the intermediate results of the algorithm in Fig. 3 always equal to $g^{(k_{n-1}\cdots k_i)_2+s}$ with $s = -1$, -2, or -3. When an attacker has retrieved the $(n-1)$th to $(i+2)$th bit of the exponent, he can detect the occurrence of $g^{4(k_{n-1}\cdots k_{i+2})_2-3}$ by DPA, because this values only occurs when $(k_{i+1}\,k_i)_2 = 0$. (This value will not appear in the intermediate results when $(k_{i+1}\,k_i)_2 = 1$, 2, or 3.) Similarly, he can detect the occurrence of $g^{4(k_{n-1}\cdots k_{i+2})_2+2}$ because it only occurs when $(k_{i+1}\,k_i)_2 = 3$.

This attack can be avoid by only selecting $s = -1$ or -2 when $(k_{i+1}\,k_i)_2 = 0$ and 2 as well as only selecting $s = -2$ or -3 when $(k_{i+1}\,k_i)_2 = 1$ or 3, because $(k_{i+1}\,k_i)_2 = 0$ and 1 (2 and 3) will become indistinguishable under this attack.

Enlarging the range of the intermediate result will increase the complexity of attacks on this weakness but it will also enlarge the size of the pre-computation table.

6   Conclusions

Randomized recoding on private exponent (private key) of exponentiation computation was previously considered as possible countermeasures against some side-channel attacks. However, this category of countermeasures is not applicable for implementing systems with varying base integer like RSA.

Based on the proposed concept of pre-borrow from the higher priority digits towards the lower priority digits, a new class of inversion-free exponentiation algorithms is developed in this paper. This category of new algorithms can be applicable for implementing important cryptosystems with varying base integer during modular exponentiation computation, e.g., the RSA cryptosystem.

The proposed basic version of inversion-free exponentiation is also enhanced by combining both randomized exponent recoding and windowing technique. The result leads to a secure and efficient countermeasure of implementing cryptosystems based on exponentiation computation over $\mathbf{Z}_n$ against most side-channel attacks. The proposed algorithm can defeat the original DPA because of the varying computational process (sequence of operation) and it can also defeat SPA because of its highly regular computational behavior. Moreover, the enlarged set of possible intermediate results can increase the complexity of most advanced DPA attacks.

The computational performance is that the average number of operations (including both multiplication and squaring) for an n-bit exponent and with window size of two is $1.5n$.

References

1. Dan Boneh, Richard A. DeMillo, and Richard J. Lipton, "On the importance of checking cryptographic protocols for faults," *Advances in Cryptology – EURO-CRYPT '97*, Lecture Notes in Computer Science 1233, pp. 37–51, Springer-Verlag, 1997.
2. Jean-Sébastien Coron, "Resistance against differential power analysis for elliptic curve cryptosystems," *Cryptographic Hardware and Embedded Systems – CHES '99*, Lecture Notes in Computer Science 1717, pp. 292–302, Springer-Verlag, 1999.
3. Pierre-Alain Fouque and Frederic Valette, "The Doubling Attack - Why Upwards Is Better than Downwards," *Cryptographic Hardware and Embedded Systems – CHES '03*, Lecture Notes in Computer Science 2779, pp. 269–280, Springer-Verlag, 2003.
4. Pierre-Alain Fouque, Frédéric Muller, Guillaume Poupard, and Frédéric Valette, "Defeating Countermeasures Based on Randomized BSD Representations," *Cryptographic Hardware and Embedded Systems – CHES '04*, Lecture Notes in Computer Science 3156, pp. 312–327, Springer-Verlag, 2004.
5. JaeCheol Ha and SangJae Moon, "Randomized signed-scalar multiplication of ECC to resist power attacks," *Cryptographic Hardware and Embedded Systems – CHES '02*, Lecture Notes in Computer Science 2523, pp. 551–563, Springer-Verlag, 2002.
6. Chris Karlof and David Wagner, "Hidden Markov model cryptanalysis," *Cryptographic Hardware and Embedded Systems – CHES '03*, Lecture Notes in Computer Science 2779, pp. 17–34, Springer-Verlag, 2003.
7. Paul Kocher, "Timing attack on implementations of Diffie-Hellman, RSA, DSS, and other systems," *Advanced in Cryptology – CRYPTO '96*, Lecture Notes in Computer Science 1109, pp. 104–113, Springer-Verlag, 1996.
8. Paul Kocher, Joshua Jaffe, and Benjamin Jun, "Differential power analysis," *Advanced in Cryptology – CRYPTO '99*, Lecture Notes in Computer Science 1666, pp. 388–397, Springer-Verlag, 1999.
9. Thomas S. Messerges, Ezzy A. Dabbish, and Robert H. Sloan, "Power Analysis Attacks of Modular Exponentiation in Smartcards," *Cryptographic Hardware and Embedded Systems (CHES '99)*, Lecture Notes in Computer Science 1717, pp. 144–157, Springer-Verlag, 1999.

10. Katsuyuki Okeya and Kouichi Sakurai, "On insecurity of the side channel attack countermeasure using addition-subtraction chains under distinguishability between addition and doubling," *Information Security and Privacy – ACISP '02*, Lecture Notes in Computer Science 2384, pp. 420–435, Springer-Verlag, 2002.
11. Katsuyuki Okeya and Dong-Guk Han, "Side Channel Attack on Ha-Moon's Countermeasure of Randomized Signed Scalar Multiplication," *Progress in Cryptology - INDOCRYPT 2003*, Lecture Notes in Computer Science 2904, pp. 334-348, Springer-Verlag, 2003.
12. Elisabeth Oswald and Manfred Aigner, "Randomized addition-subtraction chains as a countermeasure against power attacks," *Cryptographic Hardware and Embedded Systems – CHES '01*, Lecture Notes in Computer Science 2162, pp. 39–50, Springer-Verlag, 2001.
13. Sung-Ming Yen, Seungjoo Kim, Seongan Lim, and Sangjae Moon, "A countermeasure against one physical cryptanalysis may benefit another attack," *Information Security and Cryptology – ICISC '01*, Lecture Notes in Computer Science 2288, pp. 414–427, Springer-Verlag, 2002.

Efficient Computation of Tate Pairing in Projective Coordinate over General Characteristic Fields

Sanjit Chatterjee, Palash Sarkar, and Rana Barua

Cryptology Research Group,
Applied Statistics Unit,
Indian Statistical Institute,
203, B.T. Road,
Kolkata 700 108, India
{sanjit_t, palash, rana}@isical.ac.in

Abstract. We consider the use of Jacobian coordinates for Tate pairing over general characteristics. The idea of encapsulated double-and-line computation and add-and-line computation has been introduced. We also describe the encapsulated version of iterated doubling. Detailed algorithms are presented in each case and memory requirement has been considered. The inherent parallelism in each of the algorithms have been identified leading to optimal two-multiplier algorithm. The cost comparison of our algorithm with previously best known algorithms shows an efficiency improvement of around 33% in the general case and an efficiency improvement of 20% for the case of the curve parameter $a = -3$.

Keywords: Tate pairing, Jacobian coordinate, efficient implementation.

1 Introduction

Pairing based cryptography is a new way of constructing public key protocols. Initially, bilinear maps were used for attacking the discrete logarithm problem on elliptic curve groups [14, 5]. Starting with the initial work of Joux [10], Boneh-Franklin [2], etc. design of pairing based public key protocols have received a great deal of attention from the research community. See [3] for a recent survey of such protocols.

Implementation of pairing based protocols require efficient algorithms for computing pairings. An initial breakthrough in this direction has been made in [1] and [6], which introduced some nice optimisation ideas leading to dramatic improvement in pairing computation time. Since then, there have been quite a few papers on implementation aspects. Almost all of the implementation work have focussed on Tate pairing as it is faster than Weil pairing.

OUR CONTRIBUTIONS: We consider elliptic curves over large prime fields having embedding degree 2. For such fields, we consider the use of Jacobian coordinates for Tate pairing computation. The new idea that we introduce is encapsu-

C. Park and S. Chee (Eds.): ICISC 2004, LNCS 3506, pp. 168–181, 2005.

lated double-and-line computation and encapsulate add-and-line computation. We also describe encapsulated version of iterated double and line computation. From an implementation point of view, we divide curves of the form $y^2 = x^3 + ax + b$ into three cases – small a, $a = -3$ and the case where a is a general element of the field. In each case, we present detailed algorithm for pairing computation. To the best of our knowledge, such a detailed presentation of projective coordinate algorithms have not been reported earlier. We consider the memory requirement for our algorithms, a feature which has not been seriously considered earlier.

For hardware applications having special purpose crypto co-processors, it might be desirable to consider parallel versions of the algorithms. We identify the inherent parallelism in our algorithms and two-multiplier parallel version of our algorithms are optimal with respect to the number of parallel rounds.

In comparison with earlier work, we are able to obtain approximately 33% speed-up over the best known algorithm [8] for the case of general a. In the case $a = -3$ and for non-supersingular curves, the recent work by Scott [16] provides the most efficient algorithm. In comparison, for the case $a = -3$, we are able to obtain approximately 20% speed-up over the algorithm in [16].

RELATED WORK: The important work in [1] and [6] has been mentioned before. Projective coordinates were seriously considered by Izu and Takagi [8], where mainly non-supersingular curves with large embedding degrees were considered. Further, projective coordinates in conjunction with non-supersingular curves with embedding degree 2 were also considered in the work by Scott [16] mentioned earlier. A recent work by Scott and Baretto [17] considers the issue of computing trace of pairings. This paper also describes a laddering algorithm for exponentiation in $\mathbb{F}_{p^2}$ based on Lucas sequences. For general characteristics, this exponentiation algorithm is the fastest known and has to be used with the algorithm that we develop. The algorithm proposed by Eisenträerger et. al. [4] uses the double-add trick with parabolas for fast computation of pairing in affine coordinates. There are several other works on Tate pairing computation. However, most of these work with affine coordinates and over characteristic three. Hence, they are not much relevant in the present context and therefore are not discussed here.

2   Preliminaries

We discuss background material for Tate pairing, choice of curves and NAF representation.

2.1   The Tate Pairing

We first discuss how to compute the (modified) Tate pairing. Let p be an odd prime, $\mathbb{F}_p$ the corresponding finite field with p elements, E is an elliptic curve over $\mathbb{F}_p$. Let r be a large prime divisor of $(p + 1)$, such that r is coprime to p and for some $k > 0$, $r | p^k - 1$ but $r \nmid p^s - 1$ for any $1 \leq s < k$; k is called the embedding degree (MOV degree). Suppose P is a point of order r on the elliptic curve $E(\mathbb{F}_p)$ and Q is a point of same order on the elliptic curve $E(\mathbb{F}_{p^k})$, linearly independent of P. We denote the (modified) Tate pairing of order r as $e_r(P,Q) \in \mathbb{F}_{p^k}$.

$e_r(P,Q)$ is defined in terms of divisors of a rational function. A divisor is a formal sum: $D = \sum_{P \in E} a_P \langle P \rangle$, where $P \in E(\mathbb{F}_p)$. The degree of a divisor D is $deg(D) = \sum_{P \in E} a_P$. The set of divisors forms an abelian group by the addition of their coefficients in the formal sum. Let f be a (rational) function on E, then the divisor of f, $\langle f \rangle = \sum_P ord_P(f)\langle P \rangle$, where $ord_P(f)$ is the order of the zero or pole of f at P. A divisor $D = \sum_{P \in E} a_P \langle P \rangle$ is called a principal divisor if and only if it is a divisor of degree 0 (zero divisor) and $\sum_{P \in E} a_P P = \mathcal{O}$. If D is principal then there is some function f such that $D = \langle f \rangle$. Two divisors D_1 and D_2 are said to be equivalent if $D_1 - D_2$ is a principal divisor. Let $\mathcal{A}_P$ be a divisor equivalent to $\langle P \rangle - \langle \mathcal{O} \rangle$ (similarly $\mathcal{A}_Q$). Then it is easy to see that $r\mathcal{A}_P$ is principal; thus there is a rational function f_P with $\langle f_P \rangle = r\mathcal{A}_P = r\langle P \rangle - r\langle \mathcal{O} \rangle$. The (modified) Tate pairing of order r is defined as –

$$e_r(P,Q) = f_P(\mathcal{A}_Q)^{(p^k-1)/r}.$$

To compute $f_P(\mathcal{A}_Q)$, $Q \neq \mathcal{O}$ one uses Miller's algorithm [12]. Let f_a be a (rational) function with divisor $\langle f_a \rangle = a\langle P \rangle - \langle aP \rangle - (a-1)\langle \mathcal{O} \rangle$, $a \in Z$. It can be shown that $f_{2a}(Q) = f_a(Q)^2 \cdot h_{aP,aP}(Q)/h_{2aP}(Q)$ where, $h_{aP,aP}$ is the line tangent to $E(\mathbb{F}_p)$ at aP, it intersects $E(\mathbb{F}_p)$ at the point $-2aP$, and h_{2aP} is the (vertical) line that intersects $E(\mathbb{F}_p)$ at $2aP$ and $-2aP$. Now, $\langle f_r \rangle = r\langle P \rangle - \langle rP \rangle - (r-1)\langle \mathcal{O} \rangle = \langle f_P \rangle$, since $rP = \mathcal{O}$. Given P and the binary representation of r, Miller's algorithm computes $f_P(Q) = f_r(Q)$ in $\lg r$ steps by the standard double-and-add through line-and-tangent method for elliptic curve scalar multiplication. Under the condition, $r \nmid (p-1)$ we can further have $e_r(P,Q) = f_P(Q)^{(p^k-1)/r}$ for $Q \neq \mathcal{O}$, as long as $k > 1$.

In the implementation of Tate pairing over $E(\mathbb{F}_p)$, the usual practice is to take $Q \in E(\mathbb{F}_p)$ of order r and then use a distortion map $\phi()$, to get a point $\phi(Q) \in E(\mathbb{F}_{p^k})$ of order r which is linearly independent of P. A major finding in [1] is that, for particular choices of the curve parameters and distortion map $\phi()$ we can freely multiply or divide the intermediate result of pairing computation by any $\mathbb{F}_p$ element and consequently completely ignore the denominator part in the computation of Tate pairing.

2.2   Choice of Curves

Let E_1 be the elliptic curve given by the equation

$$y^2 = x^3 + ax$$

over $\mathbb{F}_p$. E_1 is super-singular if $p \equiv 3 \bmod 4$. For these curves, the curve order is $\#E_1(\mathbb{F}_p) = p+1$ and embedding degree is $k = 2$. For such curves, a distortion map [1] is $\phi(x, y) = (-x, iy) \in \mathbb{F}_{p^2} \times \mathbb{F}_{p^2}$ with $i^2 = -1$. Let r be the integer, for which we wish to compute $e_r(,)$. Then $r|(p+1)$ and the final powering in Tate pairing computation is of the form $(p^2 - 1)/r$. As observed in [1], this implies $(p-1)$ divides the final powering exponent.

Let, E_2 be the elliptic curve given by the equation

$$y^2 = x^3 - 3x + B$$

over $\mathbb{F}_p$, $p \equiv 3 \bmod 4$. Scott in his recent paper [16] considered this form of non super-singular EC with embedding degree, $k = 2$ with $\#E_2(\mathbb{F}_p) = p + 1 - t$, where t is the trace of Frobenius [11]. It is shown in [16] that the r for which $e_r(,)$ is computed over E_2 also satisfies $r|(p+1)$ and hence again $p - 1$ divides the final powering exponent of Tate pairing.

Thus for both E_1 and E_2 the following observation holds. Since $x^{p-1} = 1$ for any $x \in \mathbb{F}_p$, this implies that we can freely multiply or divide any intermediate Tate pairing result by any nonzero element of $\mathbb{F}_p$. This has been previously used to improve the efficiency of Tate pairing computation in affine coordinates. In Section 3, we point out the importance of this observation in the context of projective coordinates.

Note that, for E_1 as well as E_2, the embedding degree is $k = 2$ and the elements of the quadratic extension field ($\mathbb{F}_{p^2}$) is represented as $a + ib$, where $a, b \in \mathbb{F}_p$ and i is the *square root* of a quadratic non-residue. For $p \equiv 3 \bmod 4$ we can further have $i^2 = -1$. Essentially, the same algorithm that we develop for E_1 also applies to E_2 by evaluating $e(,)$ at P and $(-x_Q, iy_Q)$.

2.3 Non-adjacent Form Representation

The Non-Adjacent Form (NAF) representation of an integer has been suggested for elliptic curve scalar multiplication. In this representation, the digits $\{0, \pm 1\}$ are used to represent an integer with the property that no two adjacent digits are non-zero. The advantage is that, on an average, the number of non-zero digits is one-third of the length of the representation, while it is one-half in the case of binary representation. For details of NAF representation we refer the reader to [7].

For Tate pairing applications, the representation of r should be sparse, i.e., the number of non-zero digits should be small. The NAF representation is sparser than the corresponding binary representation. Hence in our algorithms, we work entirely with the NAF representation.

3 Encapsulated Computation

In the computation of Tate pairing one needs to perform an implicit scalar multiplication of the EC point P. For this, as well as for computation of the line function $h_{aP,aP}()$ one requires to perform base field inversion. But inversion for large characteristic is quite costly. The standard method to avoid inversion is to move from affine to projective coordinate system. Among the different available projective coordinate systems, the Jacobian gives the best result. In [8] the authors suggested to take the so called *simplified Jacobian-Chudnovsky coordinate* J^s as they store (X, Y, Z, Z^2) instead of (X, Y, Z). However, we have found out that if one encapsulates EC addition/doubling with line computation then there is no need to additionally store Z^2 – one can simply work in the Jacobian coordinate. Here we give the explicit formulae required for the encapsulated computation of double/add-and-line computation. In what follows, by [M] and [S], we respectively denote the cost of one multiplication and one squaring in $\mathbb{F}_p$.

3.1   Encapsulated Point Doubling and Line Computation

Here $P = (X_1, Y_1, Z_1)$ correspond to $(X_1/Z_1^2, Y_1/Z_1^3)$ in affine coordinate. We encapsulate the computation of $2P$ given P together with the computation corresponding to the associated line.

Point Doubling: From the EC point doubling rule we have the following formula:

$$X_3' = \frac{(3X_1^2 + aZ_1^4)^2 - 8X_1Y_1^2}{4Y_1^2Z_1^2}$$

$$Y_3' = \frac{3X_1^2 + aZ_1^4}{2Y_1Z_1}\left(\frac{X_1}{Z_1^2} - X_3'\right) - \frac{Y_1}{Z_1^3}$$

$$X_3 = (3X_1^2 + aZ_1^4)^2 - 8X_1Y_1^2$$

$$Y_3 = (3X_1^2 + aZ_1^4)(4X_1Y_1^2 - X_3) - 8Y_1^4$$

$$Z_3 = 2Y_1Z_1$$

Using temporary variables, we compute:

1. $t_1 = Y_1^2$;   2. $t_2 = 4X_1t_1$;   3. $t_3 = 8t_1^2$;
4. $t_4 = Z_1^2$;   5. $t_5 = 3X_1^2 + aZ_1^4$;   6. $X_3 = t_5^2 - 2t_2$;
7. $Y_3 = t_5(t_2 - X_3) - t_3$;   8. $Z_3 = 2Y_1Z_1$.

So, we require $6[S] + 4[M]$ for EC doubling. Now consider t_5. If a is a general element of $\mathbb{F}_p$, then we have to count the multiplication $a \times (Z_1^4)$. However, if a is small, i.e., it can be represented using only a few (say ≤ 8) bits, then we do not count this multiplication. In this case, aZ_1^4 can be obtained summing Z_1^4 a total of a times. This reduces the operation count to $6[S]+3[M]$. Further, if $a = -3$, then $t_5 = 3(X_1 - Z_1^2)(X_1 + Z_1^2) = 3(X_1 - t_4)(X_1 + t_4)$ and the operation count reduces to $4[S]+4[M]$. These facts are known and can be found in [7].

Line Computation: Note that, the slope μ of $h_{P,P}$, the line through P and $-2P$, is $\mu = t_5/Z_3$. So,

$$h_{P,P}(x, y) = \left(y - \frac{Y_1}{Z_1^3}\right) - \mu\left(x - \frac{X_1}{Z_1^2}\right).$$

Hence, $h_{P,P}(-x_Q, iy_Q) = \left(y_Qi - \frac{Y_1}{Z_1^3}\right) + \mu\left(x_Q + \frac{X_1}{Z_1^2}\right).$

By defining $g_{P,P}(x,y) = (2Y_1 Z_1^3) h_{P,P}(x,y)$, we get,

$$g_{P,P}(-x_Q, iy_Q) = (2Y_1 Z_1) Z_1^2 y_Q i - 2Y_1^2 + (3X_1^2 + aZ_1^4)(Z_1^2 x_Q + X_1)$$
$$= Z_3 t_4 y_Q i - (2t_1 - t_5(t_4 x_Q + X_1))$$

The ultimate result is raised to the power $(p^2 - 1)/r$, where $r|(p+1)$ (see Section 2.1). Thus we have to compute

$$(h_{P,P}(-x_Q, iy_Q))^{(p^2-1)/r} = ((h_{P,P}(-x_Q, iy_Q))^{(p-1)})^{(p+1)/r}$$
$$= ((g_{P,P}(-x_Q, iy_Q)/(2Y_1 Z_1^3))^{(p-1)})^{(p+1)/r}$$
$$= ((g_{P,P}(-x_Q, iy_Q))^{(p-1)})^{(p+1)/r}.$$

The last equation holds since $(2Y_1 Z_1^3) \in \mathbb{F}_p$ and consequently $(2Y_1 Z_1^3)^{p-1} = 1$. Thus, we can work entirely with $g_{P,P}(-x_Q, iy_Q)$ instead of $h_{P,P}(-x_Q, iy_Q)$. Since t_1, t_4 and t_5 have already been computed, $g_{P,P}(-x_Q, iy_Q)$ can be obtained using 4 additional multiplications.

Hence, *encapsulated point doubling and line computation* requires $8[M] + 6[S]$. In the case a is small, this cost is $7[M]+6[S]$ and for the case $a = -3$, this cost is $8[M]+4[S]$.

3.2   Encapsulated (Mixed) Point Addition and Line Computation

We encapsulate the computation of $P + R$ given P in affine and R in Jacobian together with the computation corresponding to the associated line.

Mixed Addition: Given $R = (X_1, Y_1, Z_1)$ and $P = (X, Y, 1)$ we compute $R + P = (X_3, Y_3, Z_3)$ as follows.

$$X_3' = \left(\frac{Y - \frac{Y_1}{Z_1^3}}{X - \frac{X_1}{Z_1^2}}\right)^2 - \frac{X_1}{Z_1^2} - X$$

$$= \left(\frac{YZ_1^3 - Y_1}{(XZ_1^2 - X_1)Z_1}\right)^2 - \frac{X_1}{Z_1^2} - X$$

$$Y_3' = \left(\frac{YZ_1^3 - Y_1}{(XZ_1^2 - X_1)Z_1}\right)(\frac{X_1}{Z_1^2} - X_3') - \frac{Y_1}{Z_1^3}$$

$$X_3 = X_3' Z_3$$
$$= (YZ_1^3 - Y_1)^2 - X_1(XZ_1^2 - X_1)^2 - X(XZ_1^2 - X_1)^2 Z_1^2$$
$$= (YZ_1^3 - Y_1)^2 - (XZ_1^2 - X_1)^2(X_1 + XZ_1^2)$$

$$Y_3 = Y_3' Z_3$$
$$= (YZ_1^3 - Y_1)((XZ_1^2 - X_1)^2 X_1 - X_3) - Y_1(XZ_1^2 - X_1)^3$$

$$Z_3 = (XZ_1^2 - X_1)Z_1$$

Using temporary variables we compute:

1. $t_1 = Z_1^2$;   2. $t_2 = Z_1 t_1$;   3. $t_3 = X t_1$;
4. $t_4 = Y t_2$;   5. $t_5 = t_3 - X_1$;   6. $t_6 = t_4 - Y_1$;
7. $t_7 = t_5^2$;   8. $t_8 = t_5 t_7$;   9. $t_9 = X_1 t_7$;
10. $X_3 = t_6^2 - (t_8 + 2t_9)$;   11. $Y_3 = t_6(t_9 - X_3) - Y_1 t_8$;   12. $Z_3 = Z_1 t_5$.

Hence, we require $3[S] + 8[M]$ for EC point addition. See [7] for details.

Line Computation: Note that, the slope μ of $h_{R,P}$, the line through R and P is $\mu = t_6/Z_3$. So,
$$h_{R,P}(x,y) = (y - Y) - \mu(x - X).$$
Hence, $h_{R,P}(-x_Q, iy_Q) = (y_Q i - Y) + \mu(x_Q + X)$. Define $g(x,y)$ as $g(x,y) = Z_3 h_{R,P}(x,y)$. Thus, we get
$$g_{R,P}(-x_Q, iy_Q) = Z_3 y_Q i - (Z_3 Y - t_6(x_Q + X))$$

As explained in the case of doubling, we can simply work with $g_{R,P}$ instead of $h_{R,P}$. Since we have already computed t_6 and Z_3 during point addition, $g_{R,P}(-x_Q, iy_Q)$ can be computed using additional three multiplications. Hence, *encapsulated point addition and line computation* requires $11[M] + 3[S]$.

4   Algorithm

We consider three situations: $a = -3$; a small (i.e., multiplication by a need not be counted); and the case where a is a general element of $\mathbb{F}_p$. For the first two cases, double-and-add algorithm is considered. For the general case, we adopt an iterated doubling technique used by Izu and Takagi [8].

4.1   Double-and-Add

We slightly modify the Miller's algorithm as improved in [1]. We will call this algorithm the modified BKLS algorithm. In the algorithm the NAF representation of r is taken to be $r_t = 1, r_{t-1}, \ldots, r_0$.

Algorithm 1 (Modified BKLS Algorithm):
1. set $f = 1$ and $V = P$
2. for $i = t - 1$ downto 0 do
3. $(u, V) = \mathsf{EncDL}(V)$;
4.   set $f = f^2 \times u$;
5.   if $r_i \neq 0$, then
6. $(u, V) = \mathsf{EncAL}(V, r_i)$;
7.   set $f = f \times u$;
8.   end if;
9. end for;
10. return f;
end Algorithm 1.

The subroutine EncDL(V) performs the computation of Section 3.1 and returns $(g_{V,V}(\phi(Q)),2V)$. The subroutine EncAL(V, r_i) takes V and r_i as input. If $r_i = 1$, it performs the computation of Section 3.2 and returns $(g_{V,P}(\phi(Q)),V + P)$; if $r_i = -1$, it first negates the point $P = (\alpha, \beta)$ to obtain $P' = -P = (\alpha, -\beta)$, then it performs the computation of Section 3.2 with P' instead of P and returns $(g_{V,-P}(\phi(Q)),V - P)$. The correctness of the algorithm follows easily from the correctness of the original BKLS algorithm.

We consider the cost. The subroutine EncDL is invoked a total of t times while EncAL is invoked a total of s times where s is the Hamming weight of $r_{t-1} \ldots r_0$. The cost of updation in Line 4 is one $\mathbb{F}_{p^2}$ squaring and one $\mathbb{F}_{p^2}$ multiplication. These operations can be completed in five $\mathbb{F}_p$ multiplications (see [17]). The cost of updation in Line 7 is three $\mathbb{F}_p$ multiplications.

The cost of EncDL depends upon the value of the curve parameter a. We analyse the total cost for the following two cases.

Case $a = -3$:

- Cost of EncDL is 8[M]+4[S].
- Cost of update in line 4 is 5[M].
- Cost of EncAL is 11[M]+3[S].
- Cost of update in line 7 is 3[M].
- Total cost is $t(13[M]+4[S]) + s(14[M]+3[S])$.

Case a is small:

- Cost of EncDL is 7[M]+6[S].
- Cost of update in line 4 is 5[M].
- Cost of EncAL is 11[M]+3[S].
- Cost of update in line 7 is 3[M].
- Total cost is $t(12[M]+6[S]) + s(14[M]+3[S])$.

4.2   Iterated Doubling

In the case where we have to consider the multiplication by the curve parameter a, we employ the technique of iterated doubling to reduce the total number of operations. As before we consider the NAF representation of r. We write the NAF representation of r as

$$(r_t = 1, r_{t-1}, \ldots, r_0) = (l_s = 1, 0^{w_s-1}, l_{s-1}, \ldots, 0^{w_0}, l_0)$$

where the l_i's are ± 1. The following algorithm is an iterated doubling version of the modified BKLS algorithm described in Section 4.1. The points $P = (\alpha, \beta)$ and $Q = (x_Q, y_Q)$ are globally available.

Algorithm 2 (iterated doubling):
Input: $P = (\alpha, \beta, 1)$ in Jacobian coordinates; $Q = (x_Q, y_Q)$.
Output: $f_P(\phi(Q))$.
1. Set $f = 1$; $g = 1$;
2. $X = \alpha$; $Y = \beta$; $Z = 1$; set $R = (X, Y, Z)$;

3. for $j = s - 1$ down to 0
4. $(f, R) = \mathsf{Encldbl}(f, R, w_j)$;
5. $(g, R) = \mathsf{EncAL}(R, l_j)$;
6. $f = f \times g$;
7. end for;
8. return f;
end Algorithm 2.

The Subroutine EncAL has already been discussed in Section 4.1. We now describe Subroutine $\mathsf{Encldbl}$.

Subroutine $\mathsf{Encldbl}$
Input: $R = (X, Y, Z)$, f and w.
Output: updated f and $2^{w+1}R$.
1. $t_1 = Y^2$; $t_2 = 4Xt_1$; $t_3 = 8t_1^2$; $t_4 = Z^2$; $w = aZ^4$; $t_5 = 3X^2 + w$;
2. $A = -(2t_1 + t_5(t_4 x_Q - X))$; $X = t_5^2 - 2t_2$;
    $Y = t_5(t_2 - X) - t_3$; $Z = 2YZ$; $B = Zt_4 y_Q$;
3. $f = f^2 \times (A + iB)$;
4. for $j = 1$ to w do
5. $w = 2t_3 w$; $t_1 = Y^2$; $t_2 = 4Xt_1$; $t_3 = 8t_1^2$; $t_4 = Z^2$; $t_5 = 3X^2 + w$;
6. $A = -(2t_1 + t_5(t_4 x_Q - X))$; $X = t_5^2 - 2t_2$;
        $Y = t_5(t_2 - X) - t_3$; $Z = 2YZ$; $B = Zt_4 y_Q$;
7. $f = f^2 \times (A + iB)$;
8. end for;
9. $R = (X, Y, Z)$;
9. return (f, R);
end Subroutine $\mathsf{Encldbl}$.

Algorithm 2 is essentially the same as Algorithm 1 except for the use of iterated doubling. The technique of iterated doubling is considered to reduce computation cost but does not affect the correctness of the algorithm. We consider the cost of the algorithm. As before let the Hamming weight of $r_{t-1}, \ldots, r_0$ be s.

- Steps 5 and 6 of Algorithm 2 are invoked s times. The total cost of these two steps is $s(14[\mathsf{M}] + 3[\mathsf{S}])$.
- Step 4 of Algorithm 2 is invoked a total of s times. The cost of the jth invocation of Step 4 is computed as follows:
  - Cost of Steps 3 and 7 in $\mathsf{Encldbl}$ is $5[\mathsf{M}]$.
  - Cost of Steps 1 and 2 in $\mathsf{Encldbl}$ is $8[\mathsf{M}] + 6[\mathsf{S}]$.
  - Cost of Steps 5 and 6 in $\mathsf{Encldbl}$ is $8[\mathsf{M}] + 5[\mathsf{S}]$.
  - Total cost of jth invocation of $\mathsf{Encldbl}$ is
    $13[\mathsf{M}] + 6[\mathsf{S}] + w_j(13[\mathsf{M}] + 5[\mathsf{S}]) = 1[\mathsf{S}] + (w_j + 1)(13[\mathsf{M}] + 5[\mathsf{S}])$.
- Total cost of Algorithm 2 is
  $s(14[\mathsf{M}] + 3[\mathsf{S}]) + \sum_{j=0}^{s-1}(1[\mathsf{S}] + (w_j + 1)(13[\mathsf{M}] + 5[\mathsf{S}]))$
  $= s(14[\mathsf{M}] + 4[\mathsf{S}]) + t(13[\mathsf{M}] + 5[\mathsf{S}])$.

4.3   Memory Requirement

The memory requirement of Algorithm 1 and Algorithm 2 are similar with Algorithm 2 requiring slightly more memory. We consider the memory requirement of Algorithm 2. To find the minimum memory requirement, first note that in Algorithm 2 we have to store and update $f \in \mathbb{F}_{p^2}$ and $X, Y, Z \in \mathbb{F}_p$ – they require 5 $\mathbb{F}_p$ storage space. We also need to store $Q = (x_Q, y_Q)$. In addition, we require some temporary variables to keep the intermediate results produced in the subroutines Encldbl and EncAL. These subroutines are called one after another – we first call Encldbl and update f together with X, Y, Z, release the temporary variables and then call EncAL where these temporary variables can be reused. The maximum of the number of temporary variables required by the two subroutines determines the number of temporary variables required in Algorithm 2. We ran a computer program to separately find these requirements. Given a straight line code what the program essentially does is to exhaustively search (with some optimisations) for all possible execution paths and output the path pertaining to minimum number of temporary variables. This turns out to be 9 for Encldbl, while it is 7 for EncAL. So, at most we require to store 16 $\mathbb{F}_p$ elements.

4.4   Parallelism

We first consider the parallelism in the encapsulated computations of Sections 3.1 and 3.2. While considering parallelism, we assume that a multiplier is used to perform squaring.

First we consider the case of encapsulated double and line computation. The situation in Section 3.1 has three cases – small a, $a = -3$ and general a. For the last case we use the iterated doubling technique of Section 4.2. We separately describe parallelism for the three cases. In each case, we first need to identify the multiplications which can be performed together. This then easily leads to parallel algorithms with a fixed number of multipliers.

Small a. In this case, multiplication by a will be performed as additions. The multiplication levels are as follows.

Level 1 : $t_1 = Y_1^2$; $t_4 = Z_1^2$; X_1^2; $Z_3 = 2Y_1Z_1$; square f;
Level 2 : $t_2 = 4X_1t_1$; $t_3 = 8t_1^2$; $t_5 = 3X_1^2 + aZ_1^4$; $t_6 = t_4x_Q$; $t_7 = t_4y_Q$;
Level 3 : $-(2t_1 + t_5(t_6 - X_1))$; $X_3 = t_5^2 - 2t_2$; $Y_3 = t_5(t_2 - X_3) - t_3$; Z_3t_7;
Level 4 : update f.

Case $a = -3$. In this case, $t_5 = 3(X_1^2 - Z_1^4) = 3(X_1 - Z_1^2)(X_1 + Z_1^2)$. The multiplication levels are as follows.

Level 1 : $t_1 = Y_1^2$; $t_4 = Z_1^2$; $Z_32Y_1Z_1$; square f;
Level 2 : $t_2 = 4X_1t_1$; $t_3 = 8t_1^2$; $t_5 = 3(X_1 - t_4)(X_1 + t_4)$; $t_6 = t_4x_Q$; $t_7 = t_4y_Q$;
Level 3 : $-(2t_1 + t_5(t_6 - X_1))$; $X_3 = t_5^2 - 2t_2$; $Y_3 = t_5(t_2 - X_3) - t_3$; Z_3t_7;
Level 4 : update f.

General a. In this case, Subroutine Encldbl is used. This consists of an initial part plus computation inside the *for* loop. The parallel version of both these parts are similar and we describe the parallel version of the loop computation. The multiplication levels are as follows.

Level 1 : $w = 2t_3w$; $t_1 = Y^2$; $t_4 = Z^2$; X^2; $Z_3 = 2YZ$; square f;
Level 2 : $t_2 = 4Xt_1$; $t_3 = 8t_1^2$; $t_5 = 3X^2 + w$; $t_6 = t_4x_Q$; $t_7 = t_4y_Q$;
Level 3 : $A = -(2t_1 + t_5(t_6 - X))$; $X = t_5^2 - 2t_2$; $Y = t_5(t_2 - X) - t_3$; Z_3t_7;
Level 4 : update f.

In each of the above cases, with two multipliers the entire operation can be performed in 9 rounds and with four multipliers it can be performed in 5 rounds. Since the total number of operations is either 17 or 18 squarings and multiplications, the number of rounds is optimal for the given number of operations and given number of multipliers.

Addition. We now consider the case of encapsulated add-and-line computation. See Section 3.2 for the details of the temporary variables and the operations. Here we mainly list the multiplication/squaring operations.

Level 1 : $t_1 = Z_1^2$;
Level 2 : $t_2 = Z_1t_1$; $t_3 = Xt_1$;
Level 3 : $t_4 = Yt_2$; $t_7 = t_5^2$; $Z_3 = Z_1t_5$;
Level 4 : $t_8 = t_5t_7$; $t_9 = X_1t_7$; $X_3 = t_6^2 - (t_8 + 2t_9)$; Z_3y_Q; Z_3Y; $t_6(x_Q - X)$;
Level 5 : $Y_3 = t_6(t_9 - X_3) - Y_1t_8$; update f;

There are a total of 17 multipications including the update operation. Using two multipliers, these can be performed in 9 rounds. On the other hand, the four multiplier algorithm is sub-optimal in the number of rounds.

Thus, for parallel version of pairing computation algorithm, one obtains optimal two-multiplier algorithms for both doubling and addition. For doubling, the four multiplier algorithm is optimal, while for addition, the four multiplier algorithm is sub-optimal. However, the Hamming weight of r will be small and hence if we use four multipliers then the sub-optimal performance will be amortized over the length of the representation of r and will not be significantly reflected in the final cost analysis.

5   Comparison

For the purpose of comparison, we assume that $r = (r_t = 1, r_{t-1}, \ldots, r_0)$ is represented in NAF having length t and Hamming weight s.

The irrelevant denominator optimisation was introduced in [1]. Further, [1] uses affine representation. The total cost including point/line computation and updation is $t(1[I]+8[M]+2[S])+s(1[I]+6[M]+1[S])$, where [I] is the cost of inversion over $\mathbb{F}_p$ and is at least 30[M], see [15].

Izu-Takagi [8] uses projective coordinates for pairing computation in general characteristics for large embedding degree k. They also consider the BKLS optimisations for supersingular curves with embedding degree $k = 2$ for general a. They assume that one $\mathbb{F}_{p^k}$ multiplication takes $k^2[M]$. For $k = 2$, this can be improved to 3[M]. In the following calculation, we use this fact. Their cost for w-iterated doubling is $6w[M]+4w[S]+13w[M]+(5w+1)[S]$ and addition is $6[M]+16[M]+3[S]$. Summing over w's, the total cost comes to $t(19[M]+9[S]) +s(22[M]+4[S])$.

The recent paper by Scott [16], also proposes the use of projective coordinates in the case $a = -3$ for certain non-supersingular curves. The paper does not distinguish between multiplication and squaring. The total cost is $21t[M]+22s[M]$. In Table 1, we summarize the above costs along with the costs obtained by our algorithms for the various cases for the curve parameter a. The best case

Table 1. Cost Comparison. Note $1[I] \geq 30[M]$ [15]

Method	Cost
BKLS [1] (affine)	$t(1[I]+8[M]+2[S])+s(1[I]+6[M]+1[S])$
Izu-Takagi [8] (general a)	$t(19[M]+9[S])+s(22[M]+4[S])$
Scott [16] ($a = -3$)	$21t[M]+22s[M]$
Algorithm 1 (a small)	$t(12[M]+6[S])+s(14[M]+3[S])$
Algorithm 1 ($a = -3$)	$t(13[M]+4[S])+s(14[M]+3[S])$
Algorithm 2 (general a)	$t(13[M]+5[S])+s(14[M]+4[S])$

occurs for Algorithm 1 with $a = -3$. Also the cases for Algorithm 1 for small a and Algorithm 2 are marginally slower than the best case. However, all three of these cases are much more efficient than any of the previous algorithms. The algorithms of Izu-Takagi [8] and Scott [16] are more efficient than the basic BKLS algorithm with affine coordinates.

For Tate pairing applications, r is generally chosen so that the Hamming weight s is small. On the other hand, for a general r, the Hamming weight s is approximately $s = t/3$. In either of these two situations, we summarize the superiority of our method as follows.

- Algorithm 1 with $a = -3$ is approximately 20% faster compared to the algorithm by Scott.
- Algorithm 2 is approximately 33% faster compared to the algorithm by Izu and Takagi.

We consider the cost comparison to EC scalar multiplication. For the purpose of security, scalar multiplication has to be resistant to side channel attacks. One simple method of attaining resistance to simple power analysis is to use Coron's dummy addition using binary representation of multiplier. Under the (realistic) assumption that the length of the binary representation of the multiplier is equal to the length of the NAF representation of r for Tate pairing, the cost of dummy-addition countermeasure is $t(2[\mathrm{M}]+7[\mathrm{S}])$ for the case of $a = -3$. This cost is comparable to the cost of Algorithm 1 for $a = -3$ when s is at most around $t/8$. Again for practical situation r can usually be chosen so that $s \le t/8$. Thus the efficiency of our algorithm is almost comparable to the efficiency of simple SPA resistant EC scalar multiplication. On the other hand, there is a wide variety of techniques for EC scalar multiplication providing very efficient algorithms. Whether the cost of Tate pairing computation can be made comparable to the most efficient EC scalar multiplication is currently a challenging research problem.

6   Conclusion

In this paper, we have considered the use of Jacobian coordinates for Tate pairing computation in general characteristics. The main idea that we have introduced is encapsulated double-and-line computation and encapsulated add-and-line computation. We have also developed encapsulated version of iterated double algorithm. The algorithms are presented in details and memory requirement has been considered. Inherent parallelism in these algorithms have been identified leading to optimal two-multiplier parallel algorithms. Our algorithms lead to an improvement of around 33% over best known algorithm for the general case where the curve parameter a is an arbitrary field element. In the special case where $a = -3$, our techniques provide an efficieny improvement of around 20% over the previously best known algorithm.

References

1. P. S. L. M. Barreto, H. Y. Kim, B. Lynn and M. Scott. Efficient algorithms for pairing-based cryptosystems. In *Advances in Cryptology - Crypto'2002*, volume 2442 of *Lecture Notes in Computer Science*, pages 354-368. Spriger-Verlag, 2002.
2. D. Boneh and M. Franklin. Identity-based encryption from the Weil pairing. *SIAM Journal of Computing*, 32(3):586-615, 2003.
3. R. Dutta, R. Barua and P. Sarkar. Pairing-based cryptography : A survey. Cryptology ePrint Archive, Report 2004/064, 2004. Available from http://eprint.iacr.org/2004/064.
4. K. Eisenträger, K. Lauter and P. L. Montgomery. Fast elliptic curve arithmetic and improved Weil pairing evaluation. CT-RSA 2003, Lecture Notes in Computer Science, Vol. 2612, pp. 343-354, Springer-Verlag, (2003). (Also see http://eprint.iacr.org/2002/112).
5. G. Frey, M. Muller and H. Ruck, The Tate Pairing and the Discrete Logarithm applied to Elliptic Curve Cryptosystems, *IEEE Trans, Inf. Theory*, 45, No. 5(1999) pp. 1717-1719.
6. S. Galbraith, K. Harrison and D. Soldera. Implementing the Tate pairing. In *Algorithmic Number Theory Symposium – ANTS V*, volume 2369 of *Lecture Notes in Computer Science*, Pages 324-337. Springer-Verlag, 2002.
7. D. R. Hankerson, A. J. Menezes and S. A. Vanstone, *Guide to Elliptic Curve Cryptography*, Springer, 2004.

8. T. Izu and T. Takagi, Efficient computation of the Tate pairing for the Large MOV degree. 5th International Conference on Information Security and Cryptolgy, ICISC 2002, Lecture Notes in Computer Science, Vol. 2587, pp. 283-297, Springer-Verlag, (2003).

9. K. Itoh, M. Takenaka, N. Torii, S. Temma and Y. Kurihara, Fast implementation of public key cryptography on DSP TMS320C6201, CHES'99, LNCS 1717, pp 61-72, Springer-Verlag, 1999.

10. A. Joux. A one-round protocol for tripartite Diffie-Hellman. In *Algorithmic Number Theory Symposium – ANTS IV*, volume 1838 of *Lecture Notes in Computer Science*, Pages 385-394. Springer-Verlag, 2000.

11. A. Menezes. *Elliptic Curve Public Key Cryptosystems*. Kluwer Academic Publishers, 1993.

12. V. S. Miller. Short programs for functions on curves. Unpublished manuscript, 1986. Available from http://crypto.stanford.edu/miller/miller.pdf

13. W. Mao and K. Harrison. Divisors, bilinear pairings and pairing enabled cryptographic applications, 2003. http://hplbwww.hpl.hp.com/people/wm/research/pairing.pdf.

14. A. Menezes, T. Okamoto and S. A. Vanstone, Reducing elliptic curve logarithms to logarithms in a finite field, *IEEE Trans, Inf. Theory*, 39, No. 5(1993) 1639-1646.

15. A. Menezes, P. C. van Oorschot and S. A. Vanstone, *Handbook of Applied Cryptology*. CRC Press, 1997

16. M. Scott. Computing the Tate Pairing, manuscript accepted for presentation in *CT-RSA 2005*.

17. M. Scott and P. S. L. M. Barreto. Compressed pairings. In *Advances in Cryptology - Crypto'2004*, volume 3152 of *Lecture Notes in Computer Science*, Spriger-Verlag, 2004. Also available in cryptology ePrint archive : Report 2004/032.

On Subliminal Channels in Deterministic Signature Schemes

Jens-Matthias Bohli and Rainer Steinwandt

IAKS, Arbeitsgruppe Systemsicherheit, Prof. Dr. Th. Beth,
Fakultät für Informatik, Universität Karlsruhe, Am Fasanengarten 5,
76 131 Karlsruhe, Germany

Abstract. Subliminal channels in randomized signature algorithms like the DSA are well-known. However, much less seems to be known about this issue when dealing with deterministic schemes. Using some known signature schemes like ESIGN-D and SFLASH^{v3} as example, we illustrate the problem of subliminal channels in non-interactive deterministic signature algorithms. Based on an appropriate formalization, a deterministic variant of RSA-PSS is shown to be subliminal free.

Keywords: subliminal channels, deterministic signature schemes.

1 Introduction

Basically, a subliminal channel in a signature scheme allows a signing party to send a covert message to an authorized receiver of signed messages. Without additional knowledge, the covert message cannot be detected, i.e., the derived signatures are indistinguishable from those not containing a subliminal message. Simmons introduced subliminal channels in [Sim84] as a solution to the prisoner's problem: A warden allows two prisoners to exchange authenticated (signed) messages but monitors their communication. The prisoners seek for a way to communicate unnoticeable to the warden and can use a subliminal channel to hide a secret message in the authenticator of a "harmless" message.

The first subliminal channels in signature schemes were pointed out by Simmons in the ElGamal signature scheme [Sim85] as well as in the DSA [Sim94]. Subsequently such channels have been discovered in many signature schemes, recently by Zhang et al. [ZLK03]. Typically, for implementing a subliminal channel, random values in the signing procedure are replaced with (symmetric) encryptions of a covert message. The intended receiver knows the symmetric key used (and possibly also the signing key, if this is necessary to reconstruct the "random" choices used in the signing procedure). A seemingly obvious method to prevent subliminal communication within signatures is the use of deterministic schemes: If there are no random choices in the signing algorithm, it seems that the signature leaves no "space" for placing a subliminal message. But the deterministic generation of the signature does not mean that this signature is the only one that passes the verification. The only published work about subliminal communication in this context we are aware of, is a short paragraph in

C. Park and S. Chee (Eds.): ICISC 2004, LNCS 3506, pp. 182–194, 2005.

the meeting minutes [IEE01]: "... the verifier has no way of telling whether the signature was generated deterministically or not." The same paragraph also indicates an obvious limitation for implementing subliminal communication by means of deterministic signature schemes: "... if the same message is signed twice and produces two different signatures, the verifier knows that the signing was not deterministic."

In this contribution we suggest a formalization of subliminal channels and subliminal freeness in non-interactive signature schemes, which is formulated in terms of a game similar to those typically used in connection with provable security. As in practical applications non-interactive signature schemes without message recovery are prevalent, we do not consider interactive signature schemes or signature schemes with message recovery. The setting for (ab)uses of subliminal channels we have in mind is that the (malicious) signer has no influence on the cover message, thus ruling out steganography in the cover message. An example for such a scenario is a malicious black-box signing device that leaks the private signing key as considered in [YY96]: Upon input of the user's signing key and a cover message, the device signs the message so that the the user's signing key can be recovered from his signatures. This recovery remains unnoticeable to the legitimate signer. Other applications are digital identification cards where the issuer cannot choose the contents of the message to be signed, since it is given, e. g., by name and address.

Albeit having clear limitations—e. g., not capturing intuitively present subliminal channels in "non-scalable" schemes—the proposed definition allows for a satisfactory discussion of a significant class of subliminal communication. Section 3 exhibits (both in an informal and in a more formal manner) subliminal channels in several deterministic signature schemes, including ESIGN-D, a deterministic version of RSA-PSS, and SFLASH^{v3}. Further on, it is shown that when being combined with a suitable key verification the already mentioned deterministic version of RSA-PSS is subliminal free.

2   Subliminal Channels in Signature Schemes

A non-interactive signature scheme is typically formalized as a tuple of algorithms for key generation, signature generation and signature verification. For taking into account subliminal channels, some modifications to this definition are needed, and thus we start by fixing some appropriate terminology.

2.1   Basic Definitions

Let us first recall the basic ingredients of an ordinary signature scheme:

Definition 1. *A signature scheme* $S = (\mathsf{Gen}, \mathsf{Sig}, \mathsf{Ver})$ *is a triple of algorithms, where*

- Gen *is a probabilistic polynomial time (pptm) algorithm that takes the security parameter k as input and returns a pair of public and secret keys* (pk, sk).
- Sig *is a pptm algorithm that takes a message M and the secret key sk as input and produces a valid signature* σ *for M under sk.*
- Ver *is a deterministic polynomial time algorithm that takes a message M, a signature* σ *and the public verification key pk as input and returns* valid *if* σ *is a valid signature for M w. r. t. pk, and* invalid *otherwise.*

In a signature scheme with subliminal channel, the key generation algorithm does not only provide a private signing and a public verification key, but also a subliminal secret key which is shared between the subliminal sender and the subliminal receiver. Although we do not require the subliminal key generation to follow the steps of the original key generation, we require the distribution of (public key, secret key)-pairs to be indistinguishable from the original one. The subliminal secret key may well depend on or even contain the secret signing key: We are interested in settings, where the possibility of subliminal communication is more important than the unforgeability of the subliminal sender's signatures.

In addition to the standard signature algorithm we need an "alternative" signature algorithm that allows to hide a subliminal message inside a signature. To implement a secure subliminal channel, a warden must not be able to distinguish (with non-negligible probability) which of the two key generation and signature algorithms has been used, even when he is able to choose the messages to be signed himself. Finally, an additional algorithm for extracting subliminal messages from received signatures is needed:

Definition 2. *A* signature scheme with subliminal channel

$$S = (\mathsf{Gen}, \mathsf{Sig}, \mathsf{Ver}, \mathsf{SubGen}, \mathsf{SubSig}, \mathsf{SubVer})$$

is a tuple of algorithms, where Gen, Sig, Ver *are as in Definition 1 and*

- SubGen *is a pptm algorithm that takes the security parameter k as input and outputs a pair of public and secret keys* pk, sk *along with a subliminal secret key ssk that is to be used for subliminal communication.*
  The resulting distribution of (pk, sk)*-pairs is required to be indistinguishable from that produced by* Gen.
- SubSig *is a pptm algorithm that takes a message M, a subliminal message m, the secret key sk, and the state information s output in the last activation (initially we set* $s := ssk$*) as input and outputs new state information s along with a signature* σ *for M, such that* σ *hides the subliminal message m.*
- SubVer *is a deterministic polynomial time algorithm that takes the message M, a signature* σ*, the public verification key pk, and the subliminal secret key ssk as input.* Ver *applied to M,* σ*, and pk returns* invalid *if and only if* SubVer *returns* invalid. *For a valid signature, the subliminal message embedded by* SubSig *is recovered by means of ssk. At this, we require that in a single*

application of the signature scheme, an embedded message can be recovered with overwhelming probability, i. e.,

$$P \begin{bmatrix} (pk, sk, ssk) \leftarrow \mathsf{SubGen}(1^k), M \leftarrow \mathcal{M}, m \leftarrow \mathcal{M}_s, \\ (., \sigma) \leftarrow \mathsf{SubSig}(M, m, sk, ssk); \\ \mathsf{SubVer}(M, \sigma, pk, ssk) = (\mathsf{valid}, m) \end{bmatrix} \geq 1 - \mathrm{negl}(k),$$

with the probability being over the random choices of SubGen *and* SubSig *and the messages M and m being chosen independently from the corresponding finite message space* $\mathcal{M} = \mathcal{M}(k)$ *resp. finite subliminal message space* $\mathcal{M}_s = \mathcal{M}_s(k)$ *according to some (application) specific probability distributions. For all values of k, the subliminal message space must contain at least 2 different messages, and for all pptm algorithms W (wardens) we require*

$$\left| P[\mathbf{Exp}^{\mathrm{ward-ind-1}}(k) = 1] - P[\mathbf{Exp}^{\mathrm{ward-ind-0}}(k) = 1] \right| \leq \mathrm{negl}(k)$$

where for $b \in \{0, 1\}$ *the experiment* $\mathbf{Exp}^{\mathrm{ward-ind-}b}(\cdot)$ *is defined as follows:*

Experiment $\mathbf{Exp}^{\mathrm{ward-ind-0}}(k)$:	*Experiment* $\mathbf{Exp}^{\mathrm{ward-ind-1}}(k)$:
$(pk, sk) \leftarrow \mathsf{Gen}(1^k);$	$(pk, sk, ssk) \leftarrow \mathsf{SubGen}(1^k);$
$d \leftarrow W^{\mathcal{S}_{sk}(\cdot, \cdot)}(pk);$	$d \leftarrow W^{\mathcal{S}_{sk,s}(\cdot, \cdot)}(pk);$
return d;	*return d;*

with $\mathcal{S}_{sk}(\cdot, \cdot)$ *an oracle which on input (M, m) returns* $\sigma \leftarrow \mathsf{Sig}(M, sk)$ *and* $\mathcal{S}_{sk,s}(\cdot, \cdot)$ *an oracle which returns* σ *with* $(\sigma, s) \leftarrow \mathsf{SubSig}(M, m, sk, s)$ *and updates its internal state s (initially equal to ssk) accordingly.*

If the signature scheme is malleable, a warden may be able to derive another valid signature for a message M that has been (subliminally) signed already. If in this case the subliminal message can still be extracted or if the signature scheme is not malleable, a subliminal channel can be regarded as "robust". In this contribution, we do not dwell on the question of robustness any further, and thus omit a formal definition.

Remark 1. For implementing a subliminal channel on basis of a deterministic signature scheme, allowing SubSig to have state is crucial: Once a subliminal message $m_1 \in \mathcal{M}_s$ has been hidden in a signature of some $M \in \mathcal{M}$, this message M cannot be (re)used to hide an $m_2 \in \mathcal{M}_s \setminus \{m_1\}$.—Otherwise the use of the subliminal channel becomes obvious.

So far we only considered to transmit a subliminal message in a single application of the (subliminal) signing algorithm. We introduce the capacity of a subliminal channel to quantify the subliminal information that can be transmitted over time.

Definition 3. *Denote by* $\mathcal{M}_s = \mathcal{M}_s(k)$ *the subliminal message space of a signature scheme with subliminal channel S and by* $P_{\mathcal{M}_s}$ *a probability distribution on* $\mathcal{M}_s$. *Then the average information in a subliminal message is*

$$H(\mathcal{M}_s) = - \sum_{m \in \mathcal{M}_s} P_{\mathcal{M}_s}(m) \log_2(P_{\mathcal{M}_s}(m)) \quad (\text{measured in bit per signature}),$$

the entropy of the source of the subliminal messages. Let $\mathcal{S}_S$ be the set of sub-liminal signature schemes that are, except for the subliminal message space $\mathcal{M}_s$, identical to S.

If, for any polynomial number of executions of SubSig, we can hide a sub-liminal message in every signature with overwhelming probability[1], we refer to $C(S) := max_{\mathcal{S}_S} H(\mathcal{M}_s)$ as capacity of the subliminal channel. Similarly as in [Sim98], we call the subliminal channel broadband if its bandwidth grows at least linearly in k, i. e., $C(S) \in \Omega(k)$. We call the subliminal channel narrowband, if its bandwidth grows no more than logarithmically in k, i. e., $C(S) \in O(\log_2(k))$.

For deterministic signature schemes the situation is different since in general a message can only once be signed subliminally without being recognized by the warden. Thus, in the deterministic case we obtain a sequence $B = (B_i)_{i \in \mathbb{N}} \in \mathbb{R}^{\mathbb{N}}$, with B_i bounding the average information that can be transmitted subliminally in the i-th signature, using the probability that we get a message M that was not previously signed:

$$B_i = C(S) \cdot \sum_{M \in \mathcal{M}} (1 - P_{\mathcal{M}}(M))^{i-1} P_{\mathcal{M}}(M)$$

For some applications of signatures it is desired to prevent any subliminal communication through valid signatures once the public verification key has been fixed, and it is desirable to have signature schemes which are provably subliminal free in this sense. The following definition tries to capture this idea:

Definition 4. Let $S = (\mathsf{Gen}, \mathsf{Sig}, \mathsf{Ver})$ be a signature scheme. We call S sublimi-nal free if for all pptm algorithms SubGen, GetCoverMessage, BadSign, BadVerify we have

$$\left| P[\mathbf{Exp}^{\mathrm{signer}-1}(k) = 1] - P[\mathbf{Exp}^{\mathrm{signer}-0}(k) = 1] \right| \leq \mathrm{negl}(k).$$

At this, SubGen obeys the restrictions in Definition 2, and for $b = 0, 1$ the ex-periment $\mathbf{Exp}^{\mathrm{signer}-b}$ is defined as follows:

Experiment $\mathbf{Exp}^{\mathrm{signer}-b}$:
    $(pk, sk, ssk) \leftarrow \mathsf{SubGen}(1^k)$;
    $(M, s) \leftarrow \mathsf{GetCoverMessage}(ssk, pk)$; (with $M \in \mathcal{M}$, s aux. information)
    $\sigma \leftarrow \mathsf{BadSign}(M, b, sk, s)$; (a signature σ with $\mathsf{Ver}(M, \sigma, pk) = \mathsf{valid}$);
    $d \leftarrow \mathsf{BadVerify}(M, \sigma, pk, s)$; (a guess d for the hidden bit b)
    return d;

2.2   Encrypting Subliminal Messages

We begin by studying how to encrypt subliminal messages in probabilistic sig-nature schemes. In a sense, Remark 1 is dual to probabilistic schemes which

---

[1] Note that deterministic schemes do not satisfy this requirement.

allow superpolynomially many signatures for the same message: In the latter case, obtaining twice the same signature for the same message is suspicious.

In other words, for a probabilistic signature scheme it is important that repeatedly sending the same subliminal message does not yield the same "random" number to be used during signing. Typically, we want to replace a random bitstring r of length linear in the security parameter with the encryption of a subliminal message. And, when signing repeatedly the same message and hiding the same subliminal message, we must be able to produce enough different signatures.

Depending on the specific application context, different approaches for producing suitable ciphertexts are possible. One approach is the following: Let Δ be a fixed bitstring of superlogarithmic length which is part of the subliminal secret key ssk and let $\mathrm{MGF}(\cdot, \cdot)$ be some mask generation function which upon input of a seed value z and a positive integer ℓ outputs a pseudorandom bitstring of length ℓ—for practical purposes we may think of a construction as in [Lab02–Appendix B.2]. Then a subliminal message m of length $|m|$ is encrypted as

$$c = \Xi \| [m \oplus \mathrm{MGF}(\Delta \| \Xi, |m|)],$$

where Ξ is a uniformly chosen random bitstring of superlogarithmic length, $\|$ denotes the concatenation of bitstrings and $\oplus$ stands for bit-wise XOR.

For the subliminal receiver decryption is straightforward and, interpreting $\mathrm{MGF}(\cdot, |m|)$ as a random oracle, it is not hard to see that for a pptm warden the resulting ciphertexts are indistinguishable from uniformly chosen random $|\Xi| + |m|$-bit strings. Note that for applying this construction, the number of possible signatures of a cover message $M \in \mathcal{M}$ into which we embed the subliminal messages must grow superpolynomially in k, so that enough random choices for Ξ are available and therewith a pptm warden is not able to find out that we exhaust only a proper subset of all possible signatures.

If we restrict the warden so that he may request the oracle just once for every (cover) message M we can omit the variable secret Ξ and replace it in the input of MGF by the now for every signature distinct message M. Thereby the number of possible signatures per message does not need to grow in k at all. The situation in deterministic schemes is similar to this, since we only hide a subliminal message when signing a (cover) message for the first time.

Alternatively, we can avoid the need to have superpolynomially many signatures if we allow SubVer to be stateful and presume that the subliminal receiver can recognize the chronological order of the signatures. In this case we can use a one time pad encryption.

3   Application to Some Proposed Signature Schemes

For some signature schemes (like SFLASH^{v3} considered in Section 3.3) parameters have not been specified for an arbitrary security parameter k, and only some fixed set of parameters has been proposed. From the practical point of view this usually causes no problem, but an asymptotic description as used above is not suitable for describing an "intuitive" subliminal channel in such a scheme. Depending on what allows for a more convenient exposition, for the concrete signature schemes considered in the sequel, we make use of both more formal and more informal descriptions of subliminal channels.

3.1   A Deterministic Braid-Based Scheme

In [KCCL02] Ko et al. propose an interesting signature scheme based on braid groups. Albeit being a proposal at a more conceptual level, for our purposes it is worthwhile to take a closer look at this scheme, since it shows a nice way to hide a subliminal message as well as a noteworthy method to discover a subliminal communication.

Given a braid group B_n, two elements $x, y \in B_n$ are said to be *conjugate*, written $x \sim y$, if there is an element $a \in B_n$ such that $y = a^{-1}xa$. Given $x, y \in B_n$ the conjugacy decision problem (CDP) is to decide if $x \sim y$, and the conjugator search problem (CSP) is to find an element $a \in B_n$ such that $y = a^{-1}xa$.

Now, the signature scheme from [KCCL02] we want to look at is based on the gap assumption that for the underlying braid group the CDP is feasible whereas the CSP is infeasible. Further on, we need a cryptographic hash function $h : \{0,1\}^* \longrightarrow B_n$ mapping a message to an element of the braid group. The secret key for the signature scheme is a braid $a \in B_n$ and the public key a CSP-hard pair $(x, x') \in B_n^2$ with $x' = a^{-1}xa$. Finally, a signature for a message $M \in \{0,1\}^*$ is given by $\sigma = a^{-1}ya$, where $y = h(M)$ is the hash value of M. The verification procedure checks if $\sigma \sim y$ and $x'\sigma \sim xy$ hold and returns valid if and only if both conjugacies are true.

By construction, this signature scheme is deterministic. Further on, it is not subliminal free according to Definition 4: Consider any message M with $y = h(M)$ not commuting with x. Then, instead of the signature $\sigma = a^{-1}ya$ the signer can also compute $\sigma' = (xa)^{-1}y(xa)$; σ' is a valid signature for M, because of $\sigma' \sim y$ and

$$x'\sigma' = a^{-1}xaa^{-1}x^{-1}yxa = a^{-1}yxa = a^{-1}x^{-1}xyxa = (xa)^{-1}xy(xa) \sim xy.$$

A subliminal receiver can easily distinguish σ and σ' if he is given the secret key a. However, the signature scheme does not provide a subliminal channel if used in this way: Given two signatures containing different subliminal bits, a warden can notice the subliminal communication by comparing the two signatures. Namely, if the signature scheme is used correctly for both computing a signature σ_1 of M_1 and σ_2 of M_2, with $y_1 = h(M_1)$ and $y_2 = h(M_2)$ we obtain $\sigma_1\sigma_2 = a^{-1}y_1aa^{-1}y_2a \sim y_1y_2$. On the other hand, if the first signature was chosen to be $\sigma_1' = (xa)^{-1}y_1(xa)$, then $\sigma_1'\sigma_2 = a^{-1}x^{-1}y_1xaa^{-1}y_2a = a^{-1}x^{-1}y_1xy_2a$ will generally not be conjugated to y_1y_2.

3.2   (Non-)Deterministic RSA-PSS

RSA-PSS [con03] is a signature scheme with appendix comprised of the basic RSA signature algorithm and a(n in general) randomized padding, the PSS-encoding. For our purposes, it is important to note that the PSS-encoding can also be used in a deterministic variant; below, we will see that the deterministic variant of RSA-PSS in conjunction with a key verification procedure is subliminal free in the sense of Definition 4.

We commence by recalling the main ingredients of the PSS-encoding: Next to a cryptographic hash function $H(\cdot)$, a mask generation function $MGF(\cdot, \cdot)$ is to be specified. To encode a message M, the signer chooses a random *salt* of a specified length for the randomization—there is no clear recommendation for the salt length in the PSS-specification, but the PSS-description mentions a typical length of the *salt* would be *hLen*, the length of the output of the hash function $H(\cdot)$, or 0. Since the length of the PSS-encoding is the length of the modulus n and thus the security parameter k, it is not implausible to choose the length of the salt to grow linearly in k. However, this enables the implementation of a broadband subliminal channel. To see this, recall the format of the PSS-encoding EM of a message M:

$$EM = [(0 \ldots 0 \| 1 \| salt) \oplus MGF(H_0, emLen - hLen - 1)] \| H_0 \| bc,$$

with

$$H_0 = H(\underbrace{0 \ldots 0}_{8 \text{ octets}} \| H(M) \| salt)$$

a hash value of length *hLen*, *emLen* the length of the string EM in octets, and bc an octet with the hexadecimal value bc. The signature σ is produced by applying the plain RSA signature algorithm to EM.

For signature verification, the verifier recovers the PSS-encoded message EM by applying the basic RSA algorithm. Then he has to check the correctness of the PSS-encoding. Reading off H_0 from the encoded message, he can compute $MGF(H_0, emLen - hLen - 1)$. With this, he can extract the random *salt* from the encoded message EM and verify the equality $H(0 \ldots 0 \| H(M) \| salt) = H_0$.

A broadband channel in RSA-PSS with randomized padding. As already pointed out in [BW02], in its randomized form RSA-PSS allows for a subliminal channel where the subliminal signer and verifier do not need to share parts of the secret signing key. Thus, in contrast to a well-known subliminal channel in the DSA, for RSA-PSS the subliminal receiver gains no advantage in forging a signature.

Proposition 1. *If the salt length grows linearly in the security parameter, then the RSA-PSS signature scheme provides a broadband subliminal channel.*

Proof. Already the original verification procedure recovers the complete salt value. Consequently, for implementing a subliminal channel it is sufficient to use an encryption procedure like the one described in Section 2.2, where the obtained ciphertexts are indistinguishable from random bitstrings. For implementing a broadband subliminal channel, we may choose both $|\Xi|$ and the length of the subliminal messages to grow linearly in k. $\qquad\square$

A different method for subliminal communication in RSA-PSS. By default, the public key (n, e) in an RSA-based scheme satisfies the condition $\gcd(\phi(n), e) = 1$. However, given only the pair (n, e) we are not aware of an efficient procedure which allows a warden to verify that $\phi(n)$ and e are indeed coprime. Thus, assume that a subliminal key generation procedure chooses a public verification key (n, e) such that $\gcd(\phi(n), e) = 3$ holds, i.e., for elements in $\mathbb{Z}_n^\times$ there will be three or no e-th roots available. In this case, there exists no secret exponent d with $de \equiv 1$ (mod $\phi(n)$), but computing the e-th roots is nevertheless feasible, provided they exist. Unfortunately, already among the elements in $\mathbb{Z}_n^\times$ only $1/3$ has e-th roots, and thus it is a priori not possible to sign arbitrary messages. Nevertheless, with a randomized padding there is a high probability to find a padding so that the encoded message corresponds to an element in $\mathbb{Z}_n^\times$ having e-th roots. Then the signer can compute these roots and embed one of three subliminal messages by choosing the least, medium, or greatest root as signature. For being able to receive such subliminal messages, the subliminal receiver is also equipped with an algorithm for computing e-th roots modulo n, i.e., the subliminal receiver may well forge signatures.

Subliminal-freeness of deterministic RSA-PSS with key verification. Omitting the random salt, the RSA-PSS signature scheme can be used as a deterministic signature scheme. The PSS-encoding operation is then acting as a full domain hash function. Although in this case the security proof given in [BR96] is not as "tight" as with the randomized padding, the scheme remains provably secure. To avoid a subliminal communication as described in the previous paragraph we can make use of safe primes, i.e., we choose $n = pq$ where the primes p and q are such that $(p-1)/2$ and $(q-1)/2$ are prime. It is known that a zero knowledge proof can be used to show that n is indeed the product of two safe primes [CM99]. Additionally, we require e to be an odd prime and check that $(2e+1) \nmid n$ holds. This ensures that e is coprime to $\phi(n)$, and therewith that for arbitrary $m \in \mathbb{Z}/n\mathbb{Z}$, the equation $\sigma^e \equiv m$ (mod n) has a unique solution $\sigma \in \mathbb{Z}/n\mathbb{Z}$. Thus, in summary before accepting a public key (n, e) we check that

- n is a product of two safe primes (using zero knowledge) and
- e is prime with $(2e+1) \nmid n$ (thereby ensuring the coprimality of e and $\phi(n)$).

Proposition 2. *When used with the above key verification procedure, the deterministic RSA-PSS signature scheme is subliminal free.*

Proof. To prove the proposition, it is sufficient to show that for any M and any verified RSA key, there is only one signature σ satisying the verification condition. By applying the basic RSA encryption operation to the signature σ, the verifier obtains a PSS-encoding $EM' = \sigma^e$ (mod n) of the message M and due to the key verification he knows, that there is only one σ with $\sigma^e \equiv EM'$ (mod n).[2]

---

[2] The verification procedure also checks that σ lies in the range $0, \dots, n-1$, i.e., σ is not only unique modulo n.

It has to be shown that the encoding EM' is the unique encoding EM of the message M. For this, let $emLen$ be the length of EM' which is checked to be $\lceil(\text{bitlength of } n)/8\rceil$ and $hLen$ be the output length of the used hash function. In the PSS-decoding, first the length of EM' is checked. Then the last octet has to be the hexadecimal value bc. Let $maskedDB$ denote the leftmost $enLen - hLen - 1$ octets, and H_0 be the next $hLen$ octets. The verification procedure finally outputs valid, if and only if $H_0 = H(M')$, where (due to the absence of $salt$) M' can be derived from M alone. Consequently, there is just one unique choice for H_0. It is checked that $maskedDB \oplus \text{MGF}(H_0, emLen - hLen - 1)$ is the string $0 \ldots 0 \| 1$. So $maskedDB = (0 \ldots 0 \| 1) \oplus \text{MGF}(H_0, emLen - hLen - 1)$, and thus $EM' = EM$ is uniquely determined.   □

3.3   A Deterministic Polynomial-Based Proposal: SFLASH^{v3}

SFLASH^{v2} is a an asymmetric signature algorithm which is part of the NESSIE Portfolio of recommended cryptographic primitives [con03]. By now the authors of the scheme "no longer recommend its use" and have put forward SFLASH^{v3} [CGP03]. Recently, Ding and Schmidt proposed a cryptanalysis of SFLASH^{v3} [DS04], but the (informal) subliminal channel described in the sequel nevertheless seems worth noting.

For details on SFLASH^{v3} we refer to [CGP03]; here a rough summary is sufficient: SFLASH^{v3} makes use of two finite fields $K \simeq \mathbb{F}_{128}$, $L \simeq \mathbb{F}_{128^{67}}$ along with bijections $\pi : \{0,1\}^7 \longrightarrow K$, $\varphi : K^{67} \longrightarrow L$. The non-public part of the key consists of an 80-bit string Δ and two affine bijections $s, t : K^{67} \longrightarrow K^{67}$. The corresponding public key is the function $G(X) = [(t \circ \varphi^{-1} \circ F \circ \varphi \circ s)(X)]_{0 \to 7 \cdot 55 - 1}$, where $F : L \longrightarrow L$, $\alpha \mapsto \alpha^{128^{33}+1}$, $[\cdot]_{0 \to 7 \cdot 55 - 1}$ indicates that only the first 56 (out of 67) rows are published, and $(f \circ g)(x) := f(g(x))$. By construction, $(Y_0, \ldots, Y_{55}) = G(X_0, \ldots, X_{66})$ can be expressed in the form $(P_0(X_0, \ldots, X_{66}), \ldots, P_{55}(X_0, \ldots, X_{66}))$ where each P_i is a polynomial of total degree ≤ 2 with coefficients in K. Also the last (unpublished) 11 rows of $(t \circ \varphi^{-1} \circ F \circ \varphi \circ s)(X_0, \ldots, X_{66})$ can be expressed through polynomials of degree ≤ 2, and we denote them by $P_{56}, \ldots, P_{66}$. Our subliminal key generation procedure reveals these polynomials to the intended subliminal receiver.

Computing and verifying signatures. Essentially, to sign a bitstring M the following steps are performed:

1. Without involving any non-public data, a 392-bit string V is derived from M by means of SHA-1.
2. Via $Y := (\pi([V]_{0 \to 6}), \pi([V]_{7 \to 13}), \ldots, \pi([V]_{385 \to 391}))$ the bitstring V is translated into a vector $Y \in K^{56}$, where the notation $[\cdot]_{a \to b}$ is to be understood as selecting the bits no. $a - b$.
3. Applying SHA-1 to the concatenation of V and Δ followed by reading off the first 77 bits of the hash value yields a bitstring $W = \text{SHA-1}(V \| \Delta)$. Via $R := (\pi([V]_{0 \to 6}), \pi([V]_{7 \to 13}), \ldots, \pi([V]_{70 \to 76}))$ this bitstring is translated into an element $R \in K^{11}$.

4. Now the value $X := (s^{-1} \circ \varphi^{-1} \cdot F^{-1} \circ \varphi \circ t^{-1})(Y\|R)$ is computed. Translating the 67 entries of X into a bitstring by means of π^{-1} yields the final (469-bit) signature S of M.

To verify a signature S' (of the correct length) of a message M, one uses π to translate S' into an element $X' \in K^{67}$. Evaluating the 56 public verification polynomials at X' yields an element $Y' \in K^{56}$. Now the signature S is accepted if and only if Y' coincides with the value Y derived from the message M in the same manner as in the first two steps of the signature procedure.

Hiding a message. In particular, the verification procedure of SFLASH^{v3} does not check whether the value R (computed in Step 3 of the signing process) has been constructed as specified: If we replace the third step of the signing procedure by "*3. Choose $R \in K^{11}$ as encryption of a subliminal message.*" then we again obtain a valid signature of M. Moreover, by means of $P_{56}, \ldots, P_{66}$ the (77-bit) value R—and therwith the encrypted subliminal message—can immediately be recovered from the resulting signature S: Instead of evaluating only $P_0, \ldots, P_{55}$ at the corresponding vector $X \in K^{67}$, now also $P_{56}, \ldots, P_{66}$ are evaluated at X.

3.4   A Deterministic Factoring-Based Scheme: ESIGN-D

ESIGN [con03] is an efficient signature scheme based on the difficulty of factoring and computing approximate e-th roots. The public key is a pair (n, e) where $n = p^2 q$ for k-bit primes p, q such that n is of length $3k$ and $e \geq 8$. Note that e does not need to be coprime to $\phi(n)$. The secret key is the pair (p, pq). Furthermore, a hash function $\mathrm{H}(\cdot)$ with output length $k - 1$ is given.

Computing and verifying signatures. To sign a message, the signer first chooses a random number $r < pq$ and computes $\alpha = \mathrm{H}(M) - r^e \pmod{n}$. Then he computes $w_0 = \lceil \frac{\alpha}{pq} \rceil$ and $w_1 = w_0 \cdot pq - \alpha$. If now $w_1 \geq 2^{2k-1}$ he begins again with a new random number r. The signature is $\sigma \equiv r + t \cdot pq \pmod{n}$ where $t \equiv w_0/(er^{e-1}) \pmod{p}$. The verification is done by checking if σ is an e-th root of $(\mathrm{H}(M)\|.^{2k-1}) \pmod{n}$, i.e., by checking if the k most significant bits of $\sigma^e \pmod{n}$ are $0\|\mathrm{H}(M)$.

The first security proof for the scheme was incorrect and to fix the proof [Gra02] proposes a deterministic variant of ESIGN, called ESIGN-D. ESIGN-D is identical to ESIGN up to the random choices in the signing algorithm which are replaced by the output of a mask generation function MGF($.,.$) (cf. the construction of M'Raïhi et al. [MNPV99]). Therefore the secret key contains an additional k-bit value Δ that does not affect the public key. The random inputs to the signing algorithm are now replaced by MGF($\mathrm{H}(M)\|\Delta\|i, |pq|$) for $i = 1, 2, \ldots$

A subliminal channel in ESIGN. In [KT99] a subliminal channel in ESIGN is proposed. To sign a message M and hide a message m with $|m| < |pq| - k - 1$ the signer computes an encryption of m into a random looking bitstring d of length

$|pq| - 1$. If $H(M) \cdot 2^k + d$ has no e-th roots then a different encoding of m has to be chosen. Otherwise the signature σ of M is an e-th root of $H(M) \cdot 2^k + d$ that can be computed efficiently, although the original signing algorithm is more efficient. The signature can be verified since $\sigma^e \pmod{n} \equiv H(M) \cdot 2^k + d = 0 \| H(M) \| d$, i. e., the k most significant bits of $\sigma^e \pmod{n}$ are $0 \| H(M)$. To extract the subliminal message the receiver decrypts d.

This subliminal channel can also be used in ESIGN-D, because the warden cannot distinguish the output of the mask generation function, modelled as random oracle, and true random.

4   Conclusion

The above discussion and examples illustrate that several proposed deterministic signature schemes allow for (practical) subliminal channels. Clearly, the kind of subliminal channels and subliminal freeness discussed in this contribution does not take all possible forms of subliminal communication into account (e. g., we did not look at the possibility of encoding information in the public verification key), but we think our discussion covers a significant class of subliminal communication that is of practical interest.

We think it is interesting to note that an, up to the addition of a key verification, deterministic version of RSA-PSS is both provably secure in the sense of existential unforgeability and also subliminal free. From our point of view, the phenomenon of subliminal communication through deterministic cryptographic schemes certainly deserves further exploration.

References

[BR96]   Mihir Bellare and Phillip Rogaway. The Exact Security of Digital Signatures – How to Sign with RSA and Rabin. In *Advances in Cryptology – EUROCRYPT 96*, volume 1070 of *Lecture Notes in Computer Science*, pages 399–416. Springer-Verlag, 1996.

[BW02]   Feng Bao and Xinkai Wang. Steganography of Short Messages through Accessories. In *Pacific Rim Workshop on Digital Steganography 2002 (STEG'02)*, 2002.

[CGP03]   Nicolas Courtois, Louis Goubin, and Jacques Patarin. SFLASH^{v3}, a fast asymmetric signature scheme. Cryptology ePrint Archive: Report 2003/211, 2003. Revised Specification of SFLASH, version 3.0., October 17th, 2003. Published under the URL http://eprint.iacr.org/2003/211/.

[CM99]   Jan Camenisch and Markus Michels. Proving in Zero-Knowledge that a Number Is the Product of Two Safe Primes. In *Advances in Cryptology – EUROCRYPT '99*, volume 1592 of *Lecture Notes in Computer Science*, pages 107–122. Springer-Verlag, 1999.

[con03]   NESSIE consortium. NESSIE Portfolio of recommended cryptographic primitives. At the time of writing available at https://www.cosic.esat.kuleuven.ac.be/nessie/deliverables/decision-final.pdf, 2003.

[DS04]   Jintai Ding and Dieter Schmidt. Cryptanalysis of SFlash^{v3}. Cryptology
            ePrint Archive: Report 2004/103, 2004. At the time of writing available at
            the URL http://eprint.iacr.org/2004/103/.

[Gra02]   Louis Granboulan. How to repair ESIGN. In *Security in Communication
            Networks*, volume 2576 of *Lecture Notes in Computer Science*, pages 234–
            240. Springer-Verlag, 2002.

[IEE01]   IEEE P1363 Working Group for Public-Key Cryptography Standards;
            Meeting Minutes (unapproved), May 22nd 2001.   At the time of
            writing available at the URL http://grouper.ieee.org/groups/1363/
            WorkingGroup/minutes/010522.txt.

[KCCL02]   Ki Hyoung Ko, Koo Ho Choi, Mi Sung Cho, and Jang Won Lee. New
            Signature Scheme Using Conjugacy Problem. Cryptology ePrint Archive:
            Report 2002/168, 2002. At the time of writing available at the URL http:
            //eprint.iacr.org/2002/168/.

[KT99]   Hidenori Kuwakado and Hatsukazu Tanaka. New Subliminal Channel Em-
            bedded in the ESIGN. *IEICE Trans. Fundamentals*, E82-A(10):2167–2171,
            1999.

[Lab02]   RSA Laboratories. PKCS #1 v.2.1: RSA Cryptography Standard. At the
            time of writing available at the URL ftp://ftp.rsasecurity.com/pub/
            pkcs/pkcs-1/pkcs-1v2-1.pdf, June 2002.

[MNPV99]   David M'Raïhi, David Naccache, David Pointcheval, and Serge Vaudenay.
            Computational Alternatives to Random Number Generators. In *Fifth An-
            nual Workshop on Selected Areas in Cryptography SAC'98*, volume 1556 of
            Lecture Notes in Computer Science, pages 72–80. Springer, 1999.

[Sim84]   Gustavus J. Simmons. The Prisoners' Problem and the Subliminal Chan-
            nel. In *Advances in Cryptology – CRYPTO '83*, pages 51–67. Plenum Press,
            New York and London, 1984.

[Sim85]   Gustavus J. Simmons. The Subliminal Channel and Digital Signatures. In
            Advances in Cryptology – EUROCRYPT '84, Lecture Notes in Computer
            Science, pages 364–378. Springer-Verlag, 1985.

[Sim94]   Gustavus J. Simmons. Subliminal Communication Is Easy Using the DSA.
            In *Advances in Cryptology – EUROCRYPT '93*, volume 765 of *Lecture
            Notes in Computer Science*, pages 218–232. Springer-Verlag, 1994.

[Sim98]   Gustavus J. Simmons. Results Concerning the Bandwidth of Subliminal
            Channels. *IEEE Journal on Selected Areas in Communications*, 16(4):463–
            473, 1998.

[YY96]   Adam Young and Moti Yung. The Dark Side of "Black-Box" Cryptography
            or: Should We Trust Capstone? In *Advances in Cryptology – CRYPTO '96*,
            volume 1109 of *Lecture Notes in Computer Science*, pages 89–103, 1996.

[ZLK03]   Fangguo Zhang, Byoungcheon Lee, and Kwangjo Kim. Exploring Signature
            Schemes with Subliminal Channel. In *The 2003 Symposium on Cryptogra-
            phy and Information Security; SCIS 2003*, 2003.

Threshold Entrusted Undeniable Signature*

Seungjoo Kim[1] and Dongho Won[2]

[1] ISAC Lab. (Information Security And Cryptology Lab.),
School of Information and Communication Engineering,
Sungkyunkwan University,
300 Cheoncheon-dong, Jangan-gu, Suwon-si, Gyeonggi-do 440-746, Korea
`skim@ece.skku.ac.kr`
`http://www.isac.re.kr`
[2] ICS Lab. (Information & Communications Security Lab.),
School of Information and Communication Engineering,
Sungkyunkwan University,
300 Cheoncheon-dong, Jangan-gu, Suwon-si, Gyeonggi-do 440-746, Korea
`dhwon@dosan.skku.ac.kr`
`http://dosan.skku.ac.kr`

Abstract. Entrusted undeniable signatures are like undeniable signatures, except that the disavowal protocol can only be run by a court in order to resolve a formal dispute. This paper introduces threshold entrusted undeniable signature scheme without trusted center. It is shown how the power to run a disavowal protocol of entrusted undeniable signature can be distributed to n agents such that any t of these can verify a signature. This facility is useful to solve "lie detector" problem of undeniable signatures.

Keywords: Undeniable signature, Entrusted undeniable signature, Threshold signature.

1   Introduction

Digital signatures are one of the most important techniques of modern cryptography, and have many applications in information security systems. Digital signatures are easily verified as authentic by anyone using the corresponding public key. This "self-authenticating" property is quite suitable for some uses, such as broadcast of announcements and public key certificate. But it is unsuitable for many other applications. Self-authentication makes signatures those are somewhat commercially or personally sensitive, for instance, much more valuable to the industrial spy or extortionist. Thus, self-authentication provides too much authentication for many applications.

To solve this problem, D. Chaum proposed a new type of digital signature, undeniable signature in Crypto'89 conference and proposed a zero-knowledge

---

* This work was supported by the University IT Research Center Project funded by the Korean Ministry of Information and Communication.

C. Park and S. Chee (Eds.): ICISC 2004, LNCS 3506, pp. 195–203, 2005.

undeniable signature in Eurocrypt'90 conference. [1, 2] Briefly, an undeniable signature is a signature which cannot be verified without the help of the signer. They are therefore less personal than ordinary signatures in the sense that a signature cannot be related to the signer without his/her help. However, if a signature is only verifiable with the aid of a signer, a dishonest signer may refuse to authenticate a genuine document. Undeniable signatures solve this problem by adding a new component called the disavowal protocol in addition to the normal components of signature and verification.

Also, Boyar et al. introduced convertible undeniable signatures. [3] In this schemes, release of a single bit string by the signer turns all of his signatures, which were originally undeniable signatures, into ordinary digital signatures.

In Eurocrypt'94, new compromised schemes between normal digital signatures and undeniable signatures were proposed by Chaum, called "designated confirmer signature schemes". [4] It was claimed that not only signer but also the designated third party has the ability of proving the validity of a given signature. In undeniable signatures, the signer might refuse to cooperate in either confirming or denying, he/she might claim the loss of keys for confirming or denying, or he/she might just be unavailable. Designated confirmer signatures can give the signer the protection of an undeniable signature while not letting his/her abuse that protection.

In [5, 6, 7], Kim et al. proposed a new signature scheme, called "nominative signature", that is a dual signature scheme of undeniable signature. Unlike an undeniable signature, the validity or invalidity of a nominative signature can be ascertained by conducting a protocol with the verifier. If a confirmation protocol is used, the cooperating verifier gives exponentially high certainty that the signature issued to him(her) and is valid. For an application of nominative signature, we consider the following case. Bob submits to a company his academic record (or any testimonial) which the president of his university signs. In this case, signer(called "nominator") is the president of university, verifier(called "nominee") is Bob and the third party is the company. That is, a nominative signature is very valuable for the case in which the content of signature is concerned with the verifier's privacy.

In [8, 9], Park et al. proposed another related notion, called "entrusted undeniable signature". Let's consider the following "lie detector" problem: Imagine that Alice works for a Toxins, Inc., and sends incriminating documents to a newspaper using an undeniable signature protocol. Alice can verify her signature to a newspaper reporter, but not to anyone else. However, CEO Bob suspects that Alice is the source of the documents. He demands that Alice run the disavowal protocol to clear her name, and Alice refuses. Bob maintains that the only reason Alice has to refuse is that she is guilty, and fires her. Entrusted undeniable signatures are suited to these sorts of tasks. [10] Entrusted undeniable signatures are like undeniable signatures, except that the disavowal protocol can only be run by a court, Carol, in order to resolve a formal dispute.

However, in a group-oriented society it is often desired that the power to run disavowal protocol is shared. Imagine that Bob tries to bribe Carol to reveal the signer's name! In this paper, we propose a threshold entrusted undeniable signature, in which the disavowal protocol can only be run by t jurors or more rather than one person. Thus, in our scheme, even if Bob corrupts less than t jurors, he can not run the disavowal protocol. The organization of the paper is as follows: In the next section, we will briefly review Park's entrusted undeniable signature scheme. In section 3, we propose a (t, n)-threshold entrusted undeniable signature scheme based on the discrete logarithm problem. In section 4, the security of our proposed scheme is analyzed. Finally, we summarize the benefits of our scheme in the last section of the paper.

2   Park's Entrusted Undeniable Signature

The key point of Park's entrusted undeniable signature is that a signer(and a verifier) uses a randomized private key $r \cdot x$ (and a corresponding public key $y^r = g^{r \cdot x} \bmod p$) instead of a real private key x (and a corresponding public key $y = g^x \bmod p$). In addition, to solve a dispute later, the signer commits to the value of random number r.

Assume that a large prime p, and a primitive element g, are made public, and used by a group of signers. Alice has a private key, x_A, and a public key, $y_A = g^{x_A} \bmod p$. Also, a court Carol has a RSA private key d_C, and a RSA public key (e_C, n_C).

To sign a message m, Alice computes $(sig_1 = y_A^r \bmod p, sig_2 = m^{r \cdot x_A} \bmod p)$ and a commitment $c = r^{e_C} \bmod n_C$ with a random number r, and sends (sig_1, sig_2, c) to a verifier Bob.

To verify the signature (sig_1, sig_2, c), Alice first shows that the discrete logarithm of $sig_1 = y_A^r \bmod p$ to the base y_A is equal to the e-th root of $c = r^{e_C} \bmod n_C$ in a zero-knowledge manner, and then runs Chaum's confirmation protocol with the randomized public key $y_A^r = g^{r \cdot x_A} \bmod p$.

If the (sig_1, sig_2, c) is forged signature, Alice can prove a court Carol that the given signature is not valid. To do this, Carol first recovers r from c with her private key d_C, and then runs Chaum's disavowal protocol on $sig_2^{r^{-1}} \bmod p$. Here we should note that the disavowal protocol can only be run by Carol knowing the private key d_C.

3   The Proposed Scheme

Being inspired of Stadler's verifiable encryptions [11], in this section, we describe a simple (t, n)-threshold entrusted undeniable signature scheme based on the discrete logarithm problem. There are four types of participants in our scheme: a signer Alice, a verifier Bob, a court(or combiner) Carol, and the n jurors.

Let p be a large prime so that $q = (p - 1)/2$ is also prime, and let $h \in Z_p^*$ be an element of order q. Let further G be a group of order p, and let g be a generator of G so that computing discrete logarithms to the base g is difficult.

Our scheme will make use of double exponentiation. By double exponentiation with base g and h we mean the function

$$Z_q \to G : x \mapsto g^{(h^x)}.$$

By the double discrete logarithm of $v \in G$ to the base g and h we mean the unique $x \in Z_q$ with

$$v = g^{(h^x)}$$

if such an x exists.

From now on, unless stated otherwise all the calculations to be performed in this paper will be done in the group G.

3.1   Setup

Let $(x_A \in Z_q, v_A = g^{y_A})$, where $y_A = h^{x_A} \bmod p$, be Alice's private/public key pair and $(x_J \in Z_q, y_J = h^{x_J} \bmod p)$ be a private/public key pair for $Jurors$ (For convenience, we assume that $Jurors = \{J_i | 1 \le i \le n\}$). We now show a protocol for generating $(x_J \in Z_q, y_J = h^{x_J} \bmod p)$ without the dealer.

[Protocol for generating Jurors' private/public key without delaer]

Suppose that a dealer with a random secret $x_J \in Z_q$ chooses a random polynomial such that $f(x) = x_J + a_1 x + \cdots + a_{t-1} x^{t-1}$, sends $s_i = f(i)$ to J_i secretly for $i = 1, \cdots, n$, and broadcasts $y_J = h^{x_J} \bmod p$ and $h^{a_1}, \cdots, h^{a_{t-1}} \bmod p$. This procedure is simulated by the following protocol without the dealer. [12, 13]

1. Each juror J_i picks $x_{J,i} \in_R Z_q$ at random and broadcasts $y_{J,i} = h^{x_{J,i}} \bmod p$ to all other jurors.
2. To distribute $x_{J,i}$, each J_i randomly selects a polynomial f_i of degree $t - 1$ in Z_q such that $f_i(0) = x_{J,i}$, i.e.,

$$f_i(x) = x_{J,i} + a_{i,1} x + a_{i,2} x^2 \cdots + a_{i,t-1} x^{t-1}$$

   with $a_{i,1}, \cdots, a_{i,t-1} \in_R Z_q$, and sends $f_i(j) \bmod q$ to J_j in a secure manner ($\forall j \ne i$). J_i also broadcasts the value

$$h^{a_{i,1}}, \cdots, h^{a_{i,t-1}} \bmod p.$$

3. From distributed $f_j(i)$ ($\forall j \ne i$), J_i checks whether, for each j ($j \ne i$),

$$h^{f_j(i)} \overset{?}{=} y_{J,j} \cdot (h^{a_{j,1}})^{i^1} \cdots (h^{a_{j,t-1}})^{i^{t-1}} \bmod p.$$

4. Let $H \overset{\Delta}{=} \{J_j | J_j \text{ is not detected to be cheating at step 3}\}$. Every J_i computes the share

$$s_i = f(i) = \sum_{j \in H} f_j(i)$$

   secretly, and computes

$$y_J = \prod_{j \in H} y_{J,j}, \quad h^{a_1} = \prod_{j \in H} h^{a_{j,1}}, \quad \cdots, \quad h^{a_{t-1}} = \prod_{j \in H} h^{a_{j,t-1}}$$

3.2   Signature Generation

To sign a message m of the group G, Alice computes $\{(sig_1, sig_2), (c_1, c_2)\} = \{(v_A^r, m^{y_A \cdot r}), (h^s \bmod p, r^{-1} \cdot y_J^s \bmod p)\}$ with two random numbers $r, s \in Z_q$.

3.3   Confirmation Protocol

Our confirmation protocol consists of two subprotocols. The first is a protocol for verifying that a commitment pair (c_1, c_2) encrypts the discrete logarithm of $sig_1 = v_A^r$ to the base v_A, and the second is a protocol for convincing the validity of a signature (sig_1, sig_2).

[Subprotocol 1]

The prover Alice now proves to the verifier Bob that the discrete logarithm of c_1 to the base h is identical to the double discrete logarithm of $sig_1^{c_2}$ to the bases v_A and y_J. It is based on the fact that if (c_1, c_2) is equal to $(h^s \bmod p, r^{-1} \cdot y_J^s \bmod p)$ for any $s \in Z_q$ then

$$sig_1^{c_2} = v_A^{r \cdot c_2} = v_A^{y_J^s}.$$

Here's the protocol (repeat K times):

1. Alice chooses a random number $w \in Z_q$, and computes $t_h = h^w \bmod p$ and $t_g = v_A^{y_J^w}$. Alice sends Bob t_h and t_g.
2. Bob chooses a random $b \in \{0, 1\}$, and sends it to Alice.
3. Alice computes $z = w - b \cdot s \bmod q$, and sends z to Bob.
4. Bob verifies that $t_h \stackrel{?}{=} h^z \cdot c_1^b \bmod p$, and $t_g \stackrel{?}{=} v_A^{y_J^z}$ (if $b = 0$) or $t_g \stackrel{?}{=} sig_1^{c_2 \cdot y_J^z}$ (if $b = 1$).

[Subprotocol 2]

In this protocol Bob is convinced that the signature (sig_1, sig_2) is valid, but cannot convince a third party by using a transcript of this protocol.

1. Bob chooses two random numbers, $a \in Z_p$ and $b \in Z_p$, and sends Alice $c = m^a \cdot g^b$.
2. Alice chooses a random number, $t \in Z_p$, and computes and sends to Bob: $z_1 = c \cdot g^t$ and $z_2 = (c \cdot g^t)^{y_A \cdot r}$
3. Bob sends Alice a and b, so that Alice can confirm that Bob did not cheat in step (1).
4. Alice sends Bob t, so that Bob can use (sig_1, sig_2) and reconstruct z_1 and z_2. If $z_1 = c \cdot g^t$ and $z_2 = sig_1^{(b+t)} \cdot sig_2^a$ then the signature is valid.

3.4   Disavowal Protocol

Unlike Chaum's undeniable signatures, in our scheme, only t or more jurors out of the *Jurors* group of n members, can run the disavowal protocol by decrypting the ciphertext (c_1, c_2) without revealing the private decryption key x_J. Our disavowal protocol consists of two subprotocols. The first is a protocol for the (t, n)-threshold ElGamal decryption of (c_1, c_2), and the second is a protocol for convincing the court Carol that a given signature is not valid, if it is not.

[Subprotocol 1]

Let (c_1, c_2) be a ElGamal ciphertext, and $H \subseteq Jurors$ agree to decrypt the ciphertext. For convenience, we assume that $H = \{J_i | 1 \le i \le t\}$.

1. Bob sends a court Carol $\{(sig_1, sig_2), (c_1, c_2)\}$.
2. Carol then chooses a random number $b \in Z_q$, computes $c_1^b \bmod p$, and broadcasts it to H with a proof that she knows the discrete logarithm of c_1^b to the base c_1. (This proof allows *Jurors* to know what signature is really being verified.)
3. After verifying that c_1^b has a correct form, every $J_i \in H$ computes the shadow $k_i = s_i \cdot \prod_{j=1, j \ne i}^{t} \frac{-j}{i-j} \bmod q$, raise $c_1^b \bmod p$ by $-k_i$, and sends this partial result to Carol.
4. To get $(c_1^b)^{(-x_J)}$, Carol multiplies each of the $(c_1^b)^{(-k_i)} \bmod p$ together:

$$\prod_{i=1}^{t} (c_1^b)^{(-k_i)}$$

$$= (c_1^b)^{(-\sum_{i=1}^{t} k_i)}$$

$$= (c_1^b)^{(-\sum_{i=1}^{t} s_i \prod_{j=1, j \ne i}^{t} \frac{-j}{i-j})}$$

$$= (c_1^b)^{(-\sum_{i=1}^{t} f(i) \prod_{j=1, j \ne i}^{t} \frac{0-j}{i-j})}$$

$$= (c_1^b)^{(-f(0))}$$

$$= (c_1^b)^{(-x_J)} \bmod p \text{ (by the Lagrange interpolation formula)}.$$

Carol then unblinds $(c_1^b)^{(-x_J)}$ with $b^{-1} \bmod q$, and multiplies this result with the second entry of the ciphertext c_2 to get r^{-1}.

[Subprotocol 2]

An alleged signer Alice may wish to convince Carol that a particular (sig_1, sig_2) is not a valid signature corresponding to her public key $v_A = g^{y_A}$ and message m. To do this, the alleged signer Alice cooperates in the following protocol.

1. Initially Carol chooses an integer $t \in \{0, \cdots, k\}$, where k is a mutually agreed constant and order k operations must be performed by the alleged signer Alice (In practice k might be 1023). and chooses a random number $a \in Z_p$.

2. Carol computes $(z_1, z_2, z_3, z_4) = (m, sig_2^{r^{-1}}, m^t \cdot g^a, sig_2^{r^{-1} \cdot t} \cdot v_A^a)$, and sends it to Alice.

3. Now Alice determines the value of t by trial and error:

$$(\frac{z_1^{y_A}}{z_2})^t = \frac{z_3^{y_A}}{z_4} \bmod p.$$

If no t is found, Alice uses a random value. Next Alice sends a blob, $blob(t, u)$, committing to the value of t, but hiding t until the randomly selected u is revealed.

4. Upon receiving the blob, Carol can send a.

5. Alice checks that a can be used to reconstruct the first message, and then finally provides u as the final message.

4   Security Analysis

To show that our proposed scheme solves the "lie detector" problem, we should consider two points. First, we have to check whether only the jurors can run the disavowal protocol. That is, we have to show that computing r from v_A^r and $(h^s \bmod p, r^{-1} \cdot y_J^s \bmod p)$ without x_J is hard. Second, we have to make sure that no "useful" information about r is given away in the subprotocol 1 of confirmation protocol. In below, we will prove two propositions. The main proof idea comes from Stadler's literature. [11] For convenience, we will omit the "mod p" and "mod q" markers.

Proposition 1. *Under the assumption that computing discrete logarithms in G is infeasible, and that breaking the ElGamal cryptosystem is hard, computing r from v_A^r and $(h^s \bmod p, r^{-1} \cdot y_J^s \bmod p)$ is at least as hard as solving the Decision-Diffie-Hellman problem to the base h in Z_p^*.*

Proof. Assume that there is an efficient algorithm $\mathcal{P}$ that computes r on input $(v^r, h^s, h^x, r^{-1} \cdot h^{s \cdot x})$, where $v = g^y$ and $y \in Z_p^*$, with a non-negligible probability ε over all $r \in Z_p^*$, and $s, x \in Z_q$. We now show hot to use $\mathcal{P}$ to decide whether a given triple (A, B, C) of elements in $\langle h \rangle$, is a Diffie-Hellman triple.

First, we randomly choose $y, r \in Z_p^*$, $\rho \in Z_q^*$, $\sigma, \tau \in Z_q$ and run $\mathcal{P}$ on input $((g^y)^r, \overline{A}, \overline{B}, r^{-1}\overline{C})$, where $\overline{A} = A^\rho h^\sigma$, $\overline{B} = Bh^\tau$, and $\overline{C} = C^\rho A^{\rho\tau} B^\sigma h^{\sigma\tau}$. Since the triple $(\overline{A}, \overline{B}, \overline{C})$ is a Diffie-Hellman triple if (A, B, C) is a Diffie-Hellman triple, and a random non-Diffie-Hellman triple, otherwise, the probability that $\mathcal{P}$ returns r depends on whether (A, B, C) is a Diffie-Hellman triple or not: (1) If (A, B, C) is a Diffie-Hellman triple then $\mathcal{P}$ returns r with probability ε. (2) If (A, B, C) is not a Diffie-Hellman triple then the probability that $\mathcal{P}$ returns r is negligible.

Let us assume on the contrary that $\mathcal{P}$ returns r with a non-negligible probability γ. Then the discrete logarithm of any $V \in G$ can be computed by repeatedly running $\mathcal{P}$ on input $(V(g^y)^\chi, h^\psi, h^\omega, t)$ with $y, t \in_R Z_p^*$, $\chi \in_R Z_p$, and

$\psi, \omega \in_R Z_q$ until $\mathcal{P}$ returns $\chi + \log_{(g^v)} V \bmod p$. Then the discrete logarithm of V to the base g can be computed by calculating $y \cdot (\chi + \log_{(g^v)} V) \bmod p$. Because the probability that $t/(\chi + \log_{(g^v)} V)^{-1} \bmod p \in \langle h \rangle$ is approximately $1/2$, the expected number of repetitions is $2/\gamma$.

After sufficiently many repetitions, a decision on whether (A, B, C) is a Diffie-Hellman triple can be made with arbitrarily small probability of error.

Proposition 2. *The prover in the subprotocol 1 of confirmation protocol can successfully cheat with a probability of at most 2^{-K}. The protocol is perfectly zero-knowledge.*

Proof. See [11] for the proof of Proposition 2.

5   Conclusions

In this paper, we show how to construct threshold entrusted undeniable signature scheme. Our signature scheme is like undeniable signature, except that the disavowal protocol can only be run by a court with an agreement with a group of n jurors. Thus our scheme does not have the disadvantage of Park's original scheme.

Acknowledgements

We would like to thank the anonymous referees for their valuable comments.

References

1. D.Chaum, H.V.Antwerpen, Undeniable signature, Advances in Cryptology - CRYPTO'89, Springer, Lecture Notes in Computer Science 435, August 20-24, 1989, pp.212-216.
2. D.Chaum, "Zero-knowledge undeniable signature", Advances in Cryptology - EUROCRYPT'90, Springer, Lecture Notes in Computer Science 473, May 21-24, 1990, pp.458-464.
3. J.Boyar, D.Chaum, I.Damgard, T.P.Pedersen, "Convertible undeniable signature", Advances in Cryptology - CRYPTO'90, Springer, Lecture Notes in Computer Science 537, August 11-15, 1990, pp.189-205.
4. D.Chaum, "Designated Confirmer Signatures", Advances in Cryptology - EUROCRYPT'94, Springer, Lecture Notes in Computer Science 950, May 9-12, 1994, pp.86-91.
5. S.Kim, S.Park, D.Won, "Nominative signatures", Proc. of ICEIC'95, International Conference on Electronics, Informations and Communications, August 7-12, 1995, pp.II-68-II-71.
6. S.Kim, S.Park, D.Won, "Zero-knowledge nominative signatures", Proc. of PragoCrypt'96, International Conference on the Theory and Applications of Cryptology, September 30 - October 3, 1996, pp.380-392.

7. Z.Huang, Y.Wang, "Convertible Nominative Signatures", Proc. of ACISP'04, Australasian Conference on Information Security and Privacy, Springer, Lecture Notes in Computer Science 3108, 2004, pp.348-357

8. S.Park, K.Lee, D.Won, "An entrusted undeniable signature", Proc. of JW-ISC'95, Korea-Japan Joint Workshop on Information Security and Cryptology, 1995, pp.120-126.

9. S.Park, T.Kim, Y.An, D.Won, "A provably entrusted undeniable signature", Proc. of IEEE SICON/ICIE'95, IEEE Singapore International Conference on Network/International Conference on Information Engineering, 1995, pp.644-648.

10. B.Schneier, "Applied Cryptography, Second Edition", John Wiley & Sons, 1996.

11. M.Stadler, "Publicly verifiable secret sharing", Advances in Cryptology - EUROCRYPT'96, Springer, Lecture Notes in Computer Science 1070, May 12-16, 1996, pp.190-199.

12. T.P.Pedersen, "Distributed provers with applications to undeniable signatures," Proc. Eurocrypt'91, Lecture Notes in Computer Science, LNCS 547, Springer–Verlag, 1991, pp.221-238.

13. T.P.Pedersen, "A threshold cryptosystem without a trusted party," Proc. Eurocrypt'91, Lecture Notes in Computer Science, LNCS 547, Springer–Verlag, 1991, pp.522-526.

On the Security Models of (Threshold) Ring Signature Schemes

Joseph K. Liu[1] and Duncan S. Wong[2,*]

[1] Department of Information Engineering,
The Chinese University of Hong Kong,
Shatin, Hong Kong
ksliu@ie.cuhk.edu.hk
[2] Department of Computer Science,
City University of Hong Kong,
Kowloon, Hong Kong
duncan@cityu.edu.hk

Abstract. We make fine-grained distinctions on the security models for provably secure ring signature schemes. Currently there are two commonly used security models which are specified by Rivest et al. [15] and Abe et al. [1]. They offer different levels of security. In this paper, we introduce a new but compatible model whose security level can be considered to be lying in between these two commonly used models. It is important to make fine-grained distinctions on the security models because some schemes may be secure in some of the models but not in the others. In particular, we show that the bilinear map based ring signature scheme of Boneh et al. [4], which have been proven secure in the weakest model (the one specified by Rivest et al. [15]), is actually insecure in stronger models (the new model specified by us in this paper and the one specified by Abe et al. [1]). We also propose a secure modification of their scheme for each of the two stronger models. In addition, we propose a threshold ring signature scheme using bilinear maps and show its security against adaptive adversaries in the strongest model defined in this paper. Throughout the paper, we carry out all of the security analyses under the random oracle assumption.

Keywords: Ring Signature, Security Models, Anonymity, Bilinear Maps.

1   Introduction

A ring signature scheme [15, 6, 1, 4, 17, 10] allows members of a group to sign messages on behalf of the group without revealing their identities (signer anonymity). It is also not possible to decide whether two signatures have been issued by the same group member. Different from a group signature scheme [8, 7, 2], the formation of a group is spontaneous and there is no group manager to revoke the

---

* This work was fully supported by a grant from CityU (Project No. 9040904).

C. Park and S. Chee (Eds.): ICISC 2004, LNCS 3506, pp. 204–217, 2005.

identity of the signer. That is, under the assumption that each user is already associated with a public key of some standard signature scheme, a user can form a group by simply collecting the public keys of all the group members including his own. These diversion group members can be totally unaware of being conscripted into the group.

Ring signatures could be used for whistle blowing [15], anonymous membership authentication for ad hoc groups [6] and many other applications which do not want complicated group formation stage but require signer anonymity. For example, in the whistle blowing scenario, a whistleblower gives out a secret as well as a ring signature of the secret to the public. From the signature, the public can be sure that the secret is indeed given out by a group member while cannot figure out who the whistleblower is. At the same time, the whistleblower does not need any collaboration of other users who have been conscripted by him into the group of members associated with the ring signature. Hence the anonymity of the whistleblower is ensured and the public is also certain that the secret is indeed leaked by one of the group members associated with the ring signature.

In 2002, Bresson et al. [6] extended the notion of ring signature schemes to a threshold setting and proposed the first threshold ring signature scheme. Later on, some other threshold ring signature schemes [16, 13] have been proposed. A t-out-of-n threshold ring signature scheme is defined as a ring signature scheme of which at least t corresponding private keys of n public keys are needed to produce a signature. The setup-free and signer anonymity properties of a conventional ring signature scheme are preserved in the threshold setting.

1.1   Contributions

We make fine-grained distinctions on the security models for provably secure ring signature schemes. Currently there are two commonly used security models which are specified by Rivest et al. [15] and Abe et al. [1]. They offer different levels of security. In this paper, we introduce a new but compatible model. Its security level can be considered to be lying in between these two commonly used models in such a way that it captures an attack called *group-changing attack* while it does not consider another attack called *multiple-known-signature existential forgery*.

It is important to make fine-grained distinctions on the security models because some schemes may be secure in some of the models but not in the others. In particular, we show that the bilinear map based ring signature scheme of Boneh et al. [4], which have been proven secure in the weakest model (the one specified by Rivest et al. [15]), is actually insecure in stronger models (the new model specified by us in this paper and the one specified by Abe et al. [1]). We show that their scheme is susceptible to group-changing attack and multiple-known-signature existential forgery. We also propose a secure modification of their scheme for each of the two stronger models.

In addition, we propose a threshold ring signature scheme using bilinear maps and show its security against adaptive adversaries in the strongest model defined in this paper. Our scheme is based on a partial proofs of knowledge protocol using secret sharing due to Cramer, et al. [9].

Throughout the paper, we carry out all of the security analyses under the random oracle assumption [3].

Paper Organization: In Sec. 2, we define a ring signature scheme, review the two commonly used security models due to Rivest et al. [15] and Abe et al. [1], and specify a new model whose security level lies in between the two commonly used models. In Sec. 3, we review the bilinear map based ring signature scheme of Boneh et al. [4] and show that it is insecure in the two stronger models (our new model and the one due to Abe et al.). We describe how to modify their scheme to make it secure in the two stronger models. In Sec. 4, we propose a bilinear map based t-out-of-n threshold ring signature scheme and show that it is secure against adaptive adversaries in the strongest model defined in this paper. We conclude the paper in Sec. 5.

2   Security Models

We describe the subtleties of different security models for ring signature schemes. Below is the definition of a ring signature scheme.

A ring signature scheme is a triple of algorithms $(\mathcal{G}, \mathcal{S}, \mathcal{V})$.

- $(x, P) \leftarrow \mathcal{G}(1^k)$ is a probabilistic polynomial-time algorithm which takes security parameter k and outputs private key x and corresponding public key P.
- $\sigma \leftarrow \mathcal{S}(1^k, x, L, m)$ is a probabilistic polynomial-time algorithm which takes as inputs security parameter k, private key x, a set L of n public keys which includes the one corresponding to x and message m, produces a signature σ.
- $1/0 \leftarrow \mathcal{V}(1^k, L, m, \sigma)$ is a polynomial-time algorithm which takes as inputs security parameter k, a set L of n public keys, a message m and a signature σ, returns 1 or 0 for accept or reject, respectively. We require that for any message m, any (x, P) generated by $\mathcal{G}(1^k)$ and any L that includes P,

$$\mathcal{V}(1^k, L, m, \mathcal{S}(1^k, x, L, m)) = 1.$$

For simplicity, we omit the denotation of the security parameter as one of the inputs of these algorithms in the rest of the paper.

A secure ring signature scheme should be able to thwart signature forgery under certain reasonable assumptions. Under the model of adaptive chosen message attack, existential unforgeability [11] means that given the public keys of all group members but not any of the corresponding private keys, an adversary, who can adaptively obtain valid signatures for any messages that he wishes, cannot forge a signature for any message m.

In the security model of Rivest et al. [15], the adversary who targets to forge a signature of message m subjects to the condition that m has *never* been presented to a signing oracle, denoted by $\mathcal{SO}^1$. Hence given a pair of message and ring signature denoted by (m, σ) with respect to a public-key set L, if another pair $(\tilde{m}, \tilde{\sigma})$ with respect to L is forged by the adversary (that is, $1 \leftarrow \mathcal{V}(L, \tilde{m}, \tilde{\sigma})$) such that $m \neq \tilde{m}$, then it is considered to be a successful forgery in this model. However, in a similar setup, given (L, m, σ), if the adversary obtains (L', m, σ') such that $\sigma' \neq \sigma$, then it is *not* considered to be a successful forgery. More importantly, it is not considered to be a forgery even if L and L' are different. In other words, a ring signature scheme could still be claimed to be secure under the model of Rivest et al. [15] even if anyone can alter the public-key sets associated with the ring signatures generated by the scheme. If an adversary can change the public-key set associated with a ring signature, we say that the corresponding ring signature scheme is susceptible to *group-changing attack*. In the following, we capture the security model of Rivest, et al. in Game 0.

Game 0:   Two entities: Simulator $\mathcal{M}$ and Adversary $\mathcal{A}$. Simulator $\mathcal{M}$ prepares a 'large' set of public keys $\mathcal{L} = \{P_1, \cdots, P_N\}$ using $\mathcal{G}$ where N is some polynomial of security parameter k. $\mathcal{M}$ invokes $\mathcal{A}$ by providing it $\mathcal{L}$ and other appropriate inputs such as pseudorandom coin flips. It then simulates the view of $\mathcal{A}$ by answering all hash queries and signing queries. Suppose in one successful simulation run, $\mathcal{A}$ outputs (L, m, σ) for some message m where $L \subseteq \mathcal{L}$. *Restriction is that m should not be present in $\mathcal{SO}$'s transcript.*

The security of ring signature schemes has two aspects: *unforgeability* and *anonymity*.

For unforgeability, the adversary $\mathcal{A}$ returns a set L of n public keys, a message $m \in \{0,1\}^*$ and a signature σ. $\mathcal{A}$ wins the game if: (1) $\mathcal{V}(L, m, \sigma) = 1$, (2) $L \subseteq \mathcal{L}$, (3) the public-key set L' of each signing query is a subset of $\mathcal{L}$, and (4) the restriction of Game 0 above applies. We use $\mathbf{Adv}_{\mathcal{A}}^{unf}(k)$ to denote the probability of $\mathcal{A}$ winning the game where k is the security parameter.

Definition 1 (Unforgeability). *A ring signature scheme is unforgeable if for any PPT (probabilistic polynomial-time) adversary $\mathcal{A}$, $\mathbf{Adv}_{\mathcal{A}}^{unf}(\cdot)$ is negligible.*

A function ϵ is negligible if for all polynomials π, $\epsilon(k) < 1/\pi(k)$ holds for all sufficiently large k.

For anonymity, Game 0 needs to be modified as follows. In the game, there are three stages. Stage 1: $\mathcal{A}$ returns a set L of n public keys and a message $m \in \{0,1\}^*$. Stage 2: $\mathcal{M}$ randomly picks a key $P_\pi \in_R L$, and generates a signature $\sigma \leftarrow \mathcal{S}(x_\pi, L, m)$ where x_π is the corresponding private key of P_π. $\mathcal{M}$ gives σ to $\mathcal{A}$. Stage 3: $\mathcal{A}$ returns a public key $\tilde{y}$ such that $\tilde{y} \in L$. $\mathcal{A}$ wins the game if: (1) $L \subseteq \mathcal{L}$, (2) the public-key set L' of each signing query is a subset of $\mathcal{L}$, (3) $\tilde{y} = P_\pi$, and (4) the restriction of Game 0 above applies. We use $\mathbf{Adv}_{\mathcal{A}}^{anon}(k)$ to denote the probability of $\mathcal{A}$ winning the game minus $1/n$ (that is the winning probability of making a random guess) where k is the security parameter.

---

[1] The signing oracle receives a message $\check{m}$ and a set $\check{L}$ of public keys in which each key is appropriately generated by $\mathcal{G}$ and returns a signature $\check{\sigma}$ such that $1 \leftarrow \mathcal{V}(\check{L}, \check{m}, \check{\sigma})$.

Definition 2 (Anonymity). *A ring signature scheme is anonymous if for any PPT (probabilistic polynomial-time) adversary* $\mathcal{A}$, $\mathbf{Adv}_{\mathcal{A}}^{anon}(\cdot)$ *is negligible.*

From the restriction of Game 0 above, we can see that group-changing attack, in particular against unforgeability, is not allowed. In Sec. 3, we will see that a ring signature scheme based on bilinear maps due to Boneh et al. [4] allows anyone to add an arbitrary number of public keys to the public-key set of any given ring signature. Hence the scheme is susceptible to group-changing attack although it has been proven secure in Game 0 under the random oracle model. We will explain why this is undesirable in some cases and group-changing attack should be considered and captured by the security model in those cases. By considering the group-changing attack, we introduce the following model.

Game 1:   Two entities: Simulator $\mathcal{M}$ and Adversary $\mathcal{A}$. $\mathcal{M}$ prepares a large set of public keys $\mathcal{L} = \{P_1, \cdots, P_N\}$ using $\mathcal{G}$ where N is some polynomial of security parameter k. $\mathcal{M}$ then invokes $\mathcal{A}$ by providing $\mathcal{L}$ and other appropriate inputs, just like the previous game. It simulates $\mathcal{A}$'s view by answering all hash queries and signing queries. Suppose in one successful simulation run, $\mathcal{A}$ outputs (L, m, σ) for some message m and $L \subseteq \mathcal{L}$. *Restriction is that (L, m) should not be present in $\mathcal{SO}$'s transcript.*

The specifications and formal definitions of unforgeability and anonymity are similar to that of before. We skip the details.

In Game 1, $\mathcal{A}$ is allowed to query the signing oracle with the forging message m as long as the corresponding public-key sets are different from L. Compared with Game 0 which does not allow m to be present in any of the signing queries, Game 1 is stronger. In this model, given a ring signature (L, m, σ), if the adversary obtains a forgery (L', m, σ') such that $L \neq L'$ but with the same message m, this is now considered as a successful forgery. In other words, the group-changing attack is taken into account. In Sec. 3, we will see that the scheme of [4] is insecure in Game 1.

There is still another scenario which is not considered in Game 1. Suppose one or multiple signatures of some message m and public-key set L are given. If the adversary then obtains a new signature on the same pair of m and L, this is not allowed or considered in Game 1. We call this attack the *multiple-known-signature existential forgery.*

Multiple-known-signature existential forgery exists in conventional (probabilistic) signature schemes. It is considered as a valid attack sometimes in target systems but not always. To include this attack, we extend Game 1 to the following model.

Game 2:   Two entities: Simulator $\mathcal{M}$ and Adversary $\mathcal{A}$. $\mathcal{M}$ prepares a large set of public keys $\mathcal{L} = \{P_1, \cdots, P_N\}$ as the previous games. $\mathcal{M}$ then invokes $\mathcal{A}$ by providing $\mathcal{L}$ and other appropriate inputs. It simulates $\mathcal{A}$'s view by answering all hash queries and signing queries. Suppose in one successful simulation run, $\mathcal{A}$ outputs (L, m, σ) for some message m and $L \subseteq \mathcal{L}$. *Restriction is that (L, m, σ) should not be present in $\mathcal{SO}$'s transcript.*

This model is similar to that defined by Abe et al. [1]. Compared with Game 1, this model not only allows the group-changing attack but also the multiple-known-signature existential forgery. In the rest of this paper, we study the constructions of ring signature schemes which can be proven secure in different models/games specified above under the random oracle model.

3   Bilinear Ring Signature Schemes

In [4], Boneh et al. presented a ring signature scheme based on bilinear maps. The scheme is reviewed in the following. It is provably secure in the model of Rivest et al. [15] under the random oracle model, that is, it is secure in Game 0. However, we will show that it is considered to be insecure in other games (Game 1 and Game 2).

Let G_1, G_2 and G_T be three (multiplicative) cyclic groups of prime order p. Let g_1 and g_2 be the generators of G_1 and G_2, respectively. Let $\psi : G_2 \to G_1$ be a computable isomorphism with $\psi(g_2) = g_1$. Let $e : G_1 \times G_2 \to G_T$ be a computable bilinear map with the following properties: (1) Bilinear: for all $u \in G_1$, $v \in G_2$ and $a, b \in \mathbb{Z}$, $e(u^a, v^b) = e(u,v)^{ab}$; (2) Non-degenerate: $e(g_1, g_2) \neq 1$. These properties imply that for any $u_1, u_2 \in G_1$, $v \in G_2$, $e(u_1 u_2, v) = e(u_1, v) \cdot e(u_2, v)$; and for any $u, v \in G_2$, $e(\psi(u), v) = e(\psi(v), u)$.

For each user i, $1 \leq i \leq n$, a random element $x_i \in_R \mathbb{Z}_p$ is picked as the user's private key and the user's public key is computed as $P_i = g_2^{x_i}$. Suppose π, $1 \leq \pi \leq n$, is the index of the actual signer. Let $H : \{0,1\}^* \to G_1$ be a hash function. For security analysis, H is viewed as a random oracle [3]. The signature generation for message $m \in \{0,1\}^*$ proceeds as follows.

1. For each i, $1 \leq i \leq n$ and $i \neq \pi$, randomly generate $a_i \in_R \mathbb{Z}_p$ and compute $\sigma_i = g_1^{a_i}$.
2. Solve the following system to obtain σ_π;

$$H(m) = \sigma_\pi^{x_\pi} \psi\left(\prod_{i=1, i \neq \pi}^{n} P_i^{a_i} \right)$$

3. The signature is $\sigma = (\sigma_1, \cdots, \sigma_n) \in G_1^n$.

To verify the signature, check if $e(H(m), g_2) = \prod_{i=1}^{n} e(\sigma_i, P_i)$.

The scheme is proven secure, that is, both unforgeable and anonymous, in the model of Rivest et al. [15] (i.e. Game 0) under the random oracle model. But alluded to in Sec. 5.4 of [4], their scheme also allows anyone to add an arbitrary number of public keys to the public-key set of any signature generated by their scheme. Below is an illustration.

Adding Group Members to a Given Signature
(Group-Changing Attack)
Given a bilinear signature $\sigma = (\sigma_1, \cdots, \sigma_n)$, anyone can add a public key, P_{n+1}, by carrying out the following steps.

1. For $i = 1, \cdots, n-1$, set $\sigma_i' \leftarrow \sigma_i$.
2. Set $\sigma_n' \leftarrow \sigma_n \cdot \psi(P_{n+1}^r)$ where $r \in_R \mathbb{Z}_p$.
3. Set $\sigma_{n+1}' \leftarrow \psi(P_n^{-r})$.

The new signature is $\sigma' = (\sigma_1', \cdots, \sigma_{n+1}')$. The technique can be generalized to let anyone add an arbitrary number of public keys to the public-key set of any given signature generated by their scheme.

Although this 'feature' can be regarded as a way to further 'anonymize' an existing ring signature, it is questionable on the intention of doing so without the consent of the actual signer. On the other hand, we argue that this group-changing attacker can bring more harm than good for many other application scenarios. For example, anyone can claim that he belongs to an ad hoc group by adding his own public key into all the corresponding signatures of that group, even though he is not invited to. In addition, other group members have no way to expel or deny anyone from being a member of the group. This is particularly undesirable when ring signature schemes are used in e-voting systems [12] where the set of eligible voters have to be prespecified and fixed and no one should be able to claim his eligibility without the approval of a voting authority.

In the following, we propose a simple modification of their scheme for preventing group-changing attack, that is, the modified version is secure in Game 1.

3.1   Variant 1 for Game 1

To turn the scheme into one which is secure in Game 1, we simply change all the evaluations of $H(m)$ to $H(L, m)$ where $L = \{P_1, \cdots, P_n\}$ is the public-key set associated with the message and signature pair.

Anonymity. The identity of the signer is unconditionally protected. For any algorithm $\mathcal{A}$ in Game 1, any set L of n public keys, and a random index value π, $1 \leq \pi \leq n$, the probability $\Pr[\mathcal{A}(\sigma) = \pi]$ is at most $1/n$, where σ is any ring signature on L generated with x_π.

Unforegability. We can obtain the similar result to Theorem 5.2 of [4] but in Game 1 such that forging a signature is as hard as solving the co-CDH problem [5]. The proofs are similar to that given in [4] and full details are omitted here except the following.

In Game 1, the adversary $\mathcal{A}$ is given a 'large' set $\mathcal{L}$ of public keys $P_1, \cdots, P_N$ where N is some polynomial of security parameter k, and is given oracle access to H and a signing oracle $\mathcal{SO}$. The adversary may work adaptively. When querying $\mathcal{SO}$, $\mathcal{A}$ has to specify a message m' and a public-key set L' such that $L' \subseteq \mathcal{L}$. m' can be the same as the final forged message m of $\mathcal{A}$ but L' must not be equal to the L of $\mathcal{A}$'s final forgery. That is, (L, m) of $\mathcal{A}$'s forgery must not be present in the transcript of $\mathcal{SO}$. The significance of this change in the use of $\mathcal{SO}$ allows us to show that Variant 1 is secure against group-changing attack. Essentially, it can be shown that if a probabilistic polynomial-time forger $\mathcal{A}$, which obtains adaptively a number of signatures (L_i', m, σ_i'), for $i = 1, \cdots, q_{\mathcal{SO}}$, where $q_{\mathcal{SO}}$ is polynomially bounded in $|p|$ (which is some polynomial of security parameter k), produces, with non-negligible probability, a forgery (L, m, σ) such that $L \neq L_i$ for all $i = 1, \cdots, q_{SO}$, then the co-CDH problem can be solved in polynomial time with non-negligible probability.

3.2   Variant 2 for Game 2

Variant 1 described above is secure against group-changing attack. However, it is not necessary to be secure against multiple-known-signature existential forgery. In the following, we show that given two signatures $(L, m, \hat{\sigma})$, $(L, m, \check{\sigma})$ of Variant 1, anyone can forge a new signature $(L, m, \tilde{\sigma})$.

Multiple-Known-Signature Existential Forgery
Let $\hat{\sigma} = (\hat{\sigma}_1, \cdots, \hat{\sigma}_n)$ and $\check{\sigma} = (\check{\sigma}_1, \cdots, \check{\sigma}_n)$. We have

$$H(L, m) = \prod_{i=1}^{n} \hat{\sigma}_i^{x_i} = \prod_{i=1}^{n} \check{\sigma}_i^{x_i}.$$

For any $a, b \in \mathbb{Z}_p$,

$$H(L, m) = (\prod_{i=1}^{n} \hat{\sigma}_i^{x_i})^{a/(a+b)} (\prod_{i=1}^{n} \check{\sigma}_i^{x_i})^{b/(a+b)} = \prod_{i=1}^{n} \tilde{\sigma}_i^{x_i}$$

where $\tilde{\sigma}_i = \hat{\sigma}_i^{a/(a+b)} \check{\sigma}_i^{b/(a+b)}$, $1 \leq i \leq n$. Hence a new signature $\tilde{\sigma} = (\tilde{\sigma}_1, \cdots, \tilde{\sigma}_n)$ can be obtained by picking a and b at random.

Being insecure in Game 2, it is implied that Variant 1 is insecure in the model defined by Abe et al. [1].

We now describe a further extension and call the scheme Variant 2 which is targeted to be secure in Game 2. The extension is to change all the evaluations of $H(L, m)$ to $H(L, m, r)$ where r is a random binary string of length $|p|$. That is, in each signature generation, a new $r \in_R \{0, 1\}^{|p|}$ is picked at random and a sequence of $(\sigma_1, \cdots, \sigma_n)$ is obtained in the similar way as the original scheme of [4] such that $e(H(L, m, r), g_2) = \prod_{i=1}^{n} e(\sigma_i, P_i)$. The signature becomes $\sigma = (r, \sigma_1, \cdots, \sigma_n)$.

(Discussions). Compared with the scheme due to Boneh, et al. [4], our scheme changes the evaluation of H by adding two more arguments: the public-key set L and a randomizer r. The two additional arguments are used for upgrading their scheme to a secure one in Game 2: L is added to prevent the public-key set of a given signature from being modified. r is added to prevent existential forgery of new signatures from given signatures of the same message (i.e. multiple-known-signature existential forgery). Below is the formal analysis.

In Game 2, the adversary $\mathcal{A}$ may query $\mathcal{SO}$ with (L, m) which is the pair of public-key set and message of $\mathcal{A}$'s final forgery. If an algorithm $\mathcal{F}$ runs in time at most t and completes successfully with probability at least ϵ, then $\mathcal{F}$ is said to be an (t, ϵ)-algorithm. The probability is taken over the domain and the coin tosses of $\mathcal{F}$.

Theorem 1. *Let* $\mathcal{L} = \{P_1, \cdots, P_N\}$ *be a set of public keys where* N *is some polynomial of security parameter* k. *Suppose* $\mathcal{A}$ *is a* (t', ϵ')-*algorithm, that takes as input* $\mathcal{L}$, *produces a ring signature forgery on a public-key set* L *such that* $L \subseteq \mathcal{L}$. *Then there exists an* (t, ϵ)-*algorithm that can solve the co-CDH problem where* $t \le 2t' + 2c(2N + q_H + Nq_S)$ *and* $\epsilon \ge \epsilon'^2$ *where* $\mathcal{A}$ *issues at most* q_S *signing queries and at most* q_H *hash queries, and an exponentiation operation takes time* c *to complete.*

Proof. The co-CDH problem can be solved by first solving two random instances of the following problem: Given g_1^{ab}, g_2^a (and g_1, g_2), compute g_1^b. We shall construct an algorithm $\mathcal{M}$ that solves this problem. This is easy if $a = 0$. In what follows, we assume $a \ne 0$.

$\mathcal{M}$ prepares a large set $\mathcal{L}$ of public keys $P_1, \cdots, P_N$ by picking $x_i \in_R \mathbb{Z}_p$ at random and setting $P_i = (g_2^a)^{x_i}$ for $i = 1, \cdots, N$. Algorithm $\mathcal{A}$ is given $\mathcal{L}$ and invoked. During the simulation, $\mathcal{M}$ simulates $\mathcal{A}$'s view by answering all hash and signing queries.

On a hash query, $\mathcal{M}$ picks at random $s \in_R \mathbb{Z}_p$ and returns $(g_1^{ab})^s$. On a signing query, suppose $\mathcal{A}$ issues a query for a message m' and a set L' of n' public keys such that $L' \subseteq \mathcal{L}$, $\mathcal{M}$ picks at random $r' \in_R \{0, 1\}^{|p|}$ and $\alpha_1, \cdots, \alpha_{n'} \in_R \mathbb{Z}_p$, sets $H(L', m', r')$ to $\prod_{i=1}^{n'} \psi(P_i)^{\alpha_i}$, and returns the signature $\sigma' = (r', g_1^{\alpha_1}, \cdots, g_1^{\alpha_{n'}})$. If collision occurs, that is the product has already been set to a H evaluation, another sequence of $(r', \alpha_1, \cdots, \alpha_{n'})$ will be chosen. It is repeated until no collision occurs. Since all the sequence elements are chosen at random, the chance of having collision occur is at most $\frac{q_H + q_S}{p}$ which is negligibly small. Same applies to the hash queries.

Eventually $\mathcal{A}$ outputs a forgery (L, m, σ) where $L \subseteq \mathcal{L}$, $|L| = n$, and $\sigma = (r, \sigma_1, \cdots, \sigma_n)$. By the random oracle model assumption, $\mathcal{A}$ has previously issued a hash query for (L, m, r). Since $H(L, m, r) = (g_1^{ab})^s$ for some s chosen by $\mathcal{M}$, $\mathcal{M}$ can compute g_1^b by $(\prod_{i=1}^{n} \sigma_i^{x_i})^{1/s}$.

$\mathcal{A}$ cannot distinguish between $\mathcal{M}$'s simulation and real life. Also $\mathcal{M}$ will not fail except with negligible chance (i.e. $\mathcal{A}$ has not issued a hash query for (L, m, r) before and the chance of this happens is $1/(p - q_H - q_S)$).

In one simulation of $\mathcal{A}$, $\mathcal{M}$ conducts N exponentiations to prepare $\mathcal{L}$, one exponentiation for each of $\mathcal{A}$'s hash queries, n exponentiations for each of $\mathcal{A}$'s signing queries, and n exponentiations for computing g_1^b. Since $n \le N$, its running time is at most the $\mathcal{A}$'s running time plus $c(2N + q_H + Nq_S)$.   □

4   Bilinear Threshold Ring Signature Schemes

A t-out-of-n threshold ring signature scheme ensures that any t members of a group of size n can generate a valid signature but not $t - 1$ members or less. It is a natural extension of the notion of (1-out-of-n) ring signature scheme.

The security of a threshold ring signature scheme also has two aspects: unforgeability and anonymity. The definition of anonymity can be extended similarly to the context of threshold ring signature schemes. We omit the details and focus on the theme of this paper: the security models for proofing existential unforgeability. Besides the parameters specified in the three models/games described in Sec. 2, an additional ingredient for threshold-of-t setting stems from the requirement of showing unforgeability even if $t-1$ group members are colluding.

On these $t-1$ adversaries (colluding group members), they can be *static* or *adaptive*. Static adversaries refer to a fix and predetermined set of group members. They capture a scenario where a particular set of $t-1$ group members try to forge a threshold-of-t signature. In other words, it is assumed that $t-1$ private keys are corrupted before the attack begins. This type of attacks is weaker than having adaptive adversaries.

In the model of adaptive adversaries, the group of colluders does not need to be prespecified before the attack. Instead, a set of $t-1$ group members is built up adaptively according to the needs during the attack.

These two types of adversaries can be applied to all the games defined in Sec. 2: Game 0, Game 1 and Game 2. As an example, we convert the strongest game, Game 2, to a threshold setting with adaptive adversaries below.

Game 2_A^T: Two entities: Simulator $\mathcal{M}$ and threshold-of-t adversary $\mathcal{A}$. $\mathcal{M}$ prepares a large set of public keys $\mathcal{L} = \{P_1, \cdots, P_N\}$ in a similar way to previous games. $\mathcal{M}$ invokes $\mathcal{A}$ by providing $\mathcal{L}$ and other appropriate inputs. The goal of $\mathcal{A}$ is to produce a threshold-of-t signature with respect to a public-key set L such that $L \subseteq \mathcal{L}$. Assume that $|L| \geq t$. In the simulation, $\mathcal{M}$ simulates $\mathcal{A}$'s view by answering all hash queries and signing queries. $\mathcal{A}$ can also ask for private keys corresponding to keys in $\mathcal{L}$ by sending a new type of queries called *private key corruption queries*: $\mathcal{A}$ specifies an index π in a private key corruption query and $\mathcal{M}$ returns the private key x_π of the public key P_π in $\mathcal{L}$. Suppose in one successful simulation run, $\mathcal{A}$ outputs threshold-of-t forgery (L, m, σ) for some message m and $L \subseteq \mathcal{L}$. *Restrictions are that (L, m, σ) is not present in $\mathcal{SO}$'s transcript of that particular simulation run, the public key set L' of each signing query is a subset of $\mathcal{L}$, and the total number of distinct private key revelation queries is at most $t-1$ in each simulation.*

T stands for threshold and A stands for adaptive adversaries on the notation of Game 2_A^T. In this game, the adversary adaptively corrupts up to $t-1$ private keys, i.e. it makes its decision on which private keys to corrupt based on the observed message-signature pairs and corrupted private keys.

4.1   A Bilinear Threshold Ring Signature Scheme Based on Secret Sharing

In the following, we present a bilinear threshold ring signature scheme based on the partial proofs of knowledge protocol due to Cramer, et al. [9]. We are able to provide the same level of strong evidence on its security by showing that it is existential unforegable in Game 2_A^T under the random oracle model. We use the same set of notations as in Sec. 3.

Signature Generation. To generate a t-out-of-n threshold ring signature, let $L = \{P_1, \cdots, P_n\}$ be the set of n public keys. Let $I \subseteq \{1, \cdots, n\}, |I| = t$, be the set of indices of t actual signers. Let $\bar{I} = \{1, \cdots, n\} \setminus I$. Let $H : \{0,1\}^* \to \mathbb{Z}_p$ be a hash function viewed as a random oracle. The signature generation proceeds as follows.

1. For each $i \in \bar{I}$, randomly generate $s_i, c_i \in_R \mathbb{Z}_p$, compute $z_i = e(g_1^{s_i}, g_2) \cdot e(g_1^{c_i}, P_i)$.
2. For each $i \in I$, compute $z_i = e(g_1^{r_i}, g_2)$ where $r_i \in_R \mathbb{Z}_p$.
3. Compute $c_0 \leftarrow H(L, t, m, z_1, \cdots, z_n)$.
4. Find a polynomial f of degree $n{-}t$ such that $f(i) = c_i$ for $i \in \bar{I}$ and $f(0) = c_0$.
5. For each $i \in I$, set $c_i = f(i)$ and compute $s_i = r_i - c_i x_i \bmod p$.
6. The signature is $\sigma = (s_1, \cdots, s_n, f)$.

Signature Verification. The signature $\sigma = (s_1, \cdots, s_n, f)$ is valid if

$$f(0) = H(L, t, m, e(g_1^{s_1}, g_2) \cdot e(g_1^{f(1)}, P_1), \cdots, e(g_1^{s_n}, g_2) \cdot e(g_1^{f(n)}, P_n)).$$

Theorem 2 (Anonymity). *For any algorithm $\mathcal{A}$, any set of n public keys $L = \{P_1, \cdots, P_n\}$ in which each of them is generated according to the key generation algorithm in Sec. 3, and a random subset $I \subseteq \{1, \cdots, n\}$, $|I| = t$, the probability $\Pr[\mathcal{A}(\sigma) = \pi : \pi \in I]$ is at most t/n, where σ is any threshold-of-t ring signature on L generated with private keys corresponding to users indexed by I.*

Proof is in Appendix A.

Theorem 3 (Existential Unforgeability in Game $2_{\mathsf{A}}^{\mathsf{T}}$). *Suppose a forger $\mathcal{A}$ is a PPT, which on inputs of a set $\mathcal{L}$ of N public keys, each of them is generated by the key generation algorithm described in Sec. 3, adaptively chooses up to $t - 1$ signers to corrupt, queries a signing oracle $\mathcal{SO}$ and a random oracle H polynomial times, and outputs a forged signature σ for a message $m \in \{0,1\}^*$ on a public-key set L with non-negligible probability where $L \subseteq \mathcal{L}$. Then DLP can be solved with non-negligible probability in polynomial time. All the restrictions in Game $2_{\mathsf{A}}^{\mathsf{T}}$ apply (in particular, the forgery (L, m, σ) should not be present in the signing oracle transcript).*

Proof is in Appendix B

5   Conclusions

We make fine-grained distinctions on the security models for provably secure ring signature schemes. In this paper, we introduce a new model whose security level can be considered to be lying in between the former two commonly used models due to Rivest et al. [15] and Abe et al. [1]. They offer three different levels of security. We explain the importance of making such fine-grained distinctions by showing, in particular, that the bilinear map based ring signature scheme of Boneh et al. [4], which has been proven secure in the weakest model (the one specified by Rivest et al. [15]) among the three, is insecure in stronger models (the new model specified in this paper and the one specified by Abe et al. [1]). We also propose a secure modification of their scheme for each of the two stronger models. A threshold ring signature scheme using bilinear maps is also proposed. It is provably secure against adaptive adversaries in the strongest model defined in this paper under the random oracle model.

Acknowledgements

Helpful discussions with Victor K. Wei are acknowledged.

References

1. M. Abe, M. Ohkubo, and K. Suzuki. 1-out-of-n signatures from a variety of keys. In *Proc. ASIACRYPT 2002*, pages 415–432. Springer-Verlag, 2002. Lecture Notes in Computer Science No. 2501.
2. M. Bellare, D. Micciancio, and B. Warinschi. Foundations of group signatures: Formal definitions, simplified requirements, and a construction based on general assumptions. In *Proc. EUROCRYPT 2003*, pages 614–629. Springer-Verlag, 2003. Lecture Notes in Computer Science No. 2656.
3. M. Bellare and P. Rogaway. Random oracles are practical: A paradigm for designing efficient protocols. In *Proc. 1st ACM Conference on Computer and Communications Security*, pages 62–73. ACM Press, 1993.
4. D. Boneh, C. Gentry, B. Lynn, and H. Shacham. Aggregate and verifiably encrypted signatures from bilinear maps. In *Proc. EUROCRYPT 2003*, pages 416–432. Springer-Verlag, 2003. Lecture Notes in Computer Science No. 2656.
5. D. Boneh, B. Lynn, and H. Shacham. Short signatures from the weil pairing. In *Proc. ASIACRYPT 2001*, pages 514–532. Springer-Verlag, 2003. Lecture Notes in Computer Science No. 2248.
6. E. Bresson, J. Stern, and M. Szydlo. Threshold ring signatures and applications to ad-hoc groups. In *Proc. CRYPTO 2002*, pages 465–480. Springer-Verlag, 2002. Lecture Notes in Computer Science No. 2442.
7. J. Camenisch and M. Stadler. Efficient group signature schemes for large groups. In *Proc. CRYPTO 97*, pages 410–424. Springer-Verlag, 1997. Lecture Notes in Computer Science No. 1294.
8. D. Chaum and E. Van Heyst. Group signatures. In *Proc. EUROCRYPT 91*, pages 257–265. Springer-Verlag, 1991. Lecture Notes in Computer Science No. 547.
9. R. Cramer, I. Damgård, and B. Schoenmakers. Proofs of partial knowledge and simplified design of witness hiding protocols. In *Proc. CRYPTO 94*, pages 174–187. Springer-Verlag, 1994. Lecture Notes in Computer Science No. 839.
10. Y. Dodis, A. Kiayias, A. Nicolosi, and V. Shoup. Anonymous identification in ad-hoc groups. In *Proc. EUROCRYPT 2004*. Springer-Verlag, 2004.
11. S. Goldwasser, S. Micali, and R. Rivest. A digital signature scheme secure against adaptive chosen-message attack. *SIAM J. Computing*, 17(2):281–308, April 1988.
12. J. K. Liu, V. K. Wei, and D. S. Wong. Linkable and anonymous signature for ad hoc groups. In *The 9th Australasian Conference on Information Security and Privacy (ACISP 2004)*, pages 325–335. Springer-Verlag, 2004. Lecture Notes in Computer Science No. 3108.

13. J. K. Liu, V. K. Wei, and D. S. Wong. A separable threshold ring signature scheme. In *The 6th Annual International Conference on Information Security and Cryptology (ICISC 2003)*, pages 12–26. Springer-Verlag, 2004. Lecture Notes in Computer Science No. 2971.

14. K. Ohta and T. Okamoto. On concrete security treatment of signatures derived from identification. In *Proc. CRYPTO 98*, pages 354–369. Springer-Verlag, 1998. Lecture Notes in Computer Science No. 1462.

15. R. Rivest, A. Shamir, and Y. Tauman. How to leak a secret. In *Proc. ASIACRYPT 2001*, pages 552–565. Springer-Verlag, 2001. Lecture Notes in Computer Science No. 2248.

16. D. Wong, K. Fung, J. Liu, and V. Wei. On the RS-code construction of ring signature schemes and a threshold setting of RST. In *5th Intl. Conference on Information and Communication Security (ICICS 2003)*, pages 34–46. Springer-Verlag, 2003. Lecture Notes in Computer Science No. 2836.

17. F. Zhang and K. Kim. ID-Based blind signature and ring signature from pairings. In *Proc. ASIACRYPT 2002*, pages 533–547. Springer-Verlag, 2002. Lecture Notes in Computer Science No. 2501.

A   Proof of Theorem 2

Proof. The polynomial f, with degree $n-t$, is uniquely determined by c_i for $i \in \bar{I}$ and c_0. c_i's are randomly generated and c_0 is the output of the random oracle H. Thus f can be considered as a function chosen randomly from the collection of all polynomials over $GF(p)$ with degree $n-t$. Then the distributions of c_i, for $i \in I$, are also uniform over $GF(p)$.

For $i \in \bar{I}$, s_i are chosen independently and distributed uniformly over $GF(p)$. For $i \in i$, r_i's are chosen independently and distributed uniformly over $GF(p)$. Since r_i's are independent of c_i and x_i, for all $i \in I$, s_i are also uniformly distributed over $GF(p)$.

In addition, for any fixed message m and fixed set of public keys L, we can see that $(s_1, \cdots, s_n)$ has exactly p^n possible solutions. Since the distribution of these possible solutions are independent and uniformly distributed no matter which t participating signers are, an adversary $\mathcal{A}$, even has all the private keys and unbound computing resources, has no advantage in identifying any one of the participating signers over random guessing. $\square$

B   Proof of Theorem 3

Proof. Let $\mathcal{L}$ be a set of N public keys, each of them is generated by the key generation algorithm described in Sec. 3. Suppose there exists a PPT $\mathcal{A}$ that outputs a forgery (L, m, σ) where $L \subseteq \mathcal{L}$, $m \in \{0,1\}^*$ and $\sigma = (s_1, \cdots, s_n, f)$ such that $\deg(f) = n - t$ with non-negligible probability. $\mathcal{A}$ may corrupt adaptively up to $t-1$ private keys of the public keys in $\mathcal{L}$. $\mathcal{A}$ may query a random oracle H and a signing oracle $\mathcal{SO}$. We construct from $\mathcal{A}$ a PPT $\mathcal{M}$ that solves DLP with non-negligible probability. That is, given $(G_2, g_2, p, \hat{P})$, $\mathcal{M}$ outputs an integer $\hat{x}$ such that $g_2^{\hat{x}} = \hat{P}$ with non-negligible probability.

To get $\mathcal{A}$ run properly, $\mathcal{M}$ randomly picks $x_1, \cdots, x_{N-1} \in_R \mathbb{Z}_p$ and randomly picks $\pi \in_R \{1, \cdots, N\}$. For $i = 1, \cdots, N-1$, $\mathcal{M}$ sets $P_i = g_2^{x_i}$. $\mathcal{M}$ sets $P_N = \hat{P}$. Then P_π and P_n are swapped. $\mathcal{A}$ is run by giving the set of N public keys as $L = \{P_1, \cdots, P_N\}$.

During the attack, $\mathcal{A}$ may query random oracle H and signing oracle $\mathcal{SO}$. $\mathcal{A}$ may also pick up to $t-1$ signers to corrupt. If $\mathcal{A}$ picks to corrupt key indexed by π, then the simulation fails. Since π is randomly chosen, the chance of picking π is at most $(t-1)/n$.

For each H-query, a random element from $\mathbb{Z}_p$ is chosen and returned. For each $\mathcal{SO}$-query, $\mathcal{A}$ specifies a public key set $L' = \{P_1, \cdots, P_{n'}\}$, a message m' and a set of indices $I' = \{i_1, \cdots, i_{t'}\}$ such that $1 \le i_j \le n'$ and $t' \le n'$. The set of indices specifies the actual signers. The answer is simulated as follows.

1. Randomly generate $c_0, c_i \in_R \mathbb{Z}_p$ for $i \notin I'$.
2. Construct f over $GF(p)$ such that $\deg(f) = n' - t'$ and $f(0) = c_0$, $f(i) = c_i$, for $i \notin I'$.
3. For $i \in I'$, compute $c_i = f(i)$.
4. For $i = 1, \cdots, n'$, randomly generate $s_i \in_R \mathbb{Z}_p$ and compute $z_i = e(g_1^{s_i}, g_2) \cdot e(g_1^{c_i}, P_i)$.
5. Assign c_0 as the value of $H(L', t', m, z_1, \cdots, z_{n'})$.
6. Output $(s_1, \cdots, s_n, f)$.

The simulation fails if step 6 causes collision, that is, the value of c_0 has been assigned before. This happens with probability at most q_H/p where q_H is the number of times that the random oracle H is queried by $\mathcal{A}$. Since $\mathcal{A}$ only queries H for polynomially number of times, the simulation is successful with overwhelming probability.

Let Θ, Ω be the random tapes given to the signing oracle and $\mathcal{A}$ such that $\mathcal{A}$ outputs a forged signature. Notice that the success probability of $\mathcal{A}$ is taken over the space defined by Θ, Ω and the random oracle H. The forged signature contains a polynomial f where $f(0) = H(L, t, m, z_1, \cdots, z_n)$ for $z_i = e(g_1^{s_i}, g_2) \cdot e(g_1^{f(i)}, P_i)$, $1 \le i \le n$. With probability at least $1 - 1/p$, there exists a query $H(L, t, m, z_1, \cdots, z_n)$ due to the assumption of ideal randomness of H. Split the sequence of outputs of oracle H as (H^-, c_0) where H^- corresponds to the answers to all H-queries except for c_0. By invoking $\mathcal{A}$ with (Θ, Ω, H^-) and randomly choosing c_0' ($\ne c_0$) polynomial times $\rho'(q_H)/p_{\mathcal{A}}$ where ρ' is some polynomial and $p_{\mathcal{A}}$ is the success probability of $\mathcal{A}$, $\mathcal{A}$ outputs at least one forged signature $\sigma' = (s_1', \cdots, s_n', f')$ with probability $p_{\mathcal{A}}/\rho''(q_H)$ where ρ'' is some polynomial, due to the heavy-row lemma [14]. Therefore, $\mathcal{M}$ can compute the discrete-log $\hat{x} = (s_\pi - s_\pi')/(c_\pi' - c_\pi) \bmod p$.

To see that $c_\pi' \ne c_\pi$ (that is, $f'(\pi) \ne f(\pi)$), we notice that since $f'(0) \ne f(0)$ and the degrees of f and f' are limited to $n - t$, there is at least one value j, $1 \le j \le n$, such that $f'(j) \ne f(j)$. The chance of having $j = \pi$ is $1/n$ as π is randomly chosen. Hence the probability of having $\mathcal{M}$ succeed is at least $p_{\mathcal{A}}/(n\rho''(q_H))$ which is non-negligible. $\qquad\square$

Identity Based Threshold Ring Signature

Sherman S.M. Chow*, Lucas C.K. Hui, and S.M. Yiu

Department of Computer Science,
The University of Hong Kong,
Pokfulam, Hong Kong
smchow, hui, smyiu @cs.hku.hk

Abstract. In threshold ring signature schemes, any group of t entities spontaneously conscript arbitrarily $n - t$ entities to generate a publicly verifiable t-out-of-n signature on behalf of the whole group, yet the actual signers remain anonymous. The spontaneity of these schemes is desirable for ad-hoc groups such as mobile ad-hoc networks. In this paper, we present an identity based (ID-based) threshold ring signature scheme. The scheme is provably secure in the random oracle model and provides trusted authority compatibility. To the best of authors' knowledge, our scheme is the first ID-based threshold ring signature scheme which is also the most efficient (in terms of number of pairing operations required) ID-based ring signature scheme (when $t = 1$) and threshold ring signature scheme from pairings.

Keywords: Threshold ring signature, identity-based signature, bilinear pairings, anonymity, spontaneity.

1 Introduction

Anonymity is becoming a major concern in many multi-user electronic commerce applications such as e-lotteries [15], e-cash and online games [12]. Group-oriented signature schemes [9] enable an entity of a group to produce a signature on behalf of the group. There are two major paradigms in anonymous group-oriented signature schemes: group signature and ring signature. In a group signature scheme, the group is predefined and there is a group manager that can revoke this anonymity. Ring signature scheme provides a similar feature. It does not support anonymity revocation mechanism, but no setup stage is needed to produce and distribute a group secret explicitly. Hence it enables any individual to spontaneously conscript arbitrarily $n - 1$ entities and generate a publicly verifiable 1-out-of-n signature on behalf of the whole group, yet the actual signer remains unconditionally anonymous. Threshold ring signature is the t-out-of-n threshold version where t or more entities can jointly generate a valid signature but $t - 1$ or fewer entities cannot. These schemes are getting more and more popular due to the increasing prevalence of pervasive computing applications and mobile ad-hoc networks, where ad-hoc groups are very common [7].

---

* Corresponding author.

C. Park and S. Chee (Eds.): ICISC 2004, LNCS 3506, pp. 218–232, 2005.
© Springer-Verlag Berlin Heidelberg 2005

1.1   Motivation of ID-Based Threshold Ring Signature

In traditional public key infrastructure (PKI), a user must pre-enroll the PKI or he/she cannot enjoy the cryptographic services provided by the PKI, e.g. no one can send them any encrypted message. Identity-based (ID-based) cryptography [5, 30] solves this problem: all users already have their corresponding public key before their enrollment since the public key can be derived via a public algorithm with input of a string that can uniquely identify each of them, such as an email address.

All previous threshold ring signature constructions are non ID-based, hence *real spontaneity* is not always possible: the public key of each member of the group is required to be published by the underlying PKI before it can be used to generate the signature. Removing this pre-requisite requirement motivates the construction of ID-based threshold ring signature scheme, which provide a better alternative than non-ID based solutions[1].

1.2   Related Work

Ring signature scheme was first formalized by Rivest *et al.* in [28]. After that, several other ring signature schemes (for examples [1, 20]) were proposed. Bresson *et al.* [7] extended the ring signature scheme into a threshold ring signature scheme using the concept of partitioning. Later, Wong *et al.* [34] proposed another threshold ring signature using tandem construction method. In [18], a constant-size ring signature was derived from the anonymous identification scheme proposed.

There are some threshold ring signature schemes with special properties recently. For examples, Liu *et al.* [24] introduced the concept of separability to threshold ring signature scheme, which enables the use of various flavours of public keys in a single threshold ring signature; Tsang *et al.* [32] introduced individual-linkability to threshold ring signature scheme, which enables anyone to determine if two ring signatures are signed with the help of the same signer; and Chan *et al.* [8] constructed CDS-type [17] *t*-out-of-*n* blind threshold ring signature, such that the signers do not know what exactly they are signers and cannot link which invocation of signing algorithm corresponding to which unblinded signature. In [24], a generic construction of threshold ring signature from any trapdoor-one-way type signature scheme and three-move type signature scheme is given. Yet, the authors have not illustrated the correctness and the security of this construction except the specific instantiations from RSA [27] and Schnorr signature [29].

Using bilinear pairing to construct ring signature is not a new idea. Inspired by the aggregate signature, a ring signature scheme was proposed in [6]. A technique similar to that of [6] was used to derive a new ring signature scheme

---

[1] Under the assumption that the trusted authority (the private key generator) will not reveal any information about who has requested for his/her private key and who has not.

in [35]. In [38], a ring signature was derived from the short signature proposed. A proxy ring signature was proposed in [39]. ID-based ring signature was introduced in [37] and subsequently a more efficient construction was proposed in [23]. Small inconsistencies in [37] and [23] were fixed by [2], together with a new proxy ring signature scheme from the delegation function due to [39]. Another ID-based ring signature with formally proven security was proposed in [22]. Threshold ring signature scheme from pairings was proposed in [33], but this scheme is not ID-based and has not addressed the requirement of TA (trusted authority) compatibility [36] in which not all the users join the same TA.

1.3   Our Contributions

In this paper, we present an ID-based threshold ring signature scheme. The scheme is provably secure in the random oracle model [4] and provides TA compatibility [36]. To the best of authors' knowledge, our scheme is the first ID-based threshold ring signature scheme. Our scheme is the most efficient (in terms of number of pairing operations required) ID-based ring signature scheme (when the threshold value $t = 1$) and also the most efficient threshold ring signature scheme from pairings.

1.4   Organization

The rest of the paper is organized as follows. The next section contains some preliminaries about the formal definitions of an ID-based threshold ring signature scheme, bilinear pairing as well as the Gap Diffie-Hellman group. Formal security definitions describing the adversary's capabilities and goals are presented in Section 3. Section 4 describes the proposed ID-based threshold ring signature scheme. The security and efficiency analysis of our scheme are given in Section 5. Finally, Section 6 concludes the paper.

2   Preliminaries

Before presenting our results, we give the framework of ID-based threshold ring signature schemes and review the definitions of bilinear pairing and Gap Diffie-Hellman groups.

2.1   Framework of ID-Based Threshold Ring Signature

An ID-based threshold ring signature scheme consists of four algorithms: Setup, KeyGen, Sign, and Verify.

- Setup: On an unary string input 1^k where k is a security parameter, it produces the master secret key s and the common public parameters $params$, which include a description of a finite signature space and a description of a finite message space.

- KeyGen: On an input of signer's identity $ID \in \{0,1\}^*$ and the master secret s, it outputs the signer's secret signing key S_{ID}.
- Sign: On input of a message m, a group of n users' identities $\{ID_i\}$, where $1 \leq i \leq n$, and the secret keys of t members $\{S_{ID_{i_j}}\}$, where $1 \leq i_j \leq n$, $1 \leq j \leq t$ and $t \leq n$; it outputs a (t,n) ID-based threshold ring signature σ on the message m.
- Verify: On a threshold ring signature σ, a message m, the threshold value t and the group of signers' identities $\{ID_i\}$ where $1 \leq i \leq n$ as the input, it outputs $\top$ for "true" or $\bot$ for "false", depending on whether σ is a valid signature signed by at least t members in the group $\{ID_i\}$ on a message m.

These algorithms must satisfy the standard consistency constraint of ID-based threshold ring signature schemes, i.e. if we have $\sigma = \text{Sign}(m, \{ID_i\}, \{S_{ID_{i_j}}\})$ and $|\{S_{ID_{i_j}}\}| = t$ (where $|\{S_{ID_{i_j}}\}|$ denotes the number of elements in the set $\{S_{ID_{i_j}}\}$), we must have $\text{Verify}(\sigma, \{ID_i\}, m, t) = \top$. Security requirements will be described in Section 3.

2.2   Bilinear Pairing and Gap Diffie-Hellman Groups

Bilinear pairing is an important primitive for many cryptographic schemes (for examples, $[2,5,6,11,13,14,16,21,22,23,25,33,35,36,37,38,39]$). Here, we describe some of its key properties.

Let $(\mathbb{G}_1, +)$ and $(\mathbb{G}_2, \cdot)$ be two cyclic groups of prime order q. The bilinear pairing is given as $\hat{e} : \mathbb{G}_1 \times \mathbb{G}_1 \rightarrow \mathbb{G}_2$, which satisfies the following properties:

1. *Bilinearity:* For all $P, Q, R \in \mathbb{G}_1$, $\hat{e}(P+Q, R) = \hat{e}(P, R)\hat{e}(Q, R)$, and $\hat{e}(P, Q+R) = \hat{e}(P, Q)\hat{e}(P, R)$.
2. *Non-degeneracy:* There exists $P, Q \in \mathbb{G}_1$ such that $\hat{e}(P, Q) \neq 1$.
3. *Computability:* There exists an efficient algorithm to compute $\hat{e}(P, Q) \; \forall P, Q \in \mathbb{G}_1$.

Definition 1. *Given a generator P of a group $\mathbb{G}$ and a 3-tuple (aP, bP, cP), the Decisional Diffie-Hellman problem (DDHP) is to decide whether $c = ab$.*

Definition 2. *Given a generator P of a group $\mathbb{G}$, (P, aP, bP, cP) is defined as a valid Diffie-Hellman tuple if $c = ab$.*

Definition 3. *Given a generator P of a group $\mathbb{G}$ and a 2-tuple (aP, bP), the Computational Diffie-Hellman problem (CDHP) is to compute abP.*

Definition 4. *If $\mathbb{G}$ is a group such that DDHP can be solved in polynomial time but no probabilistic algorithm can solve CDHP with non-negligible advantage within polynomial time, then we call $\mathbb{G}$ a Gap Diffie-Hellman (GDH) group.*

We assume the existence of a bilinear map $\hat{e} : \mathbb{G}_1 \times \mathbb{G}_1 \rightarrow \mathbb{G}_2$ that one can solve Decisional Diffie-Hellman Problem in polynomial time.

3   Formal Security Model

For an ID-based threshold ring signature scheme to be considered as secure, we need to consider its unforgeability and signer ambiguity.

3.1   Unforgeability of ID-Based Threshold Ring Signature

The following EUF-IDTR-CMIA2 game played between a challenger C and an adversary A formally defines the *existential unforgeability of ID-based threshold ring signature under adaptive chosen-message-and-identity attack.*

EUF-IDTR-CMIA2 Game:

Setup: The challenger C takes a security parameter k and runs the Setup to generate common public parameters *params* and also the master secret key s. C sends *params* to A.

Attack: The adversary A can perform a polynomially bounded number of queries in an adaptive manner (that is, each query may depend on the responses to the previous queries). The types of queries allowed are described below.

- Hash functions queries: A can ask for the values of the hash functions (e.g. $H(\cdot)$ and $H_0(\cdot)$ in our proposed scheme) for any input.
- KeyGen: A chooses an identity ID. C computes $\text{KeyGen}(ID) = S_{ID}$ and sends the result to A.
- Sign: A chooses a group of n users' identities $\{ID_i\}$ where $1 \leq i \leq n$, a threshold value t' where $t' \leq n$, and any message m. C outputs a (t', n) ID-based threshold ring signature σ.

Forgery: The adversary A outputs an ID-based threshold ring signature σ on message m "signed" by at least t' members ($t' \leq n$) of a group of n users $\{ID_i\}$ where $1 \leq i \leq n$. The only restriction is that $(m, \{ID_i\})$ does not appear in the set of previous Sign queries and less than t' private keys of $\{ID_i\}$ are returned by previous KeyGen queries. It wins the game if $\text{Verify}(\sigma, \{ID_i\}, m, t')$ is equal to $\top$. The advantage of A is defined as the probability that it wins.

Definition 5. *An ID-based threshold ring signature scheme is said to have the existential unforgeability against adaptive chosen-message-and-identity attacks property (EUF-IDTR-CMIA2 secure) if no adversary has a non-negligible advantage in the EUF-IDTR-CMIA2 game.*

3.2   Signer Ambiguity of ID-Based Threshold Ring Signature

Definition 6. *An ID-based threshold ring signature scheme is said to have the unconditional signer ambiguity if for any group of n users $\{ID_i\}$ where $1 \leq i \leq n$, any t' signers indexed by $\{i_j\}$, where $1 \leq i_j \leq n$, $1 \leq j \leq t'$ and $t' \leq n$, any message m and any signature σ, where $\sigma = \text{Sign}(m, \{ID_i\}, \{S_{ID_{i_j}}\})$, any verifier A (i.e. not a signer in the group $\{ID_{i_j}\}$), even with unbounded computing* resources, cannot identify any of the signer with probability better than a random guess. That is, $\mathcal{A}$ can only output any member of $\{i_j\}$ with probability no better than $\frac{t'}{n}$.

4   Our Proposed Scheme

In this section, we show how to adopt the techniques introduced in [24] with the elegancy of bilinear pairings to spawn an efficient ID-based threshold ring signature scheme with reasonable signature size.

4.1   Basic Construction

Define $\mathbb{G}_1, \mathbb{G}_2$, and $\hat{e}(\cdot, \cdot)$ as in the Section 2 where $\mathbb{G}_1$ is a GDH group. $H(\cdot)$ and $H_0(\cdot)$ are two cryptographic hash functions where $H : \{0,1\}^* \to \mathbb{G}_1$ and $H_0 : \{0,1\}^* \to \mathbb{Z}_q^*$.

Setup: TA randomly chooses $s \in_R \mathbb{Z}_q^*$, keeps it as the master secret key and computes the corresponding public key $P_{pub} = sP$. The system parameters are:

$$params = \{\mathbb{G}_1, \mathbb{G}_2, \hat{e}(\cdot, \cdot), q, P, P_{pub}, H(\cdot), H_0(\cdot)\}.$$

KeyGen: The signer with identity $ID \in \{0,1\}^*$ submits ID to TA. TA sets the signer's public key Q_{ID} to be $H(ID) \in \mathbb{G}_1$, computes the signer's private signing key S_{ID} by $S_{ID} = sQ_{ID}$. Then TA sends the private signing key to the signer.

Sign: Let L be the set of all identities of the n users. Without loss of generality, we assume user indexed by $\{1, 2, \cdots, t\}$ are the participating signers while user indexed by $\{t+1, t+2, \cdots, n\}$ are the non-participating signers. The participating signers carry out the following steps to give an ID-based threshold ring signature.

1. An arbitrary entity (which is trusted to keep the identities of the participating signers in confidential) "prepares the signature on behalf of" other entities in the group by performing the following computations: For $i \in \{t+1, \cdots, n\}$, chooses x_i and $h_i \in_R \mathbb{Z}_q^*$ and computes $U_i = x_i P - h_i P_{pub}$ and $V_i = x_i Q_{ID_i}$.
2. For $j \in \{1, \cdots, t\}$, each signer ID_j chooses $r_j \in_R \mathbb{Z}_q^*$ and computes $U_j = r_j P$.
3. Anyone in the group of t participating signers who got the knowledge of $\cup_{k=1}^{n}\{U_k\}$ computes $h_0 = H_0(L, t, m, \cup_{k=1}^{n}\{U_k\})$ and construct a polynomial f of degree $n-t$ over $\mathbb{Z}_q$ such that $f(0) = h_0$ and $f(i) = h_i$ for $t+1 \leq i \leq n$.
4. For $j \in \{1, \cdots, t\}$, each signer ID_j computes $h_j = f(j)$ and $V_j = r_j Q_{ID_j} + h_j S_{ID_j}$.
5. Anyone in the group of t participating signers who got the knowledge of $\cup_{k=1}^{n}\{V_k\}$ computes $V = \sum_{k=1}^{n} V_k$.
6. Output the signature for m and L as $\sigma = \{\cup_{k=1}^{n}\{U_k\}, V, f\}$.
   (The polynomial f only contains information for the hash values used and its inclusion will not compromise the unforgeability and the anonymity of the scheme.)

Verify: A verifier checks whether a signature $\sigma = \{\cup_{k=1}^{n}\{U_k\}, V, f\}$ for the message m is given by at least t signers from the set of users L as follows.

1. Check if the degree of polynomial f is $n-t$ and $H_0(L, t, m, \cup_{k=1}^{n}\{U_k\})$ is the constant term of f. Proceed if both conditions are true, reject otherwise.
2. For $k \in \{1, \cdots, n\}$, compute $h_k = f(k)$.
3. Check whether $\prod_{k=1}^{n} \hat{e}(Q_{ID_k}, U_k + h_k P_{pub}) = \hat{e}(P, V)$. If the equality holds, return $\top$. Otherwise, return $\bot$.

4.2   Trusted Authority Compatibility

It is quite often that different users joined different trusted authorities (TAs) in the reality. In [36], the notion of TA compatibility is introduced in the ID-based signcryption [16, 36] scenario. We extend their notion into TA compatibility in ID-based threshold ring signature. For ID-based threshold ring signature schemes, spontaneity will be affected if the intended group of signers joined different TAs. However, our scheme can be easily extended to handle this situation without compromising the spontaneity. We just need to change the equality to be checked in the verification algorithm to $\prod_{k=1}^{n} \hat{e}(Q_{ID_k}, U_k + h_k P_{pub_k}) = \hat{e}(P, V)$, where P_{pub_k} is the public key of the TA of the k-th user.

4.3   Robustness

Robustness is often desirable in group-oriented signature scheme. For a threshold ring signature scheme that does not support robustness, the misbehavior of any participating signer cannot be detected, and the final signature generated by the group of signers will be invalid even there is only one misbehaving signer. In our scheme, the partial signature $\sigma_j = \{h_j, U_j, V_j\}$ generated by the signer ID_j can be verified easily by checking whether $\hat{e}(Q_{ID_j}, U_j + h_j P_{pub}) = \hat{e}(P, V_j)$ holds.

5   Analysis of the Proposed Scheme

We analyze consistency, efficiency, existential unforgeability and signer ambiguity of our proposed scheme.

5.1   Consistency

The consistency of our basic construction can be easily verified by the following equations.

$$\hat{e}(P, V) = \hat{e}(P, \sum_{k=1}^{n} V_k)$$

$$= \prod_{i=1}^{t} \hat{e}(P, V_i) \prod_{j=t+1}^{n} \hat{e}(P, V_j)$$

$$= \prod_{j=1}^{t} \hat{e}(P, r_j Q_{ID_j} + h_j S_{ID_j}) \prod_{i=t+1}^{n} \hat{e}(P, x_i Q_{ID_i})$$

$$= \prod_{j=1}^{t} \hat{e}(P, (r_j + h_j s) Q_{ID_j}) \prod_{i=t+1}^{n} \hat{e}(x_i P, Q_{ID_i})$$

$$= \prod_{j=1}^{t} \hat{e}(Q_{ID_j}, (r_j + h_j s) P) \prod_{i=t+1}^{n} \hat{e}(Q_{ID_i}, x_i P - h_i P_{pub} + h_i P_{pub})$$

$$= \prod_{j=1}^{t} \hat{e}(Q_{ID_j}, U_j + h_j P_{pub}) \prod_{i=t+1}^{n} \hat{e}(Q_{ID_i}, U_i + h_i P_{pub})$$

$$= \prod_{k=1}^{n} \hat{e}(Q_{ID_k}, U_k + h_k P_{pub})$$

The consistency of the checking for the sake of robustness and that of our extended scheme with TA compatibility can be verified easily in a similar manner.

5.2   Efficiency

Although some research has been done in analyzing the complexity and speeding up the computation of pairing function (for examples, [3, 10, 19]), the pairing operations are still rather expensive. Our scheme is the most efficient (in terms of number of pairing operations required) ID-based ring signature scheme (when the threshold value $t = 1$). Taken into account the computational costs for signature generation and verification, [37] uses $4n - 1$ pairing operations while both of [2] and [23] use $2n + 1$ of them. While the most efficient 1-out-of-n ID-based ring signature scheme before the birth of our scheme is [22], which uses $n + 3$ pairings in total (i.e. signing and verification), our scheme only uses $n+1$ pairing operations. Although the difference is not great, our scheme can be further optimized since the multiplication of a series of pairings in Verify can be optimized by using the concept of "Miller lite" of Tate pairing presented in [31]. Moreover, the pairing operations in our scheme can be executed in parallel, which is not possible in schemes like [2, 23, 37].

The previous non ID-based threshold ring signature scheme from bilinear pairings in [33], requires $n + t$ pairing operations (or $(n + 1)t$ of them without optimization) for verification. Our scheme is more efficient since it only requires n pairing operations in verification and none of them in signing.

Considering the signature size, our scheme is also up to the state-of-the-art. Signature sizes in [7] and [34] are $O(n \lg n)$ and $O(n^t)$, respectively. We share the same order of space complexities as in [24] and [33]. However, due to the elegancy of elliptic curve, our scheme should achieve shorter signature size than [24].

5.3   Existential Unforgeability and Signer Ambiguity

The security of our proposed scheme is summarized in the following two theorems.

Theorem 1. *In the random oracle model (the hash functions are modeled as random oracles), if there is an algorithm $\mathcal{A}$ that can win the EUF-IDTR-CMIA2 game in polynomial time, then CDHP can be solved with non-negligible probability in polynomial time.*

Proof. Suppose the challenger $\mathcal{C}$ receives a random instance (P, aP, bP) of the CDHP and has to compute the value of abP. $\mathcal{C}$ will run $\mathcal{A}$ as a subroutine and act as $\mathcal{A}$'s challenger in the EUF-IDTR-CMIA2 game. During the game, $\mathcal{A}$ will consult $\mathcal{C}$ for answers to the random oracles H and H_0. Roughly speaking, these answers are randomly generated, but to maintain the consistency and to avoid collision, $\mathcal{C}$ keeps three lists to store the answers used. We assume $\mathcal{A}$ will ask for $H(ID)$ before ID is used in any other queries.

$\mathcal{C}$ gives $\mathcal{A}$ the system parameters with $P_{pub} = bP$. Note that b is unknown to $\mathcal{C}$. This value simulates the master key value for the TA in the game.

H requests and KeyGen requests: When $\mathcal{A}$ asks queries on the hash values of identities, $\mathcal{C}$ checks the list L_1, If an entry for the query is found, the same answer will be given to $\mathcal{A}$; otherwise, a value d_i from $\mathbb{Z}_q^*$ will be randomly generated and $d_i P$ will be used as the answer, the query and the answer will then be stored in the list. Note that the associated private key is $d_i bP$ which $\mathcal{C}$ knows how to compute.

The only exception is that $\mathcal{C}$ has to randomly choose one of the H queries from $\mathcal{A}$, say the k-th query, and answers $H(ID^*) = aP$ for this query. Since aP is a value in a random instance of the CDHP, it does not affect the randomness of the hash function H. Since both a and b are unknown to $\mathcal{C}$, a KeyGen request on this identity will make $\mathcal{C}$ fails.

H_0 requests: When $\mathcal{A}$ asks queries on the hash values, $\mathcal{C}$ checks the corresponding list L_2. If an entry for the query is found, the same answer will be given to $\mathcal{A}$; otherwise, a randomly generated value will be used as an answer to $\mathcal{A}$, the query and the answer will then be stored in the list.

Sign requests: $\mathcal{A}$ chooses a group of n users' identities $L = \{ID_j\}$, and a threshold value t' where $t' \leq n$ and any message m. On input of (m, L, t'), $\mathcal{C}$ outputs a (t', n) ID-based threshold ring signature σ as follows.

1. For $i \in \{0, t', t'+1, \cdots, n\}$, randomly choose $h_i \in_R \mathbb{Z}_q^*$.
2. Construct a polynomial f over $\mathbb{Z}_q$ such that the degree of f is $n - t'$ and $f(i) = h_i$ for $i = 0, t', t'+1, \cdots, n$.
3. For $j \in \{1, \cdots, t\}$, compute $h_j = f(j)$.
4. For $k \in \{1, \cdots, n\}$, randomly choose h_k and compute $U_k = x_k P - h_k P_{pub}$.
5. Compute $V = \sum_{k=1}^{n} x_k Q_{ID_k}$.
6. Assign h_0 as the value of $H_0(L, t, m, \cup_{k=1}^{n}\{U_k\})$; if collision occurs, generate another h_0 and repeat.
7. Output the signature as $\sigma = \{\cup_{k=1}^{n}\{U_k\}, V, f\}$.

Finally, $\mathcal{A}$ outputs a forged signature $\sigma = \{U, V, f\}$ that is "signed" by some t' members in the group $\{ID_i\}$, $ID^* \in \{ID_i\}$ and $\mathcal{A}$ only requested for the private key of some $t' - 1$ members in the group. If $ID^* \notin \{ID_i\}$, $\mathcal{C}$ fails.

It follows from the forking lemma [26] that if $\mathcal{A}$ is a sufficiently efficient forger in the above interaction, then we can construct a Las Vegas machine $\mathcal{A}'$ that outputs two signed messages (U, V, f) and (U, V', f'). To do so we keep all the random tapes in two invocations of $\mathcal{A}$ the same except h_0 returned by H_0 query of the forged message.

Now we consider the probability that ID^* is the chosen target of forgery. Let π be the index of ID^* in L, we need $f(\pi) \neq f'(\pi)$ to solve for CDHP. From the signing algorithm, $f(i) = f'(i)$ if ID_i is a non-participating signer; together with the fact that $f(0) \neq f'(0)$, we know that $f(j) \neq f'(j)$ if ID_j is a participating signer. Since $\mathcal{A}'$ knows $t' - 1$ private keys among the group $\{ID_i\}$, the probability that ID^* is a participating signer (and hence $(f(\pi) \neq f'(\pi))$) is $\frac{1}{n-t'+1}$.

Given the machine $\mathcal{A}'$ derived from $\mathcal{A}$, we can solve the CDHP by computing $abP = (h_\pi - h'_\pi)^{-1}(V - V')$. We calculate the probability of success of $\mathcal{C}$ as follows. For $\mathcal{C}$ to succeed, $\mathcal{A}$ did not ask a KeyGen query on ID^*. And the corresponding probability is at least $\frac{q_H - q_E}{q_H}$. Further, with a probability of

$$(n - t' + 1)\left(\frac{q_H-q_E-1}{q_H-q_E}\right)\left(\frac{q_H-q_E-2}{q_H-q_E-1}\right)\cdots\left(\frac{q_H-q_E-(n-t')}{q_H-q_E-(n-t'-1)}\right)\left(\frac{1}{q_H-q_E-(n-t')}\right) = \frac{n-t'+1}{q_H-q_E},$$

$ID^* \in \{ID_i\}$, hence the probability for using $\mathcal{A}$ to solve the CDHP is $\frac{1}{q_H}$.

$\square$

Theorem 2. *Our ID-based threshold ring signature scheme satisfies the property of unconditional signer ambiguity.*

Proof. The polynomial f with degree $n - t$ can be considered as a function chosen randomly from the collection of all polynomials over $\mathbb{Z}_q$ with degree $n - t$ since $h_{t+1}, \cdots, h_n$ are randomly generated and h_0 is the output of the random oracle H_0.

For $i \in \{t + 1, \cdots, n\}$, and for $j \in \{1, \cdots, t\}$, $\{x_i\}$ and $\{r_j\}$ are chosen independently and distributed uniformly over $\mathbb{Z}_q^*$. So $\{U_i\} \cup \{U_j\}$ and hence $\cup_{k=1}^n \{U_k\}$ are also uniformly distributed.

The polynomial f is determined by $h_{t+1}, \cdots, h_n$ and h_0, then the distributions of $h_1, \cdots, h_t$ are also uniform over the underlying range, with the fact that $\{S_{ID_j}\}$ is independent of $\{r_j\}$ and $\{h_j\}$, we say that $\{V_i\} \cup \{V_j\}$ and hence V are also uniformly distributed.

To conclude, for any fixed message m and fixed set of identities L, the distribution of $\{\cup_{k=1}^n \{U_k\}, V, f\}$ are independent and uniformly distributed no matter which t participating signers are. So we conclude that even an adversary with all the private keys corresponding to the set of identities L and unbounded computing resources has no advantage in identifying any one of the participating signers over random guessing. $\square$

6   Conclusion

In this paper, we present an ID-based threshold ring signature scheme. We prove the security of our scheme in the random oracle model [4]. Moreover, our scheme provides trusted authority compatibility [36]. To the best of authors' knowledge, our scheme is the first ID-based threshold ring signature scheme, which is also the most efficient ID-based ring signature scheme (when the threshold value $t = 1$) and threshold ring signature scheme from pairings in terms of the number of pairing operations. Due to the elegancy of bilinear pairing, signatures generated by our scheme are much shorter and simpler than signatures from other previous threshold ring signature schemes. Future research directions include devising an ID-based threshold ring signature scheme with constant signature size or making the threshold ring signature scheme works in a hierarchical setting [14].

Acknowledgement

This research is supported in part by the Areas of Excellence Scheme established under the University Grants Committee of the Hong Kong Special Administrative Region, China (Project No. AoE/E-01/99), a grant from the Research Grants Council of the Hong Kong Special Administrative Region, China (Project No. HKU/7144/03E), and a grant from the Innovation and Technology Commission of the Hong Kong Special Administrative Region, China (Project No. ITS/170/01).

References

1. Masayuki Abe, Miyako Ohkubo, and Koutarou Suzuki. 1-out-of-n Signatures from a Variety of Keys. In Yuliang Zheng, editor, *Advances in Cryptology - ASIACRYPT 2002, 8th International Conference on the Theory and Application of Cryptology and Information Security, Queenstown, New Zealand, December 1-5, 2002, Proceedings*, volume 2501 of *Lecture Notes in Computer Science*, pages 415–432. Springer, 2002.
2. Amit K Awasthi and Sunder Lal. ID-based Ring Signature and Proxy Ring Signature Schemes from Bilinear Pairings. Cryptology ePrint Archive, Report 2004/184, 2004. Available at http://eprint.iacr.org.
3. Paulo S.L.M. Barreto, Hae Y. Kim, Ben Lynn, and Michael Scott. Efficient Algorithms for Pairing-Based Cryptosystems. In Moti Yung, editor, *Advances in Cryptology: Proceedings of CRYPTO 2002 22nd Annual International Cryptology Conference Santa Barbara, California, USA, August 18-22, 2002*, volume 2442 of *Lecture Notes in Computer Science*, pages 354–368. Springer-Verlag Heidelberg, 2002.
4. Mihir Bellare and Phillip Rogaway. Random Oracles are Practical: A Paradigm for Designing Efficient Protocols. In *Proceedings of the 1st ACM Conference on Computer and Communications Security*, pages 62–73, 1993.

5. Dan Boneh and Matt Franklin. Identity-Based Encryption from the Weil Pairing. In Joe Kilian, editor, *Advances in Cryptology - CRYPTO 2001, 21st Annual International Cryptology Conference, Santa Barbara, California, USA, August 19-23, 2001, Proceedings*, volume 2139 of *Lecture Notes in Computer Science*, pages 213–229. Springer-Verlag Heidelberg, 2001.

6. Dan Boneh, Craig Gentry, Ben Lynn, and Hovav Shacham. Aggregate and Verifiably Encrypted Signatures from Bilinear Maps. In Eli Biham, editor, *Advances in Cryptology - EUROCRYPT 2003, International Conference on the Theory and Applications of Cryptographic Techniques, Warsaw, Poland, May 4-8, 2003, Proceedings*, volume 2656 of *Lecture Notes in Computer Science*, pages 416–432. Springer, 2003.

7. Emmanuel Bresson, Jacques Stern, and Michael Szydlo. Threshold Ring Signatures and Applications to Ad-hoc Groups. In Moti Yung, editor, *Advances in Cryptology - CRYPTO 2002, 22nd Annual International Cryptology Conference, Santa Barbara, California, USA, August 18-22, 2002, Proceedings*, volume 2442 of *Lecture Notes in Computer Science*, pages 465–480. Springer, 2002.

8. Tony K. Chan, Karyin Fung, Joseph K. Liu, and Victor K. Wei. Blind Spontaneous Anonymous Group Signatures for Ad Hoc Groups. In *1st European Workshop on Security in Ad-Hoc and Sensor Networks (ESAS 2004), Heidelberg, Germany, August 5-6, 2004*, volume 3313 of *Lecture Notes in Computer Science*. Springer, 2005.

9. David Chaum and Eugène van Heyst. Group Signatures. In Donald W. Davies, editor, *Advances in Cryptology - EUROCRYPT '91, Workshop on the Theory and Application of of Cryptographic Techniques, Brighton, UK, April 8-11, 1991, Proceedings*, volume 547 of *Lecture Notes in Computer Science*, pages 257–265. Springer, 1991.

10. YoungJu Choie and Eunjeong Lee. Implementation of Tate Pairing of Hyperelliptic Curves of Genus 2. In Jong In Lim and Dong Hoon Lee, editors, *Information Security and Cryptology - ICISC 2003, 6th International Conference Seoul, Korea, November 27-28, 2003, Revised Papers*, volume 2971 of *Lecture Notes in Computer Science*, pages 97–111. Springer, 2003.

11. Sherman S.M. Chow. Verifiable Pairing and Its Applications. In Chae Hoon Lim and Moti Yung, editors, *Information Security Applications, 5th International Workshop, WISA 2004, Jeju Island, Korea, August 23-25, Revised Papers*, volume 3325 of *Lecture Notes in Computer Science*, pages 173–187. Springer-Verlag, 2004.

12. Sherman S.M. Chow, H.W. Go, and Ricky W.M. Tang. Impact of Recent Advances in Cryptography on Online Game. In *Third International Conference on Application and Development of Computer Games, Hong Kong, April 26-27, 2004*, pages 68–73, 2004.

13. Sherman S.M. Chow, Lucas C.K. Hui, and S.M. Yiu. Identity Based Threshold Ring Signature. Cryptology ePrint Archive, Report 2004/179, 2004. Available at http://eprint.iacr.org.

14. Sherman S.M. Chow, Lucas C.K. Hui, S.M. Yiu, and K.P. Chow. Secure Hierarchical Identity Based Signature and its Application. In Javier Lopez, Sihan Qing, and Eiji Okamoto, editors, *Information and Communications Security, 6th International Conference, ICICS 2004, Malaga, Spain, October 27-29, 2004, Proceedings*, volume 3269 of *Lecture Notes in Computer Science*, pages 480–494. Springer-Verlag, 2004.

15. Sherman S.M. Chow, Lucas C.K. Hui, S.M. Yiu, and K.P. Chow. An e-Lottery Scheme using Verifiable Random Function.   In *3rd Workshop on Internet Communications Security (WICS 2005) of Computational Science and Its Applications - ICCSA 2005, International Conference, Singapore, May 9-12, 2005, Proceedings*, Lecture Notes in Computer Science. Springer-Verlag, 2005. To Appear.

16. Sherman S.M. Chow, S.M. Yiu, Lucas C.K. Hui, and K.P. Chow. Efficient Forward and Provably Secure ID-Based Signcryption Scheme with Public Verifiability and Public Ciphertext Authenticity. In Jong In Lim and Dong Hoon Lee, editors, *Information Security and Cryptology - ICISC 2003, 6th International Conference Seoul, Korea, November 27-28, 2003, Revised Papers*, volume 2971 of *Lecture Notes in Computer Science*, pages 352–369. Springer, 2003.

17. Ronald Cramer, Ivan Damgard, and Berry Schoenmakers. Proofs of Partial Knowledge and Simplified Design of Witness Hiding Protocols. In Yvo Desmedt, editor, *Advances in Cryptology - CRYPTO '94, 14th Annual International Cryptology Conference, Santa Barbara, California, USA, August 21-25, 1994, Proceedings*, volume 839 of *Lecture Notes in Computer Science*, pages 174–187. Springer, 1994.

18. Yevgeniy Dodis, Aggelos Kiayias, Antonio Nicolosi, and Victor Shoup. Anonymous Identification in Ad Hoc Groups. In Christian Cachin and Jan Camenisch, editors, *Advances in Cryptology - EUROCRYPT 2004, International Conference on the Theory and Applications of Cryptographic Techniques, Interlaken, Switzerland, May 2-6, 2004, Proceedings*, volume 3027 of *Lecture Notes in Computer Science*, pages 609–626. Springer, 2004.

19. Steven D. Galbraith, Keith Harrison, and David Soldera. Implementing the Tate Pairing. In Claus Fieker and David R. Kohel, editors, *Algorithmic Number Theory, 5th International Symposium, ANTS-V, Sydney, Australia, July 7-12, 2002, Proceedings*, volume 2369 of *Lecture Notes in Computer Science*, pages 324–337. Springer, 2002.

20. Chong Zhi Gao, Zheng an Yao, and Lei Li. A Ring Signature Scheme Based on the Nyberg-Rueppel Signature Scheme. In Jianying Zhou, Moti Yung, and Yongfei Han, editors, *Applied Cryptography and Network Security, First International Conference, ACNS 2003, Kunming, China, October 16-19, 2003 Proceedings*, volume 2846 of *Lecture Notes in Computer Science*, pages 169–175. Springer, 2003.

21. Javier Herranz and Germán Sáez. Distributed Ring Signatures for Identity-Based Scenarios. Cryptology ePrint Archive, Report 2004/190, 2004. Available at http://eprint.iacr.org.

22. Javier Herranz and Germán Sáez. New Identity-Based Ring Signature Schemes. In Javier Lopez, Sihan Qing, and Eiji Okamoto, editors, *Information and Communications Security, 6th International Conference, ICICS 2004, Malaga, Spain, October 27-29, 2004, Proceedings*, volume 3269 of *Lecture Notes in Computer Science*, pages 27–39, Malaga, Spain, October 2004. Springer-Verlag. Preliminary version available at Cryptology ePrint Archive, Report 2003/261.

23. Chih-Yin Lin and Tzong-Chen Wu. An Identity-based Ring Signature Scheme from Bilinear Pairings. Cryptology ePrint Archive, Report 2003/117, 2003. Available at http://eprint.iacr.org.

24. Joseph K. Liu, Victor K. Wei, and Duncan S. Wong. A Separable Threshold Ring Signature Scheme. In Jong In Lim and Dong Hoon Lee, editors, *Information Security and Cryptology - ICISC 2003, 6th International Conference Seoul, Korea, November 27-28, 2003, Revised Papers*, volume 2971 of *Lecture Notes in Computer Science*, pages 352–369. Springer, 2003.
25. Joseph K. Liu and Duncan S. Wong. On the Security Models of (Threshold) Ring Signature Schemes. In *Information Security and Cryptology - ICISC 2004, 7th International Conference Seoul, Korea, December 2-3, 2004, Revised Papers*, Lecture Notes in Computer Science, Seoul, Korea, December 2004. Springer-Verlag.
26. David Pointcheval and Jacques Stern. Security Arguments for Digital Signatures and Blind Signatures. *Journal of Cryptology: The Journal of the International Association for Cryptologic Research*, 13(3):361–396, 2000.
27. Ronald L. Rivest, Adi Shamir, and Leonard M. Adleman. A Method for Obtaining Digital Signatures and Public-Key Cryptosystems. *Communications of the ACM*, 26(1):96–99, January 1983.
28. Ronald L. Rivest, Adi Shamir, and Yael Tauman. How to Leak a Secret. In Colin Boyd, editor, *Advances in Cryptology - ASIACRYPT 2001, 7th International Conference on the Theory and Application of Cryptology and Information Security, Gold Coast, Australia, December 9-13, 2001, Proceedings*, volume 2248 of *Lecture Notes in Computer Science*, pages 552–565. Springer, 2001.
29. Claus-Peter Schnorr. Efficient Signature Generation by Smart Cards. *Journal of Cryptology: The Journal of the International Association for Cryptologic Research*, 4(3):161–174, 1991.
30. Adi Shamir. Identity-Based Cryptosystems and Signature Schemes. In G. R. Blakley and David Chaum, editors, *Advances in Cryptology, Proceedings of CRYPTO 1984, Santa Barbara, California, USA, August 19-22, 1984, Proceedings*, volume 196 of *Lecture Notes in Computer Science*, pages 47–53. Springer-Verlag, 19–22 August 1985.
31. Jerome A. Solinas. ID-based Digital Signature Algorithms. Slide Show presented at 7th Workshop on Elliptic Curve Cryptography (ECC 2003), August 2003.
32. Patrick P. Tsang, Victor K. Wei, Tony K. Chan, Man Ho Au, Joseph K. Liu, and Duncan S. Wong. Separable Linkable Threshold Ring Signatures. In Anne Canteaut and Kapalee Viswanathan, editors, *Progress in Cryptology - INDOCRYPT 2004, 5th International Conference on Cryptology in India, Chennai, India, December 20-22, 2004, Proceedings*, volume 3348 of *Lecture Notes in Computer Science*, pages 384–398. Springer, 2004.
33. Victor K. Wei. A Bilinear Spontaneous Anonymous Threshold Signature for Ad Hoc Groups. Cryptology ePrint Archive, Report 2004/039, 2004. Available at http://eprint.iacr.org.
34. Duncan S. Wong, Karyin Fung, Joseph K. Liu, and Victor K. Wei. On the RS-Code Construction of Ring Signature Schemes and a Threshold Setting of RST. In Sihan Qing, Dieter Gollmann, and Jianying Zhou, editors, *Information and Communications Security, 5th International Conference, ICICS 2003, Huhehaote, China, October 10-13, 2003, Proceedings*, volume 2836 of *Lecture Notes in Computer Science*, pages 34–46. Springer, 2003.
35. Jing Xu, Zhenfeng Zhang, and Dengguo Feng. A Ring Signature Scheme Using Bilinear Pairings. In Chae Hoon Lim and Moti Yung, editors, *Information Security Applications, 5th International Workshop, WISA 2004, Revised Papers*, volume 3325 of *Lecture Notes in Computer Science*, pages 163–172, Jeju Island, Korea, August 2004. Springer-Verlag.

36. Tsz Hon Yuen and Victor K. Wei. Fast and Proven Secure Blind Identity-Based Signcryption from Pairings. In A. J. Menezes, editor, *Topics in Cryptology - CT-RSA 2005, The Cryptographers' Track at the RSA Conference 2005, San Francisco, CA, USA, Febrary 14-18, 2005, Proceedings*, volume 3376 of *Lecture Notes in Computer Science*, San Francisco, CA, USA, February 2005. Springer. To Appear. Also available at Cryptology ePrint Archive, Report 2004/121.
37. Fangguo Zhang and Kwangjo Kim. ID-Based Blind Signature and Ring Signature from Pairings. In Yuliang Zheng, editor, *Advances in Cryptology - ASIACRYPT 2002, 8th International Conference on the Theory and Application of Cryptology and Information Security, Queenstown, New Zealand, December 1-5, 2002, Proceedings*, volume 2501 of *Lecture Notes in Computer Science*, pages 533–547. Springer, 2002.
38. Fangguo Zhang, Rei Safavi-Naini, and Willy Susilo. An Efficient Signature Scheme from Bilinear Pairings and Its Application. In Feng Bao, Robert H. Deng, and Jianying Zhou, editors, *Public Key Cryptography - PKC 2004, 7th International Workshop on Theory and Practice in Public Key Cryptography, Singapore, March 1-4, 2004*, volume 2947 of *Lecture Notes in Computer Science*, pages 277–290. Springer, 2004.
39. Fangguo Zhang, Reihaneh Safavi-Naini, and Chih-Yin Lin. New Proxy Signature, Proxy Blind Signature and Proxy Ring Signature Schemes from Bilinear Pairings. Cryptology ePrint Archive, Report 2003/104, 2003. Available at http://eprint.iacr.org.

Appendix

The work described in this paper was first publicly available in [13]. In an independent and more or less concurrent work [21], an ID-based ring signature scheme for general access structure was proposed. However, their scheme is inefficient for t-out-of-n threshold access structure; the space complexity of the signature and the time complexities of signing and verification are all in $O(n^t)$. Subsequently, they proposed a different construction specific to threshold access structure, which requires $2n - t$ pairing operations in signing and $2n$ of them in verification.

At ICISC 2004, a new bilinear threshold ring signature was proposed [25]. Since the structure of the threshold scheme in [21] and that of the scheme in [25] are similar, they share the same number of pairing operations in signing and verification.

To conclude, our scheme is still the most efficient ID-based threshold ring signature and the most efficient threshold ring signature scheme from pairings.

Batch Verifications with ID-Based Signatures

HyoJin Yoon[1,*], Jung Hee Cheon[1,*] and Yongdae Kim[2,**]

[1] ISaC and Department of Mathematical Sciences,
Seoul National University, Korea
{jin25,jhcheon}@math.snu.ac.kr
[2] Department of Computer Science,
University of Minnesota - Twin Cities, USA
kyd@cs.umn.edu

Abstract. An identity (ID)-based signature scheme allows any pair of users to verify each other's signatures without exchanging public key certificates. With the advent of Bilinear maps, several ID-based signatures based on the discrete logarithm problem have been proposed. While these signatures have an advantage in the fact that the system secret can be shared by several parties using a threshold scheme (thereby overcoming the security problem of RSA-based ID-based signature schemes), they all share the same efficiency disadvantage. To overcome this, some schemes have focused on finding ways to verify multiple signatures at the same time (i.e. the batch verification problem). While they had some success in improving efficiency of verification, each had a slightly diversified definition of *batch verification*. In this paper, we propose a taxonomy of batch verification against which we analyze security of well-known ID-based signature schemes. We also propose a new ID-based signature scheme that allows for all types of multiple signature batch verification, and prove its security in random oracle model.

Keywords: ID-based signatures, Batch verifications.

1   Introduction

In 1984, Shamir proposed a new model for public key cryptography, called identity (ID)-based encryption and signature schemes. The goal was to simplify key management procedures of certificate-based public key infrastructures (PKIs) [27]. Since then, several ID-based encryption and signature schemes. based on integer factorization problem, have been proposed [9, 29, 30, 21]. While these ID-based signatures have improved key management and key recovery, their disadvantage lies in the fact that the signer's key is shared with the private

---

* The first and second authors were supported in part by Korea Telecom.
** The third author is supported in part by DTC Intelligent Storage Consortium at the University of Minnesota. Part of this work was done while the third author was visiting Seoul National University in 2003.

C. Park and S. Chee (Eds.): ICISC 2004, LNCS 3506, pp. 233–248, 2005.

key generator [13, 10]. This problem can be alleviated using signatures based on the discrete logarithm problem (DLP) instead, since in this case the secret key can be shared by several parties using a threshold scheme. Several ID-based signatures with these properties that use pairings in elliptic curves have been proposed [14, 25, 8].

In spite of several advantages of ID-based signatures schemes based on pairings, they suffer from an efficiency problem that puts restrictions on their use in applications: Their signature verifications are between ten and two hundred times slower than those of DSS or RSA [1]. This problem may be critical in some applications such as electronic commerce and banking service, in which one server may have to verify many signatures simultaneously. To improve the efficiency of performance for multiple signature verification, many researchers have studied so called batch verification.

Even so, each proposed approach has a different definition of batch verification. We classify multiple signatures (i.e. input of batch verification) into the following three types, according to the number of signers and messages:

Type 1. multiple signatures on a single message generated by multiple signers.
Type 2. multiple signatures on multiple messages generated by a single signer.
Type 3. multiple signatures on multiple messages generated by multiple signers, where each message is signed by a distinct user.

Type 1 signature was traditionally classified as multisignature and has been studied for a long time [16, 23, 24, 20, 6]. Due to its simplicity, it allows for very efficient batch verification. *Type 2* batch verification proposals centered around batch RSA [11, 2] and have been a topic of research since late 80's. Compression of multiple RSA signatures of type 2 into one signature is also called condensed RSA [19]. More precisely, our discussion deals with different notion of batch verification, called *screening* [2]. That is, we only want to determine whether *the signer has at some point authenticated the text* rather than verifying if each string provided is the valid signature corresponding to the message. Recently, Boneh *et al.* proposed aggregate signatures (BGLS scheme) using bilinear maps, in which multiple signatures are aggregated into a single signature [3]. They allow for batch verification of *type 3*, but the efficiency gain is almost half of usual verifications. We note that there have been many efforts that aim at speeding up simultaneous verifications of modular exponentiations for DSA signatures [22, 18, 2, 7]. These approaches are *independent* of specific signature schemes, but the efficiency gain over the sum of individual verifications is limited. On the other hand, our approach can give significant improvement in efficiency.

Our Contributions. In this paper, we discuss batch verifications of ID-based signatures according to the above taxonomy. **(1)** We discuss security of batch verification of type 2 in the Cha-Cheon scheme. We provide a loose security reduction of batch verification of type 2 in Cha-Cheon scheme to the *computational Diffie-Hellman problem* (CDHP). It is the same as in the Hess scheme [14]. **(2)** We show that previous signature schemes are not secure in batch verification of

Type 1 or 3. **(3)** We propose a new ID-based signature scheme that is secure in batch verification of Type 1 and 3 and provide security proof under random oracle model.

Organizations. The rest of the paper is organized as follows: In Section 2, we introduce hard problems which our scheme relies upon. In Section 3, we analyze previously proposed ID-based signatures in batch verification of Type 2. We also discuss why those ID-based signature schemes fail to provide secure batch verification of Type 3 and Type 1. In Section 4, we propose a new ID-based signature scheme admitting secure batch verification of Type 3 and Type 1, provide proof of security and discuss the batch verification of each type. We conclude in Section 5.

2   Preliminary

2.1   Bilinear Maps

Consider an additive cyclic group G of prime order ℓ and a cyclic multiplicative group V. Let $e\colon G \times G \to V$ be a map which satisfies the following properties.

1. **Bilinear.** For any $aP, bP \in G$, $e(aP, bP) = e(P, P)^{ab}$.
2. **Non-degenerate.** If $e(P, Q) = 1_V$ for all P (or Q) in G, then Q (or P) is the identity of G, respectively.
3. **Efficient.** There exists an efficient algorithm to compute the map.

We call such a bilinear map as an admissible bilinear pairing.

The Weil pairing and Tate pairing in elliptic curve give good implementations of the admissible bilinear pairing. Let E be an elliptic curve over $\mathbb{F}_q$ where $q = p^n$ and p is a prime. For a prime ℓ and an ℓ torsion subgroup $E[\ell]$ of E, we define a Weil pairing $e_W : E[\ell] \times E[\ell] \to \mathbb{F}_{q^\alpha}^*$ for suitable α. Now let $G = E(\mathbb{F}_q)[\ell]$ and define a map $e\colon G \times G \to \mathbb{F}_{q^\alpha}^*$, where $e(P, Q) = e_W(P, \phi(Q))$ and ϕ is an automorphism over G. Then e is an efficiently computable non-degenerate bilinear map. The Tate pairing has similar properties and is more efficient than the Weil pairing. For the details, refer to [4].

2.2   Some Problems

Let G be a cyclic group of prime order ℓ and P a generator of G.

1. The decisional Diffie-Hellman Problem (DDHP) is to decide whether $c = ab$ in $\mathbb{Z}/\ell\mathbb{Z}$ for given $P, aP, bP, cP \in G$. If so, (P, aP, bP, cP) is called a valid Diffie-Hellman (DH) tuple.
2. The computation Diffie-Hellman Problem (CDHP) is to compute abP for given $P, aP, bP \in G$.

Now we define a gap Diffie-Hellman group.

Definition 1. *A group G is a gap Diffie-Hellman group if the DDHP in G can be efficiently computable and there exists no algorithm which can solve the CDHP in G with non-negligible probability within polynomial time.*

If we have an admissible bilinear pairing e in G, we can solve the DDHP in G efficiently as follows:

$$(P, aP, bP, cP) \text{ is a valid DH tuple} \Leftrightarrow e(aP, bP) = e(P, cP).$$

Hence an elliptic curve becomes an instance of a gap Diffie-Hellman group if the Weil (or the Tate) pairing is efficiently computable and the CDHP is sufficiently hard on the curve.

From now on, we assume that G is a gap Diffie-Hellman group generated by P, whose order is a large prime ℓ and all schemes are performed in the group G if not special remarks. To implement the a gap Diffie-Hellman group we consider G as a subset of elliptic curve as above with an admissible pairing e originated from the Weil pairing or Tate pairing.

2.3   ID-Based Signature Schemes and Attack Models for Batch Verifications

An ID-based signature scheme consists of four algorithms: *Setup, Extract, Signing* and *Verification*.

Setup. A key generation center (KGC) sets the system's secret key K_s that is called the master key and the system parameters Param.

Extract. For each identity ID, KGC generates the secret key D_{ID} corresponding to ID using K_s and Param.

Signing. A user with ID produces a signature (ID, m, σ) on a message m using her secret key D_{ID} and Param.

Verification. Given the signature (ID, m, σ), a verifier checks the validity of the σ using Param.

In batch verification, we replace the **Verification** process by the following process:

Batch Verification. Given multiple signatures $\sigma_1, \cdots, \sigma_k$ on messages $m_1, \cdots, m_k$ and corresponding identities $\text{ID}_1, \cdots, \text{ID}_k$, a verifier checks the validity of all signatures at once.

In the batch verification, if $m_1 = \cdots = m_k$ then we call this the batch verification of (multiple signatures) of Type 1. If $\text{ID}_1 = \cdots = \text{ID}_k$ then we call this the batch verification of Type 2. If each message is signed by distinct ID's then we call this the batch verification of Type 3.

We formalize the attack model for batch verification of Type 1, 2 and 3 in the general ID-based signature scheme. We call a forger $\mathcal{F}$ a k-batch forger of Type i, where i=1, 2, 3, when $\mathcal{F}$ executes the following game. Note that $\mathcal{F}$ performs an existential forgery under the adaptively chosen message and ID attack.

Setup. A k-batch forger $\mathcal{F}$ is given public system parameters.

Queries. $\mathcal{F}$ can access the hash, **Extract** and **Signing** oracle. $\mathcal{F}$ obtains the hash values of his queries, the secret keys of his chosen ID's and the signatures of his chosen ID's and messages.

Outputs. Finally, $\mathcal{F}$ outputs $\mathrm{ID}_1, \cdots, \mathrm{ID}_n$ and message $m_1, \cdots, m_n$ and corresponding signatures $\sigma_1, \cdots, \sigma_n$ of Type i where $n \leq k$ and $i = 1,\ 2,\ 3$.

$\mathcal{F}$ wins if the outputs pass the batch verification process of each type within polynomial time bound with non-negligible probability and there exists one index i such that the ID_i has not been queried to the **Extract** oracle and the message m_i corresponding to ID_i has not been asked to the **Signing** oracle.

3   Batch Verifications in ID-Based Signatures

In this section, we discuss the security of previous ID-based signature schemes.

3.1   Batch Verifications of Type 2 in the Cha-Cheon Scheme

The Cha-Cheon ID-based signature scheme consists of four algorithms: *Setup, Extract, Signing* and *Verification*.

Setup. Given a gap Diffie-Hellman group G with an admissible pairing e and its generator P, pick a random $s \in \mathbb{Z}/\ell\mathbb{Z}$ and set $P_{pub} = sP$. Choose two hash functions $H_1 : \{0,1\}^* \times G \to (\mathbb{Z}/\ell\mathbb{Z})^*$ and $H_2 : \{0,1\}^* \to G^*$. The system parameter is (P, P_{pub}, H_1, H_2). The master key is s.

Extract. Given an identity ID, the algorithm computes $Q_{\mathrm{ID}} = H_2(\mathrm{ID})$ and $D_{\mathrm{ID}} = sH_2(\mathrm{ID})$ and outputs D_{ID} as a private key of the identity ID.

Signing. Given a secret key D_{ID} and a message m, pick a random number $r \in \mathbb{Z}/\ell\mathbb{Z}$ and output a signature $\sigma = (m, U, h, V)$ where $U = rQ_{\mathrm{ID}}$, $h = H_1(m, U)$, and $V = (r + h)D_{\mathrm{ID}}$.

Verification. Given a signature $\sigma = (U, V)$ of a message m for an identity ID, compute $h = H_1(m, U)$. The signature is accepted if and only if $e(P, V) = e(P_{pub}, U + hQ_{\mathrm{ID}})$.

Let $\sigma_i = (m_i, U_i, h_i, V_i)$ be the signatures using the Cha-Cheon scheme signed by a single user with ID on distinct k-messages m_i, $U_i = r_iQ_{\mathrm{ID}}$, $V_i = (r_i + h_i)D_{\mathrm{ID}}$ and $h_i = H_1(m_i, U_i)$ where $i = 1, 2, \cdots k$, $Q_{\mathrm{ID}} = H_2(\mathrm{ID})$ and D_{ID} is a secret key of user. Then we can verify all k-signatures at once as follows:

- Compute $Q_{\mathrm{ID}} = H_2(\mathrm{ID})$ and $h_i = H_1(m_i, U_i)$ for all $i = 1, \cdots, k$.
- Check whether $e(P, \sum_{i=1}^{k} V_i) = e\left(P_{pub}, \sum_{i=1}^{k} U_i + (\sum_{i=1}^{k} h_i)Q_{\mathrm{ID}}\right)$ or not.

We know that the Cha-Cheon scheme is secure in gap Diffie-Hellman group in random oracle model [8]. Now we analyze the security of batch verification of Type 2 in the Cha-Cheon scheme.

Theorem 1. *Let $\mathcal{F}_0$ be k-batch forger of Type 2 which performs an existential forgery under an adaptively chosen message and ID attack against the Cha-Cheon scheme within a time bound T_0 with probability ϵ_0 in random oracle model. The forger $\mathcal{F}_0$ can ask queries to the oracles H_1, H_2, Extract and Signing at most q_{H_1}, q_{H_2}, q_E, and q_S-times, respectively. And $V_{q_{H_1},k}$ denotes k times the number of k-permutations of q_{H_1} elements, that is, $V_{q_{H_1},k} = k \cdot q_{H_1}(q_{H_1} - 1) \cdots (q_{H_1} - k + 1)$. If $\epsilon_0 \geq (12V_{q_{H_1},k} + 6(q_{H_1} + k \cdot q_S)^2)q_{H_2}/(\ell - 1)$, then the CDHP can be solved with probability $\geq 1/9$ and within running time $\leq 144823V_{q_{H_1},k}(1 + q_S)q_{H_2}T_0/\left(\epsilon_0\left(1 - \frac{1}{\ell}\right)\right)$.*

To prove the above theorem, we consider the properties of the ID-based scheme. While each secret key of user is chosen independently in the traditional public key system, all secret keys of users are mutually related in ID-based system. In fact, they are produced from one secret key of the whole system which is called the master key. Hence in ID-based setting it is reasonable to give not an specific ID but a system parameter to a forger. Using [8–Lemma 1], we can reduce the adaptively chosen ID attack to the *given* ID attack.

Now, consider the following lemma to reduce the security of batch verification of Type 2 in the Cha-Cheon scheme to the CDHP under the given ID attack model.

Lemma 1. *Let $\mathcal{F}$ be k-batch forger of Type 2 which performs an existential forgery under an adaptively chosen message and given ID attack against the Cha-Cheon scheme within a time bound T with probability ϵ in random oracle model. The forger $\mathcal{F}$ can ask queries to the oracles H_1, H_2, Extract and Signing at most q_{H_1}, q_{H_2}, q_E, and q_S-times, respectively. We assume that, within time bound T, $\mathcal{F}$ produces, with probability of success $\epsilon \geq \frac{12V_{q_{H_1},k} + 6(q_H + k \cdot q_S)^2}{\ell}$, multiple signatures $\sigma = (m_i, U_i, h_i, V_i)$, $i = 1, 2, \cdots, n$ and $n \leq k$, which pass the batch verification. Then, there is another probabilistic polynomial time Turing machine which has control over the machine obtained from $\mathcal{F}$ by simulation, and which produces another multiple signatures $\sigma'_i = (m'_i, U_i, h'_i, V'_i)$, $i = 1, 2, \cdots, n$ such that $h_j \neq h'_j$, for some $j \in \{1, \cdots, n\}$ and $h_i = h'_i$ for all $i = 1, \cdots, n$ such that $i \neq j$ within time $T' = \frac{144823V_{q_{H_1},k}(1 + q_S)T}{\epsilon}$.*

The Lemma 1 can be proved using the similar method with [15–Theorem 2] except the number of signatures in output $n \in \{1, \cdots, k\}$. In [15], they deal with the ring signature, so the number of signatures i.e. random parts R_i's are fixed as the number of users in the ring. But in the batch verification of Type 2, the number of signatures which is batch verified is not fixed, it is only less than k. So to fix the number of signatures during the oracle replay, we use a random variable tuple (ω, n, f) not (ω, f) when we apply the splitting lemma. Note that the number of signatures n is included in the fixed parts when we apply the splitting lemma. Thus we need k times the original $V_{q_{H_1},k}$ in [15–Theorem 2].

Proof (of theorem 1). Using the above Lemma 1, we can prove the theorem 1 in the given ID attack case. We construct an algorithm $\mathcal{C}$ to solve the CDHP using the forger $\mathcal{F}$. We assume that P, aP, bP are given as the CDHP instances. The algorithm $\mathcal{C}$ simulates a real signer to get signatures which pass the batch verification of Type 2 from $\mathcal{F}$. If $\mathcal{C}$ does not fail this simulation, $\mathcal{C}$ gets multiple signatures what he wants and using the general oracle replaying technique, $\mathcal{C}$ can solve the CDHP. In Setup, the algorithm $\mathcal{C}$ fixes a target identity ID, and put $P_{pub} = aP$.

Note that ID-**Hash Query, Extract Query, Message-Hash Query**, and **Signing Query** are the same as the proof of [8–Lemma 2]. After the queries, if the simulation does not fail, the forger $\mathcal{F}$ outputs multiple signatures $\sigma_i = (m_i, U_i, h_i, V_i)$ for given ID, where $i = 1, 2, \cdots, n$ and $n \leq k$. Then $\mathcal{C}$ replays the oracles and obtains another multiple signatures $\sigma_i' = (m_i, U_i, h_i', V_i')$, $i = 1, 2, \cdots n$ using the Lemma 1. Let $V = \sum_{i=1}^{n} V_i$, $V' = \sum_{i=1}^{n} V_i'$. Since the signatures pass the batch verification of Type 2, $\mathcal{C}$ knows that the following two equations are satisfying:

$$e(P, V) = e\left(P_{pub}, \sum_{i=1}^{n} U_i + \left(\sum_{i=1}^{n} h_i \right) Q_{\text{ID}} \right)$$

$$e(P, V') = e\left(P_{pub}, \sum_{i=1}^{n} U_i + \left(\sum_{i=1}^{n} h_i' \right) Q_{\text{ID}} \right)$$

Thus from $V = \sum_{i=1}^{n}(r_i + h_i)D_{\text{ID}}$ and $V' = \sum_{i=1}^{n}(r_i + h_i')D_{\text{ID}}$, $\mathcal{C}$ can compute $V - V' = \sum_{i=1}^{n}(h_i - h_i')D_{\text{ID}}$. Since there exist an $i \in \{1, \cdots n\}$ such that $h_i \neq h_i'$, $\mathcal{C}$ obtains $abP = D_{\text{ID}} = (V - V')/\sum_{i=1}^{n}(h_i - h_i')$. The total running time is bounded by the running time of the Lemma [15]. Thus applying [8–Lemma 1], we obtain the result of theorem 1.   □

Remark 2. *We may consider batch verification of Type 2 in the Hess scheme. In the original Hess scheme, we must compute a hash value and compare it with some value to verify. But a hash function does not have any homomorphic property, thus we cannot use directly the original Hess scheme for batch verification. Hence we slightly modify the signing and verification processes in the Hess scheme to apply the batch verification. Let the signature of a user with ID be $\sigma = (\text{ID}, m, R, h, V)$ where $h = H_1(m, U)$, $U = e(P, R)$, $V = hD_{\text{ID}} + R$ and D_{ID} is a secret key of user. Then the batch verification of Type 2 in the Hess scheme is possible and the security of them can be reduced to the CDHP similarly to the above theorem.*

The time complexity of the reduction is dominated by T_0 times k-th power of the number of H_1 hash queries over ϵ_0. That is, if k increases then the time complexity of security reduction increases exponentially. Thus the Theorem 1 gives a security proof of the batch verification in the Cha-Cheon scheme of Type 2 only when the of signature k is very small. It is the same for that of the Hess scheme.

3.2   Batch Verification of Type 3 in the Cha-Cheon Scheme

We also consider the batch verification in the Cha-Cheon [8] scheme of Type 3. However, it is not secure.

Let ID_1 be an identity of honest user $\mathcal{U}_1$ and ID_2 an identity of a 2-batch forger $\mathcal{F}$ of Type 3 in the Cha-Cheon scheme. We may assume that $\mathcal{F}$ can access to the **ID-hash** oracle and obtain $Q_1 - H_2(ID_1)$, $Q_2 = H_2(ID_2)$. Now $\mathcal{F}$ selects two random values $r_1, \tilde{r}_2$ and messages m_1, m_2, compute $U_1 = r_1 Q_1$, $h_1 = H_1(m_1, U_1)$ and

$$U_2 = \tilde{r}_2 Q_2 - h_1 Q_1 - r_1 Q_1.$$

Finally, $\mathcal{F}$ computes $h_2 = H_2(m_2, U_2)$ and $V_1 = (r_2' + h_2)D_2$, $V_2 = r_2'' D_2$, where $r_2' + r_2'' = \tilde{r}_2$, and outputs two signatures $\sigma_1 = (ID_1, m_1, U_1, h_1, V_1)$ and $\sigma_2 = (ID_2, m_2, U_2, h_2, V_2)$. Though $\mathcal{F}$ does not know the discrete log of U_2, r_2, these multiple signatures pass the batch verification of Type 3:

$$e(P_{pub}, U_1 + h_1 Q_1 + U_2 + h_2 Q_2) = e(P, r_1 D_1 + h_1 D_1 + \tilde{r}_2 D_2 - h_1 D_1 - r_1 D_1 + h_2 D_2)$$
$$= e(P, \tilde{r}_2 D_2 + h_2 D_2)$$
$$= e(P, V_1 + V_2).$$

That is, $\mathcal{F}$ pretends to generate signatures which pass the batch verification of Type 3 with the honest user U_1. Thus the batch verification of Type 3 in the Cha-Cheon scheme is not secure.

Remark 3. *In the case of the Hess scheme, because of the same reason with the previous subsection, we consider the modified Hess scheme. Similarly to the Cha-Cheon scheme, let $U_2 = e(P_{pub}, -h_1 Q_1) \cdot e(P, \widetilde{R}_2) = e(P, -h_1 D_1 + \widetilde{R}_2)$ where a random point R_2 is the same role as $U_2 = r_2 Q_{ID}$ in the Cha-Cheon scheme, then the forged signatures generated by the forger alone pass the batch verification.*

In the Cha-Cheon and Hess scheme, a random part U is used as an input of the hash function H_1. Thus although all messages are same, the hash outputs are all distinct. So the batch verification of Type 1 in the Cha-Cheon scheme is the same as that of Type 3.

4   ID-Based Signature Scheme Admitting Batch Verification of Type 3

4.1   New ID-Based Signature Scheme

This scheme consists of four algorithms: *Setup, Extract, Signing* and *Verification*.

Setup. Given a gap Diffie-Hellman group G with an admissible pairing e and its generator P, pick a random $s \in \mathbb{Z}/\ell\mathbb{Z}$ and set $P_{pub} = sP$. Choose two hash functions $H_1 : \{0,1\}^* \times G \to (\mathbb{Z}/\ell\mathbb{Z})^*$ and $H_2 : \{0,1\}^* \to G^*$. The system parameter is (P, P_{pub}, H_1, H_2). The master key is s.

Extract. Given an identity ID, the algorithm computes $Q_{\text{ID}} = H_2(\text{ID})$ and $D_{\text{ID}} = sH_2(\text{ID})$ and outputs D_{ID} as a private key of the identity ID corresponding to $Q_{\text{ID}} = H_2(\text{ID})$.

Signing. Given a secret key D_{ID} and a message m, pick a random number $r \in \mathbb{Z}/\ell\mathbb{Z}$ and output a signature $\sigma = (U, V)$ where $U = rP$, $h = H_1(m, U)$, and $V = rQ_{\text{ID}} + hD_{\text{ID}}$.

Verification. Given a signature $\sigma = (U, V)$ of a message m for an identity ID, compute $h = H_1(m, U)$. The signature is accepted if and only if $e(P, V) = e(Q_{\text{ID}}, U + hP_{pub})$.

The proposed scheme is secure under the assumption that the CDHP is hard as in the following theorem.

Theorem 2. *Let $\mathcal{F}_0$ be a forger which performs an existential forgery under an adaptively chosen message and ID attack against our ID-based scheme within a time bound T_0 with probability ϵ_0 in random oracle model. The forger $\mathcal{F}_0$ can ask queries to the oracles H_1, H_2, **Extract** and **Signing** at most q_{H_1}, q_{H_2}, q_E, and q_S-times, respectively. Assume that $\epsilon_0 \geq (10(q_S + 1)(q_S + q_{H_1})q_{H_2})/(\ell - 1)$, then the CDHP can be solved with probability $\geq 1/9$ and within running time $\leq (23q_{H_1}q_{H_2}T_0)/\left(\epsilon_0\left(1 - \frac{1}{\ell}\right)\right)$ where ℓ is a security parameter.*

Using the forking lemma [26] and [8–Lemma 1], we can prove this theorem. We discuss the rigorous proof of this theorem in the Appendix.

4.2   Security of Batch Verifications

In the our ID-based signature scheme, secure batch verification of Type 3 has possible and that of Type 2 is the same performance with the Cha-Cheon scheme. Given k signatures $(\text{ID}_1, m_1, U_1, h_1, V_1), \cdots, (\text{ID}_k, m_k, U_k, h_k, V_k)$, we can do batch verifications as follows:

- Compute $Q_i = H_2(\text{ID}_i)$ and $h_i = H_1(m_i, U_i)$ for all $i = 1, \cdots, k$. Check whether

$$e\left(P, \sum_{i=1}^{k} V_i\right) = \prod_{i=1}^{k} e\left(Q_i, U_i + h_i P_{pub}\right).$$

Now we discuss the security of batch verification of our scheme. At first we show the security of batch verification of Type 3. To reduce the adaptively chosen ID attack to the given ID attack in the case of batch verification of Type 3, we need the following lemma:

Lemma 2. *If there is a k-batch forger $\mathcal{F}_0$ of Type 3 under an adaptively chosen message and ID attack to our scheme within time bound T_0 with probability ϵ_0, then there is a k-batch forger $\mathcal{F}$ of Type 3 under an adaptively chosen* *message and given ID attack within time bound $T \leq T_0$ with the probability $\epsilon \leq \epsilon_0 \left(1 - \frac{k}{\ell}\right) \left(\frac{k}{q_{H_2}+k}\right)$, where q_{H_2} is the maximum number of queries to H_2 asked by $\mathcal{F}_0$ and k is the maximum number of signatures to be aggregated. In addition, the number of queries to hash functions, **Extract** and **Signing** asked by $\mathcal{F}_0$ are the same as those of $\mathcal{F}$.*

We show the proof of the Lemma 2 in the Appendix.

Now in random oracle model we show the security of batch verification of Type 3 under an adaptively chosen message and given ID attack.

Lemma 3. *Let $\mathcal{F}$ be k-batch forger of Type 3 which performs an existential forgery under an adaptively chosen message and given ID attack against our scheme within a time bound T with probability ϵ in random oracle model. The forger $\mathcal{F}$ can ask queries to the oracles H_1, H_2, **Extract** and **Signing** at most q_{H_1}, q_{H_2}, q_E, and q_S-times, respectively. If $\epsilon \geq (10k(q_S + 1)(q_S + k \cdot q_{H_1}))/\ell$, then the CDHP can be solved with probability $\geq 1/9$ and within running time $\leq (23q_{H_2}kT)/\epsilon$.*

Proof. We construct an algorithm $\mathcal{C}$ using the forger $\mathcal{F}$ to solve the CDHP. We assume that P, aP, bP are given as the CDHP instances. The algorithm $\mathcal{C}$ simulates a real signer to obtain signatures which pass the batch verification from $\mathcal{F}$. If $\mathcal{C}$ does not fail this simulation, he gets multiple signatures which pass the batch verification and using the general oracle replaying technique, it can solve the CDHP. In Setup, the algorithm $\mathcal{C}$ fixes a target identity ID_0, and put $P_{pub} = aP$.

Note that **ID-Hash Query**, **Extract Query**, **Message-Hash Query**, and **Signing Query** are the same as the single signature case. After the queries, if the simulation does not fail, the forger $\mathcal{F}$ outputs other $n-1$ ID's and n multiple signatures

$$\sigma_i = (m_i, U_i, h_i, V_i), \quad i = 1, 2, \cdots, n$$

where $n \leq k$.

Then $\mathcal{C}$ replays the oracles and obtains another $n'-1$ ID's and n' multiple signatures $\sigma_i' = (m_i', U_i', h_i', V_i')$, $i = 1, 2, \cdots n'$, where $n' \leq k$. By the forking lemma, the replay succeeds with the probability $\geq 1/9$ and the running time $\leq (23q_{H_2}T)/\epsilon$. Note that we may assume $h_1 \neq h_1'$ since the probability of collision of two random numbers is negligible. Since the random commitment r is fixed before the hash queries of a message, the corresponding random commitment of σ must be the same with that of σ by the forking lemma. That is, we have $ID_i = ID_j' = ID_0$ and $U_i = U_j'$ for some $i \in \{1, \cdots, n\}$ and $j \in \{1, \cdots, n'\}$ without loss of generality let $i = j' = 1$. And according to the **Extract Query**, $\mathcal{C}$ knows each secret key D_i corresponding to ID_i except that of ID_1 and by

ID-Hash Query, $\mathcal{C}$ knows discrete log of each $Q_i = H_2(\text{ID}_i)$, x_i, except that of $H_2(\text{ID}_0)$. Hence from

$$V = \sum_{i=1}^{n} V_i = \sum_{i=1}^{n} (r_i Q_i + h_i D_i) = \sum_{i=1}^{n} \{x_i(r_i P) + h_i D_i\},$$

$$V' = \sum_{i=1}^{n'} V_i' = \sum_{i=1}^{n'} \{x_i'(r_i' P) + h_i' D_i'\},$$

compute $\overline{V} = V - V' - \sum_{i=2}^{n} V_i - \sum_{i=2}^{n'} V_i' = (h_1 - h_1') D_1$ so $(h_1 - h_1')^{-1} \overline{V} = D_1 = abP$ as desired. The total running time is bounded by the running time of the forking lemma.   □

As the same reason with the Cha-Cheon scheme, the batch verification of Type 1 has the same performance with that of Type 3. In Type 2 signatures, the performance of our scheme is the same as that of the Cha-Cheon scheme.

From the Lemma 2, Lemma 3 and the Theorem 1, we obtain the following result.

Theorem 4. *Let $\mathcal{F}_0$ be a k-batch forger which performs an existential forgery under an adaptively chosen message and ID attack against our ID-based scheme with probability ϵ_0 within a time bound T_0 in random oracle model. The forger $\mathcal{F}_0$ can ask queries to the oracles H_1, H_2, Extract and Signing at most q_{H_1}, q_{H_2}, q_E, and q_S-times, respectively.*

- *In the Type 1 or 3 case, if $\epsilon_0 \geq (10(q_S + 1)(q_S + q_{H_1})(q_{H_2} + k)q_{H_2})/k(\ell - k)$, then the CDHP can be solved with probability $\geq 1/9$ and within running time $\leq (23q_{H_1}(q_{H_2} + k)T_0)/(\epsilon_0 k (1 - \frac{k}{\ell}))$.*
- *In the Type 2 case, if $\epsilon_0 \geq (12V_{q_{H_1},k} + 6(q_{H_1} + k \cdot q_S)^2)q_{H_2}/(\ell - 1)$, then the CDHP can be solved with probability $\geq 1/9$ and within running time $\leq 144823 V_{q_{H_1},k}(1 + q_S)q_{H_2}T_0/(\epsilon_0(1 - \frac{1}{\ell}))$.*

4.3   Efficiency of Batch Verifications

In this section, we compare the efficiency of verifications of k individual Cha-Cheon signatures with that of k-batch verification of Type 3 in our scheme. Here we assume that we use an elliptic curve with an admissible Tate pairing as a gap Diffie-Hellman group.

To estimate the performance of our scheme, we first present experimental results for the cost of several cryptographic primitives in Table 1. We used Miracl library v.4.8.2 [17] in P3-977 MHz with 512 Mbytes memory. In MapToPoint and Pairing, we considered a subgroup of order q in a supersingular elliptic curve E over $\mathbb{F}_p$, where p is a 512 bit prime and q is a 160 bit prime. Note that the pairing value belongs to a finite field of 1024 bits.

To verify a single Cha-Cheon signature, we need to compute two pairings, a scalar multiplication in an elliptic curve and a MapToPoint the total running

Table 1. Cost of basic operations

Function	modulus (bits)	exponent (bits)	performance (msec)
Scalar Mul. in EC	512	160	7.33
MapToPoint	512	(160)	2.42
Pairing	512	(160)	31.71

time is about 73.17ms. So the running time to verify all individual signatures signed by k distinct signers on k distinct messages is about $73k$ms. In the batch verification of Type 3 k multiple signatures using our scheme, we need to compute $k+1$ pairings, k scalar multiplications, k MapToPoints, which takes about $(41k+32)$ms. Thus if k is large, we can save about *half* of the verification time. In the batch verification of Type 2 k multiple signatures, we need to compute only two pairings, one scalar multiplication and one MapToPoint, which takes about 73ms. Thus the verification cost of the batch verification of Type 2 is almost that of a single signature.

5   Conclusion

In this paper, we classified batch verifications into three types, Type 1, 2, and 3, according to the number of signers and messages, and discussed security of previous well-known ID-based signature schemes in each type of batch verification. We have shown that the previous ID-based signature schemes are not secure in batch verifications of Type 1 and 3. We also proposed a new ID-based signature scheme admitting secure batch verification. The batch verification of Type 2 in our scheme has the same security reduction as in the previous schemes, and those of Type 1 and 3 are secure against existential forgery, under the adaptively chosen message and ID attack in random oracle model. Finally we discussed the efficiency of batch verification of Type 3 in our scheme.

References

1. P. Barreto, H. Kim, B. Lynn and M. Scott. Efficient Algorithms for Pairing-Based Cryptosystems. *Advances in Cryptology - Crypto 2002*, LNCS Vol. 2442, pp. 354–368, Springer-Verlag, 2002.
2. M. Bellare, J. Garay, and T. Rabin. Fast Batch Verification for Modular Exponentiation and Digital Signatures. *Advances in Cryptology - Eurocrypt'98*, LNCS Vol. 1403, pp. 236–250, Sringer-Verlag, 1998.
3. D. Boneh, C. Gentry, B. Lynn, and H. Shacham. Aggregate and Verifiably Encrypted Signatures from Bilinear Maps. *Advances in Cryptology - Eurocrypt 2003*, LNCS Vol. 2656, pp. 416–432, Springer-Verlag, 2003.
4. D. Boneh, B. Lynn, and H. Shacham. Short signature from the Weil pairing. *Advances in Cryptology - Asiacrypt 2001*, LNCS Vol. 2248, pp. 514–531, Springer-Verlag, 2001. The extended version is available at http://crypto.stanford.edu/~dabo/abstracts/weilsigs.html.

5. X. Boyen. Multipurpose Identity-Based Signcryption - A Swiss Army Knife for Identity-Based Cryptography. *Advances in Cryptology - Crypto 2003*, LNCS Vol. 2729, pp. 383–399, Springer-Verlag, 2003.
6. A. Boldyreva. Threshold Signatures, Multisignatures and Blind Signatures Based on the Gap-Diffie-Hellman-Group Signature Scheme. *Proceedings of PKC 2003*, LNCS Vol. 2567, pp. 31–46, Springer-Verlag, 2003.
7. C. Boyd and C. Pavlovski. Attacking and Repairing Batch Verification Schemes. *Advances in Cryptology - Asiacrypt 2000*, LNCS Vol. 1976, pp. 58–71, Springer-Verlag, 2000.
8. J. Cha and J. Cheon. An ID-based Signature from Gap-Diffie-Hellman Groups. *Public Key Cryptography - PKC 2003*, LNCS Vol. 2567, pp. 18–30, Springer-Verlag, 2003.
9. Y. Desmedt and J. Quisquater. Public-key Systems based on the Difficulty of Tampering. *Advances in Cryptology - Crypto'86*, LNCS Vol. 263, pp. 111–117, Springer-Verlag, 1987.
10. U. Feige, A. Fiat, and A. Shamir. Zero-knowledge Proofs of Identity. *J. Cryptology*, Vol. 1, pp. 77–94, 1988.
11. A. Fiat. Batch RSA. *J. Cryptology*, Vol. 10, No. 2, pp. 75–88, Springer-Verlag, 1997. A preliminary version appeared in *Advances in Cryptology - Crypto'89*, LNCS Vol. 435, pp. 175–185, Springer-Verlag, 1989.
12. F. Zhang and K. Kim, Efficient ID-based blind signature and proxy signature from bilinear pairings, *ACISP 03*, LNCS Vol. 2727, pp. 312–323, Springer-Verlag, 2003.
13. A. Fiat and A. Shamir. How to Prove Yourself: Practical Solutions to Identification and Signature Problems. *Advances in Cryptology - Crypto '86*, LNCS Vol. 263, pp. 186–194, Springer-Verlag, 1987.
14. F. Hess. Efficient Identity Based Signature Schemes Based on Pairings. *Selected Areas in Cryptography - SAC 2002*, LNCS Vol. 2595, pp. 310–324, Springer-Verlag, 2002.
15. J. Herranz and G. Sáez. Forking Lemmas in Ring Signatures' Scenario. *Progress in Cryptology - INDOCRYPT 2003*, LNCS Vol. 2904, pp. 266–279, Springer-Verlag Heidelberg, 2003.
16. K. Itakura and K. Nakamura. A Public-key Cryptosystem Suitable for Digital Multisignatures. *NEC Research and Development*, Vol. 71, pp. 1–8, 1983.
17. Shamus Software Ltd. Miracl: Multiprecision integer and rational arithmetic c/c++ library. http://indigo.ie/~mscott/.
18. D. M'Raithi and D. Naccache. Batch Exponentiation - A Fast DLP based Signature Generation Strategy. *ACM Conference on Computer and Communications Security*, pp. 58–61, ACM, 1996.
19. E. Mykletun, M. Narasimha and G. Tsudik. Providing Efficient Data Integrity Mechanisms in Outsourced Databases. *Network and Distributed System Security (NDSS)*, 2004.
20. S. Micali, K. Ohta and L. Reyzin, Accountable-subgroup Multisignatures. *On proceedings of CCS 2001*, pp. 245–254, ACM, 2001.
21. U. Maurer and Y. Yacobi. Non-interactive Public-key Cryptography. *Advances in Cryptology - Eurocrypt'91*, LNCS Vol. 547, pp. 458–460, Springer-Verlag, 1992.
22. D. Naccache, D. M'Raithi, S. Vaudenay, and D. Raphaeli. Can D.S.A be Improved? Complexity trade-offs with the Digital Signature Standard. *Advances in Cryptology - Eurocrypt'94*, LNCS Vol. 950, pp. 77–85, Springer-Verlag, 1994.
23. K. Ohta and T. Okamoto. A Digital Multisignature Scheme based on the Fiat-Shamir Scheme. *Advances in Cryptology - ASIACRYPT'91*, LNCS Vol 739. pp. 75–79, Spring-Verlag, 1991.

24. K. Ohta and T. Okamoto. Multi-signature Schemes Secure against Active Insider Attacks. *IEICE Transactions on Fundamentals of Electronics Communications and Computer Sciences*, Vol. E-82-A, No. 1, pp. 21–31, 1999.
25. K. Paterson. ID-based Signatures from Pairings on Elliptic Curves. *Electronics Letters*, Vol. 38, No. 18, pp. 1025–1026, 2002.
26. D. Pointcheval and J. Stern. Security Arguments for Digital Signatures and Blind Signatures. *J. Cryptology*, Vol. 13, No. 3, pp. 361–396, 2000. A preliminary version has appeared in *Advances in Cryptology - Eurocrypt'96*, LNCS Vol. 1070, pp. 387–398, Springer-Verlag, 1996.
27. A. Shamir. Identity-base Cryptosystems and Signature Schemes. *Advances in Cryptology - Crypto'84*, LNCS Vol. 196, pp. 47–53, Springer-Verlag, 1985.
28. C. Schnorr. Efficient Identification and Signatures for Smart Cards. *Advances in Cryptology - Crypto'89*, LNCS Vol. 435, pp. 239–252, Springer-Verlag, 1989.
29. H. Tanaka. A Realization Scheme for the Identity-based Cryptosystem. *Advances in Cryptology - Crypto'87*, LNCS Vol. 293, pp. 340–349, Springer-Verlag, 1987.
30. S. Tsuji and T. Itoh. An ID-based Cryptosystem based on the Discrete Logarithm Problem. *IEEE Journal of Selected Areas in Communications*, Vol. 7, pp. 467–473, 1989.
31. F. Zhang and K. Kim. Efficient ID-based Blind Signature and Proxy Signature from Bilinear Pairings. *ACISP 03*, LNCS Vol. 2727, pp. 312–323, Springer-Verlag, 2003.

Appendix: Security Proof

Proof of Theorem 2

Using [8–Lemma 1], we can reduce the forger $\mathcal{F}_0$ to $\mathcal{F}$ an adaptively chosen message and given ID attack within time bound $T \leq T_0$ with the probability $\epsilon \leq \epsilon_0(1 - \frac{1}{\ell})/q_{H_2}$. We construct an algorithm $\mathcal{C}$ using $\mathcal{F}$ to solve the CDHP. We assume that P, aP, and bP are given. Since the forger $\mathcal{F}$ is an adaptively chosen message attacker, he can access to the hash oracles, the extraction oracle, and the signing oracle, and ask at most q_{H_1}, q_{H_2}, q_E, and q_S queries for each oracles respectively. The algorithm $\mathcal{C}$ simulates a real signer to get a valid signature from the forger $\mathcal{F}$. If $\mathcal{C}$ does not fail this simulation, he gets a valid signature, and using the oracle replaying technique he can solve the CDHP.

We may assume the forger is well-behaved in the following sense: A forger $\mathcal{F}$ makes a **Extract** query for an ID only if an H_2 query has been made before for the ID. Also **Signing** query is made for a message m only if a H_1 queries has been made before for the m.

Then the algorithm $\mathcal{C}$ puts $P_{pub} = aP$ and performs the following game with the forger $\mathcal{F}$ for a fixed identity ID as follows:

ID-Hash Query. When $\mathcal{F}$ makes an ID-hash query ID_i, $\mathcal{C}$ gives to $\mathcal{F}$ an answer $H_2(ID_i) = bP$ if $ID_i = ID$ and $H_2(ID_i) = x_iP$ for $x_i \in_R \mathbb{Z}/\ell$ otherwise.

Extract Query. When $\mathcal{F}$ makes an extract query for ID_{i_k}, $\mathcal{C}$ gives $x_{i_k}P_{pub} = x_{i_k}(aP)$ as the secret key corresponding to $H_2(ID_{i_k})$ for an identity ID_{i_k}. Note that $\mathcal{F}$ must not ask the secret key corresponding to the $bP = H_2(ID)$.

Message-Hash Query. $\mathcal{F}$ makes q_H message-hash queries. For the j-th hash query $\mathcal{Q}_j$, $\mathcal{C}$ chooses a random value $h_j \in \mathbb{Z}/\ell$ and gives to $\mathcal{F}$ as the hash value of $\mathcal{Q}_j$ for $j = 1, \cdots q_{H_1}$ and stores them as $H_1(\mathcal{Q}_j) = h_j$.

Signing Query. If $\mathcal{F}$ asks the signature on m_{j_t} of ID_{i_t}, $\mathcal{C}$ chooses a random value $r_t \in \mathbb{Z}/\ell$ responses

$$Sign(\mathrm{ID}_{i_t}, m_{j_t}) = (\mathrm{ID}_{i_t}, m_{j_t}, U_t, h_t, V_t),$$

where $U_t = r_t P - h_t P_{pub}$ and $V_t = r_t(x_{i_t} P)$ for $t = 1, \cdots, q_S$. Since $(P, H_2(\mathrm{ID}_{i_t}), U_t + h_t P_{pub}, V_t)$ is a valid Diffie-Hellman tuple, these signatures pass the verification algorithm.

If the simulation does not fail, the forger $\mathcal{F}$ outputs a valid signature $(\mathrm{ID}, m, U, h, V)$ with probability ϵ. After a replay of the forger $\mathcal{F}$, apply the forking lemma in [26]. Then $\mathcal{C}$ obtains two valid signatures $\sigma = (\mathrm{ID}, m, U, h, V)$ and $\sigma' = (\mathrm{ID}, m, U, h', V')$ such that $h \neq h'$ with probability $\geq 1/9$ within the time $23 q_{H_1} T / \epsilon$. $\mathcal{C}$ can easily obtain the value abP from

$$\frac{(h D_{\mathrm{ID}} - h' D_{\mathrm{ID}})}{h - h'} = D_{\mathrm{ID}} = abP.$$

By the forking lemma [26] and [8–Lemma 1], we obtain the result of this theorem. $\qquad\qquad\qquad\qquad\qquad\qquad\qquad\qquad\qquad\qquad\qquad\qquad\quad$ $\square$

Proof of Lemma 2

Proof. We assume, without loss of generality, the forger $\mathcal{F}_0$ has an extract queries for any ID at most once. We consider an algorithm $\mathcal{F}$ that performs the following simulation:

Setup. $\mathcal{F}$ chooses a random number $r \in \{1, \cdots, q_{H_1}\}$. Let ID_i be the $\mathcal{F}_0$'s i-th H_2-query and $\mathrm{ID}'_i = ID$ if $i = r$ and $\mathrm{ID}'_i = \mathrm{ID}_i$ otherwise. Let $H'_2(ID_i) = H_2(ID'_i)$, **Extract'**(ID_i)=**Extract**(ID'_i) and **Signing'**(ID_i, m_i)=**Signing**(ID'_i, m_i)

Queries. If $\mathcal{F}_0$ makes the H_1, H_2 hash queries and **Extract**, **Signing** queries, then $\mathcal{F}$ computes H_1, H'_2, **Extract'** and **Signing'** as above and answers the results.

If the simulation does not fail, $\mathcal{F}_0$ outputs signatures $\sigma_i = (\mathrm{ID}^i_{out}, m_i, U_i, h_i, V_i)$ where $i = 1, \cdots k$ with probability ϵ_0. Finally, if $\mathrm{ID}^i_{out} = ID$ for some $i = 1, \cdots, k$ and $\mathrm{ID}^j_{out} \neq ID$ for all $j \neq i$ and all σ_is are valid signatures, then $\mathcal{F}$ outputs σ_i where $i = 1, \cdots, k$. Otherwise the simulation fails.

Since the output distributions of H'_2, **Extract'**, **Signing'**-queries are not distinguishable those of original ones, we know

$$\Pr[\text{For all } i \in \{1, \cdots, k\}, \ \sigma'_i\text{s are valid}] \geq \epsilon.$$

Since we consider the hash functions as the random oracles, we obtain the following result.

$$\Pr[\text{ID}_{out}^j = ID_i \text{ for some } j = 1, \cdots, k \text{ and } i = 1, \cdots q_{H_2}$$

$$| \text{ For all } i \in \{1, \cdots, k\}, \ \sigma_i'\text{s are valid}] \geq \left(1 - \frac{1}{\ell}\right)^k \geq 1 - \frac{k}{\ell}$$

Furthermore since the randomness of r, we have the following inequality.

$$\Pr[\text{ID}_{out}^i = ID_r \text{ for some } i = 1, \cdots, k \text{ and } \text{ID}_{out}^j \neq ID \text{ for some } j = 1, \cdots,$$

$$i - 1, i + 1, \cdots, k \mid \text{ID}_{out}^j = ID_i \text{ for some } j = 1, \cdots, k \text{ and } i = 1, \cdots q_{H_2}]$$

$$\geq \frac{q_{H_2} - 1}{q_{H_2}} \frac{H_{k-1}}{H_k} \geq \frac{k(q_{H_2} - 1)}{(q_{h_2} + k - 1)(q_{H_2} + k - 2)} \geq \frac{k}{2(q_{H_2} + k)}$$

Finally, summarizing these, we get the following result as desired:

$$\Pr[\text{ID}_{out}^i = ID_r = ID \text{ for some } i = 1, \cdots, k \text{ and } \text{ID}_{out}^j \neq ID \text{ for some}$$

$$j = 1, \cdots, i - 1, i + 1, \cdots, k \text{ and For all } i \in \{1, \cdots, k\},$$

$$\sigma_i'\text{s are valid}] \geq \epsilon \cdot \left(1 - \frac{1}{\ell}\right) \cdot \frac{k}{2(q_{H_2} + k)}. \qquad \square$$

A Method for Distinguishing the Two Candidate Elliptic Curves in CM Method

Yasuyuki Nogami and Yoshitaka Morikawa

Dept. of Communication Network Engineering,
Okayama University, Okayama-shi, 700-8530, Japan
{nogami, morikawa}@trans.cne.okayama-u.ac.jp

Abstract. In this paper, we first introduce a shift product-based polynomial transformation. Then, we show that the parities of $(\#E - 1)/2$ and $(\#E' - 1)/2$ are reciprocal to each other, where $\#E$ and $\#E'$ are the orders of the two candidate curves obtained at the last step of CM method algorithm. Based on this property, we propose a method to check the parity by using the shift product-based polynomial transformation. For a 160-bits prime number as the characteristic, the proposed method carries out the parity check about 20 times faster than the conventional method when 4 divides the characteristic minus 1.

Keywords: CM method, irreducible cubic polynomial, quadratic power residue/non residue.

1 Introduction

In recent years, the elliptic curve cryptosystem (ECC) has received much attention. For ECC, some attacks have been proposed[1]-[3]. From the viewpoint of security and implementation of ECC, it is said that a prime order elliptic curve is the best. In order to systematically generate prime order elliptic curves, several algorithms has been proposed[4],[5]. We can roughly classify them into two types; the one adopts a certain order counting algorithm[4],[6] and the other adopts CM method[5],[7]. This paper is related to the latter and particularly deals with no two-torsion elliptic curves defined over a prime field F_p.

Applying CM method, we can generate an elliptic curve whose order is a certain prime number. The input of the algorithm based on the CM method is the characteristic of the definition field of the curve[8], for example, then we can obtain j-invariant. According to the j-invariant, we have two candidate elliptic curves, one of these two curves has the objective order. Therefore, we must check whether the order of the curve is the objective order. In general, it is checked by picking a random point on the curve and then calculating a scalar multiplication with the rational point. This paper proposes a method that makes this check fast. In the CM method, the class polynomial computation is the most time-consuming operation[8]. Therefore, most of the conventional improvements for CM method are given for the class polynomial computation;

C. Park and S. Chee (Eds.): ICISC 2004, LNCS 3506, pp. 249–260, 2005.

however, recently Koc et al.[7] have proposed an algorithm that prepares a table of class polynomials in advance, by using this table the algorithm generates a prime order elliptic curve within several seconds. If we can distinguish the two candidate curves fast, then Koc et al. algorithm will generate prime order elliptic curves more effectively.

In this paper, we introduce a shift product-based polynomial transformation (SPPT) which are carried out by a square root calculation and a polynomial modulo operation over the prime field F_p. Then, we show that the parities of $(\#E-1)/2$ and $(\#E'-1)/2$ are reciprocal to each other, where $\#E$ and $\#E'$ are the orders of the two candidate curves obtained at the last step of CM method algorithm. Based on this property, we propose a method to check the parity by using SPPT. This parity check method does not need a scalar multiplication for a rational point, it needs a square root calculation, a polynomial modulo operation, and a quadratic power residue check instead. In addition, when 4 divides the characteristic minus 1, the proposed method does not need the square root calculation and polynomial modulo operation. It only needs a quadratic power residue check, where this check is carried out by an exponentiation in the prime field. From the experimental results, we show that the proposed method is superior to the conventional method. For a 160-bits prime number as the characteristic, the proposed method carries out the parity check about 20 times faster than the conventional method when 4 divides the characteristic minus 1. In this paper, we deal with a finite field F_q whose characteristic p is an odd prime number larger than 3. F_p denotes a prime field. $X \mid Y$ and $X \nmid Y$ mean that X divides Y and does not divide Y, respectively. $X \parallel Y$ means that X divides Y but X^2 does not divide Y. Without any additional explanation, polynomials in this paper are monic. This paper particularly focuses on the CM method for generating prime order elliptic curves.

2   Fundamentals

In this section, we go over the fundamentals of elliptic curve, quadratic residue/non-residue, and CM method.

2.1   Defining Equation

When the characteristic of F_q is not equal to 2 or 3, an elliptic curve over F_q is generally defined by

$$E(x,y) = x^3 + ax + b - y^2 = 0, \quad a, b \in F_q. \tag{1}$$

The solutions (x, y) to Eq.(1) and the point at infinity denoted by $\mathcal{O}$ are called F_q-rational points when the coordinates of x and y lie in F_q. F_q-rational points on the elliptic curve form an additive Abelian group. In this paper, we denote this group and its order by $E(F_q)$ and $\#E(F_q)$, respectively. The following parameter t is called the trace of elliptic curve $E(F_q)$;

$$t = q + 1 - \#E(F_q). \tag{2}$$

2.2   No Two-Torsion Curve

The necessary and sufficient condition for an elliptic curve to have no two-torsion points is that $E(x, 0)$ given from its defining equation is irreducible over F_q. A *two-torsion point* P means that $2P = \mathcal{O}$, where $\mathcal{O}$ plays a role of the unity in the Abelian group $E(F_q)$. It is necessary for a prime order elliptic curve to have no two-torsion points, and Nogami et al. algorithm[6] uses this necessary condition for generating prime order elliptic curves. In what follows, we consider no two-torsion curves.

2.3   The Order of Elliptic Curve

Let us rewrite the defining equation Eq.(1) as

$$y^2 = E(x, 0) = x^3 + ax + b, \quad a, b \in F_q. \tag{3}$$

For an arbitrary element $i \in F_q$, if $E(i, 0)$ is a quadratic residue in F_q, then the following two rational points on the curve are given;

$$\left(i, \pm\sqrt{E(i, 0)} \right), \tag{4}$$

where $E(i, 0) \neq 0$ because $E(F_q)$ is a no two-torsion curve. Therefore, let N be the number of quadratic residues in the following set;

$$\{x | x = E(i, 0), \forall i \in F_q\}, \tag{5}$$

the order $\#E(F_q)$ is given by

$$\#E(F_q) = 2N + 1, \tag{6}$$

where 1 shown in the right-hand side of the above equation corresponds to the point at infinity $\mathcal{O}$. From Eq.(6), N is written as

$$N = \frac{\#E(F_p) - 1}{2}. \tag{7}$$

2.4   Quadratic Residue/Non-residue

For a non-zero element $c \in F_q$, we can check whether c is a quadratic residue (QR) or quadratic non-residue (QNR) in F_q as follows;

$$c^{(q-1)/2} = \begin{cases} 1 & \text{when } c \text{ is a QR} \\ -1 & \text{when } c \text{ is a QNR} \end{cases}. \tag{8}$$

The product of two non-zero QRs and that of two QNRs become QRs in F_q. On the other hand, the product of a QR and a QNR becomes a QNR in F_q.

2.5   CM Method Algorithm

We refer to the following algorithm[8] as the conventional CM method algorithm for generating a prime order elliptic curve;

Input : Characteristic p
Output : Prime order elliptic curve $E(x, y)$ over F_p

Step1 : Find a smallest D along with t such that $4p = t^2 + Ds^2$, $D, t, s \in Z$.
Step2 : Check whether one of the orders $\#E_{\pm}(F_p) = p + 1 \pm t$ is a prime. If not a prime, then find another D along with t at **Step1**.
Step3 : Construct the class polynomial $H_D(x)$.
Step4 : Find a root $j \in F_p$ of $H_D(x)$, then set $k = j/(1728 - j) \in F_p$.
Step5 : Check whether the order of $E(x, y) = x^3 + 3kx + 2k - y^2$ is $\#E_-(F_p)$ or $\#E_+(F_p)$ by picking a random point P on $E(x, y)$ and computing $\#E_+(F_p)P$. If the order is the objective order, then output the defining equation $E(x, y)$. Otherwise, output the twisted curve of $E(x, y)$.

The calculations from Step1 to Step4 are too time-consuming steps, therefore several improvements have been proposed[5],[9]. As compared to these steps, Step5 is carried quite fast; however, Koc et al.[7] canceled these time-consuming calculations by preparing a table of class polynomials, for example, in advance. In this paper, we propose an improvement for Step5.

3   Main Idea

In what follows, we consider a prime field F_p as the definition field. We first introduce a shift product-based polynomial transformation(SPPT). Then, we show a relation between SPPT and cubic polynomial $E(x, 0)$.

3.1   Shift Product-Based Polynomial Transformation(SPPT)

Let us consider an irreducible polynomial $f(x)$ over F_p of degree m written as

$$f(x) = \sum_{i=0}^{m} f_i x^i, \ f_i \in F_p, \ f_m = 1. \tag{9}$$

If $f_{m-1} = 0$ and $p \nmid m$, as shown in App.A there exists an irreducible polynomial $\tilde{f}(x)$ that satisfies Eqs.(10).

$$\tilde{f}(x^p - x) = \prod_{i=0}^{p-1} f(x + i), \tag{10a}$$

$$\tilde{f}(x) = \sum_{i=0}^{m} \tilde{f}_i x^i, \ \tilde{f}_i \in F_p, \ \tilde{f}_m = 1, \ \tilde{f}_{m-1} = 0. \tag{10b}$$

These irreducible polynomials $f(x)$ and $\tilde{f}(x)$ hold one to one relation, *see* App.A. We consider a shift product-based polynomial transformation(SPPT) as follows;

$$\text{SPPT} : f(x) \rightarrow \tilde{f}(x). \tag{11a}$$

3.2   For Defining Equation $E(x, y)$

Let us consider $f(x) = E(x, 0)$ in **Sec.3.1**. Noting that $E(x, 0)$ is an irreducible cubic polynomial over F_p in this paper, as shown in Eq.(10a), there exists a cubic irreducible polynomial $\tilde{E}(x, 0)$ over F_p that satisfies

$$\tilde{E}(x^p - x, 0) = \prod_{i=0}^{p-1} E(x + i, 0). \tag{12}$$

By substituting $x = 0$ into Eq.(12), we have

$$\tilde{E}(0, 0) = \prod_{i=0}^{p-1} E(i, 0). \tag{13}$$

From Eq.(13), we have

$$\tilde{E}(0, 0)^{(p-1)/2} = \prod_{i=0}^{p-1} E(i, 0)^{(p-1)/2}. \tag{14}$$

By using the number N defined in **Sec.2.3** and substituting Eq.(8), we obtain

$$\tilde{E}(0, 0)^{(p-1)/2} = (-1)^{p-N}. \tag{15}$$

Noting that this paper deals with an odd prime number as the characteristic p, from Eq.(14) we have the following property;

Property 1. Let $\tilde{E}(0, 0)$ be the constant term of a cubic irreducible polynomial $\tilde{E}(x, 0)$ over F_p that satisfies Eq.(12), and let N be the number of QRs in the set Eq.(5). Then, N is odd if and only if $\tilde{E}(0, 0)$ is a QR in F_p.

3.3   Implementation of SPPT

From **Prop.1** and Eq.(15), if we have the constant term $\tilde{E}(0, 0)$, then we can check the parity of N by checking whether or not $\tilde{E}(0, 0)$ is a QR in F_p. In order to obtain the constant term $\tilde{E}(0, 0)$, in this section we consider how to determine the cubic irreducible polynomial $\tilde{E}(x, 0)$ over F_p that satisfies Eq.(12), that is how to implement SPPT introduced in **Sec.3.1**.

Let ω and τ be zeros of $f(x)$ and $\tilde{f}(x)$ introduced in **Sec.3.1**, respectively, then we have $\tau = \omega^p - \omega$ from Eq.(10a). Therefore, $\tilde{f}(x)$ is the minimal polynomial of $\omega^p - \omega$ with respect to F_p, where we should note that ω and τ belong to F_{p^3}

but not to F_p. As shown in App.B, from $f(x)$ given by Eq.(9) with $m = 3$, we obtain the following two candidates of $\tilde{f}(x)$;

$$\tilde{f}_\pm(x) = x^3 + 3f_1 x \pm \sqrt{D(f)}, \quad D(f) = -(4f_1^3 + 27f_0^2), \tag{16}$$

where $D(f)$ is the discriminant of $f(x)$. In addition, when $f(x)$ is irreducible, the discriminant $D(f)$ becomes a QR in F_p[10], therefore we can calculate its square roots in F_p by Smart's square root calculation algorithm[8], for example. Since $f(x)$ and $\tilde{f}(x)$ satisfy Eqs.(10), we can distinguish them as follows;

$$\tilde{f}(x) = \begin{cases} \tilde{f}_+(x) & \text{when } f(x) \text{ divides } \tilde{f}_+(x^p - x) \\ \tilde{f}_-(x) & \text{when } f(x) \text{ divides } \tilde{f}_-(x^p - x) \end{cases}. \tag{17}$$

For checking Eq.(17), we need polynomial modulo operations for testing whether $f(x)$ divides $\tilde{f}_+(x^p - x)$ or $\tilde{f}_-(x^p - x)$.

4   Distinguishing the Two Candidate Curves in CM Method

In this section, we consider CM method[8] as an application of **Prop.**1, however, we do not introduce CM method itself into detail. In what follows, the definition field of elliptic curve is a prime field F_p.

4.1   Two Candidate Elliptic Curves in CM Method

As introduced in **Sec.**1, several algorithms for generating prime order elliptic curves have been proposed[5],[6]. In the CM method introduced in **Sec.**2.5, the input is the characteristic p and the output is the defining equation $E(x, y)$ whose order is a prime number. In what follows, for instance, we suppose that the j-invariant is not 0 or 1728. Let us consider that X is the objective order written as

$$X = p + 1 - t. \tag{18a}$$

Using CM method algorithm, we have a pair of the characteristic p and j-invariant. In other words, we obtain the following defining equation;

$$E(x, y) = x^3 + 3kx + 2k - y^2, \ k = j/(1728 - j), \ k, j \in F_p. \tag{18b}$$

Let $\#E(F_p)$ be the order of the curve defined by Eq.(18b), it is possible for $\#E(F_p)$ to be the following two numbers;

$$\#E_\pm(F_p) = p + 1 \pm t. \tag{19}$$

Only from the j-invariant, we cannot distinguish whether $\#E(F_p)$ is $\#E_-(F_p)$ or $\#E_+(F_p)$. For this problem, as shown in **Step5**, we randomly pick a F_p-rational point P on the curve Eq.(18b), then test whether or not $XP = \mathcal{O}$ by a scalar multiplication. If $\#E(F_p)$ is not X, then we consider the twist of $E(x, y)$ as

$$E'(x, y) = x^3 + 3kc^2x + 2kc^3 - y^2, \tag{20}$$

where c is a QNR in F_p.

The parities of $(\#E_{\pm}(F_p) - 1)/2$:
If the defining equation $E(x, y)$ is a no two-torsion curve, both of the order $\#E(F_p)$ and the trace t are odd. In addition, noting that p is an odd prime number in this paper, we find that $(\#E_+(F_p) - 1)/2$ is an odd number if and only if $(\#E_-(F_p) - 1)/2$ is an even number, because

$$\text{if } 2 \,||\, (p - t), \text{ then } 4 \mid (p + t), \tag{21a}$$
$$\text{if } 4 \mid (p - t), \text{ then } 2 \,||\, (p + t), \tag{21b}$$
$$\text{where } \#E_-(F_p) - 1 = p - t, \ \#E_+(F_p) - 1 = p + t = (p - t) + 2t.$$

Therefore, we have the following property;

Property 2. The parity of $(\#E_+(F_p) - 1)/2$ and that of $(\#E_-(F_p) - 1)/2$ are reciprocal to each other.

4.2   Proposed Method

In order to distinguish whether the order $\#E(F_p)$ is $\#E_+(F_p)$ or $\#E_-(F_p)$, this paper proposes the following Step5';

Step5' : For irreducible cubic polynomial $E(x, 0) = x^3 + 3kx + 2k$, calculate $\tilde{E}(x, 0)$ that satisfies Eq.(12) by using SPPT as introduced in **Sec.3.3** and then calculate the following value $T \in F_p$;

$$T = \tilde{E}(0, 0)^{(p-1)/2}. \tag{22}$$

Let $\#E(F_p)$ be the order of $E(x, y) = x^3 + 3kx + 2k - y^2$, we have

$$N = \frac{\#E(F_p) - 1}{2} = \begin{cases} \text{odd} & \text{when } T = 1 \\ \text{even} & \text{when } T = -1 \end{cases}. \tag{23}$$

Based on this relation, check whether $\#E(F_p)$ is $\#E_+(F_p)$ or $\#E_-(F_p)$ by using **Prop.2**. If the order $\#E(F_p)$ is the objective order, then output $E(x, y)$. If not, then output the twisted curve of $E(x, y)$.

We can apply Step5' instead of Step5 in the CM method algorithm introduced in **Sec.2.5**. In Step5', we calculate an irreducible cubic polynomial $\tilde{E}(x, 0)$ from $f(x) = E(x, 0)$ by using SPPT as introduced in **Sec.3.3**.

According to **Prop.1** and **Prop.2**, we can check the parity of N from $\tilde{E}(0, 0)$ and therefore we can distinguish whether the order $\#E(F_p)$ is $\#E_+(F_p)$ or

$\#E_-(F_p)$, where N is written as Eq.(7). If the order is not the objective order, the following twisted elliptic curve has the objective order;

$$E(x, y) = x^3 + 3kc^2x + 2kc^3 - y^2, \tag{24}$$

where c is a QNR in F_p.

when 4 divides $p - 1$:
Step5 in the conventional algorithm needs a scalar multiplication for a rational point as introduced in **Sec.4.1**. On the other hand, **Step5'** only depends on whether or not $\tilde{E}(0,0)$ is a QR in F_p, as shown in Eq.(16) and Eq.(17), **Step5'** needs a square root calculation, a polynomial modulo operation for SPPT, and a quadratic power residue check Eq.(22). In other words, we need to distinguish whether $\tilde{f}(x)$ given by SPPT is $\tilde{f}_+(x)$ or $\tilde{f}_-(x)$. However, when 4 divides $p - 1$, we can easily check whether or not $\tilde{E}(0,0)$ is a QR in F_p by

$$\tilde{E}(0,0)^{(p-1)/2} = \left(\pm\sqrt{-(108k^3 + 108k^2)}\right)^{(p-1)/2}$$
$$= \left(-108k^3 - 108k^2\right)^{(p-1)/4}. \tag{25}$$

Therefore, the result of Eq.(25) does not depend on the sign $\pm$. Consequently, when 4 divides $p - 1$, we can easily check the parity of N without a square root calculation and polynomial modulo operation for SPPT. It only needs an exponentiation in the prime field F_p. As compared to a scalar multiplication for a rational point, we can carry out the calculation of the right-hand side of Eq.(25) much faster.

When 4 divides $p - 1$, we use the following Step5" instead of Step5 in the CM method algorithm introduced in **Sec.2.5**;

Step5" : For irreducible cubic polynomial $E(x,0) = x^3 + 3kx + 2k$, calculate the following value $T \in F_p$;

$$T = \left(-108k^3 - 108k^2\right)^{(p-1)/4}. \tag{26}$$

Then, let $\#E(F_p)$ be the order of $E(x, y) = x^3 + 3kx + 2k - y^2$, we have

$$N = \frac{\#E(F_p) - 1}{2} = \begin{cases} \text{odd} & \text{when } T = 1 \\ \text{even} & \text{when } T = -1 \end{cases}. \tag{27}$$

Based on this relation, check whether $\#E(F_p)$ is $\#E_+(F_p)$ or $\#E_-(F_p)$ by using **Prop.2**. If the order $\#E(F_p)$ is the objective order, then output $E(x, y)$. If not, then output the twisted curve of $E(x, y)$.

We should note that Eq.(26) does not need any square root calculations because 4 divides $p - 1$.

4.3   Experimental Result

Table 4.3 shows the computation times of Step5, Step5', and Step5" on average. We used Pentium4(2.63GHz) with C language and NTL(A library for doing number theory)[11]. For scalar multiplications of Step5, we used binary method[8].

Table 1. Comparison between **Step5**, **Step5'**, and **Step5"**

[unit:ms]

	characteristic p	Step5	Step5'	Step5"
$4 \nmid (p-1)$	$2^{160}+7$	149	45.3	–
	$2^{180}+15$	112	52.2	–
	$2^{200}+235$	131	64.4	–
	$2^{220}+463$	242	79.5	–
	$2^{240}+115$	177	88.4	–
$4 \mid (p-1)$	$2^{160}+357$	113	–	4.83
	$2^{180}+193$	139	–	5.57
	$2^{200}+697$	170	–	6.59
	$2^{220}+217$	163	–	8.05
	$2^{240}+325$	180	–	8.99

* CPU:PentiumIII 846MHz, We used NTL[11].

From the table, we find that the computation times of Step5' and Step5" are faster than that of Step5. Especially, when the characteristic p is a 160-bits prime number, the computation time of Step5" is about 20 times faster than that of Step5. Therefore, as the input of CM method algorithm, the authors recommend to choose the characteristic p such that $4 \mid (p-1)$.

In CM method, the class polynomial computation is the most time-consuming operation[8]. Therefore, most of the conventional improvements for CM method are given for the class polynomial computation; however, Koc et al[7] algorithm prepares a table of class polynomials in advance, by using this table the algorithm generates a prime order elliptic curve within several seconds. By incorporating our proposed method, Koc et al. algorithm will become about 10% faster. We can say that the proposed algorithm is enough practical.

5   Conclusion

In this paper, we have introduced a shift product-based polynomial transformation (SPPT) that was carried out by a square root calculation and a polynomial modulo operation. Then, we show that the parities of $(\#E-1)/2$ and $(\#E'-1)/2$ are reciprocal to each other, where $\#E$ and $\#E'$ are the orders of the two candidate curves obtained at the last step of CM method algorithm. Based on this property, we proposed a method to check the parity by using SPPT. This parity check method does not need a scalar multiplication for a rational point, it needs a square root calculation, a polynomial modulo operation, and a quadratic power residue check instead. When 4 divides the characteristic minus 1, the proposed method does not need the square root calculation and polynomial modulo operation. It only needs a quadratic power residue check. For a 160-bits prime number as the characteristic, the proposed method can carry out the parity check about 20 times faster than the conventional method when 4 divides the characteristic minus 1.

References

1. T.Sato, and K.Araki, "Fermat Quotients and the Polynomial Time Discrete Lot Algorithm for Anomalous Elliptic Curve," Commentarii Math. Univ. Sancti. Pauli, vol47, No.1, pp.81-92, 1998.
2. G.Frey and H.Rück, "A Remark Concerning m-Divisibility and the Discrete Logarithm in the Divisor Class Group of Curves," Math. Comp. **62** , pp.865-874, 1994.
3. P.Gaudry, F.Hess, and N.Smart, "Constructive and destructive facets of Weil descent on elliptic curves," Hewlett Packard Tech. Report HPL-2000-10 , 2000.
4. K.Horiuchi et al., "Construction of Elliptic Curves with Prime Order and Estimation of Its Complexity," IEICE Trans., J82-A, no.8, pp.1269-1277, 1999.
5. E.Konstantinou, Y.Stamatiou, and C.Zaroliagis, "On the construction of prime order elliptic curves," Indocrypto2003, LNCS 2904, pp.309-322, 2003.
6. Y.Nogami and Y.Morikawa, "Fast Generation of Elliptic Curves with Prime Order over $F_{p^{2c}}$," Proc. of Workshop on Coding and Cryptography 2003, pp.347-356, 2003.
7. E.Savas, T.Schmidt, and C.Koc, "Generating Elliptic Curves of Prime Order," CHES2001, LNCS2162, pp.142-158, 2001.
8. I.Blake, G.Seroussi, and N.Smart, Elliptic Curves in Cryptography, LNS 265, Cambridge University Press, 1999.
9. Class polynomials of CM-fields,
http://www.exp-math.uni-essen.de/zahlentheorie/classpol/class.html
10. T.Hiramoto, Y.Nogami, and Y.Morikawa, "A Fast Algorithm to Test Irreducibility of Cubic Polynomial over $GF(P)$," IEICE Trans. J84-A no.5, 2000.
11. A Library for doing Number Theory, http://www.shoup.net/ntl/.
12. R.Lidl and H.Niederreiter, Finite Fields, Encyclopedia of Mathematics and Its Applications, Cambridge University Press, 1984.

A   A Relation Between $f(x)$ and $\tilde{f}(x)$

Let τ and ω be zeros of $\tilde{f}(x)$ and $\tilde{f}(x^p - x)$, then we have

$$\tau = \omega^p - \omega, \tag{28a}$$

$$\tau^p = \omega^{p^2} - \omega^p, \tag{28b}$$

$$\tau^{p^2} = \omega^{p^3} - \omega^{p^2}, \tag{28c}$$

$$\vdots \tag{28d}$$

$$\tau^{p^{m-1}} = \omega^{p^m} - \omega^{p^{m-1}}. \tag{28e}$$

By adding these equations, we have

$$\omega^{p^m} - \omega = \tau + \tau^p + \tau^{p^2} + \cdots + \tau^{p^{m-1}} = \mathrm{Tr}(\tau) = -\tilde{f}_{m-1} = 0, \tag{29}$$

$$\text{where } \mathrm{Tr}(x) = x + x^p + x^{p^2} + \cdots + x^{p^{m-1}}.$$

Therefore, noting that τ belongs to F_{p^m} but not to its proper subfield, we find that ω also belongs to F_{p^m} but not to its proper subfield because ω satisfies Eq.(28a). In addition, we find that $\omega + 1, \omega + 2, \cdots, \omega + (p-1)$ are also zeros of $\tilde{f}(x^p - x)$. In the case of $p \nmid m$, among these zeros, there exists an element that satisfies $\mathrm{Tr}(x) = 0$[12]. Supposing that ω satisfies $\mathrm{Tr}(\omega) = 0$ and denoting the minimal polynomial of ω by $\tilde{f}(x)$, we have $\tilde{f}(x)$ that satisfies Eq.(10b).

Next, let us consider an irreducible cubic polynomial $f(x)$ of degree m over F_p whose zero ω satisfies $\mathrm{Tr}(\omega) = 0$ and $\omega \in F_{p^m}$. Then, we have

$$\prod_{i=0}^{p-1} f(x+i) = \prod_{i=0}^{p-1} (x - \omega + i)(x - \omega^p + i) \cdots (x - \omega^{p^{m-1}} + i)$$

$$= \prod_{i=0}^{m-1} (x - \omega^{p^i})(x - \omega^{p^i} + 1) \cdots (x - \omega^{p^i} + (p-1))$$

$$= \prod_{i=0}^{m-1} (x^p - x - (\omega^p - \omega)^{p^i}). \tag{30}$$

If $\tau = \omega^p - \omega$ belongs to a proper subfield $F_{p^r}, r \mid m$ of F_{p^m}, then we have

$$\tau + \tau^p + \cdots \tau^{p^{r-1}} = (\omega^p - \omega) + (\omega^{p^2} - \omega^p) + \cdots (\omega^{p^r} - \omega^{p^{r-1}}) = \omega^{p^r} - \omega. \tag{31}$$

Since ω does not belongs to the proper subfield F_{p^r}, $\omega^{p^r} - \omega$ is not equal to 0. On the other hand, $\tau + \tau^p + \cdots \tau^{p^{r-1}}$ is the sum of all conjugates of τ with respect to F_p, therefore the sum becomes an element in F_p. Let $c \in F_p$ be the element, we have

$$c + c^{p^r} + c^{p^{2r}} + \cdots + c^{p^{m'r}}$$

$$= (\omega^{p^r} - \omega) + (\omega^{p^r} - \omega)^{p^r} + (\omega^{p^r} - \omega)^{p^{2r}} + \cdots + (\omega^{p^r} - \omega)^{p^{m'r}}, \tag{32}$$

then we have

$$(m' + 1)c = \omega^{p^m} - \omega = 0, \tag{33}$$

where $m' = m/r - 1$. This paper deals with the case that the characteristic p does not divide the extension degree m, therefore c must be 0 because $p \nmid (m' + 1)$. Consequently, τ does not belong to the proper subfield F_{p^r} and therefore $f(x)$ is an irreducible polynomial of degree m over F_p whose zero $\tau = \omega^p - \omega$ satisfies

$$-f_{m-1} = \mathrm{Tr}(\tau) = \mathrm{Tr}(\omega^p - \omega) = \mathrm{Tr}(\omega)^p - \mathrm{Tr}(\omega) = 0. \tag{34}$$

Then, we have SPPT and the one to one relation between $f(x)$ and $\tilde{f}(x)$ is shown.

B   Proof of Eq.(16)

Based on the relation $\tau = \omega^p - \omega$, we have the following equations;

$$\tilde{f}_2 = -(\tau + \tau^p + \tau^{p^2}) = 0, \tag{35a}$$

$$\tilde{f}_1 = \tau\tau^p + \tau^p\tau^{p^2} + \tau^{p^2}\tau = 3f_1, \tag{35b}$$

$$\tilde{f}_0 = -\tau\tau^p\tau^{p^2} = A - B, \tag{35c}$$

$$A = \omega\omega^{2p} + \omega^p\omega^{2p^2} + \omega^{p^2}\omega^2, \tag{35d}$$

$$B = \omega^2\omega^p + \omega^{2p}\omega^{p^2} + \omega^{2p^2}\omega. \tag{35e}$$

Since $A + B = 3f_0$ and $AB = f_1^3 + 9f_0^2$, we obtain

$$\tilde{f}_0 = A - B = \pm\sqrt{-(4f_1^3 + 27f_0^2)}. \tag{36}$$

As shown above, we can easily obtain two candidates of $\tilde{f}_0$ without any calculations in the extension field F_{p^3} to which ω belongs.

Generating Prime Order Elliptic Curves: Difficulties and Efficiency Considerations*

Elisavet Konstantinou[1,2], Aristides Kontogeorgis[3],
Yannis C. Stamatiou[1,3,4], and Christos Zaroliagis[1,2]

[1] Computer Technology Institute, P.O. Box 1122, 26110 Patras, Greece
[2] Dept of Computer Eng. & Informatics, Univ. of Patras, 26500 Patras, Greece
[3] Dept of Mathematics, Univ. of the Aegean, Karlovassi, 83200, Samos, Greece
[4] Joint Research Group (JRG) on Communications and Information Systems
Security (Univ. of the Aegean and Athens Univ. of Economics and Business)
{konstane, zaro}@ceid.upatras.gr,{kontogar, stamatiu}@aegean.gr

Abstract. We consider the generation of prime order elliptic curves
(ECs) over a prime field $\mathbb{F}_p$ using the Complex Multiplication (CM)
method. A crucial step of this method is to compute the roots of a special
type of class field polynomials with the most commonly used being the
Hilbert and Weber ones, uniquely determined by the CM discriminant
D. In attempting to construct prime order ECs using Weber polynomials
two difficulties arise (in addition to the necessary transformations of the
roots of such polynomials to those of their Hilbert counterparts). The
first one is that the requirement of prime order necessitates that $D \equiv 3$
(mod 8), which gives Weber polynomials with degree three times larger
than the degree of their corresponding Hilbert polynomials (a fact that
could affect efficiency). The second difficulty is that these Weber polynomials do not have roots in $\mathbb{F}_p$. In this paper we show how to overcome
the above difficulties and provide efficient methods for generating ECs of
prime order supported by a thorough experimental study. In particular,
we show that such Weber polynomials have roots in $\mathbb{F}_{p^3}$ and present a
set of transformations for mapping roots of Weber polynomials in $\mathbb{F}_{p^3}$
to roots of their corresponding Hilbert polynomials in $\mathbb{F}_p$. We also show
how a new class of polynomials, with degree equal to their corresponding Hilbert counterparts (and hence having roots in $\mathbb{F}_p$), can be used
in the CM method to generate prime order ECs. Finally, we compare
experimentally the efficiency of using this new class against the use of
the aforementioned Weber polynomials.

Keywords: Elliptic Curve Cryptosystems, Generation of Prime Order
Elliptic Curves, Complex Multiplication, Class Field Polynomials.

---

* This work was partially supported by the Action IRAKLITOS (Fellowships for Research in the University of Patras) with matching funds from EC and the Greek
Ministry of Education.

C. Park and S. Chee (Eds.): ICISC 2004, LNCS 3506, pp. 261–278, 2005.

1   Introduction

The generation of elliptic curves (ECs) with good security properties has been one of the central considerations in Elliptic Curve Cryptography. One of the most efficient methods that can be employed for the construction of ECs with specified order is the *Complex Multiplication* (CM) method [1, 17]. Briefly, the CM method starts with the specification of a discriminant value D, the determination of the order p of the underlying prime field and the order m of the EC. It then computes a special polynomial, called *Hilbert* polynomial, which is uniquely determined by D and locates one of its roots modulo p. This root can be used to construct the parameters of an EC with order m over the field $\mathbb{F}_p$. A major drawback of Hilbert polynomials is that their coefficients grow very large with D and hence possess high computational demands. In order to eliminate this drawback, an alternative class of polynomials with much smaller coefficients, called *Weber* polynomials, can be used instead. The issue with Weber polynomials, however, is that their roots (modulo p) cannot be used to construct directly the parameters of the EC but they first have to be transformed into the roots of their corresponding Hilbert polynomials.

The CM method is not by itself adequate for applications that require robust ECs against cryptanalytic attacks. It turns out that the *properties* of the order of an EC play a central role in establishing cryptanalytic robustness. One way to establish robustness is to generate ECs whose order satisfies a certain number of properties designed to guard against the currently known most effective attacks [18, 24, 25]. An equally important alternative to cryptographic strength (see e.g., [26]) requires that the order of the generated EC is a prime number. Note that in certain applications it is necessary to have ECs of prime order [6]. Prime order ECs defined in various fields were also treated in [2, 16, 20, 23].

In this paper we follow the latter approach and study the use of the CM method for generating ECs of prime order in $\mathbb{F}_p$. Although ECs with no restrictions on their order may be generated more efficiently using a point counting (such as Schoof's [28]) algorithm[1], the requirement of prime order can severely change the situation. Point counting algorithms first choose the parameters of the EC and then compute its order. If this order is found non-prime, then another set of EC parameters is generated and the process is repeated. This can be seen, approximately, as sampling from the set of ECs of prime order (for a fixed p). There is well supported theoretical and experimental evidence [11] that this probability is, asymptotically, $\frac{c_p}{\log p}$, where c_p is a constant depending on p and satisfying $0.44 \leq c_p \leq 0.62$. Thus, it appears that prime orders are not especially favored by the point counting approach, as also noted in [11]. CM, on the other hand, starts with a prime number (the order of the EC) and *then* constructs the parameters thus avoiding this averse prime order probability.

In attempting to construct prime order ECs using Weber polynomials two additional difficulties arise. The first one is that the prime order requirement

---

[1] There are cases where point counting algorithms can be very inefficient compared to the CM method, e.g., when p is large and the discriminant value is small.

necessitates that $D \equiv 3 \pmod 8$, which in turn results in Weber polynomials with degree three times larger than the degree of their corresponding Hilbert polynomial. The second and most crucial difficulty is that such Weber polynomials (used for the construction of prime order ECs) do not have roots in $\mathbb{F}_p$ for certain values of p, as it is shown in Section 3.

Our work addresses the difficulties outlined above with an eye to applications and the practitioner's needs. We pay particular attention to support our theoretical findings with a thorough experimental study, thus shedding more light in the use of polynomials for the efficient generation of prime order ECs using the CM method, and providing guidance to the practitioner with respect to the resolution of these difficulties. In particular, we make the following contributions: (i) We show that Weber polynomials defined on values of $D \equiv 3 \pmod 8$ and used in the CM method for generating ECs of prime order have roots in the extension field $\mathbb{F}_{p^3}$ and not in $\mathbb{F}_p$. (ii) We present a set of simplified transformations that map the roots of the Weber polynomials in $\mathbb{F}_{p^3}$ to the roots of their corresponding Hilbert polynomials in $\mathbb{F}_p$. This implies that the particular Weber polynomials can be used to generate prime order ECs with the CM method. (iii) We show how a new class of polynomials can be used in the CM method for generating prime order ECs. The advantage of these polynomials is that they have the same degree with their corresponding Hilbert polynomials and hence have roots in $\mathbb{F}_p$. (iv) We perform a comparative experimental study regarding the efficiency of the CM method using the aforementioned Weber polynomials against using the new class of polynomials. Although it may seem that the use of Weber polynomials is inefficient due to their high degree and the fact that their roots lie in $\mathbb{F}_{p^3}$ (which requires operations with polynomials of degree 2), we provide experimental evidence which demonstrates that this is not always the case.

We would like to note that the case $D \equiv 3 \pmod 8$ can also be useful for the generation of ECs that do not necessarily have prime order [29] or for the generation of special curves, such as MNT curves [19, 20]. This makes our analysis for class polynomials with such discriminants even more useful.

The rest of the paper is organized as follows. In Section 2 we review some basic definitions and facts about ECs, the CM method, the Hilbert polynomials, and discuss some of their properties relevant to the generation of ECs. In Section 3 we present properties of Weber polynomials with $D \equiv 3 \pmod 8$ and describe their use in the CM method. In Section 4 we elaborate on the construction of a new class of polynomials that can also be used in the CM method. Finally, in Section 5 we present our experimental results concerning the efficiency of the CM method using Weber polynomials against using the new class of polynomials.

2   A Brief Overview of Elliptic Curve Theory and Complex Multiplication

This section contains a brief introduction to elliptic curve theory, to the Complex Multiplication method for generating prime order elliptic curves and to the Hilbert class field polynomials. Our aim is to facilitate the reading of the sections that follow. For full coverage of the necessary concepts and terms, the interested reader may consult [5]. Also, the proofs of certain theorems require basic knowledge of algebraic number theory and Galois theory. The interested reader is referred to [8, 31, 32] for definitions not given here due to lack of space.

2.1   Preliminaries of Elliptic Curve Theory

An *elliptic curve* defined over a finite field $\mathbb{F}_p$, $p > 3$ and prime, is denoted by $E(\mathbb{F}_p)$ and contains the points $(x, y) \in \mathbb{F}_p$ (in affine coordinates) that satisfy the equation (in $\mathbb{F}_p$)

$$y^2 = x^3 + ax + b, \tag{1}$$

with $a, b \in \mathbb{F}_p$ satisfying $4a^3 + 27b^2 \neq 0$. The set of these points equipped with a properly defined point addition operation and a special point, denoted by $\mathcal{O}$ and called *point at infinity* (zero element for the addition operation), forms an Abelian group. This is the *Elliptic Curve group* and the point $\mathcal{O}$ is its identity element (see [5, 30] for more details on this group).

The *order*, denoted by m, is the number of points that belong in $E(\mathbb{F}_p)$. The numbers m and p are related by the *Frobenius trace* $t = p + 1 - m$. Hasse's theorem (see e.g., [5, 30]) implies that $|t| \leq 2\sqrt{p}$. Given a point $P \in E(\mathbb{F}_p)$, its *order* is the smallest positive integer n such that $nP = \mathcal{O}$. By Langrange's theorem, the order of a point $P \in E(\mathbb{F}_p)$ divides the order m of the group $E(\mathbb{F}_p)$. Thus, $mP = \mathcal{O}$ for any $P \in E(\mathbb{F}_p)$ and, consequently, the order of a point is always less than or equal to the order of the elliptic curve.

Two of the most important quantities of an elliptic curve $E(\mathbb{F}_p)$ defined through Eq. (1) are the *curve discriminant* Δ and the *j-invariant*: $\Delta = -16(4a^3 + 27b^2)$ and $j = -1728(4a)^3/\Delta$. Given a j-invariant $j_0 \in \mathbb{F}_p$ (with $j_0 \neq 0, 1728$) *two* ECs can be constructed. If $k = j_0/(1728 - j_0) \bmod p$, one of these curves is given by Eq. (1) by setting $a = 3k \bmod p$ and $b = 2k \bmod p$. The second curve (the *twist* of the first) is given by the equation

$$y^2 = x^3 + ac^2 x + bc^3 \tag{2}$$

with c any quadratic non-residue of $\mathbb{F}_p$. If m_1 and m_2 denote the orders of an elliptic curve and its twist respectively, then $m_1 + m_2 = 2p + 2$ which implies that if one of the curves has order $p + 1 - t$, then its twist has order $p + 1 + t$, or vice versa (see [5–Lemma VIII.3]).

2.2   The Complex Multiplication Method

As stated in the previous section, given a j-invariant one may readily construct an EC. Finding a suitable j-invariant for a curve that has a given order m can be accomplished through the theory of *Complex Multiplication* (CM) of elliptic curves over the rationals. This method is called the *CM method* and in what follows we will give a brief account of it.

By Hasse's theorem, $Z = 4p - (p+1-m)^2$ must be positive and, thus, there is a unique factorization $Z = Dv^2$, with D a square free positive integer. Therefore

$$4p = u^2 + Dv^2 \tag{3}$$

for some integer u that satisfies the equation

$$m = p + 1 \pm u. \tag{4}$$

The negative parameter $-D$ is called a *CM discriminant for the prime p*. For convenience throughout the paper, we will use (the positive integer) D to refer to the CM discriminant. The CM method uses D to determine a j-invariant. This j-invariant in turn, will lead to the construction of an EC of order $p+1-u$ or $p+1+u$.

The method works as follows. Given a prime p, the smallest D is chosen for which there exists some integer u for which Eq. (3) holds. If neither of the possible orders $p+1-u$ and $p+1+u$ is suitable for our purposes, the process is repeated with a new D. If at least one of these orders is suitable, then the method proceeds with the construction of the *Hilbert polynomial* (uniquely defined by D) and the determination of its roots modulo p. Any root of the Hilbert polynomial can be used as a j-invariant. From this the corresponding EC and its twist can be constructed as described in Section 2.1. In order to find which one of the curves has the desired suitable order ($m = p + 1 - u$ or $m = p + 1 + u$), the method uses Langrange's theorem as follows: it repeatedly chooses points P at random in each EC until a point is found in one of the curves for which $mP \neq \mathcal{O}$. This implies that the curve we seek is the other one. It turns out that the most time consuming part of the CM method is the construction of the Hilbert polynomial. These polynomials have very large coefficients and their construction requires the use of high precision floating point arithmetic with complex numbers.

We now turn to the generation of prime order ECs. If m should be a prime number, then it is obvious that u should be odd. It is also easy to show that D should be congruent to 3 (mod 8) and v should be odd, too. In this paper, we follow a variant of the CM method for the construction of prime order elliptic curves. We first determine a discriminant $D \equiv 3$ (mod 8) and then we construct the two prime numbers p and m. The most trivial way to do this, is by choosing at random odd integers u and v and then check whether p and m are prime using Eq. (3) and Eq. (4). Next, a Weber polynomial corresponding to the discriminant value D is constructed and we locate a root of it. This root, however, cannot lead to the construction of the j-invariant directly, since j-invariants are roots of the Hilbert polynomials. Therefore, we must transform this root to a root of the corresponding (constructed with the same discriminant) Hilbert polynomial. The necessary transformations are given in Section 3.

2.3   Hilbert Polynomials

Every CM discriminant D defines a unique Hilbert polynomial, denoted by $H_D(x)$. Given a positive D, the Hilbert polynomial $H_D(x) \in \mathbb{Z}[x]$ is defined as

$$H_D(x) = \prod_\tau (x - j(\tau)) \tag{5}$$

for values of τ satisfying $\tau = (-\beta + \sqrt{-D})/2\alpha$, for all integers α, β, and γ such that (i) $\beta^2 - 4\alpha\gamma = -D$, (ii) $|\beta| \le \alpha \le \sqrt{D/3}$, (iii) $\alpha \le \gamma$, (iv) $\gcd(\alpha, \beta, \gamma) = 1$, and (v) if $|\beta| = \alpha$ or $\alpha = \gamma$, then $\beta \ge 0$. The 3-tuple of integers $[\alpha, \beta, \gamma]$ that satisfies these conditions is called a *primitive, reduced quadratic form* of $-D$, with τ being a root of the quadratic equation $\alpha z^2 + \beta z + \gamma = 0$. Clearly, the set of primitive reduced quadratic forms of a given discriminant is finite. The quantity $j(\tau)$ in Eq. (5) is called *class invariant* and is defined as follows. Let $z = e^{2\pi\sqrt{-1}\tau}$ and $h(\tau) = \frac{\Delta(2\tau)}{\Delta(\tau)}$, where $\Delta(\tau) = \eta(\tau)^{24} = z \left(1 + \sum_{n \ge 1} (-1)^n \left(z^{n(3n-1)/2} + z^{n(3n+1)/2}\right)\right)^{24}$. Then, $j(\tau) = \frac{(256h(\tau)+1)^3}{h(\tau)}$.

Let h be the number of primitive reduced quadratic forms, which determines the *degree* (or *class number*) of $H_D(x)$. Then, the bit precision required for the generation of $H_D(x)$ can be estimated (see [17]) by

$$\text{H-Prec}(D) \approx \frac{\ln 10}{\ln 2}(h/4 + 5) + \frac{\pi\sqrt{D}}{\ln 2} \sum_\tau \frac{1}{\alpha}$$

with the sum running over the same values of τ as the product in Eq. (5). Hilbert polynomials have roots roots modulo p under certain conditions stated in the following theorem.

Theorem 1. *A Hilbert polynomial $H_D(x)$ with degree h has exactly h roots modulo p if and only if the equation $4p = u^2 + Dv^2$ has integer solutions and p does not divide the discriminant $\Delta(H_D)$ of the polynomial.*

Proof. Let H_K be the Hilbert class field of the imaginary quadratic field $K = \mathbb{Q}(\sqrt{-D})$, and let $\mathcal{O}_{H_K}$ and $\mathcal{O}_K$ be the rings of algebraic integers of H_K and K respectively.

Let p be a prime such that $4p = u^2 + Dv^2$ has integer solutions. Then, according to [8–Th. 5.26] p splits completely in H_K. Let $H_D(x) \in \mathbb{Z}[x]$ be the Hilbert polynomial with root the real algebraic integer $j(\tau)$. Proposition 5.29 in [8] implies that $H_D(x)$ has a root modulo p if and only if p splits in H_K and does not divide its discriminant[2] $\Delta(H_D)$. But since $\frac{\mathcal{O}_{H_K}}{p\mathcal{O}_{H_K}}/\mathbb{F}_p$ is Galois, $H_D(x)$ has not only one root modulo p, but h distinct roots modulo p. $\square$

There are finitely many primes dividing the discriminant $\Delta(H_D)$ of the Hilbert polynomial and infinitely many primes to choose. In elliptic curve cryptosystems the prime p is at least 160 bits. Therefore, an arbitrary prime almost certainly does not divide the discriminant.

---

[2] For a definition of the discriminant of a polynomial see [7].

3   The CM Method Using Weber Polynomials

In this section we define Weber polynomials for discriminant values $D \equiv 3$ (mod 8) and prove that they do not have roots in $\mathbb{F}_p$ for certain primes p, but they do have roots in the extension field $\mathbb{F}_{p^3}$. We then discuss their efficiency when used in the CM method, and present a transformation that maps roots of Weber polynomials in $\mathbb{F}_{p^3}$ into the roots of their Hilbert counterparts in $\mathbb{F}_p$.

3.1   Weber Polynomials and Their Roots in Finite Fields

Weber polynomials are defined using the Weber functions (see [1, 13]):

$$f(y) = q^{-1/48} \prod_{r=1}^{\infty} (1 + q^{(r-1)/2}) \qquad f_1(y) = q^{-1/48} \prod_{r=1}^{\infty} (1 - q^{(r-1)/2})$$

$$f_2(y) = \sqrt{2}\; q^{1/24} \prod_{r=1}^{\infty} (1 + q^r) \qquad \text{where } q = e^{2\pi y \sqrt{-1}}.$$

The Weber polynomial $W_D(x) \in \mathbb{Z}[x]$ for $D \equiv 3$ (mod 8) is defined as

$$W_D(x) = \prod_{\ell} (x - g(\ell)) \qquad (6)$$

where $\ell = \frac{-b + \sqrt{-D}}{a}$ satisfies the equation $ay^2 + 2by + c = 0$ for which $b^2 - ac = -D$ and (i) $\gcd(a, b, c) = 1$, (ii) $|2b| \le a \le c$, and (iii) if either $a = |2b|$ or $a = c$, then $b \ge 0$. Let $\zeta = e^{\pi \sqrt{-1}/24}$. The class invariant $g(\ell)$ for $W_D(x)$ is defined by

$$g(\ell) = \begin{cases} \zeta^{b(c-a-a^2c)} \cdot f(\ell) & \text{if } 2 \nmid a \text{ and } 2 \nmid c \\ -(-1)^{\frac{a^2-1}{8}} \cdot \zeta^{b(ac^2-a-2c)} \cdot f_1(\ell) & \text{if } 2 \nmid a \text{ and } 2 \mid c \\ -(-1)^{\frac{c^2-1}{8}} \cdot \zeta^{b(c-a-5ac^2)} \cdot f_2(\ell) & \text{if } 2 \mid a \text{ and } 2 \nmid c \end{cases} \qquad (7)$$

if $D \equiv 3$ (mod 8) and $D \not\equiv 0$ (mod 3), and

$$g(\ell) = \begin{cases} \frac{1}{2}\zeta^{3b(c-a-a^2c)} \cdot f^3(\ell) & \text{if } 2 \nmid a \text{ and } 2 \nmid c \\ -\frac{1}{2}(-1)^{\frac{3(a^2-1)}{8}} \cdot \zeta^{3b(ac^2-a-2c)} \cdot f_1^3(\ell) & \text{if } 2 \nmid a \text{ and } 2 \mid c \\ -\frac{1}{2}(-1)^{\frac{3(c^2-1)}{8}} \cdot \zeta^{3b(c-a-5ac^2)} \cdot f_2^3(\ell) & \text{if } 2 \mid a \text{ and } 2 \nmid c \end{cases} \qquad (8)$$

if $D \equiv 3$ (mod 8) and $D \equiv 0$ (mod 3).

For these cases of the discriminant ($D \equiv 3$ (mod 8)), the Weber polynomial $W_D(x)$ has degree three times larger than the degree of its corresponding Hilbert polynomial $H_D(x)$. An upper bound for the precision requirements of Weber polynomials for both cases of D was presented in [16] and is equal to $3h + \frac{\pi\sqrt{D}}{24\ln 2}\sum_{\ell}\frac{1}{a}$ for $D \not\equiv 0$ (mod 3) and to $3h + \frac{\pi\sqrt{D}}{8\ln 2}\sum_{\ell}\frac{1}{a}$ for $D \equiv 0$ (mod 3). The sum runs over the same values of ℓ as the product of Eq. (6) and $3h$ is the degree of the Weber polynomial (h is the degree of the corresponding Hilbert polynomial).

Consider the modular function

$$\Phi_2(x,j) = (x-16)^3 - jx \tag{9}$$

where j is a class invariant for the Hilbert polynomial. The three roots of the equation $\Phi_2(x,j) = 0$ are the powers f^{24}, $-f_1^{24}$ and $-f_2^{24}$ of the Weber functions. A transformation (used in the CM method) from roots of Weber polynomials to roots of Hilbert polynomials was presented in [16], and is derived from the modular equation $\Phi_2(x,j) = 0$. The transformation for $D \not\equiv 0 \pmod 3$ is

$$R_H = \frac{(2^{12}R_W^{-24} - 16)^3}{2^{12}R_W^{-24}} \tag{10}$$

and for $D \equiv 0 \pmod 3$ is

$$R_H = \frac{(2^4 R_W^{-8} - 16)^3}{2^4 R_W^{-8}} \tag{11}$$

where R_W is a root of $W_D(x)$ and R_H is a root of $H_D(x)$. To use these transformations we have to locate R_W on a specific field, an issue not addressed in [16].

In the rest of this section we will show that when u,v are odd numbers and $D \equiv 3 \pmod 8$, then $W_D(x)$ does not have roots modulo p, but its roots belong to the extension field $\mathbb{F}_{p^3}$ (recall that the order $m = p+1\pm u$ of the elliptic curve can be prime only if u is odd, which means that in Eq. (3) v must be odd, too).

Theorem 2. *If the equation $4p = u^2 + Dv^2$ has a solution and u,v are odd integers, then the Weber polynomial $W_D(x)$ with degree $3h$ $(D \equiv 3 \pmod 8)$ has no roots modulo p.*

Proof. Given an integer c, let $\left(\frac{c}{2}\right)$ be the *Kronecker* symbol. From [22–Th. 3.1] we conclude that if $\left(\frac{-Dv^2}{2}\right) = -1$, then the polynomial $\Phi_2(x,j) \pmod p$ is irreducible modulo p. This means that if we could prove that $\left(\frac{-Dv^2}{2}\right) = -1$, then the equation $\Phi_2(x,j) = 0 \pmod p$ would have no roots $x \pmod p$ for a given $j \pmod p$. This j will be a root of Hilbert polynomial modulo p, which we know from Theorem 1 that always exists. But if there is no $x \pmod p$ that satisfies the equation $\Phi_2(x,j) = 0 \pmod p$, then the Weber polynomial cannot have a root modulo p either. If it had, then according to the transformations there would also exist an $x \pmod p$ which is a contradiction. We must prove now that $\left(\frac{-Dv^2}{2}\right) = -1$. Using the Kronecker symbol we know that $\left(\frac{-Dv^2}{2}\right) = -1$ if $-Dv^2$ is odd and $-Dv^2 \equiv \pm 3 \pmod 8$. We will show that $Dv^2 \equiv 3 \pmod 8$. Clearly, since $D \equiv 3 \pmod 8 = 8d_1 + 3$ and $v = 2v_1 + 1$ is odd, then Dv^2 is also odd. We have $Dv^2 = (8d_1 + 3)(2v_1 + 1)^2 = (8d_1 + 3)(4v_1^2 + 4v_1 + 1)$. That is, $Dv^2 \equiv 3(4v_1^2 + 4v_1 + 1) \pmod 8$ and because $v_1^2 + v_1$ is even then it is easily seen that $Dv^2 \equiv 3 \pmod 8$ which completes the proof. $\square$

The next theorem establishes the main result of this section.

Theorem 3. *If the equation $4p = u^2 + Dv^2$ has a solution with u, v odd integers, then the Weber polynomial $W_D(x)$ has h monic irreducible factors of degree 3 modulo p. Thus, the polynomial has $3h$ roots in the extension field $\mathbb{F}_{p^3}$.*

Proof. We have proved in Theorem 2 that the Weber polynomial does not have roots modulo p if u, v are odd numbers and that the polynomial $\Phi_2(x, j)$ is irreducible modulo p. This means that $\Phi_2(x, j) = 0$ has three roots $x \in \mathbb{F}_{p^3}$ for a root $j \in \mathbb{F}_p$ of the Hilbert polynomial. According to Eq. (10) and Eq. (11), $x = 2^{12} R_W^{-24}$ if $D \not\equiv 0 \pmod 3$, and $x = 2^4 R_W^{-8}$ if $D \equiv 0 \pmod 3$. Thus, there are at least three roots of the Weber polynomial that correspond to a root $j \in \mathbb{F}_p$ of the Hilbert polynomial, and which are either in $\mathbb{F}_{p^3}$ or in an extension field of greater degree (at most 72 if $D \not\equiv 0 \pmod 3$ and at most 24 if $D \equiv 0 \pmod 3$).

Let $R_{W,j}$ be a root of the Weber polynomial that corresponds to a root j of the Hilbert polynomial. Let $f_j(x)$ be the minimal polynomial of $R_{W,j} \pmod p$. The degree of this polynomial will be at least 3, because the root $R_{W,j}$ is at least in $\mathbb{F}_{p^3}$. Then, the Weber polynomial can be written as

$$W_D(x) = \prod_j f_j(x) \pmod p. \tag{12}$$

Since the degree of the Weber polynomial is $3h$ and the roots j modulo p of the Hilbert polynomial are h (see Theorem 1) we have that every minimal polynomial $f_j(x)$ will have degree 3. Thus, Weber polynomials have h irreducible cubic factors. Every factor has 3 roots in $\mathbb{F}_{p^3}$, which means that there are totally $3h$ roots in $\mathbb{F}_{p^3}$. $\qquad\square$

3.2   The Use of Weber Polynomials in the CM Method

In this subsection we will elaborate on the use of Weber polynomials for the generation of prime order ECs. The idea is that we replace Hilbert polynomials with Weber polynomials and then try to compute a root of the Hilbert polynomial from a root of its corresponding Weber polynomial. To compute the desired Hilbert root, we proceed in three stages. First, we construct the corresponding Weber polynomial. Second, we compute its roots in $\mathbb{F}_{p^3}$. Finally, we transform the Weber roots to the desired Hilbert roots in $\mathbb{F}_p$. The first stage is accomplished using the definition of Weber polynomials in Section 3.1. To compute a root of $W_D(x)$ in $\mathbb{F}_{p^3}$, we have to find an irreducible factor (modulo p) of degree 3 of the polynomial. This is achieved using Algorithm 3.4.6 from [7]. The irreducible factor has 3 roots in $\mathbb{F}_{p^3}$ from which it suffices to choose one, in order to accomplish the third stage.

Suppose that $x^3 + ax^2 + bx + c$ is an irreducible factor modulo p of the Weber polynomial. From this irreducible factor, we can compute three roots (one suffices for the CM method) of the Weber polynomial if we have already defined the reduction polynomial of the extension field $\mathbb{F}_{p^3}$. We simply set the reduction polynomial to be equal to the irreducible factor $x^3 + ax^2 + bx + c$ and then a root of the Weber polynomial would be just x.

Let us see an example: if $W_{403}(x) = x^6 - 12x^5 - 26x^4 + 4x^3 + 36x^2 + 20x + 4$ and $p = 7221076618803527297111165735009$ then a factor of the Weber polynomial modulo p is $x^3 + 5308419983557319593310936611138x^2 + 2654209991778659796655\ 46830567x + 7221076618803527\ 29711165735007$. Note that 403 is not divisible by 3 and $7221076618803527297111165735007 = p - 2 \equiv -2 \pmod{p}$.

The following lemma allows us to determine the constant term of the irreducible factor and consequently to simplify the roots' transformation as we will see later.

Lemma 1. *Let $x^3 + ax^2 + bx + c$ be an irreducible factor (modulo p) of the Weber polynomial with $D \equiv 3 \pmod 8$. Then, the following hold: (i) if $D \equiv 0 \pmod 3$, then $c = -1$; (ii) if $D \not\equiv 0 \pmod 3$, then $c = -2$.*

Proof. The constant term of the Weber polynomial is equal to $(-1)^h$ for the first case of D and $(-2)^h$ for the second case (see [14]). The Galois group of the extension H_K/K operates on the roots modulo p of $H_D(x)$, and therefore on the cubic irreducible factors of $W_D(x)$ (every root of $H_D(x)$ corresponds to three roots of $W_D(x)$ and thus to a cubic irreducible factor). Since every element in this Galois group induces the identity on $\mathbb{F}_p$, all cubic factors of $W_D(x)$ will have the same constant term. Because the constant term of a monic polynomial is equal to the product of the constant terms of its monic irreducible factors, it can be easily seen that $c = -1$ for the first case of D and $c = -2$ for the second. $\square$

We are now ready to present the transformations for mapping a Weber root in $\mathbb{F}_{p^3}$ to its corresponding Hilbert root in $\mathbb{F}_p$. Suppose that $R_W = x$ is a root of a Weber polynomial $W_D(x)$ in the extension field $\mathbb{F}_{p^3}$. The calculations in the transformations must be in $\mathbb{F}_{p^3}$ with reduction polynomial $x^3 + ax^2 + bx + c$, since R_W is a root in $\mathbb{F}_{p^3}$.

The transformations may seem quite complicated because of the arithmetic operations that take place in the extension field, but they can be simplified due to Lemma 1. Consider the case $D \not\equiv 0 \pmod 3$ for which an irreducible factor of the Weber polynomial is equal to $x^3 + ax^2 + bx - 2$. Then, $R_W^{-24} = x^{-24} = \left(\frac{x^2+ax+b}{x(x^2+ax+b)}\right)^{24} = \left(\frac{x^2+ax+b}{2}\right)^{24}$. This means that $2^{12}R_W^{-24} = \frac{(x^2+ax+b)^{24}}{2^{12}}$. Substituting it to Eq. (10) we finally have:

$$R_H = \frac{((x^2 + ax + b)^{24} - 2^{16})^3}{2^{24}(x^2 + ax + b)^{24}}. \tag{13}$$

Similarly, for $D \equiv 0 \pmod 3$ the transformation becomes:

$$R_H = \frac{2^8((x^2 + ax + b)^8 - 1)^3}{(x^2 + ax + b)^8}. \tag{14}$$

The nominator and the denominator of the two transformations are elements of $\mathbb{F}_{p^3}$. However we know that R_H is in $\mathbb{F}_p$ and we can find its value dividing only the leading coefficients of these two elements modulo p. To illustrate the above transformations, consider again the Weber polynomial W_{403}. Let p be a prime as in the previous example, and let the reduction polynomial be the factor of the $W_{403}(x)$ presented also in the previous example. Then, $((x^2 + ax + b)^{24} - 2^{16})^3 = 485216670393361675137940525358x^2 + 498390024660218217560914441491x + 437505083747867349301080018378$ and $(x^2 + ax + b)^{24} = 372203635398289746 518033419220x^2 + 193471851293797158505478806686x + 10581862220484269 1408284289782$. The root R_H of the Hilbert polynomial is equal to $\frac{485216670393361675137940525358}{2^{24}372203635398289746518033419220}$ (mod p) $= 188541528108458443856585415294$.

4   The CM Method Using a New Class of Polynomials

Even though Weber polynomials have much smaller coefficients than Hilbert polynomials and can be computed very efficiently, the fact that their degree for $D \equiv 3$ (mod 8) is three times larger than the degree of the corresponding Hilbert polynomials can be a potential problem, because it involves computations in extension fields. Moreover, the computation of a cubic factor modulo p in a polynomial with degree $3h$ is more time consuming than the computation of a single root modulo p of a polynomial with degree h.

To alleviate these problems, we can use in the CM method a relatively new class of polynomials which have degree h like Hilbert polynomials. In particular, two types of polynomials can be constructed in $\mathbb{Z}[x]$ using two families of η-products: $m_l(z) = \frac{\eta(z/l)}{\eta(z)}$ [21] for an integer l, and $m_{p_1,p_2}(z) = \frac{\eta(z/p_1)\eta(z/p_2)}{\eta(z/(p_1 p_2))\eta(z)}$ [10], where p_1, p_2 are primes such that $24|(p_1 - 1)(p_2 - 1)$. We will refer to the minimal polynomials of these products (powers of which generate the Hilbert class field and are called class invariants like $j(\tau)$) as $M_{D,l}(x)$ and $M_{D,p_1,p_2}(x)$, respectively, where D is the discriminant used for their construction.

The polynomials are obtained from these two families by evaluating their value at a suitably chosen system of quadratic forms. Once a polynomial is computed, we can use the modular equations $\Phi_l(x, j) = 0$ or $\Phi_{p_1,p_2}(x, j) = 0$, in order to compute a root modulo p of the Hilbert polynomial from a root modulo p of the $M_{D,l}(x)$ or the $M_{D,p_1,p_2}(x)$ polynomial, respectively. In this section we will construct polynomials using only the m_l family for prime values of l, in particular for $l = 3, 5, 7, 13$. The reason is that only for these values of l the modular equations have degree 1 in j. For all other values of l or for the m_{p_1,p_2} family, the degree in j is at least 2 (which makes the computations more "heavy"), the coefficients of the modular equations are quite large (which makes their use less efficient) and moreover, the computation of $m_{p_1,p_2}(z)$ involves the computation of four η-products and not two like $m_l(z)$.

In order to construct the polynomial $M_{D,l}(x)$ with $l = 3, 5, 7, 13$, we used Theorem 2 from [9] which for our purposes boils down to the following statement.

Theorem 4. [9] Let $l \in \{3, 5, 7, 13\}$ and $D > 0$ a discriminant such that $l|D$. Choose the power m_l^e as specified in Table 1. Assume $Q = [A, B, C]$ is a primitive quadratic form of discriminant D with $\gcd(A, l) = 1$, $\gcd(A, B, C) = 1$ and

$B^2 \equiv -D$ (mod $4l$). If $\tau_Q = \frac{-B+\sqrt{-D}}{2A}$, then the minimal polynomial of $m_l^e(\tau_Q)$ has integer coefficients and can be computed from an l-system.

Table 1. Class invariants for different values of l

l	class invariant
3	m_3^{12}
5	m_5^{6}
7	m_7^{4}
13	m_{13}^{2}

An l-system is a system $S = \{(A_i, B_i, C_i)\}_{1 \leq i \leq h}$ of representatives of the reduced primitive quadratic forms of a discriminant $-D$ such that $B_i^2 - 4A_iC_i = -D$, $\gcd(A_i, l) = 1$ and $B_r \equiv B_s$ (mod $2l$) for all $1 \leq r, s \leq h$. For a more formal definition see [27].

Although the construction of $M_{D,l}(x)$ polynomials is explained in [9, 21, 22], the required computation of the primitive forms is not provided. In the following, we provide all the details for computing these forms, which we also used in our implementation. Possibly there are alternative ways to generate the same polynomial $M_{D,l}(x)$ with other, equivalent forms.

For the construction of the polynomials $M_{D,l}(x)$, and according to Theorem 4, the condition $B_r \equiv B_s$ (mod $2l$) can be replaced by the condition $B_i^2 \equiv -D$ (mod $4l$) and because $D \equiv 0$ (mod l), we can write $B_i = l + 2lk_i \equiv l$ (mod $2l$) for an integer $k_i \geq 1$. In particular, $M_{D,l}(x) = \prod_{\tau_Q}(x - m_l^e(\tau_Q))$ where $Q = [A_i, B_i, C_i]$ is a primitive form satisfying the conditions $\gcd(A_i, l) = 1$, $B_i = l + 2lk_i$ and $\tau_Q = \frac{-B_i+\sqrt{-D}}{2A_i}$. The set of forms $[A_i, B_i, C_i]_{1 \leq i \leq h}$ can be computed from the set of the reduced primitive quadratic forms $[\alpha, \beta, \gamma]$ that are used for the construction of $H_D(x)$.

A form $[A_i, B_i, C_i]$ can be computed from a reduced primitive quadratic form $[\alpha, \beta, \gamma]$ using (at most) two transformations from [27–Prop. 3]. The first one transforms a form $[a, b, c]$ to an equivalent (having the same discriminant $-D$) form $[a, b + 2ak, c + bk + ak^2]$ for an integer k and the second transforms a form $[a, b, c]$ to an equivalent form $[a + bn + cn^2, b + 2cn, c]$ for an integer n. In order to compute a form $[A_i, B_i, C_i]$ we first transform a reduced primitive form $[\alpha, \beta, \gamma]$ to a form $[\alpha_1, \beta_1, \gamma_1]$ such that β_1 and γ_1 are divided by l, using the first transformation. This means that we choose an integer k such that $\beta_1 = \beta + 2\alpha k \equiv 0$ (mod l) and $\gamma_1 = \gamma + \beta k + \alpha k^2 \equiv 0$ (mod l). If $\alpha \equiv 0$ (mod l), we just set $\alpha_1 = \gamma$ and $\gamma_1 = \alpha$, and we do not apply the transformation ($\beta_1 = \beta \equiv 0$ (mod l), because $D \equiv 0$ (mod l)). After this transformation, we use the second transformation from [27] to compute the final form $[A_i, B_i, C_i]$ from $[\alpha_1, \beta_1, \gamma_1]$. Thus, $A_i = \alpha_1 + \beta_1 n + \gamma_1 n^2$, $B_i = \beta_1 + 2\gamma_1 n$ and $C_i = \gamma_1$ for an integer n such that $A_i > B_i > C_i$.

It is easy to see why this process yields a form that satisfies the desired conditions. The requirement $A_i > B_i > C_i$ exists because our experiments showed that it is necessary for the proper construction of the polynomial $M_{D,l}(x)$. For example, for $D = 51$ the reduced forms are $[1, 1, 13], [3, 3, 5]$ and the corresponding forms $[A_i, B_i, C_i]$ for $l = 3$ are $[67, 63, 15], [11, 9, 3]$.

The invariants $m_l^e(\tau)$ are related with $j(\tau)$ through the modular equation $\Phi_l(m_l^e(\tau), j(\tau)) = 0$, based on the definitions of $\Phi_l(x, j)$ for the different values of l given in Table 2.

Table 2. Modular functions for different values of l

l	$\Phi_l(x, j)$
3	$(x + 27)(x + 3)^3 - jx$
5	$(x^2 + 10x + 5)^3 - jx$
7	$(x^2 + 13x + 49)(x^2 + 5x + 1)^3 - jx$
13	$(x^2 + 5x + 13)(x^4 + 7x^3 + 20x^2 + 19x + 1)^3 - jx$

Theorem 5. *A polynomial $M_{D,l}(x)$ has h roots modulo p if and only if the equation $4p = u^2 + Dv^2$ has an integer solution and p does not divide the discriminant $\Delta(M_{D,l})$ of the polynomial.*

Proof. It follows the same lines as that of Theorem 1. We know that the class invariants m_l^e generate the Hilbert class field, and therefore Proposition 5.29 from [8] hold. This implies that $M_{D,l}(x)$ has a root modulo p when $4p = u^2 + Dv^2$ has an integer solution, and since $\frac{\mathcal{O}_{H_K}}{p\mathcal{O}_{H_K}}/\mathbb{F}_p$ is Galois, the polynomial $M_{D,l}(x)$ has h distinct solutions modulo p. ☐

The polynomials $M_{D,l}(x)$ can be used in the CM method in a more straightforward way, compared to that of Weber polynomials for the case of prime order elliptic curves. Since $M_{D,l}(x)$ has roots R_M modulo p, we use an algorithm for their computation (for example Berlekamp's algorithm [4]) and then we can compute the roots R_H modulo p of the corresponding Hilbert polynomial $H_D(x)$ from the modular equation $\Phi_l(R_M, R_H) = 0$.

We finally note that the precision required for the construction of the $M_{D,l}(x)$ polynomials is approximately $\frac{1}{l}$H-Prec(D) [9].

5   Implementation and Experimental Results

All of our implementations were made in ANSI C using the (ANSI C) GNUMP [12] library for high precision floating point arithmetic and also for the generation and manipulation of integers of unlimited precision. The implementation includes the construction of the Hilbert, Weber and $M_{D,l}(x)$ polynomials, algorithms for the computation of roots modulo p of a polynomial, algorithms for the computation of a cubic factor of a polynomial modulo p, and of course all the steps of the CM method for the generation of prime order elliptic curves. All implementations and experiments have been carried out on a Pentium III (933 MHz) running Linux and equipped with 256 MB of main memory.

Our experiments first focused on the bit precision and the time requirements needed for the construction of Weber and $M_{D,l}(x)$ polynomials with $D \equiv 3$ (mod 8). We also conducted experiments with Hilbert polynomials and we noticed, as expected, that their construction is much less efficient than the construction of Weber or $M_{D,l}(x)$ polynomials for all values of D and l. For this reason we do not report on these polynomials here (experimental studies regarding Hilbert and other polynomials can be found e.g., in [3, 15]). Concerning Weber polynomials we used discriminants $D \not\equiv 0$ (mod 3). We avoid discriminants $D \equiv 0$ (mod 3) because the precision requirements are greater than those of the case $D \not\equiv 0$ (mod 3). We have considered various values of D and h and report on our experimental results in Figure 1 and Figure 2. We noticed, as the theory dictates, that the precision required for the construction of Weber polynomials $W_D(x)$ is less than the precision required for the construction of $M_{D,l}(x)$ polynomials for all the values of l that we examined (in Section 4 we explained why we consider these particular values of l). Among the $M_{D,l}(x)$ polynomials the least precision is required for the construction of $M_{D,13}(x)$, followed by the construction of $M_{D,7}(x)$, followed by the construction of $M_{D,5}(x)$. The greatest requirements in precision are set by the $M_{D,3}(x)$ polynomials.

The same ordering can be observed in the construction time. For Figure 2 (time in seconds) we used the same values of D as in Figure 1 and also in this figure the differences among the polynomials are very clear. We observed that the time for the construction of $M_{D,l}(x)$ depends not only on the precision requirements of the polynomials, but also on the convergence rate of η-products.

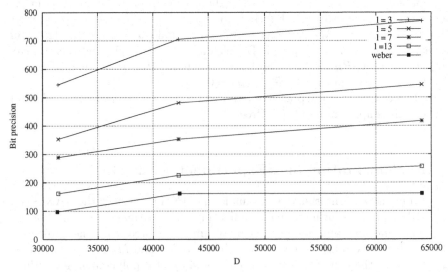

Fig. 1. Bit precision for the construction of class polynomials

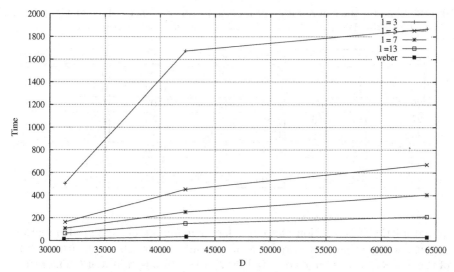

Fig. 2. Time requirements for the construction of class polynomials

The greater the l, the slower the convergence. This is why in Figure 2 the differences do not seem to be analogous with the differences in Figure 1. This favors Weber polynomials, as the η-products in their construction converge faster than any of the $M_{D,l}(x)$ polynomials, making their generation even more efficient.

The coefficients of the Weber polynomials are also smaller than the coefficients of the $M_{D,l}(x)$ polynomials, following the same relative order with precision and time. However, the disadvantage of Weber polynomials is that their degree is three times larger than the degree of the $M_{D,l}(x)$ polynomials. Therefore, the space required for the storage of a Weber polynomial $W_D(x)$ can be larger than the space required for the storage of $M_{D,13}(x)$ or $M_{D,7}(x)$. Actually, it turns out that $M_{D,l}(x)$ polynomials can be even more advantageous when it comes to storage requirements as our experiments showed. Suppose that $M_{D,l}(x) = x^h + M_1 x^{h-1} + ... + M_{h-1}x + M_h$ and h is even. We noticed that every coefficient M_i of $M_{D,l}(x)$ is divisible by l. Moreover, when $l = 13$, then $M_h = 13^{h/2}$ and $\frac{M_{h-i}}{M_i} = 13^{h/2-i}$ for $1 \leq i \leq (h/2 - 1)$. For $l = 7$, $M_h = 7^h$, $\frac{M_{h-i}}{M_i} = 7^{h-2i}$; for $l = 5$, $M_h = 5^{3h/2}$, $\frac{M_{h-i}}{M_i} = 5^{3h/2-3i}$; and finally for $l = 3$ we have $M_h = 3^{3h}$, $\frac{M_{h-i}}{M_i} = 3^{3h-6i}$. Using these properties of the $M_{D,l}(x)$ polynomials, we can reduce the space required for their storage (if someone wants to store them for subsequent use).

This is not the only advantage of $M_{D,l}(x)$ against $W_D(x)$. The large degree of the Weber polynomials is a disadvantage for the time efficiency of the CM method, because the time for finding a cubic factor of the polynomial can be much larger than the time for finding a single root modulo p of a polynomial with three times smaller degree. In Table 3 we report on the time (in seconds) that is

Table 3. Time for the computation of a cubic factor of Weber polynomials and of a linear factor of the $M_{D,l}(x)$ polynomials, together with their construction time

D	h	l	T_W	C_W	T_M	C_M
403	2	13	0.12	0.63	0.01	0.38
1027	4	13	0.40	1.31	0.02	0.36
2035	8	5	1.53	2.35	0.07	1.31
2795	12	13	3.88	3.60	0.13	2.12
4403	20	7	13.12	5.15	0.44	8.71
5603	22	13	16.97	6.94	0.50	8.38
6995	32	5	41.05	9.64	1.72	36.03
22435	32	5	41.05	17.80	1.72	72.94

required for the computation of a cubic factor modulo p of $W_D(x)$, denoted by T_W, and the time that is required for the computation of a linear factor modulo p of the $M_{D,l}(x)$ polynomials denoted by T_M, for various values of l. The prime p has size 160 bits. C_W and C_M is the time required for the construction of the $W_D(x)$ and the $M_{D,l}(x)$ polynomials, respectively. The degree of $W_D(x)$ is $3h$. Note that $C_W + T_W$ (resp. $C_M + T_M$) is the time that mostly dominates and differentiates the use of polynomials (Weber versus $M_{D,l}(x)$) in the CM method, since the time for the other steps of the method is practically independent of the polynomials used.

We observe from Table 3 that $C_W + T_W$ is almost always larger than $C_M + T_M$, implying that the use of Weber polynomials is more time consuming than the use of the $M_{D,l}(x)$ polynomials. However, we also observed that in some cases when D increases, h is of moderate size and $l \in \{3, 5\}$, the construction of the $M_{D,l}(x)$ polynomials may become less efficient (cf. last line of Table 3) and the total time of the CM method with these polynomials can be larger than the time required by the method when their corresponding Weber polynomials are used.

In conclusion, the type of polynomial that one should use depends on the particular application. If the main focus is on time or precision regarding the construction of the polynomials, then Weber polynomials should be preferred. If the focus is on fast and frequent generation of ECs and which implies storage of polynomials for subsequent use in the CM method, then the $M_{D,l}(x)$ polynomials $(l \neq 3)$ must be preferred. Finally, if the class polynomials are computed online with the CM method, then the selection of the proper polynomial depends on the value of D and h. Notice though, that Weber polynomials can be constructed for any value of $D \equiv 3 \pmod 8$, while $M_{D,l}(x)$ polynomial add a restriction for D, demanding that $D \equiv 0 \pmod l$.

References

1. A.O.L. Atkin and F. Morain, Elliptic curves and primality proving, *Mathematics of Computation* 61(1993), pp. 29-67.

2. H. Baier, Elliptic Curves of Prime Order over Optimal Extension Fields for Use in Cryptography, in *Progress in Cryptology* – INDOCRYPT 2001, LNCS Vol. 2247 (Springer-Verlag, 2001), pp. 99-107.

3. H. Baier, Efficient Algorithms for Generating Elliptic Curves over Finite Fields Suitable for Use in Cryptography, PhD Thesis, Dept. of Computer Science, Technical Univ. of Darmstadt, May 2002.

4. E. R. Berlekamp, Factoring polynomials over large finite fields, *Mathematics of Computation* 24(1970), pp. 713-735.

5. I. Blake, G. Seroussi, and N. Smart, *Elliptic curves in cryptography* , London Mathematical Society Lecture Note Series 265, Cambridge University Press, 1999.

6. D. Boneh, B. Lynn, and H. Shacham, Short signatures from the Weil pairing, in *ASIACRYPT 2001*, LNCS 2248, pp. 514-532, Springer-Verlag, 2001.

7. H. Cohen, *A Course in Computational Algebraic Number Theory*, Graduate Texts in Mathematics, **138**, Springer-Verlag, Berlin, 1993.

8. D. A. Cox, *Primes of the form $x^2 + ny^2$*, John Wiley and Sons, New York, 1989.

9. A. Enge and F. Morain, Comparing invariants for class fields of imaginary quadratic fields, in *Algebraic Number Theory* – ANTS V, Lecture Notes in Computer Science Vol. 2369, Springer-Verlag, pp. 252-266, 2002.

10. A. Enge and R. Schertz, Constructing elliptic curves from modular curves of positive genus, Preprint, 2003.

11. S. Galbraith and J. McKee, The probability that the number of points on an elliptic curve over a finite field is prime, *Journal of the London Mathematical Society*, 62(2000), no. 3, pp. 671-684.

12. GNU multiple precision library, edition 3.1.1, September 2000. Available at: http://www.swox.com/gmp.

13. IEEE P1363/D13, *Standard Specifications for Public-Key Cryptography*, 1999. http://grouper.ieee.org/groups/1363/tradPK/draft.html.

14. E. Kaltofen and N. Yui, Explicit construction of the Hilbert class fields of imaginary quadratic fields by integer lattice reduction. Research Report 89-13, Rensselaer Polytechnic Institute, May 1989.

15. E. Konstantinou, Y. Stamatiou, and C. Zaroliagis, On the Efficient Generation of Elliptic Curves over Prime Fields, in *Cryptographic Hardware and Embedded Systems* – CHES 2002, Lecture Notes in Computer Science Vol. 2523, Springer-Verlag, pp. 333-348, 2002.

16. E. Konstantinou, Y.C. Stamatiou, and C. Zaroliagis, On the Construction of Prime Order Elliptic Curves, in *Progress in Cryptology* – INDOCRYPT 2003, Lecture Notes in Computer Science Vol. 2904, Springer-Verlag, pp. 309-322, 2003.

17. G.J. Lay and H. Zimmer, Constructing Elliptic Curves with Given Group Order over Large Finite Fields, in *Algorithmic Number Theory* – ANTS-I, Lecture Notes in Computer Science Vol. 877, Springer-Verlag, pp. 250-263, 1994.

18. A. J. Menezes, T. Okamoto and S. A. Vanstone, Reducing elliptic curve logarithms to a finite field, *IEEE Trans. Info. Theory*, 39(1993), pp. 1639-1646.

19. A. Miyaji, M. Nakabayashi, and S. Takano, Characterization of Elliptic Curve Traces under FR-reduction, in *International Conference on Information Security and Cryptology* – ICISC 2000, Lecture Notes in Computer Science Vol. 2015, Springer-Verlag, pp. 90-108, 2001.

20. A. Miyaji, M. Nakabayashi, and S. Takano, New explicit conditions of elliptic curve traces for FR-reduction, *IEICE Transactions on Fundamentals*, E84-A(5):1234-1243, 2001.

21. F. Morain, Modular curves and class invariants, Preprint, June 2000.

22. F. Morain, Computing the cardinality of CM elliptic curves using torsion points, Preprint, October 2002.
23. Y. Nogami and Y. Morikawa, Fast generation of elliptic curves with prime order over $F_{p^{2c}}$, in *Proc. of the International workshop on Coding and Cryptography*, March 2003.
24. G. C. Pohlig and M. E. Hellman, An improved algorithm for computing logarithms over $GF(p)$ and its cryptographic significance, *IEEE Trans. Info. Theory*, 24 (1978), pp. 106-110.
25. T. Satoh and K. Araki, Fermat quotients and the polynomial time discrete log algorithm for anomalous elliptic curves, *Comm. Math. Univ. Sancti Pauli*, 47(1998), pp. 81-91.
26. E. Savaş, T.A. Schmidt, and Ç.K. Koç, Generating Elliptic Curves of Prime Order, in *Cryptographic Hardware and Embedded Systems* – CHES 2001, LNCS Vol. 2162 (Springer-Verlag, 2001), pp. 145-161.
27. R. Schertz, Weber's class invariants revisited, *Journal de Théorie des Nombres de Bordeaux* 4, pp. 325-343, 2002.
28. R. Schoof, Counting points on elliptic curves over finite fields, *J. Theorie des Nombres de Bordeaux*, 7(1995), pp.219-254.
29. M. Scott and P. S.L.M. Barreto, Generating more MNT elliptic curves, Cryptology ePrint Archive, Report 2004/058, 2004.
30. J. H. Silverman, *The Arithmetic of Elliptic Curves*, Springer-Verlag, GTM 106, 1986.
31. I. Stewart, *Galois Theory*, Third Edition, Chapman & Hall/CRC, Boca Raton, FL, 2004.
32. I. Stewart and D. Tall, *Algebraic Number Theory*, Second Edition, Chapman & Hall, London, 1987.

New Families of Hyperelliptic Curves with Efficient Gallant-Lambert-Vanstone Method

Katsuyuki Takashima

Information Technology R&D Center,
Mitsubishi Electric Corporation,
5-1-1, Ofuna, Kamakura, Kanagawa 247-8501, Japan
takasima@iss.isl.melco.co.jp

Abstract. The Gallant-Lambert-Vanstone method [14] (GLV method for short) is a scalar multiplication method for elliptic curve cryptography (ECC). In WAP WTLS[47], SEC 2[42], ANSI X9.62[1] and X9.63[2], several domain parameters for applications of the GLV method are described. Curves with those parameters have efficiently-computable endomorphisms. Recently the GLV method for hyperelliptic curve (HEC) Jacobians has also been studied.

In this paper, we discuss applications of the GLV method to curves with real multiplication (RM). It is the first time to use RM in cryptography. We describe the general algorithm for using such RM, and we show that some genus 2 curves with RM have enough effciency to be used in the GLV method as in the previous CM case.

Keywords: Public Key Cryptography, Elliptic Curve Cryptography, Hyperelliptic Curve Cryptography, Scalar Multiplication, GLV Method.

1 Introduction

Presently elliptic curve cryptography (ECC) and small genus (≤ 3) hyperelliptic curve cryptography (HECC) arouse much interest because of its higher efficiency in comparison with RSA. Thus efforts to improve the efficiency of those cryptosystems have been continued.

Widely used elliptic curves for ECC are classified into binary-field ones and prime-field ones. Moreover in the both categories, there exist randomly chosen ones and more efficiently computable ones ([1, 2, 42, 47]). For binary-field ones, very special EC parameters have been known as Koblitz curves for great efficiency for a long time ([24]). They use the Frobenius endomorphism for efficient computation.

In case of prime-field parameters, a method of efficient computation in [14] (Gallant-Lambert-Vanstone method, GLV method for short) can be used. Some efficiently-computable endomorphisms are used also in this case. In SEC 2[42], such curves are also called "Koblitz curves" like binary-field ones. Here we call such parameters over non-binary fields "GLV$_{nbf}$ parameters". Some GLV$_{nbf}$ EC parameters are included in the parameter list of WAP WTLS[47], SEC 2[42], ANSI X9.62[1] and X9.63[2] etc. All those are of type 2 in Table 1 in section 3.

C. Park and S. Chee (Eds.): ICISC 2004, LNCS 3506, pp. 279–295, 2005.
© Springer-Verlag Berlin Heidelberg 2005

Recently genus 2 HECC has got comparable perfomance to ECC of the same security level ([3, 26, 29]). In [38, 27], more efficient HEC parameters (not so-called Koblitz curves) are studied in the binary-field case.

So far only CM (complex multiplication) HECs are considered for HEC GLV$_{nbf}$ parameters like in the EC case. However, other endomorphisms can be used for HEC of genus > 1. Corresponding studies have not yet been reported. Hence in this paper we describe fast endomorphisms other than CM on Jacobian of genus 2 curves to enlarge the applicability of the GLV method. They are called real multiplication (RM), and we report that those are efficient enough to be used in the GLV method as in the CM case. It is the first time to use RM in cryptography.

As the RM is not induced from an automorphism of the curve, the calculation is new, and explained in section 4. The descriptions and analysis of the algorithm are divided into several subsections. Firstly, in subsection 4.1, we describe the algorithm using the point-sum representation of divisors for a curve in a 3-parameter family. Based on that, in subsection 4.2, we describe the general algorithm without solving quadratic equations (Algorithm 1). In subsection 4.3, due to Hashimoto [18] (see also [17]), we simplify the algorithm on a curve $C_{\beta,\gamma}$ (with 2 parameters β, γ), and show that an application of the endomorphism takes at most 1.5 times as long as a doubling on Jacobians. Thus in subsection 4.4, we show the endomorphism can be used in the GLV method as in the CM case. In subsection 4.5, through example RM curves, we point out that they don't have some special properties of previous CM GLV curves.

We treat $\frac{-1\pm\sqrt{5}}{2}$-multiple endomorphisms in [6, 17, 20] mainly in this paper, and we call curves with those endomorphisms on their Jacobians "Brumer-Hashimoto (BH) curves" (with 3 parameters) and "Mestre-Hashimoto (MH) curves" (with 2 parameters). For details, see section 4.

In this paper, firstly, we briefly review some necessary basic notions in section 2, and previous examples of GLV curves in section 3. As is mentioned above, we investigate BH (and MH) curves and their endomorphisms on Jacobians in section 4. And in section 5, we see (a variant of) the Algorithm 1 is also applicable for curves with $\sqrt{2}$ multiplication treated in [4, 21].

2   Basic Notions

We deal with a genus 2 curve $C : Y^2 = f(X)$ defined over a finite field $\mathbb{F}_q$ of order q where the characteristic p of $\mathbb{F}_q$ is not 2.

2.1   Mumford Representation

In this subsection, we review the representation of divisors on a genus g curve $C : Y^2 = f(X)$, so-called Mumford representation. A reduced divisor is represented by a pair of polynomials $(u(X), v(X)) \in \overline{\mathbb{F}}_q[X]^2$ s.t. $u(X)|v(X)^2 - f(X)$, $\deg v(X) < \deg u(X) \le g$ and $u(X)$ is monic. Jac$_C$ consists of all reduced divisors, and $\mathrm{Jac}_C(\mathbb{F}_q)$ consists of all reduced divisors $(u(X), v(X))$ whose coefficients are in $\mathbb{F}_q$. We call this group $\mathrm{Jac}_C(\mathbb{F}_q)$ Jacobian group in this paper.

2.2   GLV Method

Here we survey the GLV method for elliptic curve groups $E(\mathbb{F}_q)$ or Jacobian groups $\mathrm{Jac}_C(\mathbb{F}_q)$ (where the minimal polynomial of the below ψ is quadratic). We assume E or Jac_C has an efficiently-computable endomorphiam ψ. Let the restriction of ψ to a cyclic subgroup G ($\subset E(\mathbb{F}_q)$ or $\mathrm{Jac}_C(\mathbb{F}_q)$) correspond to the multiplications by $\lambda(\in [1, n-1]$, where $n := \sharp G$). For a point P (or a divisor class $[\mathcal{D}]$) $\in G$ to calculate kP (or $k[\mathcal{D}]$), firstly we calculate k_1, k_2 s.t.

$$k = k_1 + k_2\lambda \quad k_1, k_2 = O(\sqrt{n}) . \tag{1}$$

And applying "simultaneous multiple points scalar multiplication" (see [44] for example) to $P, \lambda P(= \psi(P))$ (or $[\mathcal{D}], \lambda[\mathcal{D}]$), we obtain kP (or $k[\mathcal{D}]$) respectively. Therefore, on average a speedup by a factor of 16/9 is obtained over the best general methods for scalar multiplication. We call this GLV method.

It is clear that if we can find smaller k_1, k_2, we can calculate scalar multiplication more efficiently. Therefore, in [14, 37, 25, 40] they proposed several method of calculating "smaller" k_1, k_2 (in other words, decomposition of k).

Ciet et al. [9] proposed a speedup using ψ-adic expansion of scalar instead of ordinary binary expansion. And, in [10] they also proposed countermeasures against DPA (Differential Power Analysis) in using the GLV method.

3   Previous Examples of GLV Curves (Not Using Frobenius Endomorphisms)

Previously proposed fast endomorphisms ψ for the GLV method except for Frobenius endomorphisms are divided into 3 types as in Table 1. Especially in the case of genus 1 and 2, those have the following special properties.

1. A (hyper-)elliptic curve of type 1 has an automorphism ψ. In particular in the genus 2 case (i.e. $C : Y^2 = X^5 + a_2X^3 + a_4X$), its Jacobian varieties are isogeneous to a product of elliptic curves over the degree 6 extension field of $\mathbb{F}_q$ (see [22]). We can verify it by using the criterion in [15] section 4 (see [5, 13, 28, 43] also).
2. A Curve of type 2 has also an automorphism ψ. There exists only one curve of genus 1 (or 2) of type 2 up to isomorphism over $\overline{\mathbb{F}}_q$ respectively.
3. CM ψ's of elliptic curves of type 3 are not automorphisms of curves. However, there exist only a few isomorphism classes of the curves.

That is, previously proposed ψ's of genus 1 or 2 have either of two special properties that "The Jacobian varieties are split (over the degree 6 extension field)" or "There exist only a few isomorphism (isogeny) classes".

Table 1. Examples of endomorphisms ψ except for Frobenius end (CM)

type	source	curve equation	$\mathbb{Q}$-coeff. char. poly. of ψ
1.	example 3 in [14]	$Y^2 = X^3 + aX$	$T^2 + 1$
	example 3 in [37]	$Y^2 = X^5 + aX$	$T^4 + 1$
	example 4 in [37]	$Y^2 = X^7 + aX$	$T^4 - T^2 + 1$
	example 2 in [37]	$Y^2 = X^{2g+1} + a_2 X^{2g-1} +$ $\cdots a_{2g-2}X^3 + a_{2g}X$	$T^2 + 1$
2.	example 4 in [14]	$Y^2 = X^3 + b$	$T^2 + T + 1$
	example 5 in [37] (see also [7, 41])	$Y^2 = X^{2g+1} + b$	$T^{2g} + T^{2g-1} + \cdots + T + 1$
3.	example 5 in [14]	$Y^2 = X^3 - \frac{3}{4}X^2 - 2X - 1$	$T^2 - T + 2$
	example 6 in [14]	$Y^2 = 4X^3 - 30X - 28$	$T^2 + 2$

In this paper, we propose the use of RM on genus 2 Jacobian in the GLV method. Those provide various absolutely simple (non-isomorphic) Jacobian varieties applicable for the GLV method (for comparison of computational amount with the previous curves, see section 4.4).

4   Brumer-Hashimoto (BH) Curves and Mestre-Hashimoto (MH) Curves

At first in this section, we describe a HEC of genus 2 whose Jacobian admits the action of the ring $\mathbb{Z}[\frac{1+\sqrt{5}}{2}]$ as its endomorphism ring. The curves have been studied by G. Humbert, J. -F. Mestre, and K. Hashimoto [17, 18, 20, 23, 32, 33] etc. We describe the explicit calculation method of a RM ψ on the Jacobian group.

Here we briefly summarize (a part of) previous researches of the genus 2 curves with the above property from number-theoretical viewpoint. Mestre obtained a 2-parameter family ($\{C_{\beta,\gamma}\}$ in subsection 4.3)[32, 33] based on classical works by Humbert [23] etc. Brumer has described a 3-parameter family which contains the above family as a subfamily ([6], see also [8] p.164 and the last comment in [32]). Hashimoto reconstructed curves of this family as the normal form $C_0(a, b, c)$ (equation (2)) based on the descent theory in geometric Galois theory. He also obtained explicit formulas (3) and (4) for the RM ([20]), and pointed out that they are very simple for $C_{\beta,\gamma}$ ([17, 18], see formulas (9) in subsection 4.3). Therefore, we call a general curve in the family $\{C_{\beta,\gamma}\}$ "Mestre-Hashimoto (MH) curve" and that in $\{C_0(a, b, c)\}$ "Brumer-Hashimoto (BH) curve" in this paper.[1] It is the first time to use such RM curves in cryptography.

---

[1] As is shown in [19, 20], the family $\{C_0(a, b, c)\}$ contains a subfamily of curves with split Jacobian (see also section 3). In this paper, we don't call them BH curves.

BH curves are given by the following equation. Those are parametrized by constants $a, b, c \in \mathbb{F}_q$ that are given by some rational expressions $a = a(s,t,z), b = b(s,t,z), c = c(s,t,z)$ of $s, t, z \in \mathbb{F}_q$.

$$C_0(a,b,c) : Y^2 = X^6 - (4 + 2b + 3c)X^5 + (2 + 2b + b^2 - ac)X^4$$
$$- (6 + 4a + 6b - 2b^2 + 5c + 2ac)X^3 + (1 + b^2 - ac)X^2 + (2 - 2b)X + c + 1 . \quad (2)$$

On that curve, an algebraic correspondence $X_\psi \subset C_0(a,b,c) \times C_0(a,b,c)$ is described as follows[20]. In the following equations (3) and (4) of X_ψ, coordinates of $C_0(a,b,c) \times C_0(a,b,c)$ are given by (X_1, Y_1, X_2, Y_2). We define $J_s(X) := (X - s)(sX + s - 1)((s+1)X - 1)$ for the above constant $s \in \mathbb{F}_q$.

$$A(X_1, X_2) := \tau_1 X_1^2 X_2^2 + \tau_2 X_1 X_2 (X_1 + X_2)$$
$$+ \tau_3 (X_1 + X_2)^2 + \tau_4 X_1 X_2 + \tau_5 (X_1 + X_2) + \tau_6 = 0 , \quad (3)$$
$$J_s(X_2) Y_1 = J_s(X_1) Y_2 . \quad (4)$$

Here τ_ℓ $(\ell = 1, \ldots, 6)$ are constants given by some polynomials of s, t, z. The endomorphism ψ on $\mathrm{Jac}_{C_0(a,b,c)}$ defined by X_ψ satisfies the equation $\psi^2 + \psi - 1 = 0$ (1 : identity map) [20].

Therefore, $\psi (\in \mathrm{End}(\mathrm{Jac}_{C_0(a,b,c)}))$ is a fundamental unit in the real quadratic field, that is an automorphism on the Jacobian. When genus $g > 1$ and finite field characteristic $p > g + 1$, except for some very special cases, the order of automorphism group of HEC is less than or equal to $84(g - 1)$ ([11]). Therefore, we can verify that the RM ψ is not induced by any automorphism on HEC in general. In fact, for HEC C_1 in Table 5 in subsection 4.5, the order of ψ (in $\mathbb{F}_n^*$) is 10442112734830750699l (> 84). For other curves in the Table, the order > 84 also. This is a new property of the RM ψ compared to the previous CM's for cryptographic use.

We consider the equation (2) whose RHS has a zero point in $\mathbb{F}_q$ only. Therefore, (2) is transformed to $Y^2 = f(X)$ whose RHS has degree 5. We use it as defining equation of $C_0(a,b,c)$. Thus equations (3) and (4) are transformed to $A^*(X_1, X_2) = 0$ and $J_s^*(X_2) Y_1 = J_s^*(X_1) Y_2$ respectively. Here $A^*(X_1, X_2)$ is a polynomial with $\deg_{X_1} A^* \leq 2$ and $\deg_{X_2} A^* \leq 2$, and $J_s^*(X)$ is a rational expression whose numerator and denominator have degree ≤ 3 respectively.

In the following, for a general element $(u_1(X), v_1(X))$ (i.e. $\deg u_1(X) = 2$, $\deg v_1(X) = 1$) in $\mathrm{Jac}_{C_0(a,b,c)}(\mathbb{F}_q)$ s.t.

$$u_1(X) = X^2 + u_{1,1}X + u_{1,2} , \quad v_1(X) = v_{1,1}X + v_{1,2} \quad (5)$$

where $u_{1,\ell}, v_{1,\ell} \in \mathbb{F}_q$ $(\ell = 1, 2)$, we show ψ calculation algorithms. In subsection 4.1, the algorithm using point-sum representation of diviors is described. And in subsection 4.2, that using Mumford representation (subsection 2.1) without solving several quadratic equations $A^*(x_{1,1}, X_2) = 0$, $A^*(x_{1,2}, X_2) = 0$ etc. are described.

4.1   Algorithm for ψ over $\overline{\mathbb{F}}_q$ (Using Point-Sum Representation)

ψ calculation algorithm on $\mathrm{Jac}_C(\overline{\mathbb{F}}_q)$ is as follows. In this subsection, we don't treat the last reduction (in a linear equivalence class) step of the divisors. Here we describe the calculation algorithm from a divisor $\mathcal{D}_1$ to the divisor $\mathcal{D}_2$ determined by the algebraic correspondence X_ψ.

$$\psi: \quad \mathrm{Jac}_C(\overline{\mathbb{F}}_q) \quad \longrightarrow \quad \mathrm{Jac}_C(\overline{\mathbb{F}}_q)$$
$$\left[\mathcal{D}_1 := P_{1,1} + P_{1,2}\right] \mapsto \left[\mathcal{D}_2 := (P_{2,1} + P_{2,2}) + (P_{2,3} + P_{2,4})\right] \tag{6}$$

Here we set $P_{i,j} := (x_{i,j}, y_{i,j})$ $(x_{i,j}, y_{i,j} \in \overline{\mathbb{F}}_q, i = 1, 2, j = 1, \ldots, 4)$. $x_{2,1}, x_{2,2}$ are two roots of $A^*(x_{1,1}, X_2) = 0$ in variable X_2, and $x_{2,3}, x_{2,4}$ are two roots of $A^*(x_{1,2}, X_2) = 0$ in variable X_2. And $y_{2,j} = \frac{y_{1,1}}{J_s^*(x_{1,1})} J_s^*(x_{2,j})$ for $j = 1, 2$, and $y_{2,j} = \frac{y_{1,2}}{J_s^*(x_{1,2})} J_s^*(x_{2,j})$ for $j = 3, 4$. When $\mathcal{D}_2$ is represented by two polynomials $u_2(X)$ and $v_2(X)$, $u_2(X) = \prod_{j=1,\ldots,4}(X - x_{2,j})$ and $v_2(X)$ is the unique cubic polynomial such that $v_2(x_{2,j}) = y_{2,j}(j = 1, \ldots, 4)$. If $[\mathcal{D}_1] \in \mathrm{Jac}_C(\mathbb{F}_q)$, as is shown in subsection 4.2, $[\mathcal{D}_2] \in \mathrm{Jac}_C(\mathbb{F}_q)$. Setting $d(X) := \frac{v_1(X)}{J_s^*(X)}$, the cubic polynomial whose value at $x_{2,1}, x_{2,2}$ is $d_1(:= d(x_{1,1}))$, and that at $x_{2,3}, x_{2,4}$ is $d_2(:= d(x_{1,2}))$ is represented as follows using Lagrangian interpolation formula (i.e. $L(X)$ is the cubic polynomial s.t. $d_1 = L(x_{2,1}), d_1 = L(x_{2,2})$, $d_2 = L(x_{2,3}), d_2 = L(x_{2,4})$).

$$L(X) := d_1 H_1 + d_2 H_2 \quad \text{s.t.}$$
$$H_1(X) := \prod_{j\in\{2,3,4\}}\{(X - x_{2,j})/(x_{2,1} - x_{2,j})\} + \prod_{j\in\{1,3,4\}}\{(X - x_{2,j})/(x_{2,2} - x_{2,j})\}, \tag{7}$$
$$H_2(X) := \prod_{j\in\{1,2,4\}}\{(X - x_{2,j})/(x_{2,3} - x_{2,j})\} + \prod_{j\in\{1,2,3\}}\{(X - x_{2,j})/(x_{2,4} - x_{2,j})\}.$$

At last, we obtain $v_2(X) = L(X) J_s^*(X) \bmod u_2(X)$.

4.2   Algorithm for ψ over $\mathbb{F}_q$ (Using Mumford Representation)

In this subsection, we describe the same calculation in the subsection 4.1 without solving several quadratic equations (Algorithm 1). In Algorithm 1, a subroutine ($ExpFundSym_2$) that rewrites symmetric expressions to expressions in terms of fundamental symmetric polynomials is used several times. Hence we define the subroutine $ExpFundSym_2$ as follows.

Inputs are a symmetric rational expression $\mu = \mu(X_1, X_2)$ and rational expressions $\omega_1(X_3)$ and $\omega_2(X_3)$. By using relations $\sigma_1 = X_1 + X_2$ and $\sigma_2 = X_1 X_2$, we transform $\mu(X_1, X_2)$ to the rational expression $\nu(\sigma_1, \sigma_2)$ w.r.t. σ_1, σ_2. Next we substitute ω_1, ω_2 to σ_1, σ_2 in $\nu(\sigma_1, \sigma_2)$ respectively. That is a rational expression $\xi(X_3)$ of X_3, which is the output of $ExpFundSym_2$ (i.e. $\xi(X_3) = \nu(\omega_1, \omega_2)$ $= ExpFundSym_2(X_1, X_2, X_3, \mu, \omega_1, \omega_2)$).

Example. When $\mu(X_1, X_2) = X_1^2 + X_2^2$, $\omega_1 = X_3 + 1$ and $\omega_2 = X_3 - 1$, we calculate $ExpFundSym_2(X_1, X_2, X_3, \mu, \omega_1, \omega_2)$ as follows.

At first, $\nu(\sigma_1, \sigma_2) = \sigma_1^2 - 2\sigma_2$. By substituting ω_1 and ω_2, $\nu(\omega_1, \omega_2) = \omega_1^2 - 2\omega_2$ $= (X_3 + 1)^2 - 2(X_3 - 1) = X_3^2 + 3$. Therefore, $\xi(X_3) = X_3^2 + 3$ is the output of $ExpFundSym_2(X_1, X_2, X_3, \mu, \omega_1, \omega_2)$.

Using the $ExpFundSym_2$, we give the calculation algorithm from a divisor $\mathcal{D}_1 = (u_1(X), v_1(X))$ in the formula (5) to $\mathcal{D}_2$ in the diagram (6).

Algorithm 1 RM ψ on $\mathrm{Jac}_{C_0(a,b,c)}$ (without linear equivalence reduction)

Input : $\mathcal{D}_1 = (u_1(X), v_1(X)) \in \mathrm{Jac}_{C_0(a,b,c)}(\mathbb{F}_q)$
Output : $\mathcal{D}_2 = (u_2(X), v_2(X)) \in \mathrm{Jac}_{C_0(a,b,c)}(\mathbb{F}_q)$
[**Calculation of $u_2(X)$**]
1: $\tilde{u}_2(X_{1,1}, X_{1,2}, X_2) \leftarrow A^*(X_{1,1}, X_2)A^*(X_{1,2}, X_2)$
2: Using notation $\tilde{u}_2(X_{1,1}, X_{1,2}, X_2) = \sum_{\ell=0,\dots,4} \tilde{\rho}_{4-\ell}(X_{1,1}, X_{1,2})X_2^\ell$,
   $\rho_\ell \leftarrow ExpFundSym_2(X_{1,1}, X_{1,2}, \cdot, \tilde{\rho}_\ell, -u_{1,1}, u_{1,2})$ $(\ell = 0, \dots, 4)$
3: $u_2(X) \leftarrow X^4 + \sum_{\ell=0,\dots,3}(\rho_{4-\ell}/\rho_0)X^\ell$
[**Calculation of $v_2(X)$**]
4: Using notation $A^*(X_{1,j}, X_2) = \alpha_0(X_{1,j})X_2^2 + \alpha_1(X_{1,j})X_2 + \alpha_2(X_{1,j})$ $(j = 1, 2)$,
   set $\delta_1(X_{1,j}) \leftarrow -\frac{\alpha_1(X_{1,j})}{\alpha_0(X_{1,j})}$, $\delta_2(X_{1,j}) \leftarrow \frac{\alpha_2(X_{1,j})}{\alpha_0(X_{1,j})}$ $(j = 1, 2)$.
5: $\tilde{H}_1(X, X_{1,1}, X_{2,3}, X_{2,4}) \leftarrow ExpFundSym_2(X_{2,1}, X_{2,2}, X_{1,1}, H_1, \delta_1(X_{1,1}), \delta_2(X_{1,1}))$.
6: $\tilde{\tilde{H}}_1(X, X_{1,1}, X_{1,2}) \leftarrow ExpFundSym_2(X_{2,3}, X_{2,4}, X_{1,2}, \tilde{H}_1, \delta_1(X_{1,2}), \delta_2(X_{1,2}))$.
7: Setting $\tilde{L}(X, X_{1,1}, X_{1,2}) \leftarrow d(X_{1,1})\tilde{\tilde{H}}_1(X, X_{1,1}, X_{1,2}) + d(X_{1,2})\tilde{\tilde{H}}_1(X, X_{1,2}, X_{1,1})$,
   $L(X) \leftarrow ExpFundSym_2(X_{1,1}, X_{1,2}, X, \tilde{L}, -u_{1,1}, u_{1,2})$.
8: $v_2(X) \leftarrow L(X)J_s^*(X) \bmod u_2(X)$.

[**Calculation of $u_2(X)$**]

By substituting $X_{1,1}$ and $X_{1,2}$ to X_1 in the LHS of the equation (3), we obtain two quadratic polynomials $A^*(X_{1,1}, X_2)$ and $A^*(X_{1,2}, X_2)$ w.r.t. X_2 respectively. We set $\tilde{u}_2(X_{1,1}, X_{1,2}, X_2) := A^*(X_{1,1}, X_2)A^*(X_{1,2}, X_2)$. Then $\tilde{u}_2$ is symmetric w.r.t. $X_{1,1}$ and $X_{1,2}$. Therefore, defining $\tilde{\rho}_{4-\ell}(X_{1,1}, X_{1,2})$ as in Algorithm 1 step 2 (i.e. $\tilde{u}_2 = \sum_{\ell=0,\dots,4}\tilde{\rho}_{4-\ell}(X_{1,1}, X_{1,2})X_2^\ell$), these polynomials $\tilde{\rho}_{4-\ell}(X_{1,1}, X_{1,2})$'s are symmetric. Therefore, for $\ell = 0, \dots, 4$, we can set $\rho_\ell := ExpFundSym_2$ $(X_{1,1}, X_{1,2}, \cdot, \tilde{\rho}_\ell, -u_{1,1}, u_{1,2})$ $(\in \mathbb{F}_q)$. Finally we have $u_2(X) = X^4 + \sum_{\ell=0,\dots,3}$ $(\rho_{4-\ell}/\rho_0) X^\ell$ (here, the numerators and denominators of ρ_ℓ are represented by polynomials of $u_{1,1}$, $u_{1,2}$ of degree at most 2).

[**Calculation of $v_2(X)$**]

We substitute indeterminates $X_{2,j}$ $(j = 1, \dots, 4)$ to variables $x_{2,j}$ $(j = 1, \dots, 4)$ in H_r $(r = 1, 2)$ in the formulas (7) respectively. Initially we set H_r the obtained rational expressions $H_r(X, X_{2,1}, X_{2,2}, X_{2,3}, X_{2,4})$ respectively. The total degrees w.r.t. $X_{2,j}$ of the numerators and denominators of H_r are at most 5.

Both H_r's $(r = 1, 2)$ are symmetric w.r.t. $X_{2,1}$ and $X_{2,2}$ respectively, and also symmetric w.r.t. $X_{2,3}$ and $X_{2,4}$ respectively. Hence they are expressed by polynomials of coefficients of polynomials $A^*(X_{1,1}, X_2)$, $A^*(X_{1,2}, X_2)$ of X_2. In the following, $\delta_\ell(X_{1,j}) \leftarrow \frac{\alpha_\ell(X_{1,j})}{\alpha_0(X_{1,j})}$ $(j = 1, 2)$. Next, for $r = 1, 2$, we calculate

$$\tilde{H}_r \leftarrow ExpFundSym_2(X_{2,1}, X_{2,2}, X_{1,1}, H_r, \delta_1(X_{1,1}), \delta_2(X_{1,1})),$$

$$\tilde{\tilde{H}}_r \leftarrow ExpFundSym_2(X_{2,3}, X_{2,4}, X_{1,2}, \tilde{H}_r, \delta_1(X_{1,2}), \delta_2(X_{1,2})).$$

However, by the permutation (13)(24) (product of two transpositions) of indices j in the formulas (7), $H_1(X)$ changes to $H_2(X)$ and vice versa. Hence $\tilde{H}_1(X, X_{1,2}, X_{1,1}) = \tilde{H}_2(X, X_{1,1}, X_{1,2})$. Consequently, as in Algorithm 1, we calculate $\tilde{H}_1(X, X_{1,1}, X_{1,2})$ only. Thus, as $\tilde{L}$ in Algorithm 1 is invariant under exchange between $X_{1,1}, X_{1,2}$ (two roots of $u_1(X) = 0$), coefficients of $\tilde{L}$ are expressed by polynomials of coefficients of $u_1(X)$. At last, we obtain $v_2(X)$ as $L(X)J_s^*(X) \bmod u_2(X)$.

We investigate the degrees of the above polynomials. The degrees of the numerators and denominators of δ_ℓ's are at most 2. Therefore, the degrees w.r.t. $X_{1,1}, X_{1,2}$ of the numerators and denominators of $\tilde{H}_r$ are at most 20. As the degrees of the numerators and denominators of $d(X)$ are at most 4, the degrees w.r.t. $u_{1,1}, u_{1,2}$ of the numerators and denominators of L are at most 24. Those w.r.t. $v_{1,1}, v_{1,2}$ are at most 1.

[Efficient Implementation]

In the above algorithms, we have shown the feasibility of calculation of the RM ψ without solving quadratic equations. However, the above degree estimates show that the direct implementation of Algorithm 1 seems inefficient.

Practically, based on Algorithm 1, we should take an "explicit formulas" approach like [26, 29]. That is, by writing up the formulas on finite field operation level, we can reduce the amount of the calculation of ψ more. In [26, 29], we see that explicit formulas (implementations) for the main (or dominant) cases are important practically. Similarly, in case of the calculation of ψ, the analysis of the main case is important. Hence, in the next subsection, for a subfamily $\{C_{\beta,\gamma}\}$ (with simpler equations of the algebraic correspondence) of $\{C_0(a, b, c)\}$, such main case is analysed.

4.3   Mestre-Hashimoto (MH) Curves : BH Curves with More Efficient Endomorphisms

In this subsection, specializing parameters a, b, c of $\{C_0(a, b, c)\}$, we obtain a subfamily $\{C_{\beta,\gamma}\}$ ([17, 18, 20, 32, 33]) with much more efficiently computable ψ than those used in the previous subsections. We call the general curve "Mestre-Hashimoto (MH) curve". The algorithms for the dominant case are described in Tables 2 and 3.

We substitute -1 to the parameter c in the formula (2). And we divide the obtained RHS of the formula (2) by X. Next we substitute $\frac{2-\beta}{2}$ to the parameter b and $-\frac{\beta^2}{4}+2\beta+\gamma-4$ to the parameter a in the obtained degree 5 polynomial. By using the polynomial, we obtain a parametrized genus 2 HEC $C_{\beta,\gamma}$ [17, 18, 32, 33]. Here $\beta(\neq 0), \gamma \in \mathbb{F}_q$ and "the discriminant of the RHS" $\neq 0$.

$$C_{\beta,\gamma} : Y^2 = \beta X^5 - (\beta + \gamma - 3)X^4 + (\beta^2 - 3\beta + 5 - 2\gamma)X^3 - \gamma X^2 + (\beta - 3)X - 1 . \quad (8)$$

For cryptographic purposes, we consider only the case that β has a square root in $\mathbb{F}_q$ in the following. Then the RHS can be transformed to a monic poynomial.

Table 2. Algebraic correspondence calculation algorithm

Input : $u_1(X)$, $v_1(X)$ $(\in \mathbb{F}_q[X])$

Output : $u_2(X)$, $v_2(X)$ $(\in \mathbb{F}_q[X])$

Step	Expression	Cost
1	$w_1 \leftarrow u_{1,1}^2,\ w_2 \leftarrow u_{1,1}u_{1,2},\ w_3 \leftarrow u_{1,2}^2,\ w_4 \leftarrow u_{1,1}v_{1,1},$ $w_5 \leftarrow u_{1,1}v_{1,2},\ w_6 \leftarrow u_{1,2}v_{1,1},\ w_7 \leftarrow u_{1,2}v_{1,2},\ w_8 \leftarrow \beta u_{1,1},$ $w_9 \leftarrow \beta u_{1,2},\ w_{10} \leftarrow \beta v_{1,1}$	2S,8M
2	$D_u \leftarrow w_9^2,\ D_v \leftarrow w_1 - 2w_2 + w_3 - u_{1,1} - w_9 + u_{1,2},\ D_{u,v} \leftarrow D_u D_v,$ $D_{u,v}^{\text{inv}} \leftarrow 1/D_{u,v},\ D_u^{\text{inv}} \leftarrow D_v D_{u,v}^{\text{inv}},\ D_v^{\text{inv}} \leftarrow D_u D_{u,v}^{\text{inv}},\ w = w_5 - w_6$	1I,1S,3M
3	$u_{2,1}^* \leftarrow w_8(u_{1,1} + w_9 - u_{1,2}) - 2w_9,\ u_{2,1} \leftarrow D_u^{\text{inv}} u_{2,1}^*$	2M
4	$u_{2,2}^* \leftarrow (w_8 - w_9 + \beta - 1)u_{1,1} + \beta w_9 - 4w_9 + u_{1,2} + 1,\ u_{2,2} \leftarrow D_u^{\text{inv}} u_{2,2}^*$	3M
5	$u_{2,3}^* \leftarrow w_8 - 2u_{1,1} - 2w_9 + 2u_{1,2} + 2,\ u_{2,3} \leftarrow D_u^{\text{inv}} u_{2,3}^*$	1M
6	$u_{2,4}^* \leftarrow -u_{1,1} + u_{1,2} + 1,\ u_{2,4} \leftarrow D_u^{\text{inv}} u_{2,4}^*$	1M
7	$v_{2,1}^* \leftarrow w_9(w - v_{1,2}),\ v_{2,1} \leftarrow D_v^{\text{inv}} v_{2,1}^*$	2M
8	$w^* \leftarrow (-w_{10} + 2v_{1,1} - v_{1,2})u_{1,2},\ v_{2,2}^* \leftarrow w_8 w - w_4 + w^* + v_{1,2},$ $v_{2,2} \leftarrow D_v^{\text{inv}} v_{2,2}^*$	3M
9	$v_{2,3}^* \leftarrow -w_4 + (\beta - 1)w_5 + w^* + w_6 + 2v_{1,2},\ v_{2,3} \leftarrow D_v^{\text{inv}} v_{2,3}^*$	2M
10	$v_{2,4}^* \leftarrow -w + v_{1,2},\ v_{2,4} \leftarrow D_v^{\text{inv}} v_{2,4}^*$	1M
total		1I,3S,26M

However, we notice that the following formulas (11) can be applied verbatimly' to the non-monic equation case (8) because of an invariance of the formulas (9). See the remark before the formulas.

For $C_{\beta,\gamma}$, the algebraic correspondence (3) and (4) become simply as follows due to Hashimoto [18] (see also [17]). Here the correspondence $X_\psi \subset C_{\beta,\gamma} \times C_{\beta,\gamma}$ and the coordinates of $C_{\beta,\gamma} \times C_{\beta,\gamma}$ are given as (X_1, Y_1, X_2, Y_2). Those are invariant under the transformation $(X_1, Y_1, X_2, Y_2) \mapsto (X_1, \beta^{-1/2}Y_1, X_2, \beta^{-1/2}Y_2)$.

$$\beta(X_1 X_2)^2 - (\beta - 1)(X_1 X_2) + (X_1 + X_2) + 1 = 0 ,$$
$$X_2(X_2 + 1)Y_1 = X_1(X_1 + 1)Y_2 . \tag{9}$$

According to the formulas (9), Algorithm 1 simplifies to the algorithm in Table 2 based on the concise formulas (11). We describe the calculation algorithm (or formulas) of $(u_3(X), v_3(X)) := \psi(u_1(X), v_1(X))$ when $\deg u_1 = 2$ and constants D_u, D_v defined below $\neq 0$ (those are conditions of the dominant case). We express the divisor $\mathcal{D}_2$ in the diagram (6) as follows.

$$u_2(X) = X^4 + u_{2,1}X^3 + u_{2,2}X^2 + u_{2,3}X + u_{2,4} ,$$
$$v_2(X) = v_{2,1}X^3 + v_{2,2}X^2 + v_{2,3}X + v_{2,4} . \tag{10}$$

The coefficients $u_{2,\ell}, v_{2,\ell}, \in \mathbb{F}_q,\ \ell = 1, \ldots, 4$ in formulas (10) are determined by formulas(9) as follows.

$$D_u u_{2,1} = \beta((u_{1,1} + (\beta - 1)u_{1,2})u_{1,1} - 2u_{1,2}) ,$$
$$D_u u_{2,2} = (\beta u_{1,1} - \beta u_{1,2} + \beta - 1)u_{1,1} + (\beta^2 - 4\beta + 1)u_{1,2} + 1 ,$$
$$D_u u_{2,3} = (\beta - 2)u_{1,1} - 2(\beta - 1)u_{1,2} + 2 , \quad D_u u_{2,4} = -u_{1,1} + u_{1,2} + 1 ,$$
$$D_v v_{2,1} = \beta u_{1,2}(w - v_{1,2}) ,$$
$$D_v v_{2,2} = u_{1,1}(\beta w - v_{1,1}) + ((-\beta + 2)v_{1,1} - v_{1,2})u_{1,2} + v_{1,2} , \qquad (11)$$
$$D_v v_{2,3} = (-v_{1,1} + (\beta - 1)v_{1,2})u_{1,1} + ((-\beta + 3)v_{1,1} - v_{1,2})u_{1,2} + 2v_{1,2} ,$$
$$D_v v_{2,4} = -w + v_{1,2} .$$
$$(D_u = \beta^2 u_{1,2}^2 , \quad D_v = (u_{1,1} - u_{1,2})^2 - u_{1,1} - (\beta - 1)u_{1,2} , \quad w = u_{1,1}v_{1,2} - u_{1,2}v_{1,1} .)$$

First we calculate some intermediates D_u, D_v, D_u^{-1} and D_v^{-1} (see [26, 46]). After that, the RHS of the above formulas (11) are calculated and multiplied by D_u^{-1} or D_v^{-1}. As we see in Table 2, we complete the calculation with 3 finite field squarings, 23 multiplications and 2 inversions (S, M and I in Table 2 mean squaring, multiplication and inversion respectively).

Finally, $(u_3(X), v_3(X)) \in \text{Jac}_{C_{\beta,\gamma}}(\mathbb{F}_q)$ s.t. $\deg v_3(X) < \deg u_3(X) \le 2$ is obtained by reducing a divisor defined by $(u_2(X), v_2(X))$ in a linearly equivalent class. $(u_3(X), v_3(X))$ is $\psi(u_1(X), v_1(X))$ (For reduction in the linearly equivalent class, see [31, 35] for example).

Table 3. Algebraic correspondence calculation algorithm (projective coordinate)

Input : $U_1(X)$, $V_1(X)$ ($\in \mathbb{F}_q[X]$) and Z ($\in \mathbb{F}_q$)
Output : $U_2(X)$, $V_2(X)$ ($\in \mathbb{F}_q[X]$) and D_U, D_V ($\in \mathbb{F}_q$)

Step	Expression	Cost
1	$W_1 \leftarrow U_{1,1}^2$, $W_2 \leftarrow U_{1,1}U_{1,2}$, $W_3 \leftarrow U_{1,2}^2$, $W_4 \leftarrow U_{1,1}V_{1,1}$, $W_5 \leftarrow U_{1,1}V_{1,2}$, $W_6 \leftarrow U_{1,2}V_{1,1}$, $W_7 \leftarrow U_{1,2}V_{1,2}$, $W_8 \leftarrow \beta U_{1,1}$, $W_9 \leftarrow \beta U_{1,2}$, $W_{10} \leftarrow \beta V_{1,1}$, $Z_0 \leftarrow Z^2$, $Z_1 \leftarrow ZU_{1,1}$, $Z_2 \leftarrow ZU_{1,2}$, $Z_3 \leftarrow ZV_{1,1}$, $Z_4 \leftarrow ZV_{1,2}$, $Z^* \leftarrow ZW_9$	3S,13M
2	$D_U \leftarrow W_9^2$, $D_U \leftarrow Z_0 D_U$, $D_V \leftarrow W_1 - 2W_2 + W_3 - Z_1 - Z^* + Z_2$, $D_V \leftarrow ZZ_0 D_V$, $W \leftarrow W_5 - W_6$	1S, 3M
3	$U_{2,1} \leftarrow W_8(U_{1,1} + W_9 - U_{1,2}) - 2Z^*$	1M
4	$U_{2,2} \leftarrow (W_8 - W_9 + \beta Z - Z)U_{1,1} + (\beta - 4)Z^* + Z_2 + Z_0$	3M
5	$U_{2,3} \leftarrow Z(W_8 - 2U_{1,1} - 2W_9 + 2U_{1,2} + 2Z)$	1M
6	$U_{2,4} \leftarrow -Z_1 + Z_2 + Z_0$	
7	$V_{2,1} \leftarrow W_9(W - Z_4)$	1M
8	$W \leftarrow (-W_{10} + 2V_{1,1} - V_{1,2})U_{1,2}$, $V_{2,2} \leftarrow W_8 W + (-W_4 + W)Z + V_{1,2}Z_0$	4M
9	$V_{2,3} \leftarrow Z(-W_4 + (\beta - 1)W_5 + W + W_6 + 2Z_4)$	2M
10	$V_{2,4} \leftarrow Z(-W + Z_4)$	1M
total		4S, 29M

In [26], to avoid finite field inversion, two types of projective coordinates $\mathcal{P}$ and $\mathcal{N}$ are proposed. When a divisor (5) is expressed as

$$u_1(X) = X^2 + (U_{1,1}/Z)X + (U_{1,2}/Z) , \quad v_1(X) = (V_{1,1}/Z)X + (V_{1,2}/Z) \qquad (12)$$

where $U_{1,\ell}, V_{1,\ell}, Z \in \mathbb{F}_q$ ($\ell = 1, 2$), 5-component coordinate $[U_{1,1}, U_{1,2}, V_{1,1}, V_{1,2}, Z]$ is called $\mathcal{P}$ coordinate system. By using $\mathcal{P}$ coordinate system, we can remove finite field inversions in the ψ computation as is seen in Table 3.

For $(u_3(X), v_3(X))$ $(= \psi(u_1(X), v_1(X)))$, the same representation (12) with one denominator is used. However, we use the following expression (13) with different denominators D_U and D_V for the intermediate divisor $\mathcal{D}_2$ to compute ψ efficiently. Using Algorithm 2 in Appendix, we can obtain the desired divisor $\mathcal{D}_3$ in the $\mathcal{P}$ coordinate system. That is, we express the divisor (10) as

$$
\begin{aligned}
u_2(X) &= U_2(X)/D_U = X^4 + (U_{2,1}/D_U)X^3 \\
&\quad + (U_{2,2}/D_U)X^2 + (U_{2,3}/D_U)X + (U_{2,4}/D_U) , \\
v_2(X) &= V_2(X)/D_V = (V_{2,1}/D_V)X^3 + (V_{2,2}/D_V)X^2 \\
&\quad + (V_{2,3}/D_V)X + (V_{2,4}/D_V)
\end{aligned}
\tag{13}
$$

where $U_{2,\ell}, V_{2,\ell}, D_U, D_V \in \mathbb{F}_q$. Then we can calculate $U_{2,\ell}, V_{2,\ell}, D_U, D_V$ by 4 finite field squarings and 29 multiplications (see Table 3).

The amounts of finite field operations in reduction (in a linearly equivalence class) step are 2 squarings, 9 multiplications and 1 inversion in the affine ($\mathcal{A}$) coordinate case, and 3 squarings and 20 multiplications in the projective ($\mathcal{P}$) coordinate case (see [35] and Appendix).

According to [26], a doubling on the Jacobian group are completed by the costs in the right of Table 4 (see also [45, 46]). And the amounts of finite field operations in ψ calculation are tabulated in the left of Table 4. The upper data are those without the reduction step and the lower data are total amounts for ψ. Therefore, the cost of ψ in the $\mathcal{P}$ coordinates is comparable to 1.3 times that of a doubling in $\mathcal{P}$ on the Jacobian. Using the coordinate system $\mathcal{N}$ in [26], the amount of operations changes a little, hence, we can compute ψ with enough efficiency like the $\mathcal{P}$ coordinate system.

Table 4. Amounts of finite fields operations

Operation	Cost
$\psi\mathcal{P} = \mathcal{P}$	4S, 29M
	7S, 49M
$\psi\mathcal{A} = \mathcal{A}$	1I, 3S, 26M
	2I, 5S, 35M

Operation	Cost
$2\mathcal{N} = \mathcal{P}$	7S, 38M
$2\mathcal{P} = \mathcal{P}$	6S, 38M
$2\mathcal{N} = \mathcal{N}$	7S, 34M
$2\mathcal{P} = \mathcal{N}$	6S, 34M
$2\mathcal{A} = \mathcal{A}$	1I, 5S, 22M

Consequently, we can say that, even using any coordinate systems $\mathcal{A}, \mathcal{P}$, and $\mathcal{N}$, the efficiency is at most 1.5 times that of a doubling. By that, ψ on $\text{Jac}_C(\mathbb{F}_q)$ is efficient enough to be used in the GLV method.

4.4   Complexity Comparison with the CM Case

In the genus 2 case, ψ's proposed so far are automorphisms (Table 1), and are much more efficiently computable. However, in the GLV method, computation of "simultaneous multiple points scalar multiplication" is a dominant step. Therefore, the difference of computation amount of RM ψ and that of CM only lead to the difference less than $1/100$ of total amount of scalar multiplication. Therefore, we can say that we can achieve almost the same efficiency using the RM ψ as in the CM case.

We can't obtain a meaningful advantage when we apply RM ψ in ψ-adic expansion in [9] because the method is very effective when an efficient endomorphism is much faster than doubling. Moreover it should be noted that the idea in [9] only work if the norm of the endomorphism is larger than 1, therefore it can't be applied directly to the RM presently.

DPA countermeasures in [10] are also applicable in the RM case with a small speed loss as in [10] because that is a randomization of decomposition (1) of a scalar k.

4.5 Examples of MH Curves Without Special Properties of Previous CM Curves

As in Table 5, we've obtained random curves with prime orders $n := \sharp \mathrm{Jac}_{C_{\beta,\gamma}}(\mathbb{F}_q)$ (where $q := p^l$). All were obtained using functions of Magma [30], and satisfy security conditions of Rück [39] and Frey-Rück [12] (i.e. $q^r \not\equiv 1 \bmod n (1 \leq r \leq 10^4)$). C_2 and C_3 are defined over extension fields of prime degree over prime fields, and we can check that the use of these curves avoid Weil descent attacks. Also we verified they are absolutely simple using Howe-Zhu's criteria ([22] Theorem 6), and we can verify easily that the MH family in subsection 4.3 contains curves with parametrized absolute invariants which are different from each other.

Table 5. Example parameters of curves

C_1	p	28902710783
	(β, γ)	(16698884424, 5332701667)
	n	835369018786460055929 (70 bit prime)
	λ	2787250845667990059271
C_2	(p, l)	(3,53)
	(β, γ)	([0, 1, 1, 0, 0, 0, 0, 0, 0, 2, 0, 1, 0, 0, 2, 0, 1, 0, 1, 2, 2, 2, 0, 2, 2, 1, 2, 2, 2, 0, 0, 1, 1, 2, 0, 0, 2, 2, 2, 2, 1, 0, 2, 0, 0, 0, 2, 1, 0, 1, 0, 2, 2], [1, 1, 1, 2, 2, 1, 2, 0, 1, 1, 0, 2, 2, 2, 2, 0, 2, 1, 1, 2, 0, 2, 2, 2, 0, 1, 1, 1, 0, 0, 2, 2, 0, 2, 0, 0, 2, 2, 1, 0, 2, 0, 1, 0, 2, 1, 1, 1, 2, 1, 1, 1, 2])
	n	3757102126137643244470356093792773054616024998999281 (169 bit prime)
	λ	538849826253218010328217440902343648491608841115411
C_3	(p, l)	(5,37)
	(β, γ)	([2, 3, 4, 0, 0, 4, 1, 3, 1, 3, 4, 1, 0, 4, 4, 1, 1, 2, 1, 4, 4, 0, 2, 3, 1, 3, 2, 0, 1, 0, 0, 3, 1, 4, 0, 1, 1], [3, 1, 0, 0, 4, 0, 3, 0, 1, 0, 2, 0, 1, 3, 1, 1, 1, 1, 1, 2, 0, 0, 0, 4, 3, 4, 0, 2, 3, 4, 2, 1, 0, 3, 2, 3, 3])
	n	5293955920341054807488920035801349672258069275409741 (172 bit prime)
	λ	3356990070572605689466802561620623416584134941956091

That is, the "random" RM curves are not limited to ones with some "special" properties mentioned in section 3.

C_1 is a prime field curve (i.e. $q = p$). According to [16], using optimized programs for such curves, we can obtain curves with 160 bit prime n in practical time. A minimal polynomial of indeterminate θ defining a finite field for C_2 is $\theta^{53} + 2\theta^4 + 2\theta^3 + 2\theta^2 + 1$. In Table 5, $[a_1, \ldots, a_l]$'s representing coefficients β, γ mean $\sum_{i=0}^{l-1} a_{l-i}\theta^i \in \mathbb{F}_q$. For C_3, the polynomial is $\theta^{37} + 4\theta^2 + 3\theta + 3$, and β, γ are represented as for those of C_2 also. For all curves, β has a square root in $\mathbb{F}_q$ (see subsection 4.3).

To estimate the efficiency of the GLV method using ψ, we decomposed random 10^4 k's to k_1, k_2 as in the formula (1) by the methods in [25, 40]. And we have investigated the average of the bitlengths of $\max\{|k_1|, |k_2|\}$. The method in [40] is that in section 5 of that paper, not the generalized one in section 8 (using LLL) because ψ satisfies the quadratic equation $\psi^2 + \psi - 1 = 0$.

The mean values are 33.5 for C_1, 82.7 for C_2, and 84.6 for C_3. Those are less than $\log_2(n)/2$ respectively, therefore we verified experimentally the "decomposition of k" methods in [25, 40] for CM's of EC are effective also for the RM ψ (satisfying the quadratic equation).

5   RM Curves of Genus 2 in [4, 21]

In this section, we describe another candidate RM [4, 21] for the GLV method. We refer a fact from [4].

Fact. Let $\Delta, \mathcal{U}, \mathcal{V} \in \mathbb{F}_q^*, \mathcal{W} \in \mathbb{F}_q$. A genus 2 RM curve $C : Y^2 = f(X)$ where coefficients of f are explicitly given by rational expressions of $\Delta, \mathcal{U}, \mathcal{V}$ and $\mathcal{W}$ in [4] has an endomorphism ψ on Jac$_C$ satisfying $\psi^2 = 2$, which is defined by the following formulas (14) (where coordinate of $C \times C$ are given by (X_1, Y_1, X_2, Y_2)).

$$\sum_{\ell=0}^{2} \sum_{m=0}^{2} \Phi_{\ell,m} X_1^\ell X_2^m = 0 \ , \quad Y_2 = Y_1 \frac{(\mathcal{V} - 2)(\Psi_1(X_1)X_2 + \Psi_2(X_1))}{(X_1^2 + \mathcal{U}X_1 + \mathcal{V})(\sum_{\ell=0}^{2} \Phi_{\ell,2} X_1^\ell)^2} \ . \quad (14)$$

Here $\Phi_{\ell,m}$ are constants given by polynomials of $\mathcal{U}, \mathcal{V}$, and $\Psi_1(X), \Psi_2(X)$ are polynomials of $\mathcal{U}, \mathcal{V}, X$ of degree 3 w.r.t. X.

We can calculate the RM ψ like the calculation based on Algorithm 1 because "the formulas (3) and (4)" and "the formulas (14)" have similar forms. As in subsection 4.5, for this curve family, we obtained almost prime order Jacobian groups for appropriate random parameters $\Delta, \mathcal{U}, \mathcal{V}, \mathcal{W}$.

6   Conclusions

In this paper, we discussed applications of the GLV method to curves of genus 2 with real multiplication (RM). It is the first time to use RM in cryptography.

And we have shown that for Mestre-Hashimoto (MH) curves, the application of ψ takes at most 1.5 times as long as a doubling on the Jacobian group. Hence the GLV method with the RM ψ's can achieve the almost same efficiency as that with CM endomoprhisms. And for some concrete parameters in subsection 4.5, the effectivity of previously known "decomposition of k" methods are verified experimentally.

We have verified that the general curve in MH family doesn't have special properties of previous families of CM GLV curves of genus ≤ 2.

As we have seen in sections 4 and 5, Algorithm 1 can be applied for wide range of RM's. Future studies include both of improving RM computations in this paper and searching for other applications of Algorithm 1. Especially, the optimal costs in the left one of Table 4 in various coordinate systems should be pursued intensively.

Acknowledgement. We thank Claus Diem, Mitsuru Kawazoe, Tanja Lange, Kazuto Matsuo, Yasuyuki Sakai, and the anonymous refrees of ICISC 2004 for their useful comments.

References

1. ANSI X 9.62, "*American National Standard for Financial Services - Public Key Cryptography for the Financial Services Industry : The Elliptic Curve Digital Signature Algorithm (ECDSA)*," American National Standard Institute, 1998.
2. ANSI X 9.63, "*American National Standard for Financial Services - Public Key Cryptography for the Financial Services Industry : Key Agreement and Key Transport Using Elliptic Curve Cryptography*," American National Standard Institute, 2001.
3. R. M. Avanzi, "Aspects of hyperelliptic curves over large prime fields in software implementations," *Cryptographic Hardware and Embedded Systems - CHES 2004*, Springer Verlag, (2004), 148–162.
4. P. R. Bending, "Curves of genus 2 with $\sqrt{2}$ multiplication," available at `http://www.math.uiuc.edu/Algebraic-Number-Theory/`.
5. O. Bolza, "On binary sextics with linear transformations into themselves," *Amer. J. Math.*, **10**, (1888), 47–70.
6. A. Brumer, "The rank of $J_0(N)$," *Astérisque* **228** (1995), 41–68.
7. J. P. Buhler, N. Koblitz, "Lattice basis reduction, Jacobi sums, and hyperelliptic cryptosystems," *Bulletin of the Australian Math. Soc.*, **57** (1998), 147–154.
8. J. W. S. Cassels and E. V. Flynn, "*Prolegomena to a middlebrow arithmetic of curves of genus 2*," London Math. Soc. Lecture Notes Ser. 230, Cambridge Univ. Press, 1996.
9. M. Ciet, T. Lange, F. Sica and J.-J. Quisquater, "Improved algorithms for efficient arithmetic on elliptic curves using fast endomorphism," *Advances in Cryptology - Eurocrypt 2003*, Springer Verlag, (2003), 388–400.
10. M. Ciet, J.-J. Quisquater and F. Sica, "Preventing differential analysis in GLV elliptic curve scalar multiplication," *Cryptographic Hardware and Embedded Systems - CHES 2002*, Springer Verlag, (2002), 540–550.

11. I. Duursma, P. Gaudry, F. Morain, "Speeding up the discrete log computation on curves with automorphisms," *Advances in Cryptology - Asiacrypt '99*, Springer Verlag, (1999), 103–121.

12. G. Frey and H.G. Rück, "A remark concerning m-divisibility and the discrete logarithm in the divisor class group of curves," *Math. Comp.*, **62** (1994), 865–874.

13. E. Furukawa, M. Kawazoe and T. Takahashi, "Counting points for hyperelliptic curves of type $y^2 = x^5 + ax$ over finite prime fields," *Selected Areas in Cryptography - SAC 2003*, Springer Verlag (2003), 26–41.

14. R. P. Gallant, J. L. Lambert, S. A. Vanstone, "Faster point multiplication on elliptic curves with efficient endomorphisms," *Advances in Cryptology - Crypto 2001*, Springer Verlag, (2001), 190–200.

15. P. Gaudry and É. Schost, "On the invariants of the quotients of the Jacobian of a curve of genus 2," *Applied Algebra, Algebraic Algorithms and Error-Correcting Codes AAECC-14*, Springer-Verlag, (2001) 373–386.

16. P. Gaudry and É. Schost, "Construction of secure random curves of genus 2 over prime fields," *Advances in Cryptology - Eurocrypt 2004*, Springer Verlag, (2004), 239–256.

17. K. Hashimoto, "Construction of a real multiplication by algebraic correspondences of genus 2 curves," Proceedings of RIMS **942** *Deformations of Group Schemes and Number Theory*, Kyoto, Suuriken Koukyuuroku, 1996 (In Japanese).

18. K. Hashimoto, "Abelian surfaces of GL(2)-type : Their construction and modularity," an intensive course at Tokyo Metropolitan University, unpublished, 1998, the abstract in Japanese is available at `http://www.sci.metro-u.ac.jp/math/seminar/number/number-page.html`.

19. K. Hashimoto, "Q-curves of degree 5 and jacobian surfaces of GL_2-type," *Manuscripta Math.* , **98** (1999), 165–182.

20. K. Hashimoto, "On Brumer's family of RM-curves of genus two," *Tohoku Math. J.*, **52** (2000), 475–488.

21. E. W. Howe, "Infinite families of pairs of curves over $\mathbb{Q}$ with isomorphic Jacobians," available at `arXiv:math.AG/0304471`.

22. E. W. Howe and H. J. Zhu, "On the existence of absolutely simple abelian varieties of a given dimension over an arbitrary field," *J. Number Th.*, **92** (2002), 139–163.

23. G. Humbert, "Sur les fonctions abéliennes singulières," *Œuvres de G. Humbert 2, pub par les soins de Pierre Humbert et de Gaston Julia*, Paris, Gauthier-Villars (1936), 297–401.

24. N. Koblitz, "CM-curves with good cryptographic properties," *Advances in Cryptology - Crypto 91*, Springer Verlag, (1991), 279–287.

25. D. Kim and S. Lim, "Integer decomposition for fast scalar multiplication on elliptic curves," *Selected Areas in Cryptography - SAC 2002*, Springer Verlag, (2002), 13–20.

26. T. Lange, "Formulae for arithmetic on genus 2 hyperelliptic curves," to appear in *J. AAECC - Applicable Algebra in Engineering, Communication and Computing*, already available on line (and at `http://www.ruhr-uni-bochum.de/itsc/tanja/preprints.html`).

27. T. Lange and M. Stevens, "Efficient doubling on genus 2 curves over binary fields," *Selected Areas in Cryptography - SAC 2004*, Springer Verlag, (2004).

28. F. Leprévost and F. Morain, "Revêtement de courbes elliptiques à multiplication complexe par des courbes hyperelliptiques et sommes de caractères," *J. Number Th.*, **64** No. 2, (1997), 165–182.

29. K. Matsuo, J. Chao and S. Tsujii, "Fast genus two hyperelliptic curve cryptosystem," *Technical Report ISEC 2001-31*, IEICE Japan, 2001.
30. The Magma Computational Algebra System, v. 2.11, http://magma.maths.usyd.edu.au/.
31. A. J. Menezes, Y.-H. Wu, and R. J. Zuccherato, "An elementary introduction to hyperelliptic curves," appendix of *Algebraic Aspects of Cryptography* by N. Koblitz, Springer Verlag, (1998), 155–178.
32. J.-F. Mestre, "Courbes hyperelliptiques à multiplications réelles," *C. R. Acad. Sci. Paris., t. **307**, Série I.* (1988), 721–724.
33. J.-F. Mestre, "Familles de courbes hyperelliptiques à multiplications réelles," *Arithmetic Algebraic Geometry*, Birkhäuser, (1991), 193–208.
34. D. Mumford, *Tata Lectures on Theta II*, Birkhäuser, 1984.
35. K. Nagao, "Improving group law algorithms for Jacobians of hyperelliptic curves," *ANTS-IV*, Springer Verlag, (2000), 439–447.
36. Y.-H. Park, S. Jeong, C. Kim, and J. Lim, "An alternate decomposition of an integer for faster point multiplication on certain elliptic curves," *Public Key Cryptography - PKC 2002*, Springer Verlag, (2002), 323–334.
37. Y.-H. Park, S. Jeong, and J. Lim, "Speeding up point multiplication on hyperelliptic curves with efficiently-computable endomorphisms," *Advances in Cryptology - Eurocrypt 2002*, Springer Verlag, (2002), 197–208.
38. J. Pelzl, T. Wollinger and C. Paar, "Special hyperelliptic curve cryptosystems of genus two: Efficient arithmetic and fast implementation," In *Embedded Cryptographic Hardware : Design and Security*, 2004.
39. H.-G. Rück, "On the discrete logarithm in the divisor class group of curves," *Math. Comp.* **68** (1999), 805–806.
40. F. Sica, M. Ciet, J.-J. Quisquater, "Analysis of the Gallant-Lambert-Vanstone method based on efficient endomorphisms : Elliptic and hyperelliptic curves," *Selected Areas in Cryptography - SAC 2002*, Springer Verlag, (2002), 21–36.
41. J. Sato, N. Matsuda, J. Chao, S. Tsujii, "Construction of large genus hyperelliptic curves over prime fields," *Symposium on Cryptography and Information Security - SCIS 97* 12C, IEICE Japan (In Japanese).
42. Standards for Efficient Cryptography, *"SEC2 : Recommended Elliptic Curve Domain Parameters,"* version 1.0, 20 September 2000.
43. T. Shaska and H. Völklein, "Elliptic subfields and automorphisms of genus 2," *Algebra, Arithmetic and Geometry with Applications. Papers from Shreeram S. Abhyanker's 70th Birthday Conference*, Springer Verlag, (2004), 687–707.
44. J. A. Solinas, "Low-weight binary representations for pairs of integers," CACR technical report CORR 2001-41, available at http://www.cacr.math.uwaterloo.ca/.
45. H. Sugizaki, K. Matsuo, J. Chao and S. Tsujii "An extension of Harley addition algorithm for hyperelliptic curves over finite fields of characteristic two," *Technical Report ISEC 2002-9 (2002-5)*, IEICE Japan, (2002), 49–56.
46. M. Takahashi, "Improving Harley algorithms for Jacobians of genus 2 hyperelliptic curves," *Symposium on Cryptography and Information Security - SCIS 2002*, IEICE Japan, (2002), 155–160 (In Japanese).
47. WAP WTLS, *Wireless Application Protocol Wireless Transport Layer Security Specification*, WAP Forum, April 2001, available at http://www.wapforum.org

Appendix : Reduction Step Algorithm for the Projective Coordinate

We describe the reduction step algorithm from the projective representation (13) used in subsection 4.3. In the following, all notations are the same as in subsection 4.3.

To obtain $U_{3,\ell}, V_{3,\ell}, \tilde{Z} \in \mathbb{F}_q$ ($\ell = 1, 2$) s.t.

$$u_3(X) = X^2 + (U_{3,1}/\tilde{Z})X + (U_{3,2}/\tilde{Z}) \ , \ v_3(X) = (V_{3,1}/\tilde{Z})X + (V_{3,2}/\tilde{Z}) \ ,$$

first we compute $\eta_1(X)/\eta_2(X)$ $(= D_V^2(f(X) - v_2(X)^2)/(D_U u_2(X)))$ s.t. $\eta_1(X) :=$ $D_V^2 f(X) - V_2(X)^2$ and $\eta_2(X) := U_2(X)$. In the process (Algorithm 2), we divide $\eta_1(X) = \eta_{1,0}X^6 + \ldots + \eta_{1,6}$ (not-necessarily monic) by $\eta_2(X) = \eta_{2,0}X^4 + \ldots + \eta_{2,4}$ (not-necessarily monic). Here this division can be completed very efficiently according to [35]. Coefficients of $\tilde{\eta}_1(X)$, $\tilde{\tilde{\eta}}_1(X)$ etc. in Algorithm 2 are indexed similarly as $\eta_1(X)$ or $\eta_2(X)$.

Next we obtain a linear polynomial $V_3(X)$ by reducing the cubic polynomial $V_2(X)$ (not-necessarily monic) by the quadratic polynomial $U_3(X)$ (not-necessarily monic).

At last, in step 6 in Algorithm 2, we obtain the denominator $\tilde{Z}$ and equalize the denominator of $u_3(X)$ and that of $v_3(X)$.

We attached computational amounts of finite field operations to titles of calculations in Algorithm 2. Therefore, we know the total amount of finite field operations are 3S, 20M.

Algorithm 2 Reduction step from the projective representation (13)

Input $U_2(X), V_2(X), D_U, D_V, f(X)$ (= RHS of equation (2))
Output $U_3(X) = U_{3,0}X^2 + U_{3,1}X + U_{3,2}, V_3(X) = V_{3,1}X + V_{3,2}, \tilde{Z}(= U_{3,0})$.
 [Calculation of $U_3^*(X)$] 3S, 9M

1: Set $\eta_1(X) \leftarrow$ "degree ≥ 4 part of $D_V^2 f(X) - V_2(X)^2$" and $\eta_2(X) \leftarrow U_2(X)$.
2: $\tilde{\eta}_1(X) \leftarrow$ "degree ≥ 4 part of $\eta_{2,0}\eta_1(X) - \eta_{1,0}\eta_2(X)X^2$". [2]
   $\tilde{\tilde{\eta}}_1(X) \leftarrow$ "degree ≥ 4 part of $\eta_{2,0}\tilde{\eta}_1(X) - \tilde{\eta}_{1,0}\eta_2(X)X$". [2]
3: $U_{3,0}^* \leftarrow \eta_{1,0}, U_{3,1}^* \leftarrow \tilde{\eta}_{1,0}, U_{3,2}^* \leftarrow \tilde{\tilde{\eta}}_{1,0}, U_3^*(X) \leftarrow U_{3,0}^*X^2 + U_{3,1}^*X + U_{3,2}^*$.
 [Calculation of $V_3(X)$] 8M
4: $V_3^*(X) \leftarrow U_{3,0}^*V_2(X) - V_{2,1}U_3^*(X)X$. [2]
5: $V_3(X) \leftarrow U_{3,0}^*V_3^*(X) - V_{3,1}^*U_3^*(X)$. [2]
 [Correction of the coefficients of $U_3(X)$ and denominator $\tilde{Z}$] 3M
6: $U_3(X) \leftarrow D_V U_3^*(X), \tilde{Z} \leftarrow U_{3,0}$ (leading coefficient of $U_3(X)$).

---

[2] There is no need to calculate the cancelling leading coefficient.

Some Improved Algorithms for Hyperelliptic Curve Cryptosystems Using Degenerate Divisors

Masanobu Katagi[1], Toru Akishita[1], Izuru Kitamura[1], and Tsuyoshi Takagi[2]

[1] Sony Corporation, 6-7-35 Kitashinagawa Shinagawa-ku, Tokyo 141-0001, Japan
{Masanobu.Katagi, Izuru.Kitamura}@jp.sony.com
akishita@pal.arch.sony.co.jp
[2] Technische Universität Darmstadt, Fachbereich Informatik,
Hochschulstr.10, D-64289 Darmstadt, Germany
takagi@informatik.tu-darmstadt.de

Abstract. Hyperelliptic curve cryptosystems (HECC) can be good alternatives to elliptic curve cryptosystems, and there is a good possibility to improve the efficiency of HECC due to its flexible algebraic structure. Recently, an efficient scalar multiplication technique for application to genus 2 curves using a degenerate divisor has been proposed. This new technique can be used in the cryptographic protocol using a fixed base point, e.g., HEC-DSA. This paper considers two important issues concerning degenerate divisors. First, we extend the technique for genus 2 curves to genus 3 curves. Jacobian variety for genus 3 curves has two different degenerate divisors: degree 1 and 2. We present explicit formulae of the addition algorithm with degenerate divisors, and then present the timing of scalar multiplication using the proposed formulae. Second, we propose several window methods using the degenerate divisors. It is not obvious how to construct a base point D such that $\deg(D) = \deg(aD) < g$ for integer a, where g is the genus of the underlying curve and $\deg(D)$ is the degree of divisor D. We present an explicit algorithm for generating such divisors. We then develop a window-based scheme that is secure against side-channel attacks.

Keywords: hyperelliptic curve cryptosystem, scalar multiplication, degenerate divisor, window method.

1 Introduction

Hyperelliptic curve cryptosystems (HECC) are an extension of elliptic curve cryptosystems (ECC). The operand size of HECC is $1/g$-th that of ECC because of the algebraic structure of HECC. Therefore, HECC is attractive from the viewpoint of implementation. However, the addition algorithm of HECC [Can87] requires more group operations because it is more complex than the addition algorithm of ECC. Harley [Har00a, Har00b] has recently proposed a useful algorithm, which we will call the Harley algorithm. The Harley algorithm is an explicit representation of the Cantor algorithm for genus 2 curves of odd characteristic. The extension and optimization of the Harley algorithm are being

C. Park and S. Chee (Eds.): ICISC 2004, LNCS 3506, pp. 296–312, 2005.

actively investigated with the intention of making the addition algorithms of HECC comparable to those of the ECC [MCT01, SMC+02, Lan02].

Not only studies seeking to optimize the explicit formulae, but also studies to investigate side-channel attacks on HECC are being conducted. Differential Power Analysis (DPA) is a major threat to the implementation of cryptographic algorithms. Since HECC is an extension of ECC, various countermeasures against DPA can be applied to HECC. However, degenerate divisors are important with respect to the security of HECC. A degenerate divisor of a genus-g hyperelliptic curve over a finite field $\mathbb{F}_q$ is a reduced divisor D, the degree of which is smaller than g, e.g. $D = (u, v)$ such that $u = x^{g-1} + \sum_{i<g-1} u_i x^i, v = \sum_{i<g-1} v_i x^i$, where $u_i, v_i \in \mathbb{F}_q$. Even if the curve parameters or coordinates are randomized, the HECC is vulnerable to an exceptional procedure attack using degenerate divisors [Ava03, KKA+04]. On the other hand, the degenerate divisors can be used in a positive way. Katagi et al. showed an efficient scalar multiplication using a degenerate divisor for a genus 2 hyperelliptic curve cryptosystem [KKA+04]. They also proved that the discrete logarithm problem with a degenerate base point is not substantially weaker than that with a random base point. However, to our knowledge, no explicit formula of degenerate divisors for a genus 3 hyperelliptic curve has been proposed. Moreover, the method by which to extend this method to a window-based scheme remains unclear — an algorithm that generates a pair of degenerate divisors must be considered.

1.1   Our Contribution

In this paper, we first investigate the addition formula using degenerate divisors for genus 3 hyperelliptic curves. We present some explicit addition formulae and precisely estimate their computational time. For example, an addition formula with a degenerate divisor of degree 2 (or 1) requires $1I + 52M$ (or $1I + 21M$), respectively, where I and M are the time of computing an inversion and a multiplication, respectively. Using the degenerate divisor we can achieve a faster scalar multiplication of hyperelliptic curve cryptosystems. Our estimations show that the improvement over a standard double-and-add-always algorithm is 15% (or 34%) using a divisor of degree 2 (or 1), respectively. The representation of the degenerate divisor (u, v) is shorter than that of the standard divisor, so that we achieve a compressed representation for a base point of hyperelliptic curve cryptosystems.

Next, we investigate window methods using degenerate divisors to compute scalar multiplication. In this case, we have to generate a pair of degenerate divisors (D, aD) for some integer a. First, we estimate the probability of degenerate pairs for both genus 2 and genus 3 hyperelliptic curves. We show that there exist, with a reasonable probability, pairs of degenerate divisors (D, aD) of degree 1 for genus 2 and of degree 2 for genus 3. We then show an algorithm for generating a degenerate pair for small a. From the explicit formulae described above, the divisor $aD = (u_a, v_a)$ is expressed as the polynomial of u_i, v_i over $\mathbb{F}_q$, where u_i, v_i is the component of the divisor $D = (u, v)$. We present some examples of degenerate divisors with the parameter size used for cryptographic application.

Note that the degree of this polynomial increases in the exponential time of the bit-length of a, and thus we can find a degenerate pair only for small a. Finally, we compare the efficiency and memory of the proposed schemes with those of the SPA-resistant wNAF method, which is one of the most efficient window methods [OT03]. Because of the short representation of pre-computed base points, we can achieve an efficient scalar multiplication using a smaller table size.

This paper is organized as follows: In section 2, the property of hyperelliptic curve cryptosystems is briefly reviewed. In section 3, we present the proposed scheme using degenerate divisors for a genus 3 HECC. In section 4, we propose a window-based scheme for computing scalar multiplication using the degenerate divisor. In section 5, we conclude this paper.

2   Hyperelliptic Curve Cryptosystems

In this section we review hyperelliptic curve cryptosystems related to our study.

2.1   Hyperelliptic Curves

A hyperelliptic curve C of genus g over a finite field $\mathbb{F}_q$ is defined as $y^2 + h(x)y = f(x)$, where $f(x)$ is a monic polynomial of $\mathbb{F}_q[x]$ of degree $2g + 1$ and $h(x)$ is a polynomial over $\mathbb{F}_q[x]$ of deg $h \leq g$. A point on curve C is denoted by $P = (x, y)$, and its inverse is defined as $-P = (x, -y - h(x))$. We call a point P that satisfies $P = -P$ a ramification point.

In contrast to ECC, points on a hyperelliptic curve do not form a group. Rather than points, divisors are deployed. A divisor D is a formal sum of points $\sum m_i P_i$, where $m_i \in \mathbb{Z}$. The degree of a divisor D is defined as $\sum_i m_i$. The Jacobian variety $J_c(\mathbb{F}_q)$ is defined by the quotient group $\mathbb{D}_0/\mathbb{P}$, where $\mathbb{D}_0$ is a divisor of degree 0 and $\mathbb{P}$ is a principal divisor. The principal divisor is the divisor of a rational function on C, which is a finite formal sum of the zeros and poles.

For cryptographic application, the Mumford representation is useful [Mum84]. A semi-reduced divisor can be expressed by two polynomials (u, v) of $\mathbb{F}_q[x]$.

$$u(x) = \prod_i (x - x_i)^{m_i}, \quad v(x_i) = y_i, \quad \deg v < \deg u, \quad v^2 + hv - f \equiv 0 \bmod u.$$

If $\deg u \leq g$, then the semi-reduced divisor is referred to as a reduced divisor. Elements in $J_c(\mathbb{F}_q)$ are uniquely represented as reduced divisors. We denote here the degree of the reduced divisor by $\deg(D)$.

2.2   Secure Scalar Multiplication in the Face of SPA

The fundamental operation for HECC is scalar multiplication $dD = D + \cdots + D$ (d additions), where d is an integer and D is a reduced divisor. Let $D_i = (u_i(x), v_i(x)) \in J_c(\mathbb{F}_q)$ be the reduced divisors, $i = 1, 2$. The reduced divisor D_3 of addition $D_1 + D_2$ is computed using the Cantor algorithm [Can87] [Kob89].

This operation is denoted by HECDBL if $D_1 = D_2$, or by HECADD otherwise. A standard algorithm for computing dD is as follows:

Algorithm 1. *Double-and-Add-Always Method*

Input: $d = (d_{n-1} \cdots d_1 d_0)_2,\ D \in J_c(K),\ (d_{n-1} = 1)$
Output: dD

---
1. $D[0] \leftarrow D$
2. *for i from* $n - 2$ *to* 0 *do*
3. $D[0] \leftarrow HECDBL(D[0]),\ D[1] \leftarrow HECADD(D[0],D),\ D[0] \leftarrow D[d_i]$
4. *Return(D[0])*

This double-and-add-always method computes HECADD for both $d_i = 0$ and 1. Therefore, timing attacks (TA) or Simple Power Analysis (SPA) cannot ascertain the bit hamming weight of d [Cor99].

Okeya and Takagi proposed an SPA-resistant window method for ECC [OT03], and it is also applicable to HECC. This method formulates the fixed pattern $|0 \cdots 0x|\ 0 \cdots 0x| \cdots |0 \cdots 0x|$ for some x. Though an SPA attack distinguishes HECDBL and HECADD in a scalar multiplication by measuring a single power consumption, only the identical sequence $|D \cdots DA|D \cdots DA| \cdots |D \cdots DA|$ is obtained, where D and A are denoted by HECDBL and HECADD, respectively. Therefore, the attacker cannot guess any bit information of the scalar. This method reduces the required number of HECADD, as compared with the double-and-add-always method, and thus increasing the efficiency.

2.3 Efficient Scalar Multiplication Using Degenerate Divisors

We next present an efficient scalar multiplication method using degenerate divisors, which was proposed in [KKA+04].

The Cantor algorithm is a general algorithm used for HECADD and HECDBL, but its performance is quite low. On the other hand, the Harley algorithm is an explicit representation of the Cantor algorithm in terms of the degree classification of the divisors [Har00a, Har00b]. The Harley algorithm includes the most frequent case for addition and doubling of divisors. The most frequent case is denoted by *MFCADD* and *MFCDBL*, respectively. *MFCADD* and *MFCDBL* satisfy the following conditions:

$$MFCADD : \deg(D_1) = \deg(D_2) = \deg(D_3) = g,\ D_1 \neq D_2, \gcd(u_1, u_2) = 1,$$
$$MFCDBL : \deg(D_1) = \deg(D_3) = g,\ D_1 = D_2,\ \gcd(h + 2v_1, u_1) = 1.$$

The most frequent case appears with overwhelming probability [Nag00]. Some other cases are caused by a degenerate divisor, which is defined as follows:

Definition 1. *Let C be a hyperelliptic curve over $\mathbb{F}_q$, and let $J_c(\mathbb{F}_q)$ be the Jacobian of curve C. We call a reduced divisor $D = (u, v) \in J_c(\mathbb{F}_q)$ degenerate if the degree of D is smaller than g, namely, $\deg u < g$.*

We randomly choose a degenerate divisor as the base point. The scalar multiplication using degenerate divisors is much faster than that using the standard divisors. The double-and-add-always method (Algorithm 1) in Section 2.2 is a left-to-right procedure, thus the base point D is added to $D[i]$ during scalar multiplication dD. Base point D is designated as a degenerated divisor (e.g., $\deg(D) < g$), and its addition formulae can be calculated more efficiently than the standard HECADD. In the case of genus 2 curves, the scalar multiplication using degenerate divisors is approximately 20% faster than using standard divisors if the secret scalar is 160 bits [KKA^{+}04].

The use of a degenerate divisor as the base point does not reduce the hardness of the underlying discrete logarithm problem due to the random self reducibility. The previously reported efficient scalar multiplication method with a fixed base point can be applied only to ElGamal-type encryption, the sender of Diffie-Hellman, and DSA. The scalar of these schemes is usually an ephemeral random number, and thus we are only interested in SPA. A standard countermeasure against SPA is the double-and-add-always method (Algorithm 1) in Section 2.2.

3   Proposed Algorithms for Genus 3 HECC

In this section, we present an efficient scalar multiplication using a degenerate divisor for HECC of genus 3. We deal with genus 3 HECC over a binary field $\mathbb{F}_{2^n}$, but the same discussions can be applied to genus 3 HECC over a general finite field $\mathbb{F}_q$.

3.1   Degenerate Divisor for Genus 3 HECC

For the genus 3 case, the degenerate divisors are random divisors of degree 1 or 2. We investigated the formulae with genus 3 degenerate divisors.

Let $D_1 = (u_1, v_1), D_2 = (u_2, v_2)$ be reduced divisors of the Jacobian $J_c(\mathbb{F}_{2^n})$. Denote by D_3 the addition of $D_1 + D_2$. There are exceptional group operations with degenerate divisors discussed in this paper as follows:

ExADD$^{1+3 \to 3}$: $\deg(D_1) = 1, \deg(D_2) = \deg(D_3) = 3, D_1 \neq D_2, \gcd(u_1, u_2) = 1$,
ExADD$^{1+2 \to 3}$: $\deg(D_1) = 1, \deg(D_2) = 2, \deg(D_3) = 3, D_2 = 2D_1$,
ExADD$^{2+3 \to 3}$: $\deg(D_1) = 2, \deg(D_2) = \deg(D_3) = 3, D_1 \neq D_2, \gcd(u_1, u_2) = 1$,
ExDBL$^{1 \to 2}$: $\deg(D_1) = 1, \deg(D_3) = 2, D_1 = D_2, \gcd(h, u_1) = 1$,
ExDBL$^{2 \to 3}$: $\deg(D_1) = 2, \deg(D_3) = 3, D_1 = D_2, \gcd(h, u_1) = 1$.

Table 1 shows the cost of the Harley algorithm and its degenerate versions. We re-estimated the cost of the Harley algorithm based on the previous references [Pel02]. The degenerate algorithms, moreover, are newly derived because most research has focused only on the most frequent cases of the Harley algorithm. The degenerate cases, as well as those for genus 2 HECC, are faster than the most frequent cases because the lower degree of the divisors reduces the number of field operations.

Table 1. Number of Multiplications and Inversions of the Harley Algorithm (genus 3)

Addition Formula	Cost
$MFCADD$	$1I + 78M$
$MFCDBL$	$1I + 81M$
$\mathsf{ExADD}^{1+3\to3}$	$1I + 21M$
$\mathsf{ExADD}^{1+2\to3}$	$1I + 28M$
$\mathsf{ExADD}^{2+3\to3}$	$1I + 52M$
$\mathsf{ExDBL}^{1\to2}$	$1I + 21M$
$\mathsf{ExDBL}^{2\to3}$	$1I + 53M$

3.2 Efficient Scalar Multiplication Using Degenerate Divisors

For genus 3 HECC, we can choose a divisor of degree 1 or 2 as the base point, and then $\mathsf{ExADD}^{1+3\to3}$ or $\mathsf{ExADD}^{2+3\to3}$ is used for HECADD. $\mathsf{ExADD}^{1+3\to3}$ and $\mathsf{ExADD}^{2+3\to3}$ are faster than $MFCADD$. Therefore, a scalar multiplication using a degenerate divisor is accelerated.

If we use the double-and-add-always method that is resistant to SPA, an addition chain in the scalar multiplication shows a fixed pattern which repeats HECDBL and HECADD as described in section 2.2. In particular, if $d_{n-2} = 0$, then the addition chain generates the following initial sequence for the base point D:

$$D \longrightarrow 2D \longrightarrow 3D$$
$$\longrightarrow 4D \longrightarrow 5D$$

Otherwise, if $d_{n-2} = 1$, we have the following:

$$D \longrightarrow 2D \longrightarrow 3D \longrightarrow 6D \longrightarrow 7D$$

Here, we denote D_0 as $MFCDBL$, D_1 as $\mathsf{ExDBL}^{1\to2}$, D_2 as $\mathsf{ExDBL}^{2\to3}$, A_1 as $\mathsf{ExADD}^{1+3\to3}$, A_2 as $\mathsf{ExADD}^{1+2\to3}$ and A_3 as $\mathsf{ExADD}^{2+3\to3}$. When we choose a divisor of degree 1 as the base point D, the initial pattern of the addition chain differs based on the bit d_{n-2}. The initial pattern for $d_{n-2} = 0$ is $D_1A_2D_2A_1$, but the pattern for $d_{n-2} = 1$ is $D_1A_3D_0A_1$. This means that the second most significant bit of d is vulnerable to SPA or TA. However, this vulnerability is only limited to d_{n-2} because the addition chain generates the fixed pattern D_0A_1 for other bits of d, so that an attacker cannot guess the other bits of the secret scalar.

On the other hand, when we choose a divisor of degree 2 as the base point, the initial pattern of the addition chain is $D_2A_3D_0A_3$, which is independent from the bit d_{n-2}. Thus an attacker cannot guess any bit of the secret scalar d.

We compare the computational costs of the scalar multiplication using a standard divisor and a degenerate divisor. If we choose a standard divisor, namely a divisor of degree 3, as the base point, the total cost is $159 \times ((I + 81M) + (I + 78M)) = 318I + 25281M$ for a 160-bit scalar d. Next, we estimate the case in which a divisor of degree 1 is chosen as the base point. As we pointed out, the addition chain shows a different pattern based on the bit d_{n-2}. Therefore, the total cost of scalar multiplication is $(I + 11M) + (I + 28M) + 1/2((I + 45M) +$

$(I + 81M)) + (I + 21M) + 157 \times ((I + 81M) + (I + 21M)) = 318M + 16137M$.
Thus, the scalar multiplication using a degenerate divisor of degree 1 is approximately 34% faster than that using a standard divisor under $1I = 5.2M$ (see Appendix B.1). On the other hand, if we choose a divisor of degree 2 as the base point, the first HECDBL costs $1I + 45M (\text{ExDBL}^{2 \to 3})$, and the other HECDBL and HECADD cost $1I + 81M$ $(MFCDBL)$ and $1I + 52M$ $(\text{ExADD}^{2+3 \to 3})$, respectively. Therefore, the total cost of scalar multiplication is $(1I+45M)+(1I+52M)+((1I+81M)+(1I+52M)) \times 158 = 318I+21111M$. Thus, the scalar multiplication of a degenerate divisor of degree 2 is approximately 15% faster than that of a standard divisor under $I = 5.2M$. The experimental results of these comparisons are shown in Appendix B.2.

4   Application of Window Method

In this section, we propose a more efficient method than the double-and-add-always method using degenerate divisors.

We choose a degenerate divisor D, which satisfies the condition $\deg(D) = \deg(aD) < g$ with previously known integer a, as the base point. Using a divisor that satisfies this condition, we can apply the SPA-resistant wNAF method [OT03] to the HECC scalar multiplication, as described in section 2.2. This method is faster than applying the SPA-resistant wNAF technique to the HECC using degree-g divisors under the same window size. To find such a divisor, however, is not easy because a divisor satisfying $\deg(D) < g$ exists with a very low probability.

4.1   Probability of a Pair of Degenerate Divisors (D, aD)

We investigate the probability of finding a pair of degenerate divisors (D, aD).

Lemma 2. *Assume that the base field is $K = \mathbb{F}_q$. If we choose a hyperelliptic curve C of genus g and degenerate divisors D of degree $g-1 (g > 1)$ are randomly distributed in $J_C(K)$, then the number of pairs (D, aD) satisfying $\deg(D) = \deg(aD) = g - 1$ is roughly q^{g-2}. Similarly, in the case of degenerate divisors D of degree $g - 2$ $(g > 2)$, the number of pairs (D, aD) satisfying $\deg(D) = \deg(aD) = g - 2$ is roughly q^{g-4}.*

Proof. First, the number of degenerate divisors with $g - 1$ is q^{g-1} because the number of degenerate divisors is equivalent to the number of monic polynomials of degree $g-1$. Next, $\#J_c(K)$ is roughly q^g, based on the Hasse-Weil theorem. If two degenerate divisors are selected independently, the probability that they will be a pair (D, aD) is $(1/q)^2$. In the same way, the probability of independently selecting degenerate pairs of degree $g - 2$ is $(1/q)^4$. Therefore, the number of pairs (D, aD) is $(q^g)(1/q^2) = q^{g-2}$ and $(q^g)(1/q^4) = q^{g-4}$. $\square$

From Lemma 2, the average number of pairs (D, aD) for genus 2, genus 3 with degree 1, and genus 3 with degree 2 are 1, $1/q$, and q, respectively. The probability of finding the degree 1 degenerate pairs for genus 3 is very low. Therefore, we focus on degenerate pairs for genus 2 and genus 3 with degree 2.

4.2   Algorithm for $\deg(D) = \deg(5D) = 1$ for Genus 2

In this section, we describe how to find a pair of degenerate divisors (D, aD) for genus 2 curves. First, we consider $a = 3$. From the following lemma, however, no pair that satisfies the condition $\deg(D) = \deg(3D) = 1$ exists.

Lemma 3. *Assume that $D \in J_C(K)$ is a degenerate divisor of prime order larger than 2. There exists no pair $(D, 3D)$ satisfying $\deg(D) = \deg(3D) = 1$.*

Proof. A degenerate divisor D of $\deg(D) = 1$ is represented as $D = P - P_\infty$ using P on C. $2D = D + D = P + P - 2P_\infty$. We consider $3D$, derived from $D + 2D$. If the degree of $3D$ is 1, P satisfies $P = -P$, which implies that the order of D is 2. ☐

Next, we consider $a = 5$. In the following, we describe how to find divisor D such that $\deg(D) = \deg(5D) < g$ in the case of $g = 2$.

As we noted earlier, an exhaustive search for $\deg(5D) = 1$ is difficult because there is a low probability of finding the base point D that satisfies $\deg(D) = \deg(5D) = 1$. Such a divisor D that satisfies $\deg(D) = \deg(5D) = 1$ is computed by using:

$$D \xrightarrow{\text{ExDBL}^{1 \to 2}} 2D \xrightarrow{\text{MFCDBL}} 4D \xrightarrow{\text{ExADD}^{1+2 \to 1}} 5D.$$

We first use $\text{ExDBL}^{1 \to 2}$ of a hyperelliptic curve of genus 2. $\text{ExDBL}^{1 \to 2}$ for $g = 2$ is shown in [KKA+04–Algorithm 5]. Let $D = (x + u, v)$ be a divisor of degree 1 and let $2D = (x^2 + u_1 x + u_0, v_1 x + v_0)$. The coefficients of $2D$ are represented as follows:

$$u_1 = 0, \quad u_0 = u^2,$$
$$v_1 = \frac{u^4 + f_3 u^2 + f_1 + h_1 v}{u^2 + h_1 u + h_0}, \quad v_0 = \frac{u^4 + f_3 u^2 + f_1 + h_1 v}{u^2 + h_1 u + h_0} u + v.$$

Next, we present Harley's doubling formula $MFCDBL$ using the outputs from the previous step, namely, u_0, u_1, v_0, v_1. Let $4D = (x^2 + w_1 x + w_0, z_1 x + z_0)$.

$$t_1 = (h_0 + u_0 + u_1(h_1 + u_1))(f_3 + v_1 + u_1^2) + (h_1 + u_1)(v_0 + v_1(v_1 + h_1))$$
$$t_0 = (u_1(h_0 + u_0 + u_1(h_1 + u_1)) + u_0(h_1 + u_1))(f_3 + v_1 + u_1^2)$$
$$\quad + (h_0 + u_0 + u_1(h_1 + u_1))(v_0 + v_1(v_1 + h_1))$$
$$r = u_0(u_0 + h_0 + h_1(h_1 + u_1)) + h_0(h_0 + u_0 + u_1(h_1 + u_1))$$
$$w_1 = \frac{1}{r t_1}(1 + \frac{1}{r t_1})$$
$$w_0 = \frac{r}{t_1}\frac{t_0}{t_1}(1 + \frac{t_0}{r}) + h_1 + u_1)$$
$$z_1 = (\frac{t_1}{r} + \frac{t_0}{r})(u_1 + w_1 + u_0 + w_0) + \frac{t_1}{r}(u_1 + w_1) + \frac{t_0}{r}(u_0 + w_0)$$

$$+(\frac{t_1}{r}(u_1 + w_1) + 1)w_1 + v_1 + h_1$$

$$z_0 = \frac{t_0}{r}(u_0 + w_0) + (\frac{t_1}{r}(u_1 + w_1) + 1)w_0 + v_0 + h_0$$

$5D$ is derived from $D + 4D$ via $\mathsf{ExADD}^{1+2\to2}$ for genus 2. $\mathsf{ExADD}^{1+2\to2}$ for genus 2 is shown in [Lan02–Table 2]. Since, this formula has no branch to output a degree 1 divisor, we have only to consider the case in which D and $4D$ are not coprime. The following condition applies:

$$D = P - P_\infty, \quad 4D = -P + Q - 2P_\infty \tag{1}$$

In this case, $D + 4D$ results in the degree 1 divisor $5D = Q - P_\infty$. Recall that the inverse point of $P = (u, v)$ is $-P = (u, v + h(u))$. Condition (1) yields two equations for $P = (u, v)$:

$$w(u) = w_1 u + w_0 = 0 \tag{2}$$

$$z(u) = z_1 u + z_0 = v + h(u) \tag{3}$$

Equations 2 and 3 are represented as $G_1(u, v) = 0$ and $G_2(u, v) = 0$, respectively. G_i is simplified as follows using the relation $v^2 = h(u)v + f(u)$.

$$G_i = a_i(u) + b_i(u)v = 0 \ (i = 1, 2) \tag{4}$$

Let the conjugate of $G_i(u, v)$ be $\overline{G_i} = a_i + b_i(v + h)$ [MWZ96]. The norm of $G_i(u, v)$, N_i, is a univariate polynomial function of u.

As a result, we derive that the degree of N_1 and N_2 are 53 and 72, respectively. We can easily solve these equations because univariate polynomial equations are solved in polynomial time.

We summarize this algorithm as follows:

Algorithm 2. *Finding* $\deg(D) = \deg(5D) = 1$ *for genus 2*

1. *Set the curve parameter $h_{1,0}$ and $f_{1,3}$*
2. *Solve the two univariate polynomial equations of u, $N_1(u) = 0$ and $N_2(u) = 0$.*
3. *Return $D = (u, v)$ if the two solutions have a common root else goto 1.*

We next show an example of a degenerate divisor in Appendix C.1 using Algorithm 2. We had to generate random curves for cryptographic use a few times in order to find this example. The random curves were generated using MAGMA2.11-2 [MAG].

4.3   Applying Window Method for $a = 5$

For genus 2 HECC, we cannot directly apply the SPA-resistant wNAF method [OT03] because $a \neq 3$. We will explain how to construct an SPA-resistant addition chain using the base point D such that $\deg(D) = \deg(aD)$ for $a > 3$.

We first construct a fixed pattern:

$$\underbrace{D...D}_{k_1}\underbrace{DA...DA}_{k_2},$$

where $(k_1 + k_2)$ is the bit-length of the scanned bit.

In the following we examine the situation in which $k_1 = 1$, $k_2 = 2$, and $a = 5$. We represent here all non-zero digits (i.e., $\{\pm 1, \pm 3, \pm 5, \pm 7\}$) appearing in the width-3 NAF method, as follows:

$$001 = 001 = 001 + 0d0, \qquad 00\bar{1} = 00\bar{1} = 00\bar{1} + 0d0,$$
$$003 = 011 = 001 + 010, \qquad 00\bar{3} = 0\bar{1}\bar{1} = 00\bar{1} - 010,$$
$$005 = 101 = 005 + 0d0, \qquad 00\bar{5} = \bar{1}0\bar{1} = 005 + 0d0,$$
$$007 = 111 = 005 + 010, \qquad 00\bar{7} = \bar{1}\bar{1}\bar{1} = 005 - 010,$$

where d indicates a dummy operation.

In this case, the density of non-zero digits is equal to $2/3^1$. Therefore, the total efficiency is $nECDBL + \frac{2}{3}nECADD$.

4.4   Algorithm for $\deg(D) = \deg(3D) = 2$ for Genus 3

Unlike the case of degenerate divisors for genus 2 curves, it is possible to find a pair $(D, 3D)$ satisfying $\deg(D) = \deg(3D) = 2$ for genus 3 curves because D and $3D$ become coprime. Therefore, we can directly apply the SPA-resistant wNAF method [OT03] for the base point D such that $\deg(D) = \deg(3D) = 2$. The width w is equal to 2, that is to say, the density of non-zero digits is equal to $1/2$.

In order to find such a divisor, we use the explicit formulae for computing $3D$. The divisor D that satisfies $\deg(D) = \deg(3D) = 2$ is computed using the following formulae:

$$D \xrightarrow{\text{ExDBL}^{2\rightarrow3}} 2D \xrightarrow{\text{ExADD}^{2+3\rightarrow2}} 3D$$

Note that the addition formula $\text{ExADD}^{2+3\rightarrow2}$ has a branch in step 4 in Algorithm 4. If $t_1 = 0$ in step 4, then the degree of v is reduced from 4 to 3, so that the degree of u_3 in step 6 is 2. Therefore, we utilize the condition $t_1 = 0$ in order to find the divisor D.

First, in order to find the solution of $t_1 = 0$, we must derive $2D = (x^3 + u_{12}x^2 + u_{11}x + u_{10}, v_{12}x^2 + v_{11}x + v_{10})$ as a function of $D = (x^2 + u_{21}x + u_{20}, v_{21}x + v_{20})$. We analyze the formula $\text{ExDBL}^{2\rightarrow3}$ and derive the relations D and $2D$. Next, we convert the representation from the divisor D to two points $P_1 = (x_1, y_1)$ and $P_2 = (x_2, y_2)$ on C. If we choose P_1 randomly, then the condition $t_1 = 0$ gives the equation of x_2 and y_2. Finally, we derive a univariate polynomial equation

---

[1] The minimal non-zero density of an SPA-resistant scheme with two bases D and aD ($a > 3$) is $4/7$ for $a = 13$. See Appendix.

$N_3(x_2) = 0$ using the same technique in the genus 2 case. (The degree of equation $N_3(x_2)$ is 79.)

We summarize this algorithm as follows:

Algorithm 3. *Finding* $\deg(D) = \deg(3D) = 2$ *for genus 3*

1. *Set the curve parameter* $h_{2,1,0}$ *and* $f_{0,1,2,3,4,5}$
2. *Generate random point* $P_1 = (x_1, y_1)$ *on* C
3. *Solve a univariate polynomial equation of* x_2, $N_3(x_2) = 0$
4. *Calculate* y_2 *of* P_2 *from* x_2 *if a solution exists else goto 2.*
5. *Return* $D = (x^2 + u_{21}x + u_{20}, v_{21}x + v_{20})$ *from* P_1 *and* P_2.

We show an example of a divisor in Appendix C.2 using Algorithm 3. Compared with a degenerate pair for the genus 2 case, we need not generate many curves because many candidates exist in the genus 3 with degree 2 case. The curve we selected is isomorphic to [Ver02].

4.5   Choice of Degenerate Base Point and Standard Base Point

We have proposed wNAF methods using degenerate divisors of genus 2 and 3. In this section, we compare the proposed methods with other SPA-resistant scalar multiplication methods. From the perspective that the wNAF methods use a pre-computation table, we discuss both table size and the number of group operations. In the genus 2 case, the results are shown in Table 2. The table size is represented by n bits, which is an extension degree of the base field. If we choose a degenerate divisor as the base point, then the base point is represented by $D = (x+x_0, y_0)$. The table size, therefore, must be $n \times 2$ bits. On the other hand, if we choose a standard divisor as the base point, the table size must be $2n \times 2$ bits. The number of group operations is obtained from the addition formulae in [KKA+04]. In addition, we used the relation $I = 5.8M$, which is the result of our computer experiment (see Appendix B.1). The proposed method applied to wNAF using the pair (D, aD) is 11% faster than the double-and-add-always method using a degenerate divisor. Using the same table size, the proposed method is 30% faster than the double-and-add-always method using a standard divisor. The results shown in Table 2 indicates that the proposed wNAF method is efficient for memory constrained devices. Conversely, the proposed method is not suitable for devices that have a large memory because the probability of finding multiple points of a degenerate divisor, such that $(D, aD, bD, ..)$, is low.

In the case of genus 3, the results are shown in Table 3. The number of operations is obtained from the addition formulae in Table 1. The results lead to a similar conclusion, but it is not always clear whether the degree of the degenerate divisor should be 1 or 2. The double-and-add-always method using a degree 1 divisor is faster than the wNAF method using a degree 2 divisor when $I = 5.2M$ (see Appendix B.1). In our computer experiment, the degenerate divisor of degree 1 is faster and requires less memory. This result is dependent on the computer environment for implementation. If $I > 11M$, then wNAF using a degree 2 divisor is faster.

Table 2. Degenerate base point and Standard base point (genus 2)

Base Point	Table Size (bits)	Scheme	# of operations ($I = 5.8M$)
degenerate	2n	double-and-add	$318I + 6038M(7882M)$
degenerate*	4n	wNAF ($w = 3$)	$267I + 5477M(7025M)$
standard	4n	double-and-add	$318I + 8268M(10112M)$
standard	8n	wNAF ($w = 2$)	$239I + 6293M(7679M)$
standard	16n	wNAF ($w = 3$)	$214I + 5723M(6964M)$
standard	32n	wNAF ($w = 4$)	$197I + 5239M(6381M)$

* method proposed in this paper.

Table 3. Degenerate base point and Standard base point (genus 3)

Base Point	deg(D)	Table Size (bits)	Scheme	# of operations ($I = 5.2M$)
degenerate*	1	2n	double-and-add	$318I + 16137M(17791M)$
degenerate*	2	4n	double-and-add	$318I + 21111M(22765M)$
standard	3	6n	double-and-add	$318I + 25281M(26935M)$
degenerate*	2	8n	wNAF ($w = 2$)	$239I + 17003M(18236M)$
standard	3	12n	wNAF ($w = 2$)	$239I + 19119M(20362M)$
standard	3	24n	wNAF ($w = 3$)	$214I + 17172M(18285M)$
standard	3	48n	wNAF ($w = 4$)	$197I + 15837M(16861M)$

* method proposed in this paper.

5   Summary

In this paper, we proposed explicit addition formulae with degenerate divisors for genus 3 hyperelliptic curves. The proposed technique can be applied to scalar multiplication of hyperelliptic curve cryptosystems using a fixed base point, e.g., DSA. We carefully estimated the efficiency of scalar multiplication using degenerate divisors. The proposed scheme using degenerate divisors of degree 1 or 2 attains a speed-up of approximately 34% or 15%, respectively, compared to the use of a standard divisor. In order to represent degenerate divisors, we require a shorter representation, and thus the base point can be compressed to 1/3 or 2/3.

Next, we investigated window-based methods for scalar multiplication using degenerate divisors. We estimated the probability of a pair of degenerate divisors D and aD for integer a and then presented an algorithm for generating such a pair for small a. Finally, we showed an efficiency comparison of schemes using window methods with degenerate divisors. The proposed scheme using degenerate divisors achieves a fast scalar multiplication using smaller memory than the conventional scheme using standard divisors.

Window-based methods require a pair of degenerate divisors (D, aD) for some integer a. The proposed method attempts to solve a polynomial obtained from the explicit formulae using degenerate divisors. The degree of the polynomial increases exponentially, and the polynomial can be solved for only for small a. Constructing a pair (D, aD) presents a problem when finding an algorithm for large a.

References

[Ava03]   R. Avanzi, "Countermeasures against Differential Power Analysis for Hyperelliptic Curve Cryptosystems," CHES 2003, LNCS 2779, Springer-Verlag, pp.366-381, 2003.

[Can87]   D. Cantor, "Computing in the Jacobian of a Hyperelliptic Curve," Mathematics of Computation, 48, 177, pp.95-101, 1987.

[Cor99]   J.-S. Coron, "Resistance against Differential Power Analysis for Elliptic Curve Cryptosystems," CHES '99, LNCS 1717, Springer-Verlag, pp.292-302, 1999.

[GMP]   GMP, GNU MP Library GMP. http://www.swox.com/gmp

[Har00a]   R. Harley, "Adding.text," 2000. http://cristal.inria.fr/~harley/hyper/

[Har00b]   R. Harley, "Doubling.c," 2000. http://cristal.inria.fr/~harley/hyper/

[KKA⁺04]   M. Katagi, I. Kitamura, T. Akishita, and T. Takagi, "Novel Efficient Implementations of Hyperelliptic Curve Cryptosystems using Degenerate Divisors," Cryptology ePrint Archive, 2003/203, IACR, 2004.

[Kob89]   N. Koblitz, "Hyperelliptic Cryptosystems," Journal of Cryptology, Vol.1, Springer-Verlag, pp.139-150, 1989.

[Lan02]   T. Lange, "Efficient Arithmetic on Genus 2 Hyperelliptic Curves over Finite Fields via Explicit Formulae," Cryptology ePrint Archive, 2002/121, IACR, 2002.

[MAG]   MAGMA: The Magma Computational Algebra System for Algebra, Number Theory and Geometry. http://magma.maths.usyd.edu.au/magma/

[Mum84]   D. Mumford, *Tata Lectures on Theta II*, Progress in Mathematics 43, Birkhäuser, 1984.

[MCT01]   K. Matsuo, J. Chao and S. Tsuji, "Fast Genus Two Hyperelliptic Curve Cryptosystems," Technical Report ISEC2001-31, IEICE Japan, pp.89-96, 2001.

[MWZ96]   A. Menezes, Y. Wu and R. Zuccherato, "An Elementary Introduction to Hyperelliptic Curves," Technical Report CORR 96-19, 1996. http://www.cacr.math.uwaterloo.ca/

[Nag00]   N. Nagao, "Improving Group Law Algorithms for Jacobians of Hyperelliptic Curves," ANTS-IV, LNCS 1838, Springer-Verlag, pp.439-448, 2000.

[NTL]   NTL: A Library for Doing Number Theory. http://www.shoup.net/ntl

[OT03]   K. Okeya and T. Takagi, "The Width-w NAF Method Provides Small Memory and Fast Elliptic Scalar Multiplications Secure against Side Channel Attacks," CT-RSA 2003, LNCS 2612, Springer-Verlag, pp.328-343, 2003.

[Pel02]   J. Pelzl, "Hyperelliptic Cryptosystems on Embedded Microprocessors," Diploma Thesis, Rühr-Universität Bochum, 2002.

[SMC⁺02]   T. Sugizaki, K. Matsuo, J. Chao, and S. Tsujii, "An Extension of Harley Addition Algorithm for Hyperelliptic Curves over Finite Fields of Characteristic Two," Technical Report ISEC2002-9, IEICE, pp.49-56, 2002.

[Ver02]   F. Vercauteren, "Computing Zeta Functions of Hyperelliptic Curves over Finite Fields of Characteristic 2," Crypto 2002, LNCS 2442, Springer-Verlag, pp.369-384, 2002.

A   Proposed Explicit Formulae for Genus 3

Algorithm 4. ExADD$^{2+3\rightarrow3}$

Input: $D_1 = (u_1, v_1)$, deg $u_1 = 3$, $D_2 = (u_2, v_2)$, deg $u_2 = 2$
Output: $D_3 = (u_3, v_3) = D_1 + D_2$

step	procedure	cost
1	Compute $r = \operatorname{res}(u_1, u_2)$: $w_0 \leftarrow u_{20}^2,\ w_1 \leftarrow u_{11}^2,\ w_2 \leftarrow u_{21}^2,\ w_3 \leftarrow u_{12} + u_{21}$, $w_4 \leftarrow w_0(u_{20} + u_{12}w_3),\ w_5 \leftarrow u_{21}(u_{10} + u_{11}w_3),\ w_5 \leftarrow u_{20}(w_5 + w_1)$, $w_6 \leftarrow w_3 w_2 + u_{21}u_{11},\ w_6 \leftarrow u_{10}(u_{10} + w_6),\ r \leftarrow w_4 + w_5 + w_6$	$11M$
2	Compute $ru_1^{-1} \bmod u_2 \equiv i_1 x + i_0$: $i_2 \leftarrow u_{21}u_{12},\ i_3 \leftarrow u_{21}u_{11},\ i_4 \leftarrow u_{20}u_{12}$, $i_1 \leftarrow i_2 + w_2 + u_{20} + u_{11},\ i_0 \leftarrow w_2 w_3 + i_3 + i_4 + u_{10}$	$4M$
3	Compute $t \equiv t_1 x + t_0 = r(v_1 + v_2)u_1^{-1} \bmod u_2$: $c_1 \leftarrow v_{11} + v_{21} + v_{12}u_{21},\ c_0 \leftarrow v_{20} + v_{10} + v_{12}u_{20}$, $t_2 \leftarrow i_1 c_1,\ t_3 \leftarrow i_0 c_0,\ t_1 = t_2 u_{21} + (i_1 + i_0)(c_1 + c_0) + t_2 + t_3$, $t_0 \leftarrow t_3 + t_2 u_{20}$.	$7M$
4	Compute $s = 1/r \equiv s_1 x + s_0$: $z_1 \leftarrow r t_1,\ z_2 \leftarrow 1/z_1,\ z_3 \leftarrow z_2 r,\ z_4 \leftarrow z_2 t_1,\ z_5 \leftarrow z_3 r,\ s_1 \leftarrow z_4 t_1,\ s_0 \leftarrow z_4 t_0$. **If $t_1 = 0$, goto exceptional procedure**	$1I + 6M$
5	Compute $v = su_1 + v_1 \equiv s_1 x^4 + k_3 x^3 + k_2 x^2 + k_1 x + k_0$: $t_0 \leftarrow s_0 u_{12},\ t_1 \leftarrow s_0 u_{10},\ t_2 \leftarrow s_1 u_{11}$, $k_3 \leftarrow (s_1 + s_0)(1 + u_{12}) + s_1 + t_0,\ k_2 \leftarrow t_0 + t_2 + v_{12}$, $k_1 \leftarrow (s_1 + s_0)(u_{11} + u_{10}) + t_2 + t_1 + v_{11},\ k_0 \leftarrow t_1 + v_{10}$.	$5M$
6	Compute $u_3 = s_1^{-2}(f + hv + v^2)/(u_1 u_2)$: $u_{32} \leftarrow z_5(z_5 + 1) + u_{12} + u_{21},\ t_0 \leftarrow k_3^2,\ t_1 \leftarrow u_{12}^2$, $t_2 \leftarrow z_5(z_5(f_6 + u_{12} + u_{21} + t_0 + k_3) + u_{21} + h_2 + u_{12})$, $u_{31} \leftarrow i_2 + u_{11} + u_{20} + t_1 + w_2 + t_2$, $t_3 \leftarrow (t_1 + w_2)(u_{21} + u_{12}) + i_3 + i_4 + u_{10}$, $t_4 \leftarrow i_2 + u_{20} + w_2 + u_{11} + t_1 + f_5 + (u_{21} + u_{12})(t_0 + f_6 + k_3) + k_2$, $t_4 \leftarrow z_5(t_4 + k_3 h_2) + h_2(u_{12} + u_{21}) + t_1 + w_2 + i_2 + u_{20} + u_{11} + h_1,\ t_4 \leftarrow z_5 t_4$, $u_{30} \leftarrow t_3 + t_4$.	$11M$
7	Compute $v_3 = v_{32}x^2 + v_{31}x + v_{30} \equiv su_1 + v_1 + h \bmod u_3$: $t_0 \leftarrow s_0(u_{32} + u_{12}),\ t_1 \leftarrow s_1(u_{31} + u_{11}),\ t_2 \leftarrow s_1(u_{12} + u_{32})$, $v_{32} \leftarrow t_0 + t_1 + t_2 u_{32} + u_{32} + v_{12} + h_2$, $t_4 \leftarrow s_0(u_{30} + u_{10}),\ t_5 \leftarrow (s_1 + s_0)(u_{31} + u_{11} + u_{30} + u_{10})$, $v_{31} \leftarrow t_5 + t_1 + t_4 + t_2 u_{31} + u_{31} + v_{11} + h_1$, $v_{30} \leftarrow t_4 + t_2 u_{30} + u_{30} + v_{10} + h_0$.	$8M$
total	ExADD$^{3+2\rightarrow3}$	$1I + 52M$

Algorithm 5. ExADD$^{1+3\rightarrow3}$

Input: $D_1 = (u_1, v_1)$, deg $u_1 = 3$, $D_2 = (u_2, v_2)$, deg $u_2 = 1$
Output: $D_3 = (u_3, v_3) = D_1 + D_2$

step	procedure	cost
1	Compute $r = \operatorname{res}(u_1, u_2)$: $w_0 \leftarrow u_{20}^2,\ w_1 \leftarrow w_0(u_{12} + u_{20}),\ w_2 \leftarrow u_{20}u_{11},\ r \leftarrow w_1 + w_2 + u_{10}$.	$3M$
2	Compute *inverse of* $u_1 \bmod u_2$: $inv \leftarrow 1/r$.	$1I$
3	Compute $s_0 = inv(v_1 + v_2) \bmod u_2$: $z_0 \leftarrow w_0 v_{12},\ s_0 \leftarrow inv(v_{10} + v_{20} + u_{20}v_{11} + z_0)$.	$3M$
4	Compute $u_3 = (f + hv + v^2)/(u_1 u_2),\ v = s_0 u_1 + v_1$: $u_{32} \leftarrow s_0^2 + s_0 + u_{20} + u_{12} + f_6$ $t_0 \leftarrow f_6 + s_0^2 + u_{12},\ t_1 \leftarrow u_{12}t_0,\ t_2 \leftarrow u_{20}u_{32},\ t_3 \leftarrow h_2 s_0$, $u_{31} \leftarrow t_1 + t_2 + t_3 + u_{11} + v_{12} + f_5$, $t_4 \leftarrow u_{20}(t_6 + v_{12} + f_5 + t_3 + u_{11}),\ t_5 \leftarrow v_{12}(v_{12} + u_{12} + h_2)$, $t_6 \leftarrow u_{12}(u_{12}(f_6 + u_{12}) + f_5)$, $u_{30} \leftarrow w_0 u_{32} + t_4 + t_5 + u_{12}t_0 + s_0 h_1 + t_6 + u_{10} + f_4 + v_{11}$.	$12M$
5	Compute $v_3 = v_{32}x^2 + v_{31}x + v_{30} \equiv s_0 u_1 + v_1 + h \bmod u_3$: $v_{32} \leftarrow v_{12} + h_2 + s_0(u_{12} + u_{32}) + u_{32}$, $v_{31} \leftarrow v_{11} + h_2 + s_0(u_{11} + u_{31}) + u_{31}$, $v_{30} \leftarrow v_{10} + h_2 + s_0(u_{10} + u_{30}) + u_{30}$.	$3M$
total	ExADD$^{3+1\rightarrow3}$	$1I + 21M$

B   Experimental Results

The experiment was implemented on an Intel Xeon Processor (2.80 GHz) using the Linux 2.4 (RedHat) operation system. We employed the gcc 3.3 compiler and the number theoretic library NTL5.3 [NTL] with GMP4.0 [GMP].

B.1   Timing of Field Operations

In order to obtain an appropriate relation between multiplication M and inversion I, we conducted an experiment. Let $\mathbb{F}_{2^{83}}$ be defined as $\mathbb{F}_2[t]/f(t)$, where $f(t) = t^{83} + t^7 + t^4 + t^2 + 1$ and let $\mathbb{F}_{2^{59}}$ be defined as $\mathbb{F}_2[t]/f(t)$, where $f(t) = t^{59} + t^7 + t^4 + t^2 + 1$. Table 4 shows the timing obtained in our experiment. In our computational environment, the timing ratios of the inversion by the multiplication is estimated to be $I/M = 5.8$ or 5.2 from 10^7 random samples, respectively.

Table 4. Timing of inversion and multiplication (in μs)

	mul (M)	inversion (I)	I/M
$\mathbb{F}_{2^{83}}$	0.74	4.29	5.77
$\mathbb{F}_{2^{59}}$	0.53	2.76	5.18

B.2   Scalar Multiplication Using Degenerate Divisors for Genus 3

In order to demonstrate the efficiency of the proposed algorithms, we implemented the proposed schemes in section 3. For our experiment we chose the hyperelliptic curve of genus 3 from [Ver02]. The timing of scalar multiplication is an average of 10^4 scalar multiplications. Table 5 shows that the improvement over the double-and-add-always method using a standard divisor is 7% (or 27%) using a degenerate divisor of degree 2 (or 1).

Table 5. Improved timing of scalar multiplication (genus 3)

Base Point	Timing
standard (weight 3)	32.7ms
degenerate (weight 2)	30.5ms
degenerate (weight 1)	23.8ms

C   Some Examples of D with $\deg(D) = \deg(aD)$

C.1   Example of D with $\deg(D) = \deg(5D) = 1$ for Genus 2

Finite Field $\mathbb{F}_{2^{83}}$:

$$\mathbb{F}_2[t]/f(t), \quad f(t) = t^{83} + t^7 + t^4 + t^2 + 1$$

Hyperelliptic curve C_2:

$$y^2 + (x^2 + \sum_{i=0}^{1} h_i x^i)y = x^5 + \sum_{i=0}^{3} f_i x^i,$$

$h_1 = 3b0533db4d29cf09ab889 \quad h_0 = 5b78506748b8b438bc2a1$
$f_3 = 799ce3fba76a739ca9f4d \quad f_2 = 0$
$f_1 = 5aaceb489dbd99e2b9289 \quad f_0 = 572a506ced9f3560b1acd$

Group order of the Jacobian $J_{C_2}(\mathbb{F}_{2^{83}})$:

$2 \times 4676805239455988244543155602094275148754154323469.$

Degenerate base point D:

$D = (x + x_1, y_1)$

$x_1 = 2a87b03b3d0fad48bac8 \quad y_1 = 3991e099305a0cdec6fa5$

$5D = (x + x_2, y_2)$

$x_2 = e464319c1b8b1988fb75 \quad y_2 = 51f16e7d156b43b406478$

C.2   Example of D with $\deg(D) = \deg(3D) = 2$ for Genus 3

Finite Field $\mathbb{F}_{2^{59}}$:

$$\mathbb{F}_2[t]/f(t), \quad f(t) = t^{59} + t^7 + t^4 + t^2 + 1$$

Hyperelliptic curve C_3 [Ver02]:

$$y^2 + (x^3 + \sum_{i=0}^{2} h_i x^i)y = x^7 + \sum_{i=0}^{5} f_i x^i \quad \text{over} \quad \mathbb{F}_{2^{59}}$$

$h_2 = fd34cf935a40744 \quad h_1 = ad088ab2fb72242$
$h_0 = \quad a25cacf3751cb$
$f_5 = \quad 66b8ddf5319e95 \quad f_4 = \quad 53aa4f37e88bcc$
$f_3 = f3632bafee934d5 \quad f_2 = 68170b377ee7c81$
$f_1 = \quad 69c43841ed3704 \quad f_0 = cada3798772ba56$

Group order of the Jacobian $J_{C_3}(\mathbb{F}_{2^{59}})$:

$2 \times 95780971407243394633762332360123160334059170481903949.$

Degenerate base point D:

$D = (x^2 + u_{11}x + u_{10}, v_{11}x + v_{10})$

$u_{11} = 57ef5eed895c0c4 \quad u_{10} = 27bcb5afa12dcd2$
$v_{11} = 21e335606275529 \quad v_{10} = 52e9fd32a37658c$

$3D = (x^2 + u_{21}x + u_{20}, v_{21}x + v_{20})$

$u_{21} = 1fbfc19ba156e11 \quad u_{20} = 3470af7f7fd7511$
$v_{21} = 6729f750620f0a6 \quad v_{20} = 24be045a800c3ad$

D   Proof of the Minimality

If we scan more than three bits, then schemes with lower than non-zero density can be achieved. Indeed, in the case of $k_1 = 2, k_2 = 3$, we can construct schemes with non-zero density $3/5$ for $a = 5, 7, 9, 13$. There is an SPA-resistant scheme for $k_1 = 3, k_2 = 4$, and $a = 13$. However, there is no SPA-resistant scheme if we scan larger bits. We can prove the following theorem:

Theorem 4. *The minimal non-zero density of the SPA-resistant scheme with two bases D and aD is $4/7$ for $a = 13$.*

Proof. If we try to construct an SPA-resistant scheme with a different pattern from $|D...D|DA...DA|$, then all digits appearing in the width-w NAF method can not be represented. Therefore, the SPA-resistant scheme should have the pattern $|D...D|DA...DA|$.

We have already shown a construction of the scheme with non-zero density $4/7$. In order to achieve non-zero density smaller than $4/7$, we should scan more than 11 bits. However, the largest integer $2^{12} - 1$ of 12 bits can not be represented by the pattern with $k_1 = 6$ and $k_2 = 7$. If we choose bits larger than 11, then the difference becomes larger. Consequently, there is no SPA-resistant scheme smaller than $4/7$.   □

On the Pseudorandomness of a Modification of KASUMI Type Permutations

Wonil Lee[1], Kouichi Sakurai[1], Seokhie Hong[2], and Sangjin Lee[3]

[1] Faculty of Information Science and Electrical Engineering,
Kyushu University, Fukuoka, Japan
wonil@itslab.csce.kyushu-u.ac.jp
sakurai@csce.kyushu-u.ac.jp
[2] Katholieke Universiteit Leuven, Dept. ESAT/COSIC,
Kasteelpark Arenberg 10,
B-3001 Leuven-Heverlee, Belgium
shong@esat.kuleuven.ac.be
[3] Center for Information Security Technologies (CIST),
Korea University, Seoul, Korea
sangjin@cist.korea.ac.kr

Abstract. We present a modification of KASUMI type permutations and analyze the security of it using the notion of pseudorandomness. Our modified KASUMI type permutation can be computed more efficiently than the original KASUMI type permutation. Furthermore, our results have a slightly better (same) upper bound of success probability against arbitrary attackers in the sense of (super) pseudorandomness.

Keywords: Pseudorandomness, provable security, block cipher, KASUMI.

1 Introduction

Brief history. Luby and Rackoff [4] introduced a theory for the security of block ciphers by using the notion of pseudorandomness. One of the purposes of the security analysis using the notion of pseudorandomness is to measure the security of the structures used in the block ciphers. Roughly speaking, the security of the structure is analyzed after the main functions (such as round functions in Feistel transformations) is replaced with a pseudorandom function or pseudorandom permutation. With this replacement, Luby and Rackoff showed that the three round DES type permutation is a pseudorandom permutation and the four round one is a super-pseudorandom permutation. [4]

KASUMI is a block cipher which has been adopted as a standard of 3GPP [1], where 3GPP is the body standardizing the next generation of mobile telephony. The overall structure of KASUMI is a Feistel permutation and each round function consists of two functions, FL function and FO function. Each FO function consists of a three round MISTY type permutation, where each round function

C. Park and S. Chee (Eds.): ICISC 2004, LNCS 3506, pp. 313–329, 2005.
© Springer-Verlag Berlin Heidelberg 2005

is called an FI function. And each FI function consists of a four round MISTY type permutation. See [1, 2] for details.

Recently Iwata, Yagi, and Kurosawa [2] presented results about the pseudorandomness of KASUMI for adaptive adversarial model. They first idealized KASUMI as follows.

- Each FL function is ignored.
- Each FI function is idealized by an independant (pseudo) random permutation.

They called such an idealized KASUMI a "KASUMI type permutation," and proved that the four round and six round idealized KASUMI type permutations are pseudorandom and super-pseudorandom, respectively, for adaptive adversaries.

Motivation.   The results of [2] are related to the following question:

- How to provide a construction method of a (super) pseudorandom permuation with large input size using several independent pseudorandom permutations with small input size.

More specifically, their results show that if we only consider the KASUMI-like structure which has Feistel structure as the overall structure and MISTY structure as the round structure, there exist

1. a construction method of a $4n$-bit input size pseudorandom permuation using **"twelve"** independent n-bit input size pseudorandom permutations, and
2. a construction method of a $4n$-bit input size super pseudorandom permuation using **"eighteen"** independent n-bit input size pseudorandom permutations.

The results of [2] shows that the high level structure of KASUMI block cipher can be used to get the above results.

Here, we can think a next natural question. That is, how to reduce the number of using n-bit input size pseudorandom permutations in order to obtain a $4n$-bit input size (super) pseudorandom permuation, while it preserves security of the above results.

Our contribution.   In this paper, we show that if we only consider the KASUMI-like structure, there exist

1. a construction method of a $4n$-bit input size pseudorandom permuation using **"ten"** n-bit input size pseudorandom permutations and
2. a construction method of a $4n$-bit input size super-pseudorandom permuation using **"sixteen"** n-bit input size pseudorandom permutations.

We will first define a modification of the high level structure of KASUMI block cipher and then prove our above results with the modified structure in the adaptive adversarial model.

Our modification is similar to the KASUMI type permutation except the use of just two round MISTY-type permutation as the round function of it. Details can be shown in Section 2.2. Here note that two round MISTY-type permutation can be computed parallelly (the values of each round functions can be computed simultaneously). So our modification can be computed more efficiently than the original KASUMI type permutation. Furthermore our results have slightly better (same) upper bound of success probability of arbitrary attacker in the sense of (super) pseudorandomness. A summary of our results is given by Table 1. The model of attacker and the meaning of q will be described momentarily.

Table 1. Summary of the previous results and our contributions. (# BP means the number of basic permutations and UP an upper bound of success probability.)

	Pseudorandom		Super-pseudorandom	
	# BP	UP	# BP	UP
[2]	12	$\dfrac{15}{2} \cdot \dfrac{q(q-1)}{2^n - 1}$	18	$\dfrac{9q(q-1)}{2^n - 1}$
this paper	10	$\dfrac{7q(q-1)}{2^n - 1}$	16	$\dfrac{9q(q-1)}{2^n - 1}$

2   Preliminaries

Our overall treatment follows the nicely laid out framework of Iwata, Yagi, and Kurosawa [2].

2.1   Notation

For a bit string $x \in \{0,1\}^{4n}$, we denote the first n bits of x by x_{LL}, the next n bits of x by x_{LR}, the third n bits of x by x_{RL}, and the last n bits of x by x_{LL}. That is, $x = (x_{LL}, x_{LR}, x_{RL}, x_{RR})$. For a set of l-bit strings $\{x^{(i)} | x^{(i)} \in \{0,1\}^l\}_{1 \le i \le q}$, we say $\{x^{(i)}\}_{1 \le i \le q}$ are distinct if $x^{(i)} \ne x^{(j)}$ for all $1 \le i < j \le q$.

If S is a set, $s \xleftarrow{R} S$ denotes the process of picking an element from S uniformly at random. Denote by P_n the set of all permutations over $\{0,1\}^n$, which consists of $(2^n)!$ permutations in total. For functions f and g, $g \circ f$ denotes the function $x \mapsto g(f(x))$.

2.2   A Modification of KASUMI Type Permutation

In this section, we provide the definition of our modification of KASUMI type permutations. We call it "the MKASUMI type permutation".

Definition 1 (The basic MKASUMI type permutation). *Let $x \in \{0,1\}^{4n}$. For any permutations $p_1, p_2 \in P_n$, define the basic MKASUMI type permutation $\psi_{p_1, p_2} \in P_{4n}$ as*

$$\psi_{p_1,p_2}(x) \triangleq y$$

where $y_{LL} \triangleq x_{RL}$, $y_{LR} \triangleq x_{RR}$, $y_{RL} \triangleq x_{RL} \oplus p_1(x_{RR}) \oplus p_2(x_{RL}) \oplus x_{LL}$, *and* $y_{RR} \triangleq x_{RL} \oplus p_1(x_{RR}) \oplus x_{LL}$.

Definition 2 (The r round MKASUMI type permutation). *Let $r \geq 1$ be an integer, and $p_1, p_2, ..., p_{2r} \in P_n$ be permutations. Define the r round MKA-SUMI type permutation $\psi(p_1, p_2, ..., p_{2r}) \in P_{4n}$ as*

$$\psi(p_1, p_2, ..., p_{2r}) \triangleq \psi_{p_{2r-1},p_{2r}} \circ \psi_{p_{2r-3},p_{2r-2}} \circ \cdots \circ \psi_{p_1,p_2}$$

See Fig. 1 for illustrations. In this paper, for $1 \leq i \leq q$ and $1 \leq j \leq 2r$, let $I_j^{(i)}$ denote the input to p_i when the input to ψ is $x^{(i)}$ and the output is $y^{(i)}$. Similarly, let $O_j^{(i)}$ denote the output of p_i when the input to ψ is $x^{(i)}$ and the output is $y^{(i)}$.

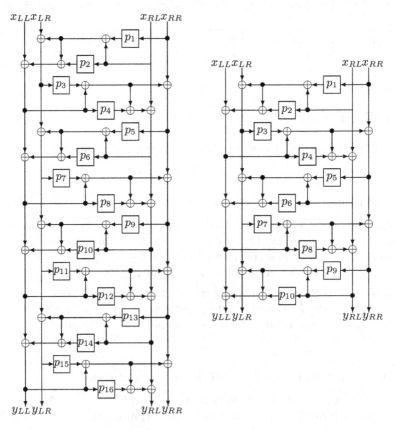

Fig. 1. Eight round MKASUMI type permutation $\psi(p_1, ..., p_{16})$ and five round MKA-SUMI type permutation $\psi(p_1, ..., p_{10})$

2.3   Pseudorandom and Super-Pseudorandom Permutations

Our adaptive adversary A is modeled as a Turing machine that has black-box access to an oracle (or oracles). The computational power of A is unlimited, but the total number of oracle calls is limited to a number q. After making at most q queries to the oracle(s) adaptively, A outputs a bit. In this paper, we assume that A never asks a query if its answer is determined by a previous query-answer pair.

The pseudorandomness of a block cipher Ψ over $\{0,1\}^{4n}$ captures its computational indistinguishability from P_{4n}, where the adversary is given access to the forward direction of the permutation. In other words, it measures security of a block cipher against adaptive chosen plaintext attack.

Definition 3 (Pseudorandomness). *Let a block cipher Ψ be a family of permutations over $\{0,1\}^{4n}$. Let A be an adversary. Then, in the sense of pseudorandomness of Ψ, A's advantage is defined by*

$$Adv_{\Psi}^{prp}(A) \triangleq |Pr(\psi \overset{R}{\leftarrow} \Psi : A^{\psi} = 1) - Pr(R \overset{R}{\leftarrow} P_{4n} : A^{R} = 1)|.$$

A^{ψ} indicates A with an oracle which, in response to a query x, returns $y \leftarrow \psi(x)$. A^{R} indicates A with an oracle which, in response to a query x, returns $y \leftarrow R(x)$.

The super-pseudorandomness of a block cipher Ψ over $\{0,1\}^{4n}$ captures its computational indistinguishability from P_{4n}, where the adversary is given access to both directions of the permutation. In other words, it measures security of a block cipher against adaptive chosen plaintext and adaptive chosen ciphertext attacks.

Definition 4 (Super-Pseudorandomness). *Let a block cipher Ψ be a family of permutations over $\{0,1\}^{4n}$. Let A be an adversary. Then, in the sense of super-pseudorandomness of Ψ, A's advantage is defined by*

$$Adv_{\Psi}^{sprp}(A) \triangleq |Pr(\psi \overset{R}{\leftarrow} \Psi : A^{\psi,\psi^{-1}} = 1) - Pr(R \overset{R}{\leftarrow} P_{4n} : A^{R,R^{-1}} = 1)|.$$

$A^{\psi,\psi^{-1}}$ indicates A with an oracle which, in response to a query $(+,x)$, returns $y \leftarrow \psi(x)$, and in response to a query $(-,y)$, returns $x \leftarrow \psi^{-1}(y)$. $A^{R,R^{-1}}$ indicates A with an oracle which, in response to a query $(+,x)$, returns $y \leftarrow R(x)$, and in response to a query $(-,y)$, returns $x \leftarrow R^{-1}(y)$.

3   Five Round MKASUMI Type Permutation Is Pseudorandom

Theorem 1. *For $1 \le i \le 10$, let $p_i \in P_n$ be a random permutation. Let $\psi = \psi(p_1, ..., p_{10})$ be a five round MKASUMI. And let $R \in P_{4n}$ be a random permutation and $\Psi \triangleq \{\psi \mid \psi = \psi(p_1, ..., p_{10}), p_i \in P_n \text{ for } 1 \le i \le 10\}$.*

Then for any adversary A that makes at most q queries in total,

$$Adv_{\Psi}^{prp}(A) \leq \frac{7q(q-1)}{2^n - 1}.$$

Proof. Let $\mathcal{O}$ be either R or ψ. The adversary A has oracle access to $\mathcal{O}$. A can make a query x and the oracle returns $y = \mathcal{O}(x)$. For the i-th query A makes to $\mathcal{O}$, define the query-answer pair $(x^{(i)}, y^{(i)}) \in \{0,1\}^{4n} \times \{0,1\}^{4n}$, where A's query was $x^{(i)}$ and the answer it got was $y^{(i)}$. Define view v of A as $v = \langle (x^{(1)}, y^{(1)}), ..., (x^{(q)}, y^{(q)}) \rangle$.

Since A is computationally unbounded, we may without loss of generality assume that A is deterministic. This implies that for every $1 \leq i \leq q$ the i-th query $x^{(i)}$ is fully determined by the first $i-1$ query-answer pairs, and the final output of A (0 or 1) depends only on v. Therefore, there exists a function $C_A(\cdot)$ such that

$$\begin{cases} C_A(x^{(1)}, y^{(1)}, ..., x^{(i-1)}, y^{(i-1)}) = x^{(i)} \text{ for } 1 \leq i \leq q \text{ and} \\ C_A(v) = A\text{'s final output.} \end{cases}$$

We say that $v = \langle (x^{(1)}, y^{(1)}), ..., (x^{(q)}, y^{(q)}) \rangle$ is a *possible* view if for every $1 \leq i \leq q$, $C_A(x^{(1)}, y^{(1)}, ..., x^{(i-1)}, y^{(i-1)}) = x^{(i)}$.

Let $v_{one} \triangleq \{v | C_A(v) = 1 \text{ and } v \text{ is possible}\}$. Further, we let v_{good} be a set of all possible view $v = \langle (x^{(1)}, y^{(1)}), ..., (x^{(q)}, y^{(q)}) \rangle$ which satisfies the following four conditions: (1) $C_A(v) = 1$, (2) $\{y_{RL}^{(i)}\}_{1 \leq i \leq q}$ are distinct, (3) $\{y_{RR}^{(i)}\}_{1 \leq i \leq q}$ are distinct, and (4) $\{x_{RL}^{(i)} \oplus x_{RR}^{(i)} \oplus y_{RL}^{(i)} \oplus y_{RR}^{(i)}\}_{1 \leq i \leq q}$ are distinct.

Evaluation of p_R. We first evaluate $p_R \triangleq Pr(R \xleftarrow{R} P_{4n} : A^R = 1)$. We have $p_R = \frac{\#\{R | A^R = 1\}}{(2^{4n})!}$. For each $v \in v_{one}$, the number of R such that

$$R(x^{(i)}) = y^{(i)} \text{ for all } 1 \leq i \leq q \tag{1}$$

is exactly $(2^{4n} - q)!$. Therefore, we have $p_R = \sum_{v \in v_{one}} \frac{\#\{R | R \text{ satisfying } (1)\}}{(2^{4n})!} = \#v_{one} \cdot \frac{(2^{4n}-q)!}{(2^{4n})!}$.

Evaluation of p_ψ. We evaluate $p_\psi \triangleq Pr(\psi \xleftarrow{R} \Psi : A^\psi = 1)$, where $\psi \xleftarrow{R} \Psi$ means that $p_i \xleftarrow{R} P_n$ for $1 \leq i \leq 10$ and then let $\psi \leftarrow \psi(p_1, ..., p_{10})$. Then we have $p_\psi = \frac{\#\{(p_1, ..., p_{10}) | A^\psi = 1\}}{((2^n)!)^{10}}$.

We have the following lemmas. A proof of Lemma 1 is given in Section 4.

Lemma 1 (Main Lemma). *For any fixed possible view $v = \langle (x^{(1)}, y^{(1)}), ..., (x^{(q)}, y^{(q)}) \rangle$ such that $\{y_{RL}^{(i)}\}_{1 \leq i \leq q}$ are distinct, $\{y_{RR}^{(i)}\}_{1 \leq i \leq q}$ are distinct, and $\{x_{RL}^{(i)} \oplus x_{RR}^{(i)} \oplus y_{RL}^{(i)} \oplus y_{RR}^{(i)}\}_{1 \leq i \leq q}$ are distinct, the number of $(p_1, ..., p_{10})$ which satisfies*

$$\psi(x^{(i)}) = y^{(i)} \text{ for } 1 \leq \forall i \leq q \tag{2}$$

is at least $(1 - \frac{11}{2} \cdot \frac{q(q-1)}{2^n - 1}) \cdot \{2^n!\}^6 \cdot \{(2^n - q)!\}^4$.

Lemma 2. $\#v_{good} \geq \#v_{one} - \frac{3}{2} \cdot \frac{q(q-1)}{2^n-1} \cdot \frac{(2^{4n})!}{(2^{4n}-q)!}$.

Proof. The proof is almost similar to the proof of [2]. So we omit it. ∎

Then from Lemma 1 and 2, we have

$$p_\psi = \sum_{v \in v_{one}} \frac{\#\{(p_1,...,p_{10})|(p_1,...,p_{10}) \text{ satisfying } (2)\}}{\{(2^n)!\}^{10}}$$

$$\geq \sum_{v \in v_{good}} \frac{\#\{(p_1,...,p_{10})|(p_1,...,p_{10}) \text{ satisfying } (2)\}}{\{(2^n)!\}^{10}}$$

$$\geq \sum_{v \in v_{good}} (1 - \frac{11}{2} \cdot \frac{q(q-1)}{2^n-1}) \cdot \frac{\{(2^n-q)!\}^4}{\{(2^n)!\}^4}$$

$$\geq (\#v_{one} - \frac{3}{2} \cdot \frac{q(q-1)}{2^n-1} \cdot \frac{(2^{4n})!}{(2^{4n}-q)!}) \cdot (1 - \frac{11}{2} \cdot \frac{q(q-1)}{2^n-1}) \cdot \frac{\{(2^n-q)!\}^4}{\{(2^n)!\}^4}$$

$$= (p_R - \frac{3}{2} \cdot \frac{q(q-1)}{2^n-1}) \cdot (1 - \frac{11}{2} \cdot \frac{q(q-1)}{2^n-1}) \cdot \frac{\{(2^n-q)!\}^4}{\{(2^n)!\}^4} \cdot \frac{(2^{4n})!}{(2^{4n}-q)!}$$

Now it is easy to see that $\frac{\{(2^n-q)!\}^4}{\{(2^n)!\}^4} \cdot \frac{(2^{4n})!}{(2^{4n}-q)!} \geq 1$ (this can be shown easily by an induction on q). Then $p_\psi \geq (p_R - \frac{3}{2} \cdot \frac{q(q-1)}{2^n-1}) \cdot (1 - \frac{11}{2} \cdot \frac{q(q-1)}{2^n-1}) \geq p_R - \frac{7q(q-1)}{2^n-1}$. Applying the same argument to $1-p_\psi$ and $1-p_R$ yields that $1-p_\psi \geq 1-p_R - \frac{7q(q-1)}{2^n-1}$, and we have $|p_\psi - p_R| \leq \frac{7q(q-1)}{2^n-1}$. ∎

From Theorem 1, it is very easy to show $\psi = \psi(p_1,...,p_{10})$ is pseudorandom even if each p_i is a pseudorandom permutation by using a standard hybrid argument [4].

4   Proof of Lemma 1

First, we need following three lemmas.

Lemma 3. Let $v = \langle (x^{(1)}, y^{(1)}),...,(x^{(q)}, y^{(q)}) \rangle$ be a fixed possible view. Then $\#\{(p_1, p_2, p_3, p_4)| \exists i, j \text{ such that } 1 \leq i < j \leq q \text{ and } I_6^{(i)} = I_6^{(j)}\} \leq \frac{q(q-1)}{2} \cdot \frac{3 \cdot \{(2^n)!\}^4}{2^n-1}$.

Proof. A proof is given in Appendix. ∎

Lemma 4. Let $v = \langle (x^{(1)}, y^{(1)}),...,(x^{(q)}, y^{(q)}) \rangle$ be a fixed possible view. Then $\#\{(p_1, p_2, p_3, p_4)| \exists i, j \text{ such that } 1 \leq i < j \leq q \text{ and } I_5^{(i)} = I_5^{(j)}\} \leq \frac{q(q-1)}{2} \cdot \frac{2 \cdot \{(2^n)!\}^4}{2^n-1}$.

Proof. If we prove the lemma in the following four cases, the proof is almost similar to the proof of Lemma 3. So we omit the details.

Case 1: $x_{RL}^{(i)} \neq x_{RL}^{(j)}$.

Case 2: $x_{RR}^{(i)} \neq x_{RR}^{(j)}$ and $x_{RL}^{(i)} = x_{RL}^{(j)}$.

Case 3: $x_{LL}^{(i)} \neq x_{LL}^{(j)}$, $x_{RL}^{(i)} = x_{RL}^{(j)}$ and $x_{RR}^{(i)} = x_{RR}^{(j)}$.

Case 4: $x_{LR}^{(i)} \neq x_{LR}^{(j)}$, $x_{LL}^{(i)} = x_{LL}^{(j)}$, $x_{RL}^{(i)} = x_{RL}^{(j)}$, and $x_{RR}^{(i)} = x_{RR}^{(j)}$.   ■

Lemma 5. *Let $v = \langle (x^{(1)}, y^{(1)}), ..., (x^{(q)}, y^{(q)}) \rangle$ be a fixed possible view such that $\{x_{RL}^{(i)} \oplus x_{RR}^{(i)} \oplus y_{RL}^{(i)} \oplus y_{RR}^{(i)}\}_{1 \leq i \leq q}$ are distinct. Then*

$$\#\{(p_1, p_2, p_3, p_4) | \; \exists \; i,j \; such \; that \; 1 \leq i < j \leq q \; and \; O_8^{(i)} = O_8^{(j)}\} \leq \frac{q(q-1)}{2} \cdot \frac{\{(2^n)!\}^4}{2^n - 1}$$

Proof.   First, we fix i and j such that $1 \leq i < j \leq q$, and consider the condition $O_8^{(i)} = O_8^{(j)}$. Now observe that $O_8^{(i)} = O_8^{(j)}$ is equivalent to the following condition:

$$p_4(I_4^{(i)}) \oplus x_{RL}^{(i)} \oplus y_{RL}^{(i)} \oplus x_{RR}^{(i)} \oplus y_{RR}^{(i)} = p_4(I_4^{(j)}) \oplus x_{RL}^{(j)} \oplus y_{RL}^{(j)} \oplus x_{RR}^{(j)} \oplus y_{RR}^{(j)}.$$

Then the number of p_4 which satisfies the above condition is at most $\frac{(2^n)!}{2^n - 1}$, since our assumption. Therefore,

$$\#\{(p_1, p_2, p_3, p_4) | \; (p_1, p_2, p_3, p_4) \; satisfies \; O_8^{(i)} = O_8^{(j)}\} \leq \frac{\{(2^n)!\}^4}{2^n - 1}$$

and since we have $\binom{q}{2}$ choice of i and j the lemma follows.   ■

Proof of Lemma 1.   (See figures in Appendix.)

Initially, $x^{(1)}, ..., x^{(q)}, y^{(1)}, ..., y^{(q)}$ are fixed.

Number of $(p_1, ..., p_4)$.   From Lemma 3, 4, and 5, the number of $(p_1, ..., p_4)$ such that:

- $I_5^{(i)} \neq I_5^{(j)}$, $I_6^{(i)} \neq I_6^{(j)}$, and $O_8^{(i)} \neq O_8^{(j)}$ for $1 \leq \forall i < \forall j \leq q$,

is at least $\{2^n!\}^4 - \frac{q(q-1)}{2} \cdot \frac{\{2^n!\}^4}{2^n - 1} - \frac{2q(q-1)}{2} \cdot \frac{\{2^n!\}^4}{2^n - 1} - \frac{3q(q-1)}{2} \cdot \frac{\{2^n!\}^4}{2^n - 1}$. Fix any $(p_1, ..., p_4)$ which satisfy these three conditions.

Number of p_5.   For any fixed i and j such that $1 \leq i < j \leq q$, the number of p_5 such that

$$p_5(I_5^{(i)}) \oplus I_6^{(i)} \oplus I_3^{(i)} = p_5(I_5^{(j)}) \oplus I_6^{(j)} \oplus I_3^{(j)},$$

which is equivalent to $I_7^{(i)} = I_7^{(j)}$, is at most $\frac{(2^n)!}{2^n - 1}$, since $I_5^{(i)} \neq I_5^{(j)}$.
    Similarly, the number of p_5 such that

$$p_5(I_5^{(i)}) \oplus I_3^{(i)} \oplus I_6^{(i)} \oplus y_{LR}^{(i)} \oplus y_{RL}^{(i)} = p_5(I_5^{(j)}) \oplus I_3^{(j)} \oplus I_6^{(j)} \oplus y_{LR}^{(j)} \oplus y_{RL}^{(j)},$$

which is equivalent to $O_9^{(i)} = O_9^{(j)}$, is at most $\frac{(2^n)!}{2^n - 1}$, since $I_5^{(i)} \neq I_5^{(j)}$.

Then the number of p_5 which satisfies:

– $I_7^{(i)} \neq I_7^{(j)}$ and $O_9^{(i)} \neq O_9^{(j)}$ for $1 \leq \forall i < \forall j \leq q$,

is at least $(2^n)! - \frac{q(q-1)\{(2^n)!\}}{2^n-1}$. Fix any p_5 which satisfy the above two conditions.

Number of p_6. For any fixed i and j such that $1 \leq i < j \leq q$, the number of p_6 such that

$$p_6(I_6^{(i)}) \oplus I_6^{(i)} \oplus O_3^{(i)} \oplus O_5^{(i)} \oplus x_{RR}^{(i)} \oplus y_{RR}^{(i)} = p_6(I_6^{(j)}) \oplus I_6^{(j)} \oplus O_3^{(j)} \oplus O_5^{(j)} \oplus x_{RR}^{(j)} \oplus y_{RR}^{(j)},$$

which is equivalent to $O_7^{(i)} = O_7^{(j)}$, is at most $\frac{(2^n)!}{2^n-1}$, since $I_6^{(i)} \neq I_6^{(j)}$.

Similarly, the number of p_6 satisfies

$$p_6(I_6^{(i)}) \oplus I_4^{(i)} \oplus I_6^{(i)} \oplus O_5^{(i)} = p_6(I_6^{(j)}) \oplus I_4^{(j)} \oplus I_6^{(j)} \oplus O_5^{(j)},$$

which is equivalent to $I_8^{(i)} = I_8^{(j)}$, is at most $\frac{(2^n)!}{2^n-1}$, since $I_6^{(i)} \neq I_6^{(j)}$.

Similarly, the number of p_6 satisfies

$$p_6(I_6^{(i)}) \oplus I_3^{(i)} \oplus I_4^{(i)} \oplus y_{LL}^{(i)} \oplus y_{LR}^{(i)} = p_6(I_6^{(j)}) \oplus I_3^{(j)} \oplus I_4^{(j)} \oplus y_{LL}^{(j)} \oplus y_{LR}^{(j)},$$

which is equivalent to $O_{10}^{(i)} = O_{10}^{(j)}$, is at most $\frac{(2^n)!}{2^n-1}$, since $I_6^{(i)} \neq I_6^{(j)}$.

Then the number of p_6 satisfies:

– $O_7^{(i)} \neq O_7^{(j)}$, $I_8^{(i)} \neq I_8^{(j)}$, and $O_{10}^{(i)} \neq O_{10}^{(j)}$ for $1 \leq \forall i < \forall j \leq q$,

is at least $(2^n)! - \frac{3q(q-1)\{(2^n)!\}}{2(2^n-1)}$. Fix any p_6 which satisfy the above three conditions.

Number of $(p_7, ..., p_{10})$. Now $p_1, ..., p_6$ are fixed in such a way that $\{I_7^{(i)}\}_{1 \leq i \leq q}$ are distinct, $\{O_7^{(i)}\}_{1 \leq i \leq q}$ are distinct, $\{I_8^{(i)}\}_{1 \leq i \leq q}$ are distinct, $\{O_8^{(i)}\}_{1 \leq i \leq q}$ are distinct, $\{O_9^{(i)}\}_{1 \leq i \leq q}$ are distinct, and $\{O_{10}^{(i)}\}_{1 \leq i \leq q}$ are distinct. We know from our condition that $\{I_9^{(i)}\}_{1 \leq i \leq q}$ are distinct and $\{I_{10}^{(i)}\}_{1 \leq i \leq q}$ are distinct. Therefore, we have exactly $(2^n - q)!$ choice of p_i for each $i = 7, 8, 9, 10$.

Completing the proof. To summarize, we have:

– at least $(1 - 3 \cdot \frac{q(q-1)}{2^n-1}) \cdot \{2^n!\}^4$ choice of $p_1, ..., p_4$.
– at least $(1 - \frac{q(q-1)}{2^n-1}) \cdot \{2^n!\}$ choice of p_5.
– at least $(1 - \frac{3}{2} \cdot \frac{q(q-1)}{2^n-1}) \cdot \{2^n!\}$ choice of p_6.
– exactly $\{(2^n - q)!\}^4$ choice of $p_7, ..., p_{10}$.

Then the number of $(p_1, ..., p_{10})$ which satisfy (2) is at least

$$(1 - 3 \cdot \frac{q(q-1)}{2^n-1}) \cdot (1 - \frac{q(q-1)}{2^n-1}) \cdot (1 - \frac{3}{2} \cdot \frac{q(q-1)}{2^n-1}) \cdot \{2^n!\}^6 \cdot \{(2^n-q)!\}^4$$

$$\geq (1 - \frac{11}{2} \cdot \frac{q(q-1)}{2^n-1}) \cdot \{2^n!\}^6 \cdot \{(2^n-q)!\}^4.$$

∎

5   Eight Round MKASUMI Type Permutation Is Super-Pseudorandom

Theorem 2. *For $1 \leq i \leq 16$, let $p_i \in P_n$ be a random permutation. Let $\psi = \psi(p_1, ..., p_{16})$ be the eight round MKASUMI. And let $R \in P_{4n}$ be a random permutation and $\Psi \triangleq \{\psi \mid \psi = \psi(p_1, ..., p_{16}), p_i \in P_n \text{ for } 1 \leq i \leq 16\}$.*
Then for any adversary A that makes at most q queries in total,

$$Adv_{\Psi}^{sprp}(A) \leq \frac{9q(q-1)}{2^n - 1}.$$

Proof. Let $\mathcal{O}$ be either R or ψ. The adversary A has oracle access to $\mathcal{O}$ and $\mathcal{O}^{-1}$. There are two types of queries A can make: either $(+, x)$ or $(-, y)$. For the i-th query A makes to $\mathcal{O}$ or $\mathcal{O}^{-1}$, define the query-answer pair $(x^{(i)}, y^{(i)}) \in \{0, 1\}^{4n} \times \{0, 1\}^{4n}$, where either A's query was $(+, x)$ and the answer it got was $y^{(i)} = \mathcal{O}(x^{(i)})$ or A's query was $(-, y)$ and the answer it got was $x^{(i)} = \mathcal{O}^{-1}(y^{(i)})$. Define view v of A as $v = \langle (x^{(1)}, y^{(1)}), ..., (x^{(q)}, y^{(q)}) \rangle$.

Since A is computationally unbounded, we may without loss of generality assume that A is deterministic. This implies that for every $1 \leq i \leq q$ the i-th query $x^{(i)}$ is fully determined by the first $i-1$ query-answer pairs, and the final output of A (0 or 1) depends only on v. Therefore, there exists a function $C_A(\cdot)$ such that

$$\begin{cases} C_A(x^{(1)}, y^{(1)}, ..., x^{(i-1)}, y^{(i-1)}) = \text{ either } (+, x^{(i)}) \text{ or } (-, y^{(i)}) \text{ for } 1 \leq i \leq q \text{ and} \\ C_A(v) = A\text{'s final output.} \end{cases}$$

We say that $v = \langle (x^{(1)}, y^{(1)}), ..., (x^{(q)}, y^{(q)}) \rangle$ is a *possible* view if for every $1 \leq i \leq q$, $C_A(x^{(1)}, y^{(1)}, ..., x^{(i-1)}, y^{(i-1)}) \in \{(+, x^{(i)}), (-, y^{(i)})\}$. Let $v_{one} \triangleq \{v | C_A(v) = 1 \text{ and } v \text{ is possible}\}$.

Evaluation of p_R. We first evaluate $p_R \triangleq Pr(R \xleftarrow{R} P_{4n} : A^{R, R^{-1}} = 1)$. We have $p_R = \#v_{one} \cdot \frac{(2^{4n} - q)!}{(2^{4n})!}$ as was done in the proof of Theorem 1.

Evaluation of p_ψ. We evaluate $p_\psi \triangleq Pr(\psi \xleftarrow{R} \Psi : A^{\psi, \psi^{-1}} = 1)$, where $\psi \xleftarrow{R} \Psi$ means that $p_i \xleftarrow{R} P_n$ for $1 \leq i \leq 16$ and then let $\psi \leftarrow \psi(p_1, ..., p_{16})$. Then we have $p_\psi = \frac{\#\{(p_1, ..., p_{16}) | A^{\psi, \psi^{-1}} = 1\}}{((2^n)!)^{16}}$.

We have the following main lemma. A proof of this lemma is given in Appendix.

Lemma 6 (Main Lemma). *For any fixed possible view $v = \langle (x^{(1)}, y^{(1)}), ..., (x^{(q)}, y^{(q)}) \rangle$, the number of $(p_1, ..., p_{16})$ which satisfies*

$$\psi(x^{(i)}) = y^{(i)} \text{ for } 1 \leq \forall i \leq q \tag{3}$$

is at least $(1 - \frac{9q(q-1)}{2^n - 1}) \cdot \{2^n!\}^{12} \cdot \{(2^n - q)!\}^4$.

Then from Lemma 6 , we have

$$p_\psi = \sum_{v \in v_{one}} \frac{\#\{(p_1, ..., p_{10})|(p_1, ..., p_{16}) \text{ satisfying } (3)\}}{\{(2^n)!\}^{16}}$$

$$\geq \sum_{v \in v_{good}} (1 - \frac{9q(q-1)}{2^n - 1}) \cdot \frac{\{(2^n - q)!\}^4}{\{(2^n)!\}^4}$$

$$= \#v_{one} \cdot (1 - \frac{9q(q-1)}{2^n - 1}) \cdot \frac{\{(2^n - q)!\}^4}{\{(2^n)!\}^4}$$

$$= p_R \cdot (1 - \frac{9q(q-1)}{2^n - 1}) \cdot \frac{\{(2^n - q)!\}^4}{\{(2^n)!\}^4} \cdot \frac{(2^{4n})!}{(2^{4n} - q)!}.$$

Since $\frac{\{(2^n - q)!\}^4}{\{(2^n)!\}^4} \cdot \frac{(2^{4n})!}{(2^{4n} - q)!} \geq 1$, $p_\psi \geq p_R \cdot (1 - \frac{9q(q-1)}{2^n - 1}) \geq p_R - \frac{9q(q-1)}{2^n - 1}$. Applying the same argument to $1 - p_\psi$ and $1 - p_R$ yields that $1 - p_\psi \geq 1 - p_R - \frac{9q(q-1)}{2^n - 1}$ and we have $|p_\psi - p_R| \leq \frac{9q(q-1)}{2^n - 1}$. ∎

6   Discussion and Concluding Remarks

In this paper, we showed that 5 round MKASUMI type permutation is pseudorandom and 8 round MKASUMI type permutation is super-pseudorandom. Until now, we found a distinguisher for 3 round MKASUMI type permutation in the notion of pseudorandomness and found a distinguisher for 4 round MKASUMI type permutation in the notion of super pseudorandomness. It is easy to make a distinguisher for 3 round MKASUMI type permutation in the notion of pseudorandomness. So we omit it. We can distinguish 4 round MKASUMI type permutation from random permutation in the notion of super pseudorandomness using just two plaintext queries and one ciphertext query as follows: Choose two distinct plaintexts $x^{(1)} = (a, b, c, d)$ and $x^{(2)} = (a', b, c, d)$ and denote the corresponding two ciphertexts by $y^{(1)} = (e, f, g, h)$ and $y^{(2)} = (e', f', g', h')$, respectively. Then it is easy to see that

(1) $O_3^{(1)} \oplus I_4^{(1)} \oplus O_3^{(2)} \oplus I_4^{(2)} = a \oplus a'$,
(2) $O_3^{(1)} \oplus I_4^{(1)} \oplus d = p_7(f) \oplus e \oplus h$,
(3) $O_3^{(2)} \oplus I_4^{(2)} \oplus d = p_7(f') \oplus e' \oplus h'$

By (1), (2), and (3), $p_7(f) \oplus p_7(f') = a \oplus a' \oplus e \oplus e' \oplus h \oplus h'$. Hence we can obtain the value of $p_7(f) \oplus p_7(f')$ since we know $a, a', e, e', h,$ and h'. Next, choose ciphertext $y^{(3)} = (e, f', g \oplus p_7(f) \oplus p_7(f'), h \oplus p_7(f) \oplus p_7(f'))$ and denote the corresponding plaintext by $x^{(3)} = (a'', b'', c'', d'')$. Then it is easy to check that $c \oplus c'' = d \oplus d''$ holds. Therefore we can make a distinguisher with this property. But the following problems still remain to be solved.

- Can 4 round MKASUMI type permutation be pseudorandom?
- Can 5, 6, and 7 round MKASUMI type permutation be super-pseudorandom?

In addition, it is still open whether 5 round original KASUMI type permutation can be super-pseudorandom.

Acknowledgements. This work was supported (in part) by the Ministry of Information & Communications, Korea, under the Information Technology Research Center (ITRC) Support Program. The first author was supported by the 21st Century COE Program 'Reconstruction of Social Infrastructure Related to Information Science and Electrical Engineering' of Kyushu University. The third author was supported by the Post-doctoral Fellowship Program of Korea Science & Engineering Foundation (KOSEF).

References

1. 3GPP TS 35.202 v 3.1.1. Specification of the 3GPP confidentiality and integrity algorithms, Document 2: KASUMI specification. Available at http://www.3gpp.org /tb/other/algorithms.htm.
2. T. Iwata, T. Yagi, and K. Kurosawa. *On the Pseudorandomness of KASUMI Type Permutations*, The Eighth Australasian Conference on Information Security and Privacy, ACISP 2003, LNCS, Vol. 2727, Springer-Verlag, pp. 130-141, 2003.
3. J. S. Kang, S. U. Shin, D. Hong, and O. Yi, *Provable security of KASUMI and 3GPP encryption mode f8*, Advances in Cryptology - ASIACRYPT 2001, LNCS 2248, Springer-Verlag, pp. 255-271, 2001.
4. M. Luby and C. Rackoff, *How to construct pseudorandom permutations from pseudoradom functions*, SIAM J. Comput., Vol. 17, No. 2, pp. 373-386, April 1988.
5. M. Matsui, *New block encryption algorithm MISTY*, Fast Software Encryption, FSE 1997, LNCS 1267, Springer-Verlag, pp. 54-68, 1997.
6. M. Naor and O. Reingold, *On the construction of pseudorandom permutations: Luby-Rackoff revised*, J. Cryptology, Vol. 12, No. 1, Springer-Verlag, pp. 29-66, 1999.
7. J. Patarin, *Pseudorandom permutations based on the DES scheme*, Proceedings of EUROCODE'90, LNCS 514, Springer-Verlag, pp. 193-204, 1990.

Appendix

Proof of Lemma 3

First we fix i and j such that $1 \leq i < j \leq q$, and consider the condition

$$I_6^{(i)} = I_6^{(j)} \tag{4}$$

in the following *four* cases:

Case 1: $x_{RR}^{(i)} \neq x_{RR}^{(j)}$. First, consider the condition

$$p_1(x_{RR}^{(i)}) \oplus x_{RL}^{(i)} \oplus x_{LL}^{(i)} = p_1(x_{RR}^{(j)}) \oplus x_{RL}^{(j)} \oplus x_{LL}^{(j)}. \tag{5}$$

The number of p_1 which satisfies (5) is at most $\frac{(2^n)!}{2^n-1}$ since $x_{RR}^{(i)} \neq x_{RR}^{(j)}$. Thus

$$\#\{(p_1,p_2,p_3,p_4)|\ (p_1,p_2,p_3,p_4) \text{ satisfies both (4) and (5)}\} \leq \frac{\{(2^n)!\}^4}{2^n-1}. \quad (6)$$

Next, consider any p_1 which does not satisfy (5), that is,

$$p_1(x_{RR}^{(i)}) \oplus x_{RL}^{(i)} \oplus x_{LL}^{(i)} \neq p_1(x_{RR}^{(j)}) \oplus x_{RL}^{(j)} \oplus x_{LL}^{(j)}. \quad (\neg 5)$$

For this p_1, we consider the condition

$$p_2(x_{RL}^{(i)}) \oplus p_1(x_{RR}^{(i)}) \oplus x_{RL}^{(i)} \oplus x_{LL}^{(i)} = p_2(x_{RL}^{(j)}) \oplus p_1(x_{RR}^{(j)}) \oplus x_{RL}^{(j)} \oplus x_{LL}^{(j)} \quad (7)$$

which is equivalent to $I_4^{(i)} = I_4^{(j)}$. Since $(\neg 5)$ holds, the number of p_2 which satisfies (7) is at most $\frac{(2^n)!}{2^n-1}$, and thus we have

$$\#\{(p_1,p_2,p_3,p_4)|\ (p_1,p_2,p_3,p_4) \text{ satisfies (4), } (\neg 5), \text{ and (7)}\} \leq \frac{\{(2^n)!\}^4}{2^n-1}. \quad (8)$$

Next, consider any p_1 which satisfies $(\neg 5)$, and any p_2 which dose not satisfy (7). That is,

$$p_2(x_{RL}^{(i)}) \oplus p_1(x_{RR}^{(i)}) \oplus x_{RL}^{(i)} \oplus x_{LL}^{(i)} \neq p_2(x_{RL}^{(j)}) \oplus p_1(x_{RR}^{(j)}) \oplus x_{RL}^{(j)} \oplus x_{LL}^{(j)}, \quad (\neg 7)$$

which is equivalent to $I_4^{(i)} \neq I_4^{(j)}$. For these p_1, p_2, and any p_3, the number of p_4 which satisfies

$$p_4(I_4^{(i)}) \oplus O_3^{(i)} \oplus I_4^{(i)} \oplus X_{RL}^{(i)} = p_4(I_4^{(j)}) \oplus O_3^{(j)} \oplus I_4^{(j)} \oplus X_{RL}^{(j)},$$

which is equivalent to (4), is at most $\frac{(2^n)!}{2^n-1}$. Thus,

$$\#\{(p_1,p_2,p_3,p_4)|\ (p_1,p_2,p_3,p_4) \text{ satisfies (4), } (\neg 5), \text{ and } (\neg 7)\} \leq \frac{\{(2^n)!\}^4}{2^n-1}. \quad (9)$$

Thus, from (6),(8), and (9), we have

$$\#\{(p_1,p_2,p_3,p_4)|\ (p_1,p_2,p_3,p_4) \text{ satisfies (4)}\} \leq \frac{3 \cdot \{(2^n)!\}^4}{2^n-1}. \quad (10)$$

Case 2: $x_{RL}^{(i)} \neq x_{RL}^{(j)}$ **and** $x_{RR}^{(i)} = x_{RR}^{(j)}$.   For any p_1, the number of p_2 which satisfies (7) is at most $\frac{(2^n)!}{2^n-1}$ since $x_{RL}^{(i)} \neq x_{RL}^{(j)}$, and thus we have

$$\#\{(p_1,p_2,p_3,p_4)|\ (p_1,p_2,p_3,p_4) \text{ satisfies (4) and (7)}\} \leq \frac{\{(2^n)!\}^4}{2^n-1}. \quad (11)$$

Next, for any p_1, any p_2 which satisfies $(\neg 7)$, and any p_3, the number of p_4 which satisfies (4) is at most $\frac{(2^n)!}{2^n-1}$. Therefore we have

$$\#\{(p_1,p_2,p_3,p_4)|\ (p_1,p_2,p_3,p_4) \text{ satisfies (4) and } (\neg 7)\} \leq \frac{\{(2^n)!\}^4}{2^n-1}. \quad (12)$$

Thus, from (11), and (12), we have

$$\#\{(p_1, p_2, p_3, p_4)|\ (p_1, p_2, p_3, p_4) \text{ satisfies } (4)\} \leq \frac{2 \cdot \{(2^n)!\}^4}{2^n - 1}. \tag{13}$$

Case 3: $x_{LL}^{(i)} \neq x_{LL}^{(j)}$, $x_{RL}^{(i)} = x_{RL}^{(j)}$, and $x_{RR}^{(i)} = x_{RR}^{(j)}$. For any p_1 and any p_2, ($\neg$7) is satisfied. Therefore, for any p_1, any p_2, and any p_3, the number of p_4 which satisfies (4) is at most $\frac{(2^n)!}{2^n-1}$. Thus we have

$$\#\{(p_1, p_2, p_3, p_4)|\ (p_1, p_2, p_3, p_4) \text{ satisfies } (4)\} \leq \frac{\{(2^n)!\}^4}{2^n - 1}. \tag{14}$$

Case 4: $x_{LR}^{(i)} \neq x_{LR}^{(j)}$, $x_{LL}^{(i)} = x_{LL}^{(j)}$, $x_{RL}^{(i)} = x_{RL}^{(j)}$, and $x_{RR}^{(i)} = x_{RR}^{(j)}$. In this case, there exists no p_1, p_2, p_3, and p_4 that satisfies (4). Therefore we have

$$\#\{(p_1, p_2, p_3, p_4)|\ (p_1, p_2, p_3, p_4) \text{ satisfies } (4)\} = 0. \tag{15}$$

Completing the proof. By taking the maximum of (10),(13),(14), and (15),

$$\#\{(p_1, p_2, p_3, p_4)|\ (p_1, p_2, p_3, p_4) \text{ satisfies } (4)\} \leq \frac{4 \cdot \{(2^n)!\}^4}{2^n - 1}$$

for any case. Finally, since we have $\binom{q}{2}$ choice of i and j, the lemma follows. ■

Proof of Lemma 6

Initially, $x^{(1)}, ..., x^{(q)}, y^{(1)}, ..., y^{(q)}$ are fixed.

Number of $(p_1, ..., p_4)$. From Lemma 3 and 4, the number of $(p_1, ..., p_4)$ such that $I_5^{(i)} \neq I_5^{(j)}$ and $I_6^{(i)} \neq I_6^{(j)}$ for $1 \leq \forall i < \forall j \leq q$, is at least $\{2^n!\}^4 - \frac{2q(q-1)}{2} \cdot \frac{\{2^n!\}^4}{2^n-1} - \frac{3q(q-1)}{2} \cdot \frac{\{2^n!\}^4}{2^n-1}$. Fix any $(p_1, ..., p_4)$ which satisfy the above two conditions.

Number of $(p_{13}, ..., p_{16})$. From Lemma 3 and 4, the number of $(p_{13}, ..., p_{16})$ such that $I_{11}^{(i)} \neq I_{11}^{(j)}$ and $I_{12}^{(i)} \neq I_{12}^{(j)}$ for $1 \leq \forall i < \forall j \leq q$, is at least $\{2^n!\}^4 - \frac{2q(q-1)}{2} \cdot \frac{\{2^n!\}^4}{2^n-1} - \frac{3q(q-1)}{2} \cdot \frac{\{2^n!\}^4}{2^n-1}$. We have used the symmetry of MKASUMI type permutation. Fix any $(p_{13}, ..., p_{16})$ which satisfy the above two conditions.

Number of p_5. For any fixed i and j such that $1 \leq i < j \leq q$, the number of p_5 such that

$$p_5(I_5^{(i)}) \oplus I_6^{(i)} \oplus I_3^{(i)} = p_5(I_5^{(j)}) \oplus I_6^{(j)} \oplus I_3^{(j)},$$

which is equivalent to $I_7^{(i)} = I_7^{(j)}$, is at most $\frac{(2^n)!}{2^n-1}$, since $I_5^{(i)} \neq I_5^{(j)}$.

Then the number of p_5 which satisfies $I_7^{(i)} \neq I_7^{(j)}$ for $1 \leq \forall i < \forall j \leq q$, is at least $(2^n)! - \frac{q(q-1)}{2} \cdot \frac{(2^n)!}{2^n-1}$. Fix any p_5 which satisfy the above condition.

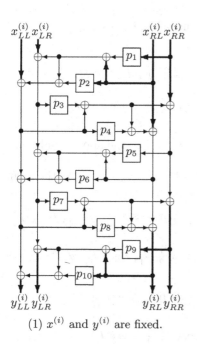

(1) $x^{(i)}$ and $y^{(i)}$ are fixed.

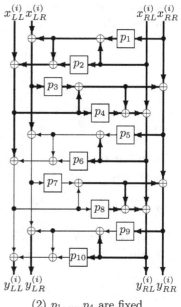

(2) $p_1, ..., p_4$ are fixed.

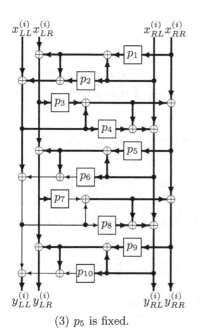

(3) p_5 is fixed.

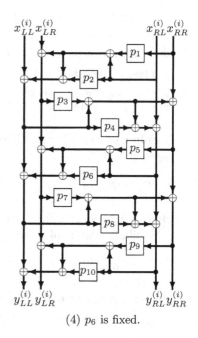

(4) p_6 is fixed.

Number of p_{11}. For any fixed i and j such that $1 \le i < j \le q$, the number of p_{11} such that

$$p_{11}(I_{11}^{(i)}) \oplus I_{12}^{(i)} \oplus I_{13}^{(i)} = p_{11}(I_{11}^{(j)}) \oplus I_{12}^{(j)} \oplus I_{13}^{(j)},$$

which is equivalent to $I_9^{(i)} = I_9^{(j)}$, is at most $\frac{(2^n)!}{2^n-1}$, since $I_{11}^{(i)} \ne I_{11}^{(j)}$.

Then the number of p_{11} which satisfies $I_9^{(i)} \ne I_9^{(j)}$ for $1 \le \forall i < \forall j \le q$, is at least $(2^n)! - \frac{q(q-1)}{2} \cdot \frac{(2^n)!}{2^n-1}$. Fix any p_{11} which satisfy the above condition.

Number of p_6. For any fixed i and j such that $1 \le i < j \le q$, the number of p_6 such that

$$p_6(I_6^{(i)}) \oplus I_6^{(i)} \oplus O_3^{(i)} \oplus O_5^{(i)} \oplus x_{RR}^{(i)} \oplus I_9^{(i)} = p_6(I_6^{(j)}) \oplus I_6^{(j)} \oplus O_3^{(j)} \oplus O_5^{(j)} \oplus x_{RR}^{(j)} \oplus I_9^{(j)},$$

which is equivalent to $O_7^{(i)} = O_7^{(j)}$, is at most $\frac{(2^n)!}{2^n-1}$, since $I_6^{(i)} \ne I_6^{(j)}$.

Similarly, the number of p_6 satisfies

$$p_6(I_6^{(i)}) \oplus I_4^{(i)} \oplus I_6^{(i)} \oplus O_5^{(i)} = p_6(I_6^{(j)}) \oplus I_4^{(j)} \oplus I_6^{(j)} \oplus O_5^{(j)},$$

which is equivalent to $I_8^{(i)} = I_8^{(j)}$, is at most $\frac{(2^n)!}{2^n-1}$, since $I_6^{(i)} \ne I_6^{(j)}$.

Similarly, the number of p_6 satisfies

$$p_6(I_6^{(i)}) \oplus I_3^{(i)} \oplus I_4^{(i)} \oplus I_{11}^{(i)} \oplus I_{12}^{(i)} = p_6(I_6^{(j)}) \oplus I_3^{(j)} \oplus I_4^{(j)} \oplus I_{11}^{(j)} \oplus I_{12}^{(j)},$$

which is equivalent to $O_{10}^{(i)} = O_{10}^{(j)}$, is at most $\frac{(2^n)!}{2^n-1}$, since $I_6^{(i)} \ne I_6^{(j)}$.

Then the number of p_6 satisfies $O_7^{(i)} \ne O_7^{(j)}$, $I_8^{(i)} \ne I_8^{(j)}$, and $O_{10}^{(i)} \ne O_{10}^{(j)}$ for $1 \le \forall i < \forall j \le q$, is at least $(2^n)! - \frac{3q(q-1)}{2} \cdot \frac{(2^n)!}{2^n-1}$. Fix any p_6 which satisfy the above three conditions.

Number of p_{12}. For any fixed i and j such that $1 \le i < j \le q$, the number of p_{12} such that

$$p_{12}(I_{12}^{(i)}) \oplus I_7^{(i)} \oplus I_{12}^{(i)} \oplus O_{11}^{(i)} \oplus O_{13}^{(i)} \oplus y_{LR}^{(i)} = p_{12}(I_{12}^{(j)}) \oplus I_7^{(j)} \oplus I_{12}^{(j)} \oplus O_{11}^{(j)} \oplus O_{13}^{(j)} \oplus y_{LR}^{(j)},$$

which is equivalent to $O_9^{(i)} = O_9^{(j)}$, is at most $\frac{(2^n)!}{2^n-1}$, since $I_{12}^{(i)} \ne I_{12}^{(j)}$.

Similarly, the number of p_{12} satisfies

$$p_{12}(I_{12}^{(i)}) \oplus I_{12}^{(i)} \oplus I_{14}^{(i)} \oplus O_{11}^{(i)} = p_{12}(I_{12}^{(j)}) \oplus I_{12}^{(j)} \oplus I_{14}^{(j)} \oplus O_{11}^{(j)},$$

which is equivalent to $I_{10}^{(i)} = I_{10}^{(j)}$, is at most $\frac{(2^n)!}{2^n-1}$, since $I_{12}^{(i)} \ne I_{12}^{(j)}$.

Similarly, the number of p_{12} satisfies

$$p_{12}(I_{12}^{(i)}) \oplus I_5^{(i)} \oplus I_6^{(i)} \oplus I_{13}^{(i)} \oplus I_{14}^{(i)} = p_{12}(I_{12}^{(j)}) \oplus I_5^{(j)} \oplus I_6^{(j)} \oplus I_{13}^{(j)} \oplus I_{14}^{(j)},$$

which is equivalent to $O_8^{(i)} = O_8^{(j)}$, is at most $\frac{(2^n)!}{2^n-1}$, since $I_{12}^{(i)} \ne I_{12}^{(j)}$.

Then the number of p_{12} satisfies $O_9^{(i)} \neq O_9^{(j)}$, $I_{10}^{(i)} \neq I_{10}^{(j)}$, and $O_8^{(i)} \neq O_8^{(j)}$ for $1 \leq \forall i < \forall j \leq q$, is at least $(2^n)! - \frac{3q(q-1)}{2} \cdot \frac{(2^n)!}{2^n - 1}$. Fix any p_{12} which satisfy the above three conditions.

Number of $(p_7, ..., p_{10})$. Now $p_1, ..., p_6, p_{11}, ..., p_{16}$ are fixed in such a way that $\{I_7^{(i)}\}_{1 \leq i \leq q}$ are distinct, $\{O_7^{(i)}\}_{1 \leq i \leq q}$ are distinct, $\{I_8^{(i)}\}_{1 \leq i \leq q}$ are distinct, $\{O_8^{(i)}\}_{1 \leq i \leq q}$ are distinct, $\{I_9^{(i)}\}_{1 \leq i \leq q}$ are distinct, $\{O_9^{(i)}\}_{1 \leq i \leq q}$ are distinct, $\{I_{10}^{(i)}\}_{1 \leq i \leq q}$ are distinct, and $\{O_{10}^{(i)}\}_{1 \leq i \leq q}$ are distinct.

Therefore, we have exactly $(2^n - q)!$ choice of p_i for each $i = 7, 8, 9, 10$.

Completing the proof To summarize, we have:

- at least $(1 - \frac{5}{2} \cdot \frac{q(q-1)}{2^n - 1})^2 \cdot \{2^n!\}^8$ choice of $p_1, ..., p_4, p_{13}, ..., p_{16}$.
- at least $(1 - \frac{1}{2} \cdot \frac{q(q-1)}{2^n - 1})^2 \cdot \{2^n!\}^2$ choice of (p_5, p_{11}).
- at least $(1 - \frac{3}{2} \cdot \frac{q(q-1)}{2^n - 1})^2 \cdot \{2^n!\}^2$ choice of (p_6, p_{12}).
- exactly $\{(2^n - q)!\}^4$ choice of $p_7, ..., p_{10}$.

Then the number of $(p_1, ..., p_{16})$ which satisfy (3) is at least

$$(1 - \frac{5}{2} \cdot \frac{q(q-1)}{2^n - 1})^2 \cdot (1 - \frac{1}{2} \cdot \frac{q(q-1)}{2^n - 1})^2 \cdot (1 - \frac{3}{2} \cdot \frac{q(q-1)}{2^n - 1})^2 \cdot \{2^n!\}^{12} \cdot \{(2^n - q)!\}^4$$

$$\geq (1 - \frac{9q(q-1)}{2^n - 1}) \cdot \{2^n!\}^{12} \cdot \{(2^n - q)!\}^4.$$

∎

Provably Secure Double-Block-Length Hash Functions in a Black-Box Model

Shoichi Hirose

Graduate School of Informatics, Kyoto University, Kyoto 606-8501, Japan
`hirose@i.kyoto-u.ac.jp`

Abstract. In CRYPTO'89, Merkle presented three double-block-length hash functions based on DES. They are optimally collision resistant in a black-box model, that is, the time complexity of any collision-finding algorithm for them is $\Omega(2^{\ell/2})$ if DES is a random block cipher, where ℓ is the output length. Their drawback is that their rates are low. In this article, new double-block-length hash functions with higher rates are presented which are also optimally collision resistant in the black-box model. They are composed of block ciphers whose key length is twice larger than their block length.

Keywords: double-block-length hash function, black-box model, block cipher.

1 Introduction

A cryptographic hash function is a function which maps an input of arbitrary length to an output of fixed length. It is one of the most important primitives in cryptography [14] and should satisfy preimage resistance, second-preimage resistance and collision resistance. Informally, preimage resistance means that, given an output, it is infeasible to obtain an input which produces the output. Second-preimage resistance means that, given an input, it is infeasible to obtain another input which produces the same output as the given input. Collision resistance means that it is infeasible to obtain two different inputs which produce the same output. For simplicity, a cryptographic hash function is called a hash function in this article.

A hash function usually consists of iteration of a compression function with fixed input/output length and is called an iterated hash function. Compression-function constructions are classified into two types: based on block ciphers and from scratch. The topic of this article is the former. It minimizes design and implementation effort with secure block ciphers. Its major drawback is slow processing speed. However, it is compensated by fast block ciphers such as AES. Furthermore, some recent work has pointed out weakness of SHA families [1, 18]. Thus, block-cipher-based hash functions may become more important.

Block-cipher-based hash functions are classified into two categories: single-block-length (SBL) and double-block-length (DBL). A SBL hash function is a

C. Park and S. Chee (Eds.): ICISC 2004, LNCS 3506, pp. 330–342, 2005.

hash function whose output length is equal to the block length. The output length of a DBL hash function is twice larger than the block length.

It is well-known that the birthday attack can find a collision of a hash function with time complexity $O(2^{\ell/2})$, where ℓ is the output length of the hash function. The block length of widely used block ciphers is 64 or 128. Thus, SBL hash functions are no longer secure in terms of collision resistance.

For DBL hash functions, many constructions have been presented [4, 7, 8, 9, 10, 12, 15]. Among them, three DBL hash functions by Merkle [15] have been shown to be optimally collision resistant in a black-box model: the time complexity of any collision-finding algorithm for them is $\Omega(2^{\ell/2})$, where ℓ is the output length. However, their rates are at most 0.276 and they are not so efficient.

In this article, DBL hash functions are proposed which are more efficient and optimally collision resistant in the black-box model. They can be represented in a simple form. They are of parallel type and their rates are $1/2$. They are based on block ciphers whose key length is twice larger than the block length. Thus, they can be constructed with AES or other previous AES candidates, which support 128-bit blocks and 256-bit keys.

The DBL hash functions proposed in this article consist of two different block ciphers to be provably secure. Though it seems their drawback, a genuine tweakable block cipher [13] will help obtain virtually two different block ciphers with different tweaks. Furthermore, it is possible to transform a DBL hash function with different block ciphers to the one with only one block cipher with slightly lower rate by the method used in MDC-2 [4].

Collision resistance as well as preimage resistance of the proposed DBL hash functions is proved in the black-box model. In this model, for the proposed DBL hash functions, second-preimage resistance can be regarded as preimage resistance for the output corresponding to the given input. In the black-box model, a block cipher is assumed to be an invertible keyed random permutation. This is an ideal but still proper assumption in that most of the attacks on block-cipher-based hash functions do not utilize the internal structure of the block ciphers. The technique in [3] is used in the security proofs in this article. It is assumed that two block ciphers are independent in our analysis.

The rest of this article is organized as follows. Section 2 includes notations, definitions and related work. In Section 3, provably secure DBL hash functions with rate $1/2$ consisting of two block ciphers are presented. Security proofs are also shown. In Section 4, it is mentioned how to construct provably secure DBL hash functions with one block cipher. A concluding remark is given in Section 5.

2   Preliminaries

2.1   Related Work

Preneel, Govaerts and Vandewalle [16] discussed the security of SBL hash functions against several attacks. They considered SBL hash functions with compression functions represented by $h_i = e(k, x) \oplus z$, where e is an (n, n) block cipher,

$k, x, z \in \{h_{i-1}, m_i, h_{i-1} \oplus m_i, v\}$ and v is a constant. They concluded that 12 out of $64 (= 4^3)$ hash functions are secure against the attacks. However, they did not provide any formal proofs.

Black, Rogaway and Shrimpton [3] presented a detailed investigation of provable security of SBL hash functions given in [16] in the black-box model. The most important result shown in their paper is that the time complexity of any collision-finding algorithm against 20 hash functions including the 12 mentioned above is $\Omega(2^{\ell/2})$, where ℓ is the output length.

Knudsen, Lai and Preneel [11] discussed the security of DBL hash functions with rate 1 based on (n, n) block ciphers. Hohl, Lai, Meier and Waldvogel [7] discussed the security of compression functions of DBL hash functions with rate $1/2$. On the other hand, the security of DBL hash functions with rate 1 based on $(n, 2n)$ block ciphers was discussed by Satoh, Haga and Kurosawa [17] and by Hattori, Hirose and Yoshida [6].

Many schemes with rate less than 1 were also presented. Merkle [15] presented three DBL hash functions based on DES with rates at most 0.276. They are optimally collision resistant in the black-box model. MDC-2 and MDC-4 [4] are also DBL hash functions based on DES with rates $1/2$ and $1/4$, respectively. Lai and Massey proposed the tandem/abreast Davies-Meyer [12]. They consist of a $(n, 2n)$ block cipher and their rates are $1/2$. It is an open question whether the four schemes are optimally collision resistant or not.

Knudsen and Preneel studied the schemes to construct secure compression functions with longer outputs from secure ones based on error-correcting codes [8, 9, 10]. It is also an open question whether optimally collision resistant compression functions are constructed by their schemes.

Recently, Black, Cochran and Shrimpton [2] showed that it is impossible to construct a highly efficient block-cipher-based hash function provably secure in the black-box model. A block-cipher-based hash function is highly efficient if it makes exactly one block-cipher call for each message block and all block-cipher calls use a single key.

2.2   Cryptographic Hash Functions

A cryptographic hash function H is a function which maps an input of arbitrary length to an output of fixed length. H should satisfy the following properties.

Preimage resistance For a given output y, it is intractable to find an input x such that $y = H(x)$.

Second-preimage resistance For a given input x, it is intractable to find an input x' such that $H(x) = H(x')$ and $x \neq x'$.

Collision resistance It is intractable to find a pair of inputs x and x' such that $H(x) = H(x')$ and $x \neq x'$.

A hash function $H : \{0, 1\}^* \rightarrow \{0, 1\}^\ell$ usually consists of a compression function $f : \{0, 1\}^\ell \times \{0, 1\}^{\ell'} \rightarrow \{0, 1\}^\ell$ and an initial value $h_0 \in \{0, 1\}^\ell$. An input m is divided into the ℓ'-bit blocks $m_1, m_2, \ldots, m_l$. Then,

$$h_i = f(h_{i-1}, m_i)$$

is computed successively for $1 \le i \le l$ and $h_l = H(m)$. H is called an iterated hash function.

Unambiguous padding is applied to m if its length is not a multiple of ℓ'. It is outside the scope of this article and is not described here.

2.3   Block Ciphers and a Black-Box Model

A block cipher with the block length n and the key length κ, $e : \{0,1\}^\kappa \times \{0,1\}^n \to \{0,1\}^n$, is called an (n, κ) block cipher. An (n, κ) block cipher is an invertible keyed permutation: $e(k, \cdot)$ is a permutation for every $k \in \{0,1\}^\kappa$, and it is easy to compute both $e(k, \cdot)$ and $e(k, \cdot)^{-1}$. The set of all (n, κ) block ciphers is denoted by $B(n, \kappa)$.

Most of the attacks on hash functions based on block ciphers do not utilize the internal structure of the block ciphers. Thus, the security of hash functions based on block ciphers is often analyzed in a black-box model, that is, under the assumption that $e(k, \cdot)$ is a random invertible permutation for each k.

In the black-box model, an encryption e and a decryption e^{-1} can be simulated by the following two oracles. An encryption oracle e returns a randomly selected ciphertext for a query which is a pair of a key and a plaintext. A decryption oracle e^{-1} returns a randomly selected plaintext for a query which is a pair of a key and a ciphertext. The oracles e and e^{-1} share a table of triplets of keys, plaintexts and ciphertexts, (k_i, x_i, y_i)'s, which are produced by the queries and the corresponding answers. Referring to the table, they randomly select an answer to a new query under the restriction that $e(k, \cdot)$ is a permutation for every k. They also add the triplet produced by the query and the answer to the table.

Without loss of generality, it is assumed that any adversary with the two oracles e and e^{-1} asks only once on a triplet of a key, a plaintext and a ciphertext obtained by a query and a corresponding answer: Once the adversary obtains (k, x, y) by a query and the answer, he just keeps it and asks neither (k, x) nor (k, y) afterward.

2.4   DBL Hash Functions

DBL hash functions with two block-cipher calls in their compression functions are discussed in the article. Let f be a compression function such that

$$(h_i, g_i) = f(h_{i-1}, g_{i-1}, m_i),$$

where $h_i, g_i, m_i \in \{0,1\}^n$ and n is the block length. f consists of f_U and f_L such that

$$\begin{cases} h_i = f_U(h_{i-1}, g_{i-1}, m_i) \\ g_i = f_L(h_{i-1}, g_{i-1}, m_i). \end{cases}$$

h_i is not fed into f_L and this kind of compression function is called the parallel type. This type of compression function is considered in this article.

Each of f_U and f_L is composed of a block cipher as follows:

$$\begin{cases} h_i = e_U(k_U, x_U) \oplus z_U \\ g_i = e_L(k_L, x_L) \oplus z_L, \end{cases}$$

where k_U, x_U, z_U and k_L, x_L, z_L are uniquely defined by h_{i-1}, g_{i-1}, m_i.

The rate r of an iterated hash function of block-cipher-based f is defined by

$$r = \frac{|m_i|}{(\# \text{ of block-cipher calls in } f) \times n}.$$

It is a measure of the efficiency of block-cipher-based hash functions.

The major difference should be noticed between the DBL hash functions previously proposed and ones proposed in the article. e_U and e_L are identical for the former, but are different for the latter.

2.5   Definitions of Security

As has been discussed in this section, the security of DBL hash functions is analyzed in the black-box model. Insecurity is quantified by success probability of an optimal resource-bounded adversary. In the black-box model, the resource is the number of the queries to encryption and decryption oracles.

For a set S, $z \leftarrow_R S$ represents random sampling from S under the uniform distribution. For a probabilistic algorithm $\mathcal{M}$, $z \leftarrow_R \mathcal{M}(x)$ means that z is an output of $\mathcal{M}$ with an input x and the output distribution is based on the random choices of $\mathcal{M}$ and the input distribution.

Collision Resistance. The following experiment $\texttt{FindColHF}(\mathcal{A}, H)$ is introduced to define the collision resistance of a DBL hash function H with two block ciphers e_U and e_L. The adversary $\mathcal{A}$ is a collision-finding algorithm of H with oracles e_U, e_U^{-1} and e_L, e_L^{-1}. Let $e_P^{\pm 1}$ represent a pair of oracles e_P and e_P^{-1} for $P \in \{U, L\}$.

> $\texttt{FindColHF}(\mathcal{A}, H)$
> $\quad e_U \leftarrow_R B(n, \kappa); \; e_L \leftarrow_R B(n, \kappa);$
> $\quad (m, m') \leftarrow_R \mathcal{A}^{e_U^{\pm 1}, e_L^{\pm 1}};$
> $\quad \text{if } m \neq m' \wedge H(m) = H(m') \text{ return } 1; \text{ else return } 0;$

$\texttt{FindColHF}(\mathcal{A}, H)$ returns 1 iff $\mathcal{A}$ finds a collision. Let $\mathbf{Adv}_H^{\text{coll}}(\mathcal{A})$ be the probability that $\texttt{FindColHF}(\mathcal{A}, H)$ returns 1. The probability is taken over the uniform distribution on $B(n, \kappa)$ and coin tosses of $\mathcal{A}$.

Definition 1 (Collision resistance of a hash function). *For $q \geq 1$, let*

$$\mathbf{Adv}_H^{\text{coll}}(q) = \max_{\mathcal{A}} \left\{ \mathbf{Adv}_H^{\text{coll}}(\mathcal{A}) \right\},$$

where $\mathcal{A}$ makes at most q queries to each of $e_U^{\pm 1}$ and $e_L^{\pm 1}$.

The following experiment $\mathtt{FindColCF}(\mathcal{A}, f, h_0)$ is introduced to define the collision resistance of a compression function f with two block ciphers e_U and e_L. h_0 is an initial value of an iterated hash function of f.

$\mathtt{FindColCF}(\mathcal{A}, f, h_0)$
$\quad e_U \leftarrow_R B(n, \kappa);\ e_L \leftarrow_R B(n, \kappa);$
$\quad ((h, m), (h', m')) \leftarrow_R \mathcal{A}^{e_U^{\pm 1}, e_L^{\pm 1}};$
$\quad \text{if } ((h, m) \neq (h', m') \wedge f(h, m) = f(h', m')) \vee f(h, m) = h_0 \text{ return } 1;$
$\quad \text{else return } 0;$

$\mathtt{FindColCF}(\mathcal{A}, f, h_0)$ returns 1 iff $\mathcal{A}$ finds a collision of f or a preimage of h_0. Let $\mathbf{Adv}_f^{\mathrm{comp}}(\mathcal{A})$ be the probability that $\mathtt{FindColCF}(\mathcal{A}, f, h_0)$ returns 1.

Definition 2 (Collision resistance of a compression function). *For $q \geq 1$, let*

$$\mathbf{Adv}_f^{\mathrm{comp}}(q) = \max_{\mathcal{A}} \left\{ \mathbf{Adv}_f^{\mathrm{comp}}(\mathcal{A}) \right\},$$

where $\mathcal{A}$ asks at most q queries to each of $e_U^{\pm 1}$ and $e_L^{\pm 1}$.

Preimage Resistance. The following experiment $\mathtt{FindPreImg}(\mathcal{A}, G)$ is introduced to define the preimage resistance of G with two block ciphers e_U and e_L. G is a hash function or a compression function.

$\mathtt{FindPreImg}(\mathcal{A}, G)$
$\quad e_U \leftarrow_R B(n, \kappa);\ e_L \leftarrow_R B(n, \kappa);\ y \leftarrow_R \{0, 1\}^\ell;$
$\quad x \leftarrow_R \mathcal{A}(y)^{e_U^{\pm 1}, e_L^{\pm 1}};$
$\quad \text{if } G(x) = y \text{ return } 1;\ \text{else return } 0;$

$\mathtt{FindPreImg}(\mathcal{A}, G)$ returns 1 iff $\mathcal{A}$ finds a preimage of G for an output y chosen randomly. Let $\mathbf{Adv}_G^{\mathrm{img}}(\mathcal{A})$ be the probability that $\mathtt{FindPreImg}(\mathcal{A}, G)$ returns 1.

Definition 3 (Preimage resistance). *For $q \geq 1$, let*

$$\mathbf{Adv}_G^{\mathrm{img}}(q) = \max_{\mathcal{A}} \left\{ \mathbf{Adv}_G^{\mathrm{img}}(\mathcal{A}) \right\},$$

where $\mathcal{A}$ makes at most q queries to each of $e_U^{\pm 1}$ and $e_L^{\pm 1}$.

Generally speaking, second-preimage resistance is stronger security requirement than preimage resistance. A preimage may have some information of another preimage which produces the same output. However, in the black-box model, for the hash functions or the compression functions considered in the subsequent sections, a preimage has no information useful to find another preimage. Thus, only preimage resistance is discussed in this article.

3   Provably Secure DBL Hash Functions with Two Block Ciphers

In this section, the security of DBL hash functions with compression functions shown in Fig. 1 is analyzed. Let f be a compression function such that $(h_i, g_i) = f(h_{i-1}, g_{i-1}, m_i)$ and

$$\begin{cases} h_i = f_U(h_{i-1}, g_{i-1}, m_i) \\ g_i = f_L(h_{i-1}, g_{i-1}, m_i). \end{cases}$$

f_U and f_L consist of $(n, 2n)$ block ciphers e_U and e_L, respectively, and are represented as follows:

$$\begin{cases} h_i = e_U(k_{U1}\|k_{U2}, x_U) \oplus z_U \\ g_i = e_L(k_{L1}\|k_{L2}, x_L) \oplus z_L, \end{cases}$$

where '$\|$' is the concatenation and $k_{U1}, k_{U2}, x_U, z_U, k_{L1}, k_{L2}, x_L, z_L \in \{0,1\}^n$ are represented by linear combinations of $h_{i-1}, g_{i-1}, m_i \in \{0,1\}^n$. Namely,

$$\begin{pmatrix} k_{U1} \\ k_{U2} \\ x_U \\ z_U \end{pmatrix} = U \begin{pmatrix} h_{i-1} \\ g_{i-1} \\ m_i \end{pmatrix}, \quad \begin{pmatrix} k_{L1} \\ k_{L2} \\ x_L \\ z_L \end{pmatrix} = L \begin{pmatrix} h_{i-1} \\ g_{i-1} \\ m_i \end{pmatrix}$$

and both U and L are 4×3 $\{0,1\}$-matrices.

3.1   Collision Resistance

In this subsection, a sufficient and simple condition of U and L is presented for an iterated hash function of f to be collision resistant.

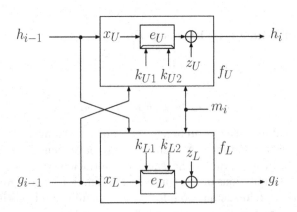

Fig. 1. A Diagram of Compression Functions with Two Block Ciphers and with Rate 1/2

The collision resistance of compression functions is focused on in the remaining part. It has been shown in [5, 15] that an iterated hash function is collision resistant if its compression function is. The following lemma states the fact in the black-box model.

Lemma 1. *[3] Let H be an iterated hash function of f. Then, for $q \geq 1$,*
$$\mathbf{Adv}_H^{\mathrm{coll}}(q) \leq \mathbf{Adv}_f^{\mathrm{comp}}(q).$$

First, a notation and a simple lemma are given for later use. For $1 \leq r \leq 4$, let $U(r)$ and $L(r)$ denote 3×3 $\{0,1\}$-matrices obtained by deleting the r-th row of U and L, respectively.

Lemma 2. *If both $U(3)$ and $U(4)$ are non-singular, then*
$$z_U \in \{x_U, x_U \oplus k_{U1}, x_U \oplus k_{U2}, x_U \oplus k_{U1} \oplus k_{U2}\}.$$

Proof. Since $U(4)$ is non-singular, z_U can be represented by a linear combination of x_U, k_{U1}, k_{U2}. On the other hand, since $U(3)$ is non-singular, z_U cannot be represented by any linear combinations of k_{U1}, k_{U2}. □

A sufficient condition is given for a compression function to be collision resistant in the following lemma.

Lemma 3. *Suppose that all of $U(3), U(4), L(3), L(4)$ are non-singular. Then, for every $1 \leq q \leq 2^{n-1} + 1$,*
$$\mathbf{Adv}_f^{\mathrm{comp}}(q) \leq q(q+1)/2^{2n-1}.$$

Proof. Let $\mathcal{A}$ be a collision-finding algorithm of f with oracles $e_U^{\pm 1}$ and $e_L^{\pm 1}$. $\mathcal{A}$ asks q queries to each of $e_U^{\pm 1}$ and $e_L^{\pm 1}$.
Since both $U(4)$ and $L(4)$ are non-singular and
$$\begin{pmatrix} k_{U1} \\ k_{U2} \\ x_U \end{pmatrix} = U(4) \begin{pmatrix} h_{i-1} \\ g_{i-1} \\ m_i \end{pmatrix}, \qquad \begin{pmatrix} k_{L1} \\ k_{L2} \\ x_L \end{pmatrix} = L(4) \begin{pmatrix} h_{i-1} \\ g_{i-1} \\ m_i \end{pmatrix},$$

the correspondence between (k_{U1}, k_{U2}, x_U) and (k_{L1}, k_{L2}, x_L) is 1-to-1. Thus, once a pair of an input and an output of e_U, $(k_{U1}, k_{U2}, x_U, y_U)$, is fixed by $\mathcal{A}$'s query to e_U or e_U^{-1} and its reply, an input to e_L, (k_{L1}, k_{L2}, x_L), is uniquely determined. Similarly, $\mathcal{A}$'s query to e_L or e_L^{-1} and its reply also uniquely determine an input to e_U.

On the other hand, it is necessary to ask a query to each of $e_U^{\pm 1}$ and $e_L^{\pm 1}$ in order to obtain a pair of an input and an output of f. The fact mentioned above implies that the correspondence between a pair of a query and a reply of $e_U^{\pm 1}$ and that of $e_L^{\pm 1}$ is 1-to-1. Hence, without loss of generality, it is assumed that $\mathcal{A}$ asks a query to an oracle and the corresponding query to the other oracle at a time.

Since $h_i = e_U(k_{U1}\|k_{U2}, x_U) \oplus z_U = y_U \oplus z_U$ and
$$z_U \in \{x_U, x_U \oplus k_{U1}, x_U \oplus k_{U2}, x_U \oplus k_{U1} \oplus k_{U2}\}$$

from Lemma 2, h_i depends both on x_U and on y_U and one of x_U and y_U is determined randomly by a reply of the oracle. Thus, h_i is randomly determined by the oracle. g_i is also randomly determined by the other oracle.

It is assumed that $z_U = x_U$ and $z_L = x_L$ in the rest of the proof. The proof is similar for the other cases.

For every $1 \leq j \leq q$, let C_j be the event such that

$$(x_{Uj} \oplus y_{Uj} = h_0 \wedge x_{Lj} \oplus y_{Lj} = g_0) \vee$$
$$\exists j' < j \, (x_{Uj} \oplus y_{Uj} = x_{Uj'} \oplus y_{Uj'} \wedge x_{Lj} \oplus y_{Lj} = x_{Lj'} \oplus y_{Lj'}),$$

where x_{Uj}, y_{Uj} and x_{Lj}, y_{Lj} correspond to the pairs of the j-th query and its reply of $e_U^{\pm 1}$ and $e_L^{\pm 1}$, respectively. Then,

$$\Pr[C_j] \leq \frac{j}{(2^n - (j-1))^2}.$$

Thus, if $q \leq 2^{n-1} + 1$, then

$$\mathbf{Adv}_f^{\mathrm{comp}}(\mathcal{A}) \leq \Pr[C_1 \vee \cdots \vee C_q] \leq \sum_{j=1}^{q} \Pr[C_j]$$

$$\leq \sum_{j=1}^{q} \frac{j}{(2^n - (j-1))^2} \leq \sum_{j=1}^{q} \frac{j}{(2^n - 2^{n-1})^2}$$

$$= \frac{q(q+1)}{2^{2n-1}}.$$

$\square$

The following theorem is led immediately from Lemmas 1 and 3.

Theorem 1. *Let H be an iterated hash function of f. Suppose that all of $U(3)$, $U(4), L(3), L(4)$ are non-singular for f. Then,*

$$\mathbf{Adv}_H^{\mathrm{coll}}(q) \leq q(q+1)/2^{2n-1}$$

for every $1 \leq q \leq 2^{n-1} + 1$.

From this theorem, any constant probability of success in finding a collision implies that $q = \Omega(2^n)$.

3.2   Preimage Resistance

Preimage resistance of iterated hash functions presented in the previous subsection is discussed here.

The following lemma shows the relationship between preimage resistance of an iterated hash function and that of its compression function. This lemma is also implicit in [19].

Lemma 4. *[3] Let H be an iterated hash function of f. Then, for $q \geq 1$,*
$$\mathbf{Adv}_H^{\mathrm{img}}(q) \leq \mathbf{Adv}_f^{\mathrm{img}}(q).$$

The preimage resistance of compression functions given in the previous subsection is presented in the following lemma.

Lemma 5. *Suppose that all of $U(3), U(4), L(3), L(4)$ are non-singular. Then, for every $g \geq 1$,*

$$\mathbf{Adv}_f^{\text{img}}(q) \leq q/(2^n - q)^2.$$

Proof. Let $\mathcal{A}$ be a preimage-finding algorithm of f with oracles $e_U^{\pm 1}$ and $e_L^{\pm 1}$. $\mathcal{A}$ asks q queries to each of $e_U^{\pm 1}$ and $e_L^{\pm 1}$. Let w be the input of $\mathcal{A}$ and $w = (w_U, w_L)$, where $w_U, w_L \in \{0,1\}^n$.

It is necessary to ask a query to each of $e_U^{\pm 1}$ and $e_L^{\pm 1}$ in order to obtain a pair of an input and an output of f. As in the proof of Lemma 3, the correspondence between a pair of a query and a reply of $e_U^{\pm 1}$ and that of $e_L^{\pm 1}$ is 1-to-1. Hence, without loss of generality, it is assumed that $\mathcal{A}$ asks a query to an oracle and the corresponding query to the other oracle at a time.

Since $h_i = y_U \oplus z_U$ and

$$z_U \in \{x_U, x_U \oplus k_{U1}, x_U \oplus k_{U2}, x_U \oplus k_{U1} \oplus k_{U2}\}$$

from Lemma 2, h_i depends both on x_U and on y_U and one of x_U and y_U is determined randomly by a reply of the oracle. Thus, h_i is randomly determined by the oracle. g_i is also randomly determined by the other oracle.

It is assumed that $z_U = x_U$ and $z_L = x_L$ in the rest of the proof. The proof is similar for the other cases.

For every $1 \leq j \leq q$, let I_j be the event such that

$$x_{Uj} \oplus y_{Uj} = w_U \wedge x_{Lj} \oplus y_{Lj} = w_L$$

where x_{Uj}, y_{Uj} and x_{Lj}, y_{Lj} correspond to the pairs of the j-th query and its reply of $e_U^{\pm 1}$ and $e_L^{\pm 1}$, respectively. Then,

$$\Pr[\mathsf{I}_j] \leq \frac{1}{(2^n - (j-1))^2}.$$

Thus,

$$\mathbf{Adv}_f^{\text{img}}(\mathcal{A}) \leq \Pr[\mathsf{I}_1 \vee \cdots \vee \mathsf{I}_q] \leq \sum_{j=1}^{q} \Pr[\mathsf{I}_j] \leq \sum_{j=1}^{q} \frac{1}{(2^n - (j-1))^2}$$

$$\leq \frac{q}{(2^n - q)^2}.$$

$\square$

The following theorem is led immediately from Lemmas 4 and 5.

Theorem 2. *Let H be an iterated hash function of f. Suppose that all of $U(3), U(4), L(3), L(4)$ are non-singular for f. Then, for every $q \geq 1$,*

$$\mathbf{Adv}_H^{\text{img}}(q) \leq \frac{q}{(2^n - q)^2}.$$

Theorem 2 implies nothing about the preimage resistance for $q \geq 2^n - 2^{n/2} + 1$. It states, however, that the success probability is (asymptotically) negligible as long as $q = c\, 2^n$ for any positive constant $c < 1$:

$$\mathbf{Adv}_H^{\mathrm{img}}(c\, 2^n) \leq \frac{c}{(1-c)^2} \frac{1}{2^n}.$$

For example, if $c = 1/2$, then $\mathbf{Adv}_H^{\mathrm{img}}(2^{n-1}) \leq 1/2^{n-1}$.

4   Provably Secure DBL Hash Functions with One Block Cipher

Let e be an (n, κ) block cipher and $n + 2 \leq \kappa$. In this section, the security of DBL hash functions with compression functions shown in Fig. 2 is analyzed. The left-side function is focused on. Let us call it f.

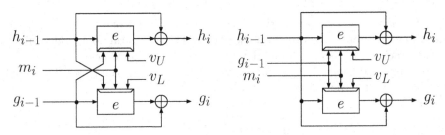

Fig. 2. Compression Functions with One Block Cipher

The compression function f is represented as follows:

$$\begin{cases} h_i = e(g_{i-1}\|m_i\|v_U, h_{i-1}) \oplus h_{i-1} \\ g_i = e(h_{i-1}\|m_i\|v_L, g_{i-1}) \oplus g_{i-1}, \end{cases}$$

where $m_i \in \{0, 1\}^\ell$ for some $1 \leq \ell < \kappa - n$, and v_U and v_L are constants in $\{0, 1\}^{\kappa - n - \ell}$ such that $v_U \neq v_L$.

Since $v_U \neq v_L$, in the black-box model, e with v_U and e with v_L can be regarded as two independent random block ciphers. Furthermore, there exists 1-to-1 correspondence between a pair of an input and an output of e with v_U and that of e with v_L.

From these observations, it is clear that the following lemma can be proved in the similar way as Lemma 3.

Lemma 6. *For the compression function f, if $1 \leq q \leq 2^{n-1} + 1$, then*

$$\mathbf{Adv}_f^{\mathrm{comp}}(q) \leq q(q + 1)/2^{2n-1}.$$

The following theorem states the collision resistance of an iterated hash function of f. This is immediately lead from Lemmas 1 and 6.

Theorem 3. *Let H be an iterated hash function of f. Then,*

$$\mathbf{Adv}_H^{\mathrm{coll}}(q) \leq q(q+1)/2^{2n-1}$$

for every $1 \leq q \leq 2^{n-1}+1$.

For preimage resistance, similarly, the following theorem is obtained.

Theorem 4. *Let H be an iterated hash function of f. Then, for $q \geq 1$,*

$$\mathbf{Adv}_H^{\mathrm{img}}(q) \leq \frac{q}{(2^n - q)^2}.$$

In the black-box model, it is sufficient that $v_U, v_L \in \{0,1\}$ and $v_U \neq v_L$. However, in practice, v_U, v_L should be longer in order to avoid weak keys and to increase independence. Suppose that ℓ_{con} be the length of v_U or v_L and $\kappa = 2n$. Then, the rate of H is $(1 - \ell_{\mathrm{con}}/n)/2$. For example, the rate is $7/16$ if $\ell_{\mathrm{con}} = n/8$.

The idea that two block ciphers are obtained from one block cipher by fixing a part of the key with different constants is found in the design of MDC-2 [4]. However, the security proof as shown above does not seem to be applied to MDC-2.

5   Conclusion

In this article, DBL hash functions provably secure in the black-box model have been presented. They are based on $(n, 2n)$ block ciphers and can be represented in a simple form. Future work is to explore more efficient DBL hash functions optimally collision resistant.

References

1. E. Biham and R. Chen. Near-collisions of SHA-0. Cryptology ePrint Archive, Report 2004/146, 2004. http://eprint.iacr.org/.
2. J. Black, M. Cochran, and T. Shrimpton. On the impossibility of highly efficient blockcipher-based hash functions. Cryptology ePrint Archive, Report 2004/062, 2004. http://eprint.iacr.org/.
3. J. Black, P. Rogaway, and T. Shrimpton. Black-box analysis of the block-cipher-based hash-function constructions from PGV. In *CRYPTO 2002 Proceedings*, pages 320–335, 2002. Lecture Notes in Computer Science 2442.
4. B. O. Brachtl, D. Coppersmith, M. M. Hyden, S. M. Matyas Jr., C. H. W. Meyer, J. Oseas, S. Pilpel, and M. Schilling. Data authentication using modification detection codes based on a public one-way encryption function, mar 1990. U. S. Patent # 4,908,861.

5. I. Damgård. A design principle for hash functions. In *CRYPTO'89 Proceedings*, pages 416–427, 1990. Lecture Notes in Computer Science 435.
6. M. Hattori, S. Hirose, and S. Yoshida. Analysis of double block length hash functions. In *9th IMA International Conference on Cryptography and Coding*, pages 290–302, 2003. Lecture Notes in Computer Science 2898.
7. W. Hohl, X. Lai, T. Meier, and C. Waldvogel. Security of iterated hash functions based on block ciphers. In *CRYPTO'93 Proceedings*, pages 379–390, 1994. Lecture Notes in Computer Science 773.
8. L. Knudsen and B. Preneel. Hash functions based on block ciphers and quaternary codes. In *ASIACRYPT'96 Proceedings*, pages 77–90, 1996. Lecture Notes in Computer Science 1163.
9. L. Knudsen and B. Preneel. Fast and secure hashing based on codes. In *CRYPTO'97 Proceedings*, pages 485–498, 1997. Lecture Notes in Computer Science 1294.
10. L. Knudsen and B. Preneel. Construction of secure and fast hash functions using nonbinary error-correcting codes. *IEEE Transactions on Information Theory*, 48(9):2524–2539, 2002.
11. L. R. Knudsen, X. Lai, and B. Preneel. Attacks on fast double block length hash functions. *Journal of Cryptology*, 11(1):59–72, 1998.
12. X. Lai and J. L. Massey. Hash function based on block ciphers. In *EUROCRYPT'92 Proceedings*, pages 55–70, 1993. Lecture Notes in Computer Science 658.
13. M. Liskov, R. L. Rivest, and D. Wagner. Tweakable block ciphers. In *CRYPTO 2002 Proceedings*, pages 31–46, 2002. Lecture Notes in Computer Science 2442.
14. A. J. Menezes, P. C. van Oorschot, and S. A. Vanstone. *Handbook of Applied Cryptography*. CRC Press, 1996.
15. R. C. Merkle. One way hash functions and DES. In *CRYPTO'89 Proceedings*, pages 428–446, 1990. Lecture Notes in Computer Science 435.
16. B. Preneel, R. Govaerts, and J. Vandewalle. Hash functions based on block ciphers: A synthetic approach. In *CRYPTO'93 Proceedings*, pages 368–378, 1994. Lecture Notes in Computer Science 773.
17. T. Satoh, M. Haga, and K. Kurosawa. Towards secure and fast hash functions. *IEICE Transactions on Fundamentals*, E82-A(1):55–62, 1999.
18. X. Wang, D. Feng, X. Lai, and H. Yu. Collisions for hash functions MD4, MD5, HAVAL-128 and RIPEMD. Cryptology ePrint Archive, Report 2004/199, 2004. http://eprint.iacr.org/.
19. R. S. Winternitz. A secure one-way hash function built from DES. In *IEEE Symposium on Security and Privacy*, pages 88–90, 1984.

Padding Oracle Attacks
on Multiple Modes of Operation[*]

Taekeon Lee[1], Jongsung Kim[1], Changhoon Lee[1],
Jaechul Sung[2], Sangjin Lee[1], and Dowon Hong[3]

[1] Center for Information Security Technologies(CIST),
Korea University, Anam Dong, Sungbuk Gu, Seoul, Korea
{imml97, joshep, crypto77, sangjin}@cist.korea.ac.kr
[2] Department of Mathematics, University of Seoul,
90 Cheonnong-Dong Dongdaemun Gu, Seoul, 130-743, Korea
{jcsung}@uos.ac.kr
[3] Information Security Technology Division,ETRI,
Taejon, Korea
{dwhong}@etri.re.kr

Abstract. In [12] Vaudenay presented side-channel attacks on the CBC encryption mode cipher under the padding oracle attack models, which enable an adversary to determine the correct message with knowledge of ciphertext. Black and Urtubia generalized these attacks in several directions, considering various padding schemes [4]. In this paper we extend these attacks to other kinds of modes of operation for block ciphers. Specifically, we apply the padding oracle attacks to multiple modes of operation with various padding schemes. As a results of this paper, 12 out of total 36 double modes and 22 out of total 216 triple modes are vulnerable to the padding oracle attacks. It means that the 12 double modes and the 22 triple modes exposed to these types of attacks do not offer the better security than single modes.

Keywords: Padding oracle attacks, Multiple modes of operation, Block ciphers.

1 Introduction

Vaudenay showed various ways to perform an efficient side-channel attack named padding oracle attack on modes of operation [12]. This attack requires an oracle which on receipt of a ciphertext, decrypts it and replies to the sender whether the padding is VALID or INVALID. Black and Urtubia generalized Vaudenay's attack to various padding schemes and the CBC mode of operation [4]. Afterwards, Paterson and Yau [9] introduced padding oracle attacks on the ISO CBC mode encryption standard and Klima and Rosa [8] showed that the CBC mode

---

[*] This work was supported by MOST research fund (M1-0326-08-0001).

C. Park and S. Chee (Eds.): ICISC 2004, LNCS 3506, pp. 343–351, 2005.

using ABYT-PAD can be attacked in the PKCS#7 format. Based on these kinds of attacks, in this paper, we evaluate the security of multiple modes of operation with various padding schemes.

Biham presented multiple modes of operation which are made of the connected several single modes (e.g. ECB, CBC, OFB, CFB, CBC^{-1}, CFB^{-1}) [2]. There exist $36(= 6^2)$ double and $216(=6^3)$ triple modes in total. Especially, the double and triple DES can be considered as modes of operation (ECB|ECB) and (ECB|ECB^{-1}|ECB), respectively.

Wagner analyzed the multiple modes of operation proposed by Biham [13]. However, his method is too unrealistic. So, Hong et al. use only known-IV chosen texts to attack many triple modes of operation which are combined with cascade operations [5]. Hong et al. presented that 123 out of 216 triple modes are analyzed with complexities less than Biham's results, and showed that the security of many triple modes decreases when the initial values are exposed. This paper analyzes all the double and triple modes of operation with various padding schemes under the padding oracle attack models, which are different from the above attack methods. As a result, 12 out of total 36 double modes and 22 out of total 216 triple modes are vulnerable to these types of attacks.

This paper is organized as follows. Preliminaries are presented in Section 2. Section 3 discusses padding oracle attacks on multiple modes of operation. Finally, we summarize our conclusion in Section 4.

2   Preliminaries

This section presents some notations and padding methods which are used throughout this paper, and then describes multiple modes of operation which are targets in our attacks.

2.1   Notations

- P : A plaintext with message and padding
- M : An unpadded message string
- L_M : The length of the message string M (bit)
- $C = (IV, C_1, C_2, \cdots, C_q)$: A ciphertext output after a mode encryption
- IV : An initial vector used in a mode
- C_i : The i-th ciphertext block, (i starts with 1)
- C_i^j : The j-th byte of i-th ciphertext block, (j starts with 0)
- n : A block size of block cipher (byte)
- e_j : A 32-bit binary string in which the j-th bit is one and the others are zeros, (j starts with 0)
- $X||Y$: The concatenation of strings X and Y
- $X \oplus Y$: The exclusive-or of strings X and Y
- $\mathcal{O}$: An oracle to distinguish whether the deciphered plaintext is correct padding or not
- VALID or INVALID : Responses of oracle, whether the deciphered plaintext is correct or incorrect

2.2   Padding Methods

Let the size of message be $(8n \times q + m)$ bit $(0 \le m < 8n, 0 \le q)$.

- CBC-PAD : If $m \ne 0$ and one byte is left to padding, we append a single 0x01 to message M, if there are two bytes of padding needed, we append 0x0202 to M, and so on. If $m = 0$ then we append one block $nn \cdots n$ to M(in case $n = 16$, then we append one block $0x1010 \cdots 10$).

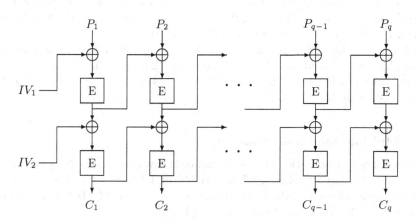

Fig. 1. Example of the Double Mode of Operation : CBC|CBC

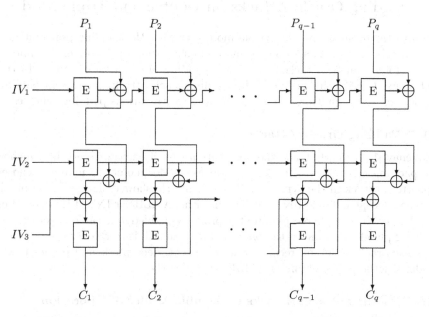

Fig. 2. Example of the Triple Mode of Operation : CFB|OFB|CBC

- ESP-PAD : If the last block is one byte left to padding, we append a single 0x01 to message M, if there are two bytes of padding needed, we append 0x0102 to M, and so on. If $m = 0$ then we append one block 0x010203 $\cdots n$ to M.
- XY-PAD : This padding method uses two distinct constant byte values X and Y. We transform M by first appending X one time, then adding the necessary number of Y values. If $m = 0$ then we append one block $XY \cdots Y$ to M.
- ISO(9797-1)-Padding method 3 : If the last block is $t(> 0)$ bits left to padding, we append t-bit string 0^t to M together with L_M to the first block, i.e., $P = L_M \| M \| 0^t$. If $m = 0$ then $P = L_M \| M$.
- ISO(10118-1)-Padding method 3 : If the last block is $t(> 0)$ bits left to padding, we append a 64-bit message length L_M from the least significant bit at the last block and padding $'100 \cdots 0'$ between message M and length L_M, i.e., $P = M \| 10 \cdots 0 \| L_M$.

2.3   Multiple Modes of Operation

Multiple modes of operation consist of multiple layers where each layer can be of the form ECB, CBC, OFB, CFB, CBC^{-1} or CFB^{-1}. Figs. 1 and 2 show examples of double and triple modes of operation, respectively. In this paper, we focus on total 36 double modes and 216 triple modes.

3   Padding Oracle Attacks on Double and Triple Modes

We investigate all double and triple modes whether the padding oracle attacks are possible or not. As a result, we accomplished that 12 double modes and 22 triple modes can be attacked under the padding oracle attack models. In this section we describe the outline of the padding oracle attack and present padding oracle attacks on multiple modes of operation with various padding schemes.

3.1   Padding Oracle Attacks

Vaudenay [12] recently made the observation that if one can ascertain somehow the padding error status, it can be used as a side-channel to mount a chosen ciphertext attack in the symmetric key setting. He showed that given an oracle $\mathcal{O}$ which accepts a ciphertext and returns either VALID or INVALID depending on whether the deciphered plaintext is properly padded, one can recover the underlying plaintext. Namely, the padding oracle attack is to recover the plaintext with the oracle which on receipt of a ciphertext, decrypts it and replies to the sender whether the padding is VALID or INVALID.

3.2   Padding Oracle Attacks on Double Modes of Operation

In the padding oracle attack model, we find the plaintext to correspond to a given ciphertext by using the padding oracle. We here present padding oracle attacks on double modes of operation. We begin by attacking CBC|CBC mode of operation using various padding methods. Our attack consists of two stages; in the first stage, we find a padding length of message and in the second, we find the plaintext to correspond to a given ciphertext.

CBC-PAD

- **Stage of Finding the Padding Length**

  Let oracle $\mathcal{O}$ and ciphertext $C = (IV_1, IV_2, C_1, C_2, \cdots, C_q)$ be given. Then, we can perform a binary search to discover the padding length of message as follows; To begin with, we manipulate C_{q-2}, i.e., we change C_{q-2} into C'_{q-2} such that the difference of C_{q-2} and C'_{q-2} is only one byte in the middle (one half) position and then require a response (VALID or INVALID) of oracle $\mathcal{O}$ with respect to $(IV_1, IV_2, C_1, \cdots, C_{q-3}, C'_{q-2}, C_{q-1}, C_q)$. At this time, if a change in any padding part of the corresponding plaintext block P_q is induced then oracle $\mathcal{O}$ will always return INVALID. On the other hands, if a change in any message part of the corresponding plaintext block P_q is induced then oracle $\mathcal{O}$ will always return VALID. In case we have the VALID response from oracle $\mathcal{O}$, we require a response of oracle $\mathcal{O}$ with respect to $(IV_1, IV_2, C_1, \cdots, C_{q-3}, C''_{q-2}, C_{q-1}, C_q)$, where the difference of C_{q-2} and C''_{q-2} is only one byte in the three fourths position. On the other hands, in case oracle $\mathcal{O}$ gives the INVALID response, we require a response of oracle $\mathcal{O}$ with respect to $(IV_1, IV_2, C_1, \cdots, C_{q-3}, C''_{q-2}, C_{q-1}, C_q)$, where the difference of C_{q-2} and C''_{q-2} is only one byte in the one fourth position. We repeatedly perform this process every a single byte of C_{q-2} and thus we can find the padding length in $log_2(n)$ queries where n is the length of a block in bytes.

- **Stage of Finding the Plaintext**

  As described above could find the length of the padding by using $log_2(n)$ queries. It follows that we know what is the padding among 0x01, 0x0202, 0x030303, $\cdots$. In order to get all the last plaintext block, we change ciphertext block C_{q-2} and then require a response of oracle $\mathcal{O}$ with respect to the altered ciphertext. For example, if the padding is of 0x0202, then we manipulate the last two bytes of the ciphertext block C_{q-2} to be appeared padding 0x0303 of the deciphered plaintext. For the altered ciphertext, we again alter C_{q-2}^{13} into $t \oplus C_{q-2}^{13}$ ($0 \le t \le 2^8 - 1$) and require responses of oracle $\mathcal{O}$ with respect to these 2^8 altered ciphertexts. In case oracle $\mathcal{O}$ responds VALID among the 2^8 queries (it can be done with about 2^7 queries on aerage), we can use the value t satisfying the VALID response to recover the underlying plaintext byte $P_q^{13}(= t \oplus 0x03)$. Refer to the Fig. 3. This method can be applied repeatedly to recover the last plaintext block P_q. Furthermore, we can use this method repeatedly to find all plaintext blocks $P_1, P_2, \cdots, P_q$. Especially, in order to find plaintext blocks P_1 and P_2 we alter IV_1 and IV_2 respectively, and require responses of oracle $\mathcal{O}$ with respect to the altered ciphertexts. We should do $128 \times n$ oracle queries to find one block of plain-

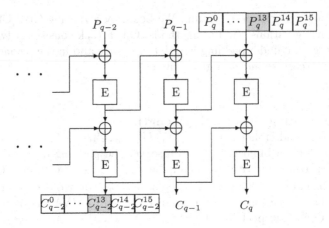

Fig. 3. Padding oracle attack : CBC|CBC

text on average. So we needs about $128 \times n \times q + log_2(n)$ oracle queries on average to find the corresponding plaintext for the given ciphertext C.

ESP-PAD, XY-PAD

With the same method, we can recover the corresponding plaintexts for a given ciphertext on the CBC|CBC mode when this mode uses ESP-PAD or XY-PAD as its padding scheme. We summarize the results in Table 1.

Table 1. Attack Complexities on Double Modes of Operation

Padding Methods	CBC-PAD	ESP-PAD	XY-PAD	ISO(9797-1) -PAD 3	ISO(10118-1) -PAD 3
Complexities (# oracle queries)	$128nq$ $+log_2(n)$	$128nq$ $+log_2(n)$	$128nq$ $+log_2(n)$	$(8n-1)(q-1)$ $+log_2(8n)+2$	$(8n-1)(q-1)$ $+log_2(8n)+2$
Double modes that can be broken by the padding oracle attacks	CBC|CBC, CBC|OFB, OFB|CBC, OFB|OFB, OFB|CFB, CFB|OFB, CFB|CFB, OFB|CFB^{-1}, CFB|CFB^{-1}, CFB^{-1}|OFB, CFB^{-1}|CFB, CFB^{-1}|CFB^{-1}			CBC|CBC, CBC|OFB, OFB|CBC, OFB|OFB	

* q : Number of plaintext/ciphertext block size
* n : A block size of the block cipher (byte)

ISO(9797-1)-Padding Method 3

– Stage of Finding the Padding Length

In this padding scheme, if a given ciphertext C has greater than or equal to 4 blocks, we can find the padding length of message by using the technique of the foregoing first stage. However, in case C has less than 4 blocks, we should use an additional technique to find the padding length of message. This is due to the fact that this padding scheme puts the message length into the first block. For the convenience, we assume that the size of a block is a 128-bit. If a given ciphertext has three blocks, i.e., $C = (IV_1, IV_2, C_1, C_2, C_3)$, then the first block with the message length will be changed when we alter the given ciphertext block C_1 as before. Therefore, we use $C' = (IV_1 \oplus e_7, IV_2, C_1, C_1, C_2, C_3)$ instead of $C = (IV_1, IV_2, C_1, C_2, C_3)$ as an oracle query. At this time, if the response of oracle is VALID then the message length L_M is 256 (Note that L_M is the bit length of the message M corresponding to C), because the message length is in $128 < L_M \leq 256$. If the response of oracle is INVALID, then we have $128 < L_M \leq 256$. Thus the response of oracle is VALID for the query $C'' = (IV_1 \oplus e_{8,7}, IV_2, C_1, C_1, C_2, C_3)$. Based on the VALID query C'', we can find the padding length of message by using the technique of the previous first stage. That is, we perform a binary search in bit for the second C_1 of C'' as described above. So, we find the length of padding in at most $log_2(128) + 2$ queries where 128 is the length of block in bit. In case a given ciphertext has two blocks, it is easy to see that the length of message can be found by the above method.

– **Stage of Finding the Plaintext**
  As stated above, we let the size of a block be a 128-bit. Following is the algorithm for recovering the plaintext.

Input : $L_M, IV_1, IV_2, C_1, C_2, \cdots, C_q$
Output : $P_i^1, P_i^2, \cdots, P_i^{127}$ the rightmost 127 bits of P_i, $1 \leq i \leq q$
      where P_i^j means the j-th bit of P_i, $(0 \leq j \leq 127)$.

If $3 \leq i \leq q$
  for $j = 127$ to 1 do
  $IV_1' = IV_1 \oplus L_M \oplus (256 + j)$
  $C' = (IV_1', IV_2, C_1, C_{i-2}, C_{i-1}, C_i)$
  $\begin{cases} b = 0, & if\ \mathcal{O}(C') = VALID \\ b = 1, & if\ \mathcal{O}(C') = INVALID \end{cases}$
  $P_i^j = b$
  If (b=1)
  $C_{i-2} = C_{i-2} \oplus e_j$ (In case $i = 3$, C_{i-2} represents only the second C_1 in C'.)
  end for

  return $P_i^1, P_i^2, \cdots, P_i^{127}$

If $i = 2$, we use $C'' = (IV_1', IV_2, C_1, IV_2, C_1, C_2)$ instead of
    $C' = (IV_1', IV_2, C_1, C_{i-2}, C_{i-1}, C_i)$ to find the plaintext
    in the above sub-algorithm.

Algorithm to find the plaintext

Using the above algorithm we can recover the corresponding plaintext for a given ciphertext except for the most significant bit P_i^0 of each block. This is due to the fact that this padding scheme does not append any strings to M when the message is the multiple of 128-bit. So we should do an exhaustive search for the $(q-1)$-bit remaining bits using encryption oracle and thus we could find all the underlying plaintext blocks. This attack needs total $127 \times (q-1) + log_2(128) + 2$ oracle queries and 2^{q-1} encryptions on average to find all the plaintext blocks. Generally, this attack requires $(8n-1) \times (q-1) + log_2(8n) + 2$ oracle queries and 2^{q-1} encryptions.

ISO(10118-1)-Padding Method 3

Using the same method used in attacking the CBC|CBC mode with the ISO(9797-1)-Padding method 3, we can attack the CBC|CBC mode with the ISO(10118-1)-Padding method 3 with the same complexity.

Results on Our Attacks: We evaluated the security of all possible double modes of operation against the padding oracle attack. As a result, 12 out of 36 double modes are vulnerable to the padding oracle attack. The attack procedure of the 12 double modes is similar to that of CBC|CBC mode and omitted here. See Table 1 for the details of our results.

3.3   Padding Oracle Attacks on Triple Modes of Operation

Similarly, we can apply the padding oracle attacks to triple modes of operation with various padding schemes. In our observation, 22 out of total 216 triple modes are vulnerable to the padding oracle attacks. See Table 2 for the details of our results.

Table 2. Attack Complexities on Triple Modes of Operation

Padding Methods	CBC-PAD	ESP-PAD	XY-PAD	ISO(9797-1) -PAD 3	ISO(10118-1) -PAD 3
Complexities (# oracle queries)	$128nq$ $+log_2(n)$	$128nq$ $+log_2(n)$	$128nq$ $+log_2(n)$	$(8n-1)(q-1)$ $+log_2(8n)+2$	$(8n-1)(q-1)$ $+log_2(8n)+2$
Triple modes that can be broken by the padding oracle attacks	CBC|CBC|CBC, CBC|CBC|OFB, CBC|OFB|CBC, CBC|OFB|OFB, OFB|CBC|CBC, OFB|CBC|OFB, OFB|OFB|CBC, OFB|OFB|OFB, OFB|OFB|CFB, OFB|OFB|CFB^{-1}, OFB|CFB|CBC, OFB|CFB|OFB, OFB|CFB|CFB, OFB|CFB|CFB^{-1}, CFB|OFB|CBC, CFB|OFB|OFB, CFB|OFB|CFB, CFB|OFB|CFB^{-1}, CFB|CFB|CBC, CFB|CFB|OFB, CFB|CFB|CFB, CFB|CFB|CFB^{-1}			CBC|CBC|CBC, CBC|CBC|OFB, CBC|OFB|CBC, CBC|OFB|OFB, OFB|CBC|CBC, OFB|CBC|OFB, OFB|OFB|CBC,OFB|OFB|OFB	

4   Conclusion

We performed padding oracle attacks on multiple modes of operation. In the case of double modes of operation, among 36 modes, we succeeded in attacking 12 modes, and in the case of triple modes of operation, among 216 modes, we succeeded in attacking 22 modes with various padding schemes under the padding oracle attack models. The results imply that 12 double modes and 22 triple modes are vulnerable to these types of attacks and they do not offer the better security than single modes. We believe that our results are valuable to analyze the security of any multiple modes of operation. Any presence of message integrity layer in encryption schemes with a padding methods would invalidate the attacks described.

References

1. R. Baldwin and R. Rivest, *The RC5, RC5-CBC, RC5-CBC-Pad, and RC5-CTS algorithms*, RFC 2040, 1996.
2. E. Biham, *Cryptanalysis of multiple modes of operation*, Journal of Cryptology, Vol. 11, No. 1, pp. 45-58, 1998.
3. E. Biham, *Cryptanalysis of triple modes of operation*, Journal of Cryptology, Vol. 12, No. 3, pp. 161-184, 1999.
4. J. Black and H. Urtubia, *Side-Channel Attacks on Symmetric Encryption Schemes: The Case for Authenticated Encryption*, In Proc. of 11th USENIX Security Symposium, San Francisco 2002, pp. 327-338, 2002.
5. D. Hong, J. Sung, S. Hong, W. Lee, S. Lee, J. Lim, and O. Yi, *Known-IV Attacks on Triple Modes of Operation of Block Ciphers*, Advances in Cryptology - ASIACRYPT 2001, LNCS 2248, pp. 208-221, Springer-Verlag, 2001.
6. ISO/IEC 9797-1: Information technology, *Security tehniques- Message Auhentication Codes (MACs). Part 1: Mechanisms using a block cipher*, 1999.
7. ISO/IEC FDIS 10118-1: Information technology, Security techniques. Hashfunctions. Part 1: General (Final Draft), 2000.
8. V. Klima and T. Rosa, *Side Channel Attacks on CBC Encrypted Messages in the PKCS#7 Format*, Cryptology ePrint Archive, Report 2003/098, 2003.
9. G. Paterson and Arnold Yau, *Padding Oracle Attacks on the ISO CBC Mode Encryption Standard*, CT-RSA 2004, LNCS 2964, pp. 305-323, Springer-Verlag, 2004.
10. J. Sung, D. Hong, and S. Lee, *Key Recovery Attacks on the RMAC, TMAC, and IACBC*, ACISP 2003, LNCS 2727, pp 265-273, Springer-Verlag, 2003.
11. J. Sung, S. Lee, J.Lim, W. Lee and O. Yi, *Concrete Security Analysis of CTR-OFB and CTR-CFB Modes of Operation*, ICISC 2001, LNCS 2288, pp.103-113, Springer-Verlag, 2001.
12. S. Vaudenay, *Security Flaws Induced by CBC Padding - Applications to SSL, IPSEC, WTLS . . .*, Advances in Cryptology - EUROCRYPT 2002, LNCS 2332, pp. 534-545, Springer-Verlag, 2002.
13. D. Wagner, *Cryptanalysis of some recently-proposed multiple modes of operation*, In FSE 1998, LNCS 1372, pp. 254-269, Springer-Verlag, 1998.

An Evolutionary Algorithm to Improve the Nonlinearity of Self-inverse S-Boxes*

Hua Chen and Dengguo Feng

State Key Laboratory of Information Security, Institute of Software,
Chinese Academy of Sciences, Beijing 100080, P.R. China
{chenhua, feng}@is.iscas.ac.cn

Abstract. Self-inverse S-Boxes have been given much attention in the design of block ciphers recently. In this paper, based on Walsh Hadamard transform of Boolean functions, an evolutionary algorithm is investigated to increase the nonlinearity of self-inverse S-Boxes. The improved S-Boxes still remain self-inverse. Under this algorithm, randomly generated self-inverse S-Boxes can easily evolve into the ones with much higher nonlinearity.

Keywords: Self-inverse, S-Box, Walsh Hadamard Transform, Nonlinearity.

1   Introduction

S-Boxes are the only nonlinear component of most block ciphers. Whether or not S-Boxes are good can heavily influence the security of the whole algorithm. So how to generate cryptographically strong S-Boxes is a critical problem in the design of block ciphers.

Many ways have been explored to construct S-Boxes. One important approach is exponent permutation in finite field, which was investigated in [12, 11]. The S-Box of AES(Advanced Encryption Standard) is an example of exponent permutations, which is constructed with inverse mapping plus affine transformation in finite field. To construct S-Boxes with the truth table is also a good avenue. For example, in [13, 14], S-Boxes were generated by using single output Boolean functions. Some other examples were introduced in [10, 15, 9, 6, 16].

To generate S-Boxes with an evolutionary way has attracted much interest in recent years. In [7] an evolutionary design with the genetic algorithm was discussed, and two evolutional strategies on 8×4 S-Boxes were provided. It was also found that, if some improvement work was applied on the middle evolutionary objects, the genetic algorithm can be more effective[7]. In [8], how to improve the nonlinearity of bijective S-Boxes was investigated. The basic idea was, the nonlinearity of bijective S-Boxes can be improved by swapping some pairs of

---

* This work was supported by Chinese Natural Science Foundation (Grant No.60373047 and 60025205) and 863 Project (Grant No. 2003AA14403).

C. Park and S. Chee (Eds.): ICISC 2004, LNCS 3506, pp. 352–361, 2005.

output vectors. Literature [3] extended the work of [8] by providing some conditions under which the nonlinearity and difference uniformity of bijective S-Boxes can be improved simultaneously. In [2] the work of [8] was combined with the genetic algorithm.

As we know, self-inverse S-Boxes have been given much attention in the design of block ciphers which have SP structure. Self-inverse S-Boxes can save the storage space in implementation because encryption and decryption use the same S-Boxes. If a block cipher also owes a self-inverse structure, it can ensure the equivalent security of both encryption and decryption. A self-inverse S-Box is also called an involution[1]. The S-Box of Anubis [1] is an involution. A recursive structure is used to construct the S-Box. In [5], how to remove the linear redundancy of finite field based involutions was investigated which also maintained a high level of several important cryptographic properties and self inverse property still holds for resulting S-Boxes. However, the method is only efficient for finite field based involutions because the good properties of resulting S-Boxes heavily rely on the original S-Boxes. In this paper, random involutions are considered, and an evolutionary algorithm is investigated to improve the nonlinearity of randomly generated self-inverse S-Boxes.

This paper is organized as follows. In Section 2 some basic theories of Boolean functions are introduced. Then the evolutionary algorithm is presented in Section 3. Under the given algorithm, the experimental results for 10000 8×8 randomly generated self-inverse S-Boxes are provided in Section 4. Finally, the conclusion remarks are presented in section 5.

2   Basic Theory

First, let $f(x) : F_2^n \longrightarrow F_2$ denote a Boolean function, and $\hat{f}(x) = (-1)^{f(x)}$ is the polarity form of $f(x)$. It is easily seen that $\hat{f}(x) = 1$ when $f(x) = 0$, and $\hat{f}(x) = -1$ when $f(x) = 1$.

A linear Boolean function can be defined as $L_w(x) = w_1 x_1 \oplus w_2 x_2 \oplus \cdots \oplus w_n x_n$, $w \in F_2^n$. Its polarity form is $\hat{L}_w(x)$. An affine Boolean function is a linear Boolean function plus a constant, which can be represented as $A_{w,c}(x) = L_w(x) \oplus c$. L_n represents the set of all linear and affine n-ary Boolean functions. The nonlinearity of a Boolean function is the minimum hamming distance to any affine function.

$\hat{F}(w) = \sum_x \hat{f}(x) \hat{L}_w(x)$ is called the Walsh Hadamard transform(WHT) on $f(x)$. Denote WH_{max} as the maximum absolute value taken by $\hat{F}(w)$. If N_f represents the nonlinearity of boolean functions, $N_f = \frac{1}{2}(2^n - WH_{max})$. Apparently, the nonlinearity can be increased by reducing WH_{max}.

A S-Box is also a Boolean function, which is a mapping from n binary inputs to m binary outputs. If $n = m$ and the outputs of a S-Box differs from each other, the S-Box is called bijective. Obviously, it is also reversible. If the inverse of a S-Box is itself, we call it a self-inverse S-Box or an involution.

Given a S-Box $S(x) : F_2^n \longrightarrow F_2^m$ (also called $n \times m$ S-Box), $N_f = min d_H(u \cdot S(x), l(x))(l \in L_n, 0 \neq u \in F_2^m)$ is the nonlinearity of S(x). d_H represents the hamming distance. Let $I(x) = y$ represent an involution from n bits to n bits with $I(y) = I^{-1}(y) = x$. Denote $I_\theta(x) = L_\theta(y)$. The Walsh Hadamard transform of $I(x)$ has a square matrix of size 2^n. The element $\hat{I}_\theta(w)$ represents the WHT of $L_\theta(y)$.

Theorem 1. *The WHT matrix of an involution $I(x)$ is symmetric.*

Proof: As we know, an involution is also a bijective S-Box. It has been proven in [8] that $\hat{I}_\theta(w) = \hat{I}_w^{-1}(\theta)$. And:

$$\hat{I}_w^{-1}(\theta) = \sum_x I_w^{-1}(x)\hat{L}_\theta(x)$$

$$= \sum_x \hat{L}_w(I^{-1}(x))\hat{L}_\theta(x)$$

$$= \sum_x \hat{L}_w(I^{-1}(I(y)))\hat{L}_\theta(x)$$

$$= \sum_x \hat{L}_w(y)\hat{L}_\theta(x)$$

$$= \hat{I}_w(\theta)$$

so $\hat{I}_\theta(w) = \hat{I}_w(\theta)$, which completes the proof.

Theorem 1 can reduce the computation complexity on the WHT matrix of an involution. Similarly to a single Boolean function, if WH_{max} is denoted as the maximum absolute value of $\hat{I}_\theta(w)$, $N_f = \frac{1}{2}(2^n - WH_{max})$.

In the next section, the hill climbing algorithm will be introduced to increase N_f of involutions with WHT matrix. Under this algorithm, bad involutions can easily evolved into good ones in nonlinearity.

3   Hill Climbing Self-inverse S-Boxes

In [8] a method (called hill climbing) is proposed to improve the nonlinearity of bijective S-Boxes. Under this method, to swap two output vectors of a S-Box can improve the nonlinearity if they satisfy some conditions. Now define $I(x) = y$ as an involution. According to the method in [8], if two different input vectors x_1 and x_2 satisfy some conditions, they are swapped. The new S-Box $I'(x)$ has the higher nonlinearity than $I(x)$. So $I'(x_1) = I(x_2), I'(x_2) = I(x_1)$. For x that satisfies $x \neq x_1$ and $x \neq x_2$, $I'(x) = I(x)$. If x_2 satisfies $I(x_2) \neq x_1$ and $I(x_2) \neq x_2$, $I'(I'(x_1)) = I'(I(x_2)) = I(I(x_2)) = x_2$. Apparently $I'(x)$ do not satisfy self-inverse property any more. In [5], two pairs of output vectors are swapped in order to maintain the self-inverse property. It is also adopted in this paper. Different from [5], nonlinearity is considered when two pairs of output vectors are swapped and the conditions are explored under which the nonlinearity can be increased.

For an involution $I(x) = y$, $\hat{I}_\theta(w)$ is an element of Walsh Hadamard transform matrix. WH_{max} is the maximum absolute value of $\hat{I}_\theta(w)$. And some sets of (w, θ) pairs are defined as follows:

$$W_1^+ = \{(w, \theta) : \hat{I}_\theta(w) = WH_{max}\}$$
$$W_1^- = \{(w, \theta) : \hat{I}_\theta(w) = -WH_{max}\}$$
$$W_2^+ = \{(w, \theta) : \hat{I}_\theta(w) = WH_{max} - 2\}$$
$$W_2^- = \{(w, \theta) : \hat{I}_\theta(w) = -WH_{max} + 2\}$$
$$W_3^+ = \{(w, \theta) : \hat{I}_\theta(w) = WH_{max} - 4\}$$
$$W_3^- = \{(w, \theta) : \hat{I}_\theta(w) = -WH_{max} + 4\}$$
$$W_4^+ = \{(w, \theta) : \hat{I}_\theta(w) = WH_{max} - 6\}$$
$$W_4^- = \{(w, \theta) : \hat{I}_\theta(w) = -WH_{max} + 6\}$$
$$W_5^+ = \{(w, \theta) : \hat{I}_\theta(w) = WH_{max} - 8\}$$
$$W_5^- = \{(w, \theta) : \hat{I}_\theta(w) = -WH_{max} + 8\}$$

Theorem 2. *Let $I(x) = y : F_2^n \longrightarrow F_2^n$ be an involution. Further, let x_1 and x_2 be distinct input vectors with corresponding outputs $y_1 = I(x_1)$ and $y_2 = I(x_2)$, which also satisfy $I(x_1) \neq x_2$ and $I(x_2) \neq x_1$. Let $I'(x)$ be an S-Box the same as $I(x)$ except that $I'(x_1) = y_2$, $I'(x_2) = y_1$, $I'(y_1) = x_2$ and $I'(y_2) = x_1$. Then the nonlinearity of $I'(x)$ will exceed that of $I(x)$ if and only if all of the following conditions are satisfied:*

(a) *For all $(w, \theta) \in W_1^+$, condition (1) or (2) must be satisfied;*
For all $(w, \theta) \in W_{2,3}^-$, neither condition (1) nor (2) is satisfied;
(1.) $L_w(x_1) \neq L_w(x_2)$
$L_\theta(y_1) = L_w(x_1)$
$L_\theta(y_2) = L_w(x_2)$
and not all the followings are true:
$L_w(y_1) \neq L_w(y_2)$
$L_\theta(x_1) = L_w(y_2)$
$L_\theta(x_2) = L_w(y_1)$

(2.) *not all the followings are true:*
$L_w(x_1) \neq L_w(x_2)$ $L_\theta(y_1) = L_w(x_2)$
$L_\theta(y_2) = L_w(x_1)$
and
$L_w(y_1) \neq L_w(y_2)$
$L_\theta(x_1) = L_w(y_1)$
$L_\theta(x_2) = L_w(y_2)$

For all $(w, \theta) \in W_1^-$, the conditions are similar to $(w, \theta) \in W_1^+$ except the signs change('=' will change to '$\neq$', '$\neq$' will change to '=').

For all $(w,\theta) \in W_{2,3}^+$, the conditions are similar to $(w,\theta) \in W_{2,3}^-$ except the signs change.

(b) For all $(w,\theta) \in W_{4,5}^+$
not all the followings are true:
$$L_w(x_1) = L_w(x_2)$$
$$L_\theta(y_1) = L_w(x_2)$$
$$L_\theta(y_2) = L_w(x_1)$$
$$L_w(y_1) \neq L_w(y_2)$$
$$L_\theta(x_1) \neq L_w(y_2)$$
$$L_\theta(x_2) \neq L_w(y_1)$$
For all $(w,\theta) \in W_{4,5}^-$, the conditions are similar to $(w,\theta) \in W_{4,5}^+$ except the signs change.

Proof: By definition we have $\hat{I}_\theta(w) = \sum_x \hat{I}_\theta(x)\hat{L}_w(x)$ for the original S-Box and $\hat{I}_\theta'(w) = \sum_x \hat{I}_\theta'(x)\hat{L}_w(x)$ for the changed S-Box by swapping the two output pairs for input pairs (x_1, x_2) and (y_1, y_2). For a (w,θ), $\Delta \hat{I}_\theta(w) = \hat{I}_\theta'(w) - \hat{I}_\theta(w)$.

As we know, the difference between the sum terms of $\hat{I}_\theta'(w)$ and that of $\hat{I}_\theta(w)$ only involves x_1, x_2, y_1, and y_2. Cancelling, we are left with:

$$\Delta \hat{I}_\theta(w) = \hat{I}_\theta'(x_1)\hat{L}_w(x_1) + \hat{I}_\theta'(x_2)\hat{L}_w(x_2) - \hat{I}_\theta(x_1)\hat{L}_w(x_1) - \hat{I}_\theta(x_2)\hat{L}_w(x_2)$$
$$+\hat{I}_\theta'(y_1)\hat{L}_w(y_1) + \hat{I}_\theta'(y_2)\hat{L}_w(y_2) - \hat{I}_\theta(y_1)\hat{L}_w(y_1) - \hat{I}_\theta(y_2)\hat{L}_w(y_2)$$
$$= \hat{I}_\theta(x_2)\hat{L}_w(x_1) + \hat{I}_\theta(x_1)\hat{L}_w(x_2) - \hat{I}_\theta(x_1)\hat{L}_w(x_1) - \hat{I}_\theta(x_2)\hat{L}_w(x_2)$$
$$+\hat{I}_\theta(y_2)\hat{L}_w(y_1) + \hat{I}_\theta(y_1)\hat{L}_w(y_2) - \hat{I}_\theta(y_1)\hat{L}_w(y_1) - \hat{I}_\theta(y_2)\hat{L}_w(y_2)$$

Since each output function is changing in either zero or two truth table positions, we have $\Delta \hat{I} \in \{-8, -4, 0, 4, 8\}$ for all (w,θ). TO increase the nonlinearity we must reduce WH_{max}. For the convenience to study the required conditions, denote $\Delta_1 = \hat{I}_\theta(x_2)\hat{L}_w(x_1) + \hat{I}_\theta(x_1)\hat{L}_w(x_2) - \hat{I}_\theta(x_1)\hat{L}_w(x_1) - \hat{I}_\theta(x_2)\hat{L}_w(x_2)$, and $\Delta_2 = \hat{I}_\theta(y_2)\hat{L}_w(y_1) + \hat{I}_\theta(y_1)\hat{L}_w(y_2) - \hat{I}_\theta(y_1)\hat{L}_w(y_1) - \hat{I}_\theta(y_2)\hat{L}_w(y_2)$. $\Delta_1 \in -4, 0, 4$, $\Delta_2 \in -4, 0, 4$.

For all $(w,\theta) \in W_1^+$, $\Delta \hat{I} = -4$ or -8 is required. It is equivalent to $\Delta_1 = -4, \Delta_2 \neq 4$ or $\Delta_1 \neq 4, \Delta_2 = -4$.

When $\Delta_1 = -4, \Delta_2 \neq 4$, (1) $-$ (4) must hold simultaneously and either of (5) $-$ (8) must not hold.

(1) $\hat{I}_\theta(x_2)\hat{L}_w(x_1) = -1$
(2) $\hat{I}_\theta(x_1)\hat{L}_w(x_2) = -1$
(3) $\hat{I}_\theta(x_1)\hat{L}_w(x_1) = 1$
(4) $\hat{I}_\theta(x_2)\hat{L}_w(x_2) = 1$

(5) $\hat{I}_\theta(y_2)\hat{L}_w(y_1) = 1$
(6) $\hat{I}_\theta(y_1)\hat{L}_w(y_2) = 1$
(7) $\hat{I}_\theta(y_1)\hat{L}_w(y_1) = -1$
(8) $\hat{I}_\theta(y_2)\hat{L}_w(y_2) = -1$

Which is equivalent to condition (1.) in (a).

When $\Delta_1 \neq 4, \Delta_2 = -4$, $(9) - (12)$ must not hold simultaneously and $(13) -$ (16) must hold simultaneously.

(9) $\hat{I}_\theta(x_2)\hat{L}_w(x_1) = 1$
(10) $\hat{I}_\theta(x_1)\hat{L}_w(x_2) = 1$
(11) $\hat{I}_\theta(x_1)\hat{L}_w(x_1) = -1$
(12) $\hat{I}_\theta(x_2)\hat{L}_w(x_2) = -1$

(13) $\hat{I}_\theta(y_2)\hat{L}_w(y_1) = -1$
(14) $\hat{I}_\theta(y_1)\hat{L}_w(y_2) = -1$
(15) $\hat{I}_\theta(y_1)\hat{L}_w(y_1) = 1$
(16) $\hat{I}_\theta(y_2)\hat{L}_w(y_2) = 1$

Which is equivalent to condition (2.) in (a).

Similarly, for all $(w, \theta) \in W_1^-$, $\Delta\hat{B} = 4$ or 8 is required. It is equivalent to $\Delta_1 = 4, \Delta_2 \neq -4$ or $\Delta_1 \neq -4, \Delta_2 = 4$. It is not difficult to prove that the conditions are similar to those for $(w, \theta) \in W_1^+$ except the signs change.

For all $(w, \theta) \in W_{2,3}^-$, in order to reduce WH_{max}, $\Delta\hat{I} = 4$ or 8 or 0. Apparently, the conditions should be the complement of those for $(w, \theta) \in W_1^+$, so both condition (1) and (2) in (a) should be satisfied. Similarly, for all $(w, \theta) \in W_{2,3}^-$, only the signs are changed from those of $(w, \theta) \in W_{2,3}^+$.

For all $(w, \theta) \in W_{4,5}^+$, in order to reduce WH_{max}, $\Delta\hat{I} \neq 8$. This requires $\Delta_1 \neq 4$ or $\Delta_2 \neq 4$, which implies condition (b). Similarly, for all $(w, \theta) \in W_{4,5}^-$, only the signs are changed from that of $(w, \theta) \in W_{4,5}^+$.

Corollary 1. $I'(x)$ *is also an involution.*

Proof: Here two cases need to be considered. One case is $x \notin \{x_1, x_2, y_1, y_2\}$, so $y = I(x) \notin \{x_1, x_2, y_1, y_2\}$. Because $I'(x)$ is same as $I(x)$ except the outputs of x_1, x_2, y_1 and y_2, $I'(x) = I(x)$ and $I'(y) = I(y)$. Hence, $I'(I'(x)) = I'(I(x)) = I(I(x)) = x$.

The other case is $x \in x_1, x_2, y_1, y_2$ and the following equations hold:

$$I'(I'(x_1)) = I'(y_2) = x_1$$
$$I'(I'(x_2)) = I'(y_1) = x_2$$
$$I'(I'(y_1)) = I'(x_2) = y_1$$
$$I'(I'(y_2)) = I'(x_1) = y_2$$

So $I'(x)$ is an involution. Which completes the proof.

For $S(x) : F_2^n \longrightarrow F_2^m$, if some $x \in F_2^n$ satisfy $S(x) = x$, we say $S(x)$ has fixed points. S-Boxes with fixed points will enhance chances to attack algorithms. So fixed points are not encouraged in the design of S-Boxes. Theorem 2 can ensure improved involutions have no fixed points if the original ones has no fixed points either.

Corollary 2. *If $I(x)$ has no fixed points, $I'(x)$ has no fixed points either.*

Proof: For $x \notin \{x_1, x_2, y_1, y_2\}$, $I'(x) = I(x) \neq x$. And because $I(x_1) \neq x_2$, $I(x_2) \neq x_1$ also holds. So:

$$I'(x_1) = I(x_2) \neq x_1$$
$$I'(x_2) = I(x_1) \neq x_2$$
$$I'(y_1) = x_2 \neq I(x_1) = y_1$$
$$I'(y_2) = x_1 \neq I(x_2) = y_2$$

Which completes the proof.

On the basis of Theorem 2, a hill climbing algorithm will be proposed to improve the nonlinearity of an involution. The basic idea is, for an involution, two output vectors that satisfy the conditions in Theorem 2 will be swapped. The procedure is repeated until the nonlinearity reaches a local maximum, which means no output pairs can be swapped to increase the nonlinearity. When the original S-Box has no fixed points, the final S-Box has no fixed points either. The algorithm can also ensure the self-inverse property of the final S-Box.

InvolutionHillClimb(I(x))

(1) Calculate the Walsh-Hadamard transform matrix of $I(x)$ and determine the maximum value WH_{max}.
(2) Find the pairs of (w, θ) which belong to the sets W_1^+, W_1^-, $W_{2,3}^+$, $W_{2,3}^-$, $W_{4,5}^+$ and $W_{4,5}^-$.
(3) Repeat Until no pair satisfies tests:
   (a) Select a input pair (x_1, x_2) and check it against the conditions in Theorem 2.
   (b) If the current pair satisfies the conditions then get the new involution $I'(x)$ and calculate the new WHT matrix, else select the next pair for testing.
(4) Output $I'(x)$.

The algorithm must stop because the nonlinearity has an upper bound. It is not possible to increase the nonlinearity infinitely. For $n \times m$ S-Boxes, the bound is $2^{n-1}(1 - 2^{-n/2})$[4]. According to Theorem 2, the maximum possible input pairs (x_1, x_2) that need to be considered should be $C_n^2/2 - n/2$. So in the algorithm, before a input pair (x_1, x_2) is found to satisfy the conditions in Theorem 2, the maximum possible input pairs that need to be checked should be $C_n^2/2 - n/2$. And in the proof of Theorem 2, it is easily seen that the minimum increment of nonlinearity is 4. So if the nonlinearity of the original involution is nl, the maximum iterative times in the algorithm should be smaller than $(C_n^2/2 - n/2) \times (2^{n-1}(1 - 2^{-n/2}) - nl)/4$.

4   Experiments and Results

$8 \times m$ S-Boxes are popularly used in the design of block ciphers, and the maximum nonlinearity is 112[4]. It is easy to construct best S-Boxes in nonlinearity with exponent permutations, but they do not satisfy self-inverse property. In this section, the experimental results on 8×8 involutions will be given.

First of all, 10000 8×8 involutions with no fixed points will be randomly generated with the following algorithm:

GenerateInvolution

1. For every input $x \in F_2^n$, set the mark value of x as 0.
2. Select a x whose mark value is 0. Set the new mark value of x is 1. Generate a random value $m \in F_2^n$ whose mark value is 0.
3. Set $I(x) = m$ and $I(m) = x$.
4. Set the new mark value of m is 1.
5. Repeat 2-4 steps until no x satisfies the condition.
6. Output the $n * n$ involution $I(x)$.

Then the hill climbing algorithm in section 3 is applied on the generated 10000 involutions. Table 1 and Figure 1 respectively give the comparison results of the original involutions and the improved ones after the hill climbing algorithm. From table 1 we can see the nonlinearity of the original randomly generated involutions is mainly distributed over $(88, 90, 92, 94, 96)$, and their propagations are 8.34% , 14.89%, 26.14%, 29.25% and 13.24%. The propagation of 98 is only 0.93%. So highly nonlinear involutions are hardly obtained by the random generation. When these random generated involutions are improved by the hill climbing algorithm, the nonlinearity of the final involutions concentrates on $(98, 100)$ and their proportions are 53.9% and 42.17%. From Figure 1 we can also see the improvement effect is apparent. Besides, most improved S-Boxes have higher S-Boxes than Anubis because its nonlinearity is only 94[1].

Table 1. Comparison results of original involutions and improved involutions

nonlinearity	original S-Box	improved S-Box
78	26	0
80	40	0
82	90	0
84	190	0
86	374	0
88	834	0
90	1489	0
92	2614	7
94	2925	12
96	1324	367
98	93	5390
100	1	4217
102	0	7

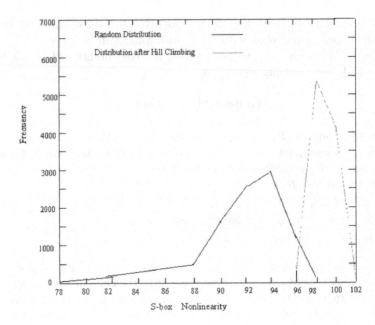

Fig. 1. The comparison curve of original involutions and improved involutions

5   Conclusion

In this paper, Walsh Hadamard transform of self-inverse S-Boxes is studied. On the basis of it, the hill climbing algorithm is introduced to improve self-inverse S-Boxes. After being improved, the self-inverse characteristics still holds. Then 10000 8×8 self-inverse sample S-Boxes are randomly generated with a given algorithm in this paper. The Experimental results show that the original involutions have been greatly improved in nonlinearity after being hill climbed. Besides, most of the improved involutions has much higher nonlinearity than the one of Anubis.

There are still some open problems for further research. The work in this paper can be combined with the genetic algorithm to generate much stronger involutions. Besides the nonlinearity, other properties of S-Boxes can be considered to be improved with the similar idea in this paper.

References

1. Paulo   S.L.M.   Barreto,   Vincent   Rijmen,   *The   ANUBIS   Block   Cipher*,http://planeta.terra.com.br/informatica/paulobarreto/AnubisPage.html.
2. H. Chen, D. Feng. An Effective Evolutionary Strategy for Bijective S-Boxes. *Proceedings of the IEEE Congress on Evolutionary Computation 2004 (CEC'04), June 20-23, 2004, Portland Oregon, pp. 2120–2123.*

3. H. Chen, D. Feng, W. Wu. An effective algorithm for improving the cryptographic properties of bijective S-Boxes. *Journal of Computer Research and Development. Vol 41 No 8 2004, pp. 1410–1414*

4. Deng-guo Feng, Wen-ling Wu. Designing And Analysis of Block Cipher. BeijingTsinghua University Press2000. 67–69

5. Joanne Fuller, William Millan, Ed Dawson. Multi-objective Optimisation of Bijective S-Boxes. *Proceedings of the IEEE Congress on Evolutionary Computation 2004 (CEC'04), June 20-23, 2004, Portland Oregon, pp. 1525–1532.*

6. K.Kim, T. Matsumoto, and H. Imai. A Recursive Construction Method of S-Boxes Satisfying Strict Avalanche Criterion. In *Advances in Cryptology - Crypto '90*, Proceedings, LNCS, volume 537 pp. 564–574. Springer-Verlag, 1991.

7. William Millan, L.Burnett,G.Carter,A.Clark, and E.Dawson. Evolutionary Heuristics for Finding Cryptographically Strong S-Boxes,ICICS'99, BerlinSpring- er-Verlag,1999,LNCS vol.1726 pp.263–274

8. William Millan. How to Improve the Nonlinearity of Bijective S-Boxes,ACISP' 98 Berlin Springer-Verlag,1998,LNCS vol.1438 pp.181–192

9. K.Nyberg. Perfect Nonlinear S-Boxes. In *Advances in Cryptology - Eurocrypt '91*, Proceedings, LNCS, volume 547, pp. 378–386. Springer-Verlag, 1991.

10. K.Nyberg. On the Construction of Highly Nonlinear Permutations. In *Advances in Cryptology - Eurocrypt '92*, Proceedings, LNCS, volume 658, pp. 92–98. Springer-Verlag, 1993.

11. J.Pieprzyk. Non-linearity of Exponent Permutations. In *Advances in Cryptology - Eurocrypt' 89*, Proceedings, LNCS, volume 434, pp. 81–92. Springer-Verlag, 1990.

12. J.Pieprzyk. Bent Permutations. In *International Conference on Finite Fields, Coding Theory and Advances in Communications.* Las Vegas, pp. 173–181, 1991.

13. J.Pieprzyk and G.Finkelstein. Permutations that Maximise Non-Linearity and their Cryptographic Significance. In *Proceedings of Fifth IFIP International Conference on Computer Security IFIP/SEC' 88*, pp. 63–74,1988.

14. J. Pieprzyk and G. Finkelstein. Towards Effective Nonlinear Cryptosystem Design. IEE Proceedings, Pt E., 135(6):325–335, November 1988.

15. A.F. Webster and S.E. Tavares. On the Design of S-Boxes. In *Advances in Cryptology - Crypto '85*, Proceedings, LNCS, volume 218, pp. 523–534. Springer-Verlag, 1996.

16. J.Seberry, X.-M. Zhang and Y. Zheng. Systematic Generation of Cryptographically Robust S-Boxes. In *Proceedings of the First ACM Conference on Computer and Communications Security.* pp. 171–182, 1994.

Identity-Based Access Control
for Ad Hoc Groups*

Nitesh Saxena, Gene Tsudik, and Jeong Hyun Yi**

School of Information and Computer Science,
University of California at Irvine,
Irvine, CA 92697, USA
{nitesh, gts, jhyi}@ics.uci.edu

Abstract. The proliferation of group-centric computing and communication motivates the need for mechanisms to provide *group access control.* Group access control includes mechanisms for admission as well as revocation/eviction of group members. Particularly in ad hoc groups, such as peer-to-peer (P2P) systems and mobile ad hoc networks (MANETs), secure group admission is needed to bootstrap other group security services. In addition, secure membership revocation is required to evict misbehaving or malicious members. Unlike centralized (e.g., multicast) groups, ad hoc groups operate in a decentralized manner and accommodate dynamic membership which make access control both interesting and challenging. Although some recent work made initial progress as far as the admission problem, the membership revocation problem has not been addressed.

In this paper, we develop an identity-based group admission control technique which avoids certain drawbacks of previous (certificate-based) approaches. We also propose a companion membership revocation mechanism. Our solutions are robust, fully distributed, scalable and, at the same time, reasonably efficient, as demonstrated by the experimental results.

Keywords: access control, ad-hoc group security, threshold signatures.

1   Introduction

Ad hoc groups are becoming increasingly popular these days. A number of peer-to-peer (P2P) systems as well as mobile ad-hoc networks (MANETs) fall into the category of ad hoc groups. These groups are characterized by two important features, (1) lack of trusted authority and (2) dynamic membership, which often implies dynamic topology. These features prompt a number of challenges for routing as well as content placement and retrieval. They also make it difficult

---

* This work was supported in part by an award from the Army Research Office (ARO) under contract W911NF0410280 and a grant from SUN Microsystems.
** Corresponding author.

C. Park and S. Chee (Eds.): ICISC 2004, LNCS 3506, pp. 362–379, 2005.

to develop effective and efficient security mechanisms. The need for security in MANETs and P2P has been widely recognized by the research community and the bulk of prior work has been in the context of traditional security services such as secure group communication (group key agreement and key management) and secure routing (in MANETs). Although these services are certainly important, another equally important issue – **group access control** – has not received due attention.

In a group setting, the traditional notion of access control is to prevent unauthorized entities from accessing group resources. However, if we consider group membership itself to be a resource, the problem of admission and revocation/eviction of group members can be viewed as a form of access control. A secure admission control is necessary to prevent unauthorized users from joining a group, i.e., accessing the "membership" resource. Without such a mechanism, group membership is open to malicious users and the group becomes vulnerable, e.g., to Sybil attacks [1]. Moreover, a group as a whole must be able to contend with the possibility of group members becoming selfish[1] or malicious/compromised. Once detected, such rogue members need to be removed from the group. This necessitates secure (and efficient) membership revocation mechanisms.

Without effective group access control mechanisms, other group security services: secure group communication [2, 3] and secure routing (in mobile ad hoc networks), such as Ariadne [4], SPINS [5], etc., are difficult to achieve.

One fairly obvious application for the type of secure distributed group access control that we envisage is in the domain of private P2P groups formed atop wide-open (essentially public) P2P systems, such as Kazaa, Morpheus or Gnutella. In fact, a recent article in the Time Magazine [6] examines the popular trend of creating so-called *"Darknets"* [7] – secure private groups within Kazaa and Morpheus – in order to escape the intensified crackdown on music and other content sharing. Another example of a somewhat futuristic, military oriented application of group access control in a mobile ad hoc network with unmanned aerial vehicles (UAVs) is described in [8].

2   Related Work and Motivation

Previous work on admission control in ad hoc groups ([9], [10], [11], and [12]) employed a menu of cryptographic techniques to perform secure group admission. The purpose is for a certain threshold of group members to make collaborative decisions regarding the admission of a prospective member and provide it with a signed group membership certificate. Among these signature schemes are: plain (RSA or DSA) signatures, accountable subgroup multi-signatures (ASM) [13],

---

[1] A selfish member is interested in obtaining service but refuses to provide it. Service can range from sharing files in a content-sharing P2P to forwarding traffic in a MANET.

threshold signatures ([14], [9]). Unfortunately, these schemes have certain drawbacks that make them unfit for practical group admission control scenarios.

Lineage Problem with Plain Signatures and Multisignatures. In particular, admission control based on either plain signatures or accountable sub-group multisignatures has a *lineage* problem. This problem occurs when a membership certificate is issued to a new member: each member (sponsor) who takes part in the admission process needs to confirm (by signing) its agreement to admit this new member. Essentially, a membership certificate has to be signed by some number of membership sponsors.[2] However, each sponsor needs to attach its own certificate to its signature on a new member's certificate in order to make group certificates universally verifiable. However, a sponsor's own certificate also has to be counter-signed by its erstwhile sponsors, and so on, and so forth. This is clearly unworkable since a member's certificate would have to be accompanied by a number of certificate chains that affirm its lineage.

Inapplicability of Known Threshold RSA Signatures. A $(t-1, n)$ threshold signature scheme [15] enables any subgroup of t members in a group to collaboratively sign a message on behalf of that group. This is achieved by secret-sharing the signature key among the group members, and allowing them to compute a signature on some message via a distributed protocol in which the members use the shares of the signature key instead of the key itself. Threshold signatures are naturally attractive for the purpose of admission control since they prevent the aforementioned lineage problem: a membership certificate can be signed by any set (of a certain size) of sponsors while the signature length is constant and identities of individual sponsors are not revealed.

Various flavors of threshold RSA signatures exist in literature that might be used to construct the admission control protocol. However, unfortunately, **none** of these schemes are directly applicable (refer to [16] for details).

In an effort to mitigate the above problem of the known threshold RSA signatures, Kong, et al. [9] proposed a new threshold RSA scheme, geared toward providing security services in mobile ad-hoc networks. Subsequently the scheme and its MANET applications were described in [9, 8], and most recently in a journal version [17]. Unfortunately, this scheme is neither robust (i.e. it can not tolerate malicious group members) nor secure. The robustness problem was first pointed out in [12]. For an explicit attack exploiting the insecurity of the scheme, the reader is referred to [18].

Limitation of Threshold DSA Signature. An alternative scheme (called *TS-DSA*) [12] is based on the threshold DSS signature scheme [19]. This scheme is robust and hence tolerates malicious insiders. However, the practicality of this scheme is questionable since, as illustrated by experiments in [11], it is very

---

[2] This number is determined by the group admission policy; common examples are a certain fraction of current members or a fixed threshold. See [10] for a detailed discussion of admission policies.

costly due to $O(t^2)$ communication among the t signers. Furthermore, *TS-DSA* remains secure as long as there are less than $\lfloor \frac{t+1}{2} \rfloor$ malicious members. In other words, for the scheme to be able to tolerate $t-1$ faults, $2t-1$ signers are required.

Our Contributions. In addition to the above discussion, the previously proposed admission control mechanisms are certificate-based, making them quite impractical[3] for mobile ad hoc networks where the amount of communication is directly related to the battery power of the mobile devices (refer to [20]). In this paper, we develop a new group admission control technique, which we refer to as *ID-GAC* (**ID**entity-based **G**roup **A**dmission **C**ontrol). As the name suggests, *ID-GAC* is an identity-based approach (and therefore more communication efficient), in contrast to previous, certificate-based schemes. Furthermore, we present a membership revocation mechanism geared to work in conjunction with *ID-GAC*.

Organization. After specifying the notation in Table 1, we define the security model for a generic group access control system in Section 3. In Sections 4 and 5, we present the new ID-based admission control protocol and demonstrate its security. Section 6 presents the membership revocation protocols. Experimental results are illustrated in Section 7 and finally, some outstanding issues are discussed in Section 8 followed by the conclusion in Section 9.

Table 1. Notation

M_i	group member i	id_i	ID for M_i
t	admission threshold	t_r	revocation threshold
n	total number of M_i-s	$\mathbb{G}_1, \mathbb{G}_2$	cyclic GDH groups of order q
A	generator of group $\mathbb{G}_1$	B	group public key
$\hat{e}$	map s.t. $\hat{e} : \mathbb{G}_1 \times \mathbb{G}_1 \rightarrow \mathbb{G}_2$	T_i	membership token for M_i
SK_i	secret key of M_i	PK_i	public key of M_i
$S_i(m)$	signature on message m	MRL	membership revocation list
SL_i	list of signers for M_i	H	hash func. such as SHA-1 or MD5
ss_i	secret share of M_i	H_1	hash func. s.t. $H_1 : \{0,1\}^* \rightarrow \mathbb{G}_1^*$
$pss_j(i)$	partial share for M_i by M_j	H_2	hash func. s.t. $H_2 : \{0,1\}^* \times \mathbb{G}_1 \rightarrow \mathbb{Z}_q^*$

3   Security Model

In this section, we define a generic security model for peer-based membership management. In other words, we describe what we mean by secure admission control and secure membership revocation, respectively. Since this work is applied in nature, the model is specified informally.

---

[3] The typical size of a group membership certificate is quite large, e.g., 5KB for 1024-bit DSA parameters.

3.1   Admission Control

A *secure admission* mechanism is a secure interactive protocol between a prospective member M_{new} and a set of current group members $\{M_i \mid 1 \leq i \leq s\}$ where $1 \leq t \leq s \leq n$. (In other words, the number of current members taking part in the admission is at least the number necessary for the admission threshold which is, in turn, no greater than the number of current members.) This protocol must satisfy the following properties:

1. COMPLETENESS. When the protocol completes (in polynomial time), M_{new} has a group membership token, if at least t out of n group members vote in favor of admission. In addition, M_{new} also acquires the membership revocation information if any, which allows it to keep track of revoked group members. *Optionally*[4], M_{new} is also in possession of a means to take part in future admission decisions.
2. TRACEABILITY. If one or more malicious sponsors do not provide correct information in the course of the admission process, M_{new} can detect, trace and publicly identify such members.[5]
3. IMPERSONATION RESISTANCE. It is computationally infeasible for anyone, who has not been successfully admitted via the admission protocol, to impersonate a genuine group member (whether to members or non-members).

3.2   Membership Revocation

A secure *membership revocation* mechanism is a protocol among the members of the group wherein any set of t_r $(\geq t)$ members attempt to revoke a current member M_r who possesses a membership token T_r and/or a secret share ss_r. This mechanism needs to satisfy the following properties:

1. COMPLETENESS. At the end of this protocol, i.e. when t_r revocation requests are lodged against M_r, the latter is unable to: (1) prove group membership and (2) participate in future admission decisions (in case M_r previously had voting/admission rights).
2. IMPERSONATION RESISTANCE. Only a genuine group member (possessing a valid membership token) can take part in the revocation process. In other words, non-group members and previously revoked group members can not lodge valid revocation requests.
3. COLLUSION RESISTANCE. At the end of the protocol, M_r (if she possessed a secret share) is unable to collaborate with any set of previously revoked members and recover the group secret x. In other words, any set of revoked share-holding members can not collude to interpolate the group secret.

---

[4] Whether a new member gets voting rights that allow it to take part in future admission decisions depends upon group admission policy.
[5] The group members misbehaving in this manner may then be revoked from the group by triggering the revocation mechanism.

Remark: As will be discussed in Section 6, every group member maintains an updated list of the revoked members which we refer to as *membership revocation list* (MRL).

4 ID-Based Group Admission Control (ID-GAC)

In this section, we present our new admission control mechanism *ID-GAC*. The mechanism is based on the threshold version [21] of BLS signature scheme [22]. The description includes: set-up, admission process and security arguments following the model in Section 3. *ID-GAC* is an identity-based mechanism since the membership token used to prove membership is derived from the group member's identity.

4.1   Setup

ID-GAC can be initialized by either: (1) a trusted dealer or (2) a group of $2t-1$ or more founding members. In either case, the dealer first initializes and generates the appropriate elliptic curve domain parameters $(p, \mathbb{F}_p, a, b, A, q)$. The elliptic curve is represented by the equation: $y^2 = x^3 + ax + b$. $\mathbb{G}_1$ is set to be a group of order q generated by A, $\mathbb{G}_2$ is a subgroup of $\mathbb{F}_{p^2}^*$ of order q, and $\hat{e} : \mathbb{G}_1 \times \mathbb{G}_1 \to \mathbb{G}_2$ is defined to be a public bilinear mapping. Also, $H_1 : \{0,1\}^* \to \mathbb{G}_1^*$ is the hash function that maps binary strings to non–zero points in $\mathbb{G}_1$. All of this information is published and all group members (as well as prospective members) are assumed to have access to it.

INITIALIZATION BY DEALER. TD selects a random polynomial $f(z) = f_0 + f_1 z + \cdots + f_{t-1} z^{t-1}$ over $\mathbb{Z}_q$ of degree $t-1$, such that the group secret is $f(0) = f_0 = x$. In order to enable verifiable secret sharing (VSS) [23], TD computes and publishes the witnesses $W_i = f_i A$ for $(i = 0, \cdots, t-1)$. The witness value $W_0 = xA$, also denoted by B, is actually the group public key. Next, for each M_i, TD computes the secret share ss_i and the identity-based membership token T_i (valid until the time exp^6) such that: $ss_i = f(id_i) \pmod{q}$ and $T_i = xH_1(id_i\|exp)$. Note that TD is not required hereafter.

SELF-INITIALIZATION BY FOUNDING MEMBERS. t or more founding members M_i-s select individual polynomials $f_i(z)$ over $\mathbb{Z}_q$ of degree $t-1$, such that $f_{i0} = x_i$. Then, using the DKG protocol [24], each M_i computes its own secret share ss_i,

---

6 Membership tokens are valid for a certain period of time. The duration of the validity period are be defined by the group policy (which is out of scope of this paper). For simplicity, we assume that all membership certificates reflect the same expiration period exp. In order to enable implicit revocation, each member M_i needs to be provided with $T_i = xH_1(id_i\|exp)$. This structure implies that T_i is not valid after the time exp. Once expired, the token needs to be renewed via the admission process. The group founding members might be provided with long(er)-term membership tokens. We assume that all nodes have reasonably synchronized clocks, within a certain skew.

such that $ss_i = \sum_{j=1}^{l} f_j(id_i) \pmod{q}$ ($l \geq 2t - 1$). Once M_i gets its share, it is rather easy to recover the secret using Lagrange interpolation. Also, the dealing process supports VSS. Now, in order to provide each member with a membership token, any set of t founding members must collaborate. For example, group members $M_2, M_3, \cdots, M_{t+1}$ may collaborate and provide M_1 with a membership token as $T_1 = \sum_{j=2}^{t+1}(ss_j l_j(0))H_1(id_1\|exp)$ $[= xH_1(id_1\|exp)]$.

4.2   Admission Process

Let n ($\geq t$) be the number of current group members. In order to be admitted to the group, a prospective member M_{new} must collect at least t votes from current group members. Figure 1 shows protocol message flows for the admission process. The goal is for M_{new} to obtain a membership token T_{new} which can then be used to prove membership. (In practice, M_{new} also needs to obtain the current membership revocation list – MRL – to keep track of revoked group members.)

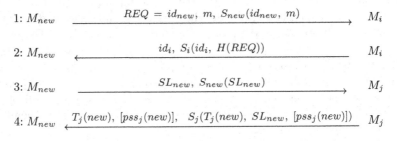

Fig. 1. *ID-GAC* Protocol

1. M_{new} sends a signed join request message m as well as its identity id_{new} to at least t current group members[7].
2. Group members who wish to participate in the admission reply with their respective id-s to M_{new} along with the signature on the previous message sent by M_{new}.
3. M_{new} picks (perhaps, at random) t sponsors M_j-s, forms a signer list SL_{new} which contains the id-s of t responders, signs it, and sends it to each M_j.
4. Each signing member M_j sends back to M_{new} the partial membership token $T_j(new)$ and (optionally) the partial share of the secret $pss_j(new)$ such that

$$T_j(new) = (ss_j \cdot l_j(0))H_1(id_{new}\|exp) \quad \text{and}$$

$$pss_j(new) = ss_j \cdot l_j(id_{new}) + r_j \pmod{q}$$

---

[7] In order to secure the protocol against common *replay* attacks [25], we note that it is necessary to include timestamps, nonces and protocol message identifiers. However, in order to keep our description simple, we omit these values.

$$\text{where,} \quad l_j(x) = \prod_{i=1}^{t} \frac{x - id_i}{id_j - id_i}.$$

M_{new} is also provided with a signed copy of the current MRL. Note that the Lagrange coefficients $l_j(id_{new})$-s are publicly known, and therefore, M_{new} can derive ss_j from $pss_j(new)$. This can be prevented using the *shuffling* technique proposed in [9] by adding extra random value r_j-s to each share. These r_j-s are secret random values and must sum up to zero by construction. They must be securely shared among the t sponsoring M_j-s.

5. Finally, M_{new} calculates its secret share ss_{new} (if provided) and the membership token T_{new} by adding up the values obtained in the last step. The share acquisition and membership token acquisition procedures are discussed next.

4.3   Membership Token Acquisition

The membership token acquisition procedure is also performed as part of the 4-th message of the above protocol. Each sponsor M_j computes a partial membership token $T_j(new)$ for M_{new} and then sends it to M_{new}. (It is important to note that the partial membership token $T_j(new)$ is actually a BLS [22] signature on id_{new} using $(ss_j \cdot l_j(0))$ as the secret key for M_j.) Then, M_{new} computes its membership token T_{new} by summing up $T_j(new)$ $(j = 1, \cdots, t)$ and verifies the correctness by checking $\hat{e}(B, H_1(id_{new} \| exp)) = \hat{e}(A, T_{new})$. This equation can be easily shown to be correct using the properties of the bilinear map $\hat{e}$.

4.4   Share Acquisition

The share acquisition procedure – whereby M_{new} obtains its share ss_{new} from sponsors – is performed as part of the 4-th message in the *ID-GAC* protocol (as shown in Figure 1).

The t sponsoring members (M_j-s) compute the shuffled partial share $pss_j(new)$ for M_{new} as $pss_j(new) = ss_j \cdot l_j(id_{new}) + r_j$ (mod q) using the shuffling technique [9]. Each M_j sends $pss_j(new)$ to M_{new}. Then, M_{new} computes its share ss_{new} by summing up $pss_j(new)$ $(j = 1, \cdots, t)$ and verifying the correctness using VSS [23].

By verifying the secret share and the membership token as described above, M_{new} is assured of possessing correct credentials. Armed with the membership token T_{new}, M_{new} can prove membership. Also, using a secret share ss_{new}, M_{new} can use the group secret x in collaboration with other $(t-1)$ group members and take part in future admission protocols. In Section 5, we discuss two schemes that can be used to prove membership.

Next, we discuss the security of the proposed *ID-GAC* scheme.

4.5   Security Considerations

In this section, we argue the security of the proposed scheme, based on the security model of Section 3.

1. COMPLETENESS. This property follows by inspection. At the end of the protocol, M_{new} receives the membership token T_{new} which is verified as: $\hat{e}(B, H_1(id_{new}\|exp)) = \hat{e}(A, T_{new})$. Using T_{new}, M_{new} can prove membership. M_{new} also receives a secret share ss_{new} which is verified using VSS as: $ss_{new}A = \sum_{i=0}^{t-1} id_{new}{}^i W_i$. Using ss_{new}, M_{new} can take part in future admission decisions and can also recover the group secret x in collaboration with any other $t-1$ members. Of course, M_{new} can obtain these credentials in polynomial time.

2. TRACEABILITY. In case the verification of T_{new} and/or ss_{new} fails, M_{new} must identify (trace) sponsors that sent invalid partial token(s) and/or partial secret share(s). To verify each partial secret share, M_{new} can perform the VSS procedure. Correctness of each membership token share $T_j(new)$ can be verified as follows:

$$\hat{e}(T_j(new), A) = \hat{e}(H_1(id_{new}\|exp), l_j(0) \sum_{i=0}^{t-1} id_j{}^i W_i).$$

If the above verification fails, M_{new} concludes that M_j is cheating.

3. IMPERSONATION RESISTANCE. In order to make sure that M_{new} is communicating with only genuine group members, it can verify the partial credentials as in TRACEABILITY above and in the process can trace any impersonating non-member.

5   Proving Membership

We employ two schemes that can be used by a group member to prove membership. We consider proving membership to internal parties (members) and external parties (non-members). The internal membership proof (IMP) is a pairing-based secret handshake scheme proposed in [26]. The external membership proof (EMP) is the identity-based signature scheme in GDH groups as proposed in [27].

6   Membership Revocation

In addition to implicitly revoking membership tokens via expiration (see footnote 6), rogue members need to be explicitly revoked, e.g., for reasons of selfishness, maliciousness or compromise.

A secure membership revocation mechanism should satisfy all properties outlined in Section 3.2. One trivial solution is to have a set of t share holding group members (revokers) collaborate and renew membership tokens for all members, except the one being revoked. The revokers also need to update the secret shares using proactivity in case the revoked member possessed an old share. This approach is clearly very inefficient.

In this section we present a practical revocation mechanism based on Membership Revocation Lists (MRLs). This is analogous to the simple and widespread certificate revocation technique (CRLs [28]) used in traditional PKIs. However, unlike CRLs, our solution is fully distributed.

6.1   MRL Update

Upon every revoke operation, each group member needs to update its copy of the MRL. Also, an entry needs to be removed from the MRL once the corresponding member's membership token expires.

6.2   Membership Validation

If and when a user V receives a signed message from another user M_u claiming to be the group member (of a group with public key B), V needs to first verify the validity of M_u's membership and then the signature (using EMP signature verfication). The validation procedures will be different (internal or external) based on whether V is a group member or a non-member.

Internal Membership Validation. If V is a group member, validation involves only a lookup of its MRL and checking the status of M_u. If MRL contains no entry corresponding to id_u, V concludes that M_u is a member in good standing.

External Membership Validation. An non-member V needs to perform the following protocol to validate M_u's status.

1. V sends to the group a membership validation request for M_u's membership.
2. Share holding group members interested in answering this request, reply with their IDs to V.
3. V waits for at least t such responses, creates a signers' list SL_v containing the IDs of the interested members and sends it to them.
4. These members then look up the status of M_u in their respective MRLs and provide V with a signed response.
5. V verifies the signature on the status of M_u.

The above validation procedure is quite similar to the admission process described in Section 4.2. Moreover, it could also be viewed as a distributed version of the Online Certificate Status Protocol ($OCSP$) [29] in the context of certificate revocation.

6.3   Revocation Process

To revoke an allegedly malicious or misbehaving member M_r, any current member M_a can bootstrap the revocation process. The following steps need to be performed.

1. M_a broadcasts a revocation request message referencing M_r, using EMP signature generation.
2. All other group members M_j-s ($j \neq r$) perform the internal membership validation for M_a as in Section 6.2. [8] They then perform the MRL update

---

[8] Although this is concerned with the group revocation policy, here we assume that a member which is "under-review" can also lodge valid revocation requests.

(as discussed in Section 6.1) to add M_a to the revoker list for M_r. The status of M_r is then set to "under-review". Once the number of revokers in this list reach t_r, the status of M_r is updated to "revoked".

3. If M_r is the $\hat{t}^{(th)}$ ($\hat{t} \leq t-1$) member to be revoked[9], the t_r revokers collaborate and update the shares using the proactive method. If less than t out of t_r revokers possessed the secret shares, the group founding members need to perform the share update.

4. All available group members M_i-s ($i \neq r$) possessing voting rights, then contact the revokers (or founding members) and renew their shares.

6.4   Security Considerations

In this section, we argue that the above revocation mechanism is secure based on the security model sketched in Section 3.2.

1. COMPLETENESS. By inspection: As soon as t_r valid revocation requests are lodged against M_r, each group member records them in its local copy of the MRL and sets the status corresponding to id_r as "revoked". Now, M_r can not prove membership via either IMP or EMP protocol and/or take part in the admission process since its membership validation (as described in Section 6.2) would fail.

2. IMPERSONATION RESISTANCE. As part of Step 1 of the revocation process above, a revocation request submitted by M_a against M_r is signed using EMP signing. Upon receiving such a request, each group member (except M_a) first validates M_a's status by doing an MRL lookup and then validates M_a's membership by verifying the signature on the request message (via EMP verification). The former guarantees that M_a is itself not revoked and the latter ensures that M_a is indeed a group member. Therefore, it is impossible for a revoked member and computationally infeasible for a non-member to lodge valid revocation requests. Thus, the proposed revocation mechanism is impersonation resistant.

3. COLLUSION RESISTANCE. The revocation procedure involves the secret share updates atleast after every $t-1$ members are revoked. Therefore, the share of the tth revoked member will not correspond to the shares of $t-1$ previously revoked members in yielding the group secret using the polynomial interpolation. This implies that no set of revoked members can collude.

6.5   Discussion

The MRL-based solution requires group members to synchronize in order to maintain up-to-date MRLs. However, it is certainly unrealistic to expect all members to be on line all of the time. Any member can establish the freshness of its MRL by performing a procedure similar to membership validation (in

---

[9] How many revocation operations trigger a share update is determined by the group revocation policy.

Section 6.2). A set of t interested members (possessing the secret shares) respond with the complete and signed MRL, as opposed to the membership validation procedure, where they respond with the status of a particular member. This procedure can also be performed periodically.

Discrepancies in MRLs might arise for a number of reasons. A group member might go off-line or become unreachable temporarily. Such events are common in asynchronous groups, such as most P2P systems and MANETs. More generally, a group can become partitioned due to some network event, e.g., a router failure. Suppose a partition occurs and a group is split into two subgroups G_A and G_B. The two subgroups operate independently and their MRLs evolve separately. If at a later time, G_A and G_B merge back into a single group, the two MRLs: MRL_A and MRL_B, need to be appropriately merged. This particular scenario presents a major challenge since the techniques described above will not work. Consider what happens if, while the group is partitioned, members of G_A decide to revoke their counterparts in G_B, and vice versa. (Not surprisingly, this remains a major item for future work.)

7   Performance Analysis

We now present and discuss the performance measurement results for the proposed *ID-GAC* admission and eviction techniques. In particular, we describe our experience with the implementation of these schemes and experiments in P2P and MANET settings, focusing on the respective costs of admission, traceability, membership proofs and revocation. We also compare our results with the previously proposed DSA-based admission mechanism [12], [11], wherever applicable.

7.1   Implementation

The *ID-GAC* library is built using OpenSSL [30] and MIRACL [31] (optimized using Comba method) libraries. The latter was needed to implement various identity-based functions. Currently, *ID-GAC* consists of approximately 10,000 lines of C/C++ source code and supports Linux 2.4.

We used the elliptic curve E defined by the equation: $y^2 = x^3 + 1$ over $\mathbb{F}_p$ with $p > 3$ a prime satisfying $p = 2 \pmod 3$ and q being a prime factor[10] of $p + 1$. The size of q is set to be 160 bits and p is a 512-bit prime. The group $\mathbb{G}_1$ is a subgroup of points generated by A such that $A \in E(\mathbb{F}_p)$. The group $\mathbb{G}_2$ is a subgroup of $\mathbb{F}_{p^2}^*$ of order q. The bilinear map $\hat{e} : \mathbb{G}_1 \times \mathbb{G}_1 \to \mathbb{G}_2$ is the well-known Tate pairing. Note that the pairing value belongs to finite field of 1024 bits.

7.2   Basic Operations

To estimate the performance of *ID-GAC*, we first present the costs of the primitive operations in Table 2. For measuring the costs of basic operations in *ID-*

---

[10] By Euler's theorem, q must divide $\#E(\mathbb{F}_p)$. For the curve $y^2 = x^3 + 1$, $\#E(\mathbb{F}_p) = p + 1$.

GAC, we used a machine with an Intel P3 800MHz processor and 384MB memory. All experiments were repeated 1,000 times for each measurement in order to get fairly accurate average results.

7.3   Experimental Setup

We now describe the experimental setup used for the experiments in both P2P (Gnutella [32]) and MANET settings. We used two laptops and two PDAs, each configured with 802.11*b* in ad-hoc mode. For routing purposes, we used the Optimized Link State Routing Protocol (OLSR) [33]. The two laptops (running Linux 2.4) are equipped with 800/900-MHz P-III processors and 384/256MB RAM, respectively. The two PDAs (running Linux *Familiar*) each have a 400MHz XSCALE processor and 64MB RAM. In all experiments, we ran equal number of processes (current group members) on both the laptops. The PDAs were used for routing purposes only.

Table 2. Costs of Primitive Operations (P3-800MHz)

Function	modulus (bits)	exponent (bits)	average time (msec)
DSA sign	1024	160	4.78
DSA verify	1024	160	5.74
Map-to-point ($H_1(\cdot)$)	512	160	3.13
BLS sign	512	160	8.91
Pairing	512	160	37.24

Note that both P2P and MANET experiments were done on the equipment with same computing power. The main purpose of our experiments is to measure the computation and communication costs in wireless as well as wired networks and to demonstrate the practicality of ID-based admission and revocation mechanisms. For *ID-GAC* experiments, the modulus size ($|p|$) was set to 512-bits and 1024-bit modulus was used for *TS-DSA* experiments[11].

Remark: In all experiments below, we used a member/authorizer paradigm. A *member*, in this context, is a group member who has no voting/admission rights (i.e., no secret share), whereas an *authorizer* has them. Since their respective costs sometimes vary substantially, they are graphed separately.

7.4   Node Admission

Figure 2 shows the admission cost with varying threshold (for both member and authorizer) for *ID-GAC* and *TS-DSA* schemes in (a) MANET and (b) Gnutella

---

[11] Computing discrete log in $\mathbb{F}_{p^2}$ is sufficient for computing discrete log in $\mathbb{G}_1$. Therefore, for proper security of discrete log in $\mathbb{F}_{p^2}$ the prime p should be at least 512-bits long (so that the group size is at least 1024-bits long). This will ensure that the GDH problem remains sufficiently hard.

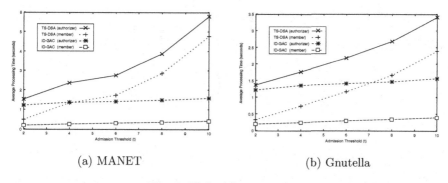

(a) MANET   (b) Gnutella

Fig. 2. Node Admission Cost

Table 3. Bandwidth Comparison

	TS-DSA	ID-GAC
Admission (member)	$(2t - 1) * |GMC_j|$	$t * |T_j|$
Admission (authorizer)	$(2t - 1) * |GMC_j| + t * |pss_j|$	$t * (|T_j| + |pss_j|)$

experiments. The admission costs also include the verification of the membership token and/or the secret share.

As shown in Figure 2, *ID-GAC* exhibits appreciably better performance that *TS-DSA* in MANET, as well as in P2P setting. The results imply that the amount of communication in *TS-DSA* contributes significantly to the overall cost of admission, although computation-wise it is still quite efficient (see Table 2). The communication overhead for *TS-DSA* is even higher in MANET, than in Gnutella experiments. This is clearly due to the error-prone, low-bandwidth wireless channel.

7.5 Bandwidth Consumption

Table 3 compares the respective bandwidth costs. (Refer to Table 1 for notation.) While *TS-DSA* uses certificates, *ID-GAC* is identity-based which obviates any need for explicit membership certificates. Certificate size is relatively large, e.g., 5KB with 1024-bit DSA parameters. For example, if a prospective member wants to join the group as a member, $(2t - 1) * 5K * 8$ bits must be transfered, whereas, only $t * 512$ bits are needed in *ID-GAC*.

Also, it is well known that, in many small devices (such as low-end MANET nodes or sensors) sending a single bit is roughly equivalent to adding $1,000$ 32-bit numbers, in terms of battery power consumption [20]. For example, in case of $t = 3$, the bandwidth cost with *TS-DSA* is about 133 times higher than *ID-GAC*. In other words, *TS-DSA* consumes 133 times more energy than *ID-GAC* for the communication. Hence, we expect *ID-GAC* to be more suitable for MANET scenarios which often involves power-constrained devices.

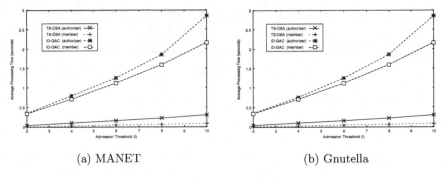

| (a) MANET | (b) Gnutella |

Fig. 3. Traceability Cost

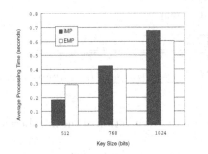

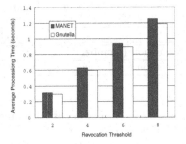

Fig. 4. Membership Proof Cost   **Fig. 5.** Membership Revocation Cost

7.6   Traceability

Traceability costs are presented in Figure 3. Due to the costly computation of Tate pairings, *ID-GAC* performs poorly, as compared to *TS-DSA*.

However, since the misbehavior in the admission protocol leads ultimately to the eviction of the corresponding group member, we argue that traceability is a *rare exceptional* measure; thus we consider its costs to be relatively unimportant.

7.7   Membership Proof

We now discuss measurement results for the membership proofs in *ID-GAC*, as outlined in Section 5. The respective costs of *IMP* and *EMP* computations (including both signing and verifying) are shown in Figure 4, with varying key sizes. Recall that *IMP* is needed if two members want to authenticate each other secretly and establish a shared secret key; in contrast, *EMP* can (also) be used to prove membership to outsiders.

7.8   Revocation

The graph in Figure 5 represents the average cost needed to revoke a particular group member for the varying threshold using the mechanism described in

Section 6. In this context, the threshold (t_r) is the revocation threshold. Costs include: generation of signed revocation requests, broadcast to the group and validation of these requests. Delay between two consecutive revocation requests is not taken into account.

8   Discussion and Further Improvement

In this section, we discuss some of the outstanding issues concerning *ID-GAC*, focusing primarily on the performance.

The bilinear mapping operation is an elegant procedure that forms the basis for the verification in the admission process and the membership proof. However, it is an expensive computation which dominates the overall running time of the protocol. Moreover, it is only because of the Tate pairing operation in our protocol that the key size, i.e. the size of the prime p needs to be at least 512-bits; although it is well-known that a 160-bits ECC system is as secure as a 1024-bits RSA based system. Barreto, et al. [34] suggest some modifications and optimizations for the Tate pairing operation, most of which are implemented in the MIRACL library [31] that we used. In order to further improve performance, we need to parallelize these operations and pre-compute as much as possible. As discussed in Section 4.5, the traceability of malicious sponsors is an optional procedure that involves verification of individual sponsor's signatures, which costs two pairings per verification. This cost can be significantly lowered by using pre-computations as described in [31].

The choice of the elliptic curve being used can certainly influence the overall cost of the scheme. BLS short signature scheme [22] uses a supersingular curve defined by the equation $y^2 = x^3 + 2x \pm 1$ over $\mathbb{F}_{3^l}$ where l is a positive exponent. Barreto, et al. [34] specify the cost of generating a BLS signature on such a curve (defined in $\mathbb{F}_{3^{97}}$) to be just 3.57 ms on a P3 1GHz machine, after some optimizations and preprocessing. This seems like a significant improvement over the costs incurred in our measurements which were based on a different curve defined in a prime field. Therefore, porting our protocol for these curves appears to be an attractive way to reduce costs. Since the signature generation here is cheaper though the verification is still costly, this in fact could be an ideal candidate for an admission control mechanism, where one would prefer the computation load to be high on the prospective member, much lesser so on the current group members. In addition, these signatures are very short, only 154-bits in length, which will give rise to short membership tokens in *ID-GAC*; another seemingly attractive prospect.

9   Conclusion

In this paper, we proposed *ID-GAC*, an identity-based scheme for secure admission control in dynamic ad hoc groups along with a distributed membership revocation mechanism based on the membership revocation lists. *ID-GAC* borrows ideas from threshold secret sharing and ID-based cryptography. As demonstrated by extensive experimentation in both P2P and MANET settings, the proposed scheme is far more efficient than the previously proposed solution, *TS-DSA*[11], [12]. The measurement results and performance analysis further indicate that *ID-GAC* is even more applicable in MANET devices where bandwidth and battery power are of prime concern.

References

1. Douceur, J.R.: The Sybil Attack. In: International Workshop on Peer-to-Peer Systems (IPTPS'02). (2002)
2. Steiner, M., Tsudik, G., Waidner, M.: Key Agreement in Dynamic Peer Groups. In: IEEE Transactions on Parallel and Distributed Systems. (2000)
3. Steiner, M., Tsudik, G., Waidner, M.: Cliques: A new approach to group key agreement. IEEE Transactions on Parallel and Distributed Systems (2000)
4. Hu, Y.C., Perrig, A., Johnson, D.B.: Ariadne: A secure on-demand routing protocol for ad hoc networks. In: Proceedings of the Eighth ACM International Conference on Mobile Computing and Networking (Mobicom 2002). (2002)
5. Perrig, A., Szewczyk, R., Wen, V., Culler, D., Tygar, J.D.: Spins: Security protocols for sensor networks. In: Mobile Computing and Networking. (2001)
6. Hamilton, A.: Playing in the dark: The heat is on, and music swappers are taking their business underground. TIME Magazine (2003)
7. Biddle, P., England, P., Peinado, M., Willman, B.: The darknet and the future of content distribution. In: ACM Workshop on Digital Rights Management. (2002)
8. Kong, J., Luo, H., Xu, K., Gu, D.L., Gerla, M., Lu, S.: Adaptive Security for Multi-level Ad-hoc Networks. In: Journal of Wireless Communications and Mobile Computing (WCMC). Volume 2. (2002) 533–547
9. Kong, J., Zerfos, P., Luo, H., Lu, S., Zhang, L.: Providing Robust and Ubiquitous Security Support for MANET. In: IEEE 9th International Conference on Network Protocols (ICNP). (2001)
10. Kim, Y., Mazzocchi, D., Tsudik, G.: Admission Control in Peer Groups. In: IEEE International Symposium on Network Computing and Applications (NCA). (2003)
11. Saxena, N., Tsudik, G., Yi, J.H.: Admission Control in Peer-to-Peer: Design and Performance Evaluation. In: ACM Workshop on Security of Ad Hoc and Sensor Networks (SASN). (2003) 104–114
12. Narasimha, M., Tsudik, G., Yi, J.H.: On the Utility of Distributed Cryptography in P2P and MANETs: The Case of Membership Control. In: IEEE 11th International Conference on Network Protocol (ICNP). (2003) 336–345
13. Ohta, K., Micali, S., Reyzin, L.: Accountable Subgroup Multisignatures. In: ACM Conference on Computer and Communications Security. (2001) 245–254
14. Gennaro, R., S.Jarecki, H.Krawczyk, T.Rabin: Robust Threshold DSS Signatures. In Maurer, U., ed.: EUROCRYPT '96. Number 1070 in LNCS, IACR (1996) 354–371
15. Desmedt, Y., Frankel, Y.: Threshold cryptosystems. In Brassard, G., ed.: CRYPTO '89. Number 435 in LNCS, IACR (1990) 307–315
16. Saxena, N., Tsudik, G., Yi, J.H.: Access Control in Ad Hoc Groups. In: International Workshop on Hot Topics in Peer-to-Peer Systems (HOT-P2P). (2004)
17. Luo, H., Kong, J., Zerfos, P., Lu, S., Zhang, L.: URSA: Ubiquitous and Robust Access Control for Mobile Ad Hoc Networks, available online at http://www.cs.ucla.edu/wing/publication/publication.html. In: IEEE/ACM Transactions on Networking (ToN), to appear. (2004)

18. Jarecki, S., Saxena, N., Yi, J.H.: An Attack on the Proactive RSA Signature Scheme in the URSA Ad Hoc Network Access Control Protocol. In: ACM Workshop on Security of Ad Hoc and Sensor Networks (SASN). (2004)

19. Gennaro, R., Jarecki, S., Krawczyk, H., Rabin, T.: Robust and Efficient Sharing of RSA Functions. In Koblitz, N., ed.: CRYPTO '96. Number 1109 in LNCS, IACR (1996) 157–172

20. Barr, K., Asanovic, K.: Energy Aware Lossless Data Compression. In: International Conference on Mobile Systems, Applications, and Services (MobiSys). (2003)

21. Boldyreva, A.: Efficient threshold signatures, multisignatures and blind signatures based on the gap-diffie-hellman-group signature scheme. In: Proceedings of International Workshop on Practice and Theory in Public Key Cryptography. Volume 2567 of LNCS. (2003) 31–46

22. Boneh, D., Lynn, B., Shacham, H.: Short Signatures from the Weil Pairing. In Boyd, C., ed.: ASIACRYPT'01. Number 2248 in LNCS, IACR (2001) 514–532

23. P.Feldman: A Practical Scheme for Non-interactive Verifiable Secret Sharing. In: 28th Symposium on Foundations of Computer Science (FOCS). (1987) 427–437

24. Gennaro, R., Jarecki, S., Krawczyk, H., Rabin, T.: Secure Distributed Key Generation for Discrete-Log based Cryptosystems. In: Eurocrypt 99. (1999)

25. Menezes, A.J., van Oorschot, P.C., Vanstone, S.A.: Handbook of Applied Cryptography. CRC Press series on discrete mathematics and its applications. (1997) ISBN 0-8493-8523-7.

26. Balfanz, D., Durfee, G., Shankar, N., Smetters, D., Staddon, J., Wong, H.C.: Secret Handshakes from Pairing-Based Key Agreements. In: IEEE Symposium on Security and Privacy. (2003) 180–196

27. Cha, J., Cheon, J.: An ID-based signature from Gap-Diffie-Hellman Groups. In: Proceedings of International Workshop on Practice and Theory in Public Key Cryptography. Volume 2567 of LNCS. (2003) 18–30

28. Kocher, P.C.: On Certificate Revocation and Validation. In: Proceedings of International Conference on Financial Cryptography. Volume 1465 of LNCS. (1998)

29. Myersand, M., Ankney, R., Malpani, A., Galperin, S., Adams, C.: Online Certificate Status Protocol - OCSP . RFC 2560, IETF (1999)

30. OpenSSL Project: ⟨http://www.openssl.org/⟩

31. MIRACL Library: ⟨http://indigo.ie/~mscott/⟩

32. The Gnutella Protocol Specification v0.4: http://www.clip2.com/Gnutella Protocol04.pdf

33. OLSR: ⟨http://hipercom.inria.fr/olsr/⟩

34. Barreto, P.S.L.M., Kim, H.Y., Lynn, B., Scott, M.: Efficient Algorithms for Pairing-based Cryptosystems. In Yung, M., ed.: CRYPTO '02. Number 2442 in LNCS, IACR (2002) 354–369

Mobile Mixing[*]

Marcin Gogolewski[2,**], Mirosław Kutyłowski[2], and Tomasz Łuczak[1]

[1] Faculty of Mathematics and Computer Science,
Adam Mickiewicz University, Umultowska 87, 61-614 Poznań, Poland
{marcing, tomasz}@amu.edu.pl
[2] Institute of Mathematics, Wrocław University of Technology,
Wybrzeże Wyspiańskiego 27, 50-370 Wrocław, Poland
Miroslaw.Kutylowski@pwr.wroc.pl

Abstract. We consider a process during which encoded messages are processed through a network; at one step a message can be delivered only to a neighbor of the current node; at each node a message is recoded cryptographically so that an external observer cannot link the messages before and after re-coding. The goal of re-coding is to hide origins of the messages from an adversary who monitors the traffic. Recoding becomes useful, if at least two messages simultaneously enter a node – then the node works like a mix server.

We investigate how long the route of messages must be so that traffic analysis does not provide any substantial information for the adversary. Anonymity model we consider is very strong and concerns distance between a priori probability distribution describing origins of each message, and the same probability distribution but conditioned upon the traffic information. We provide a rigid mathematical proof that for a certain route length, expressed in terms of mixing time of the network graph, variation distance between the probability distributions mentioned above is small with high probability (over possible traffic patterns).

While the process concerned is expressed in quite general terms, it provides tools for **proving** privacy and anonymity features of many protocols. For instance, our analysis extends results concerning security of an anonymous communication protocol based on onion encoding – we do not assume, as it is done in previous papers, that a message can be sent directly between arbitrary nodes. However, the most significant application now might be proving immunity against traffic analysis of RFID tags with universal re-encryption performed for privacy protection.

Keywords: anonymous communication, traffic analysis, Markov chain, variation distance, rapid mixing, onion protocol, RFID tag.

[*] Partially supported by KBN grants 0 T00A 003 23 and 1 P03A 025 27 in years 2002-2004.
[**] Partially done while the author was affiliated with Adam Mickiewicz University at Poznań.

C. Park and S. Chee (Eds.): ICISC 2004, LNCS 3506, pp. 380–393, 2005.

1   Introduction

Contemporary cryptography provides techniques for encoding messages so that an unauthorized party cannot read or modify (without detection) a protected message. However, in many practical situations we would like to prevent an adversary not only from being able to read/modify messages, but also from learning that a message has been transferred from one party to another. This information might be of fundamental value for instance for business negotiations. Hiding communication could be also highly desirable in private sphere and for consumer protection. The problem is particularly acute in public networks, where physical means of privacy protection have their limitations.

Despite their practical significance, technologies of anonymous communication and privacy enhancing still require a lot of attention. Although many proposals have been already made, they are often immune only against certain types of attacks. Even worse, rigid security analysis is usually missing – it is replaced by case-by-case arguments, why simple or widely known attacks do not work. On the other hand, in this area privacy guarantees are of an extraordinary value. Simple intuitions might be misleading, for instance if an adversary with huge resources applies some very sophisticated stochastic tools which we are not even aware of. However, for many nice-looking protocols relatively simple attacks have been found (as it has occured for [10]).

There are many protocols aimed for anonymous communication. Most of them are based on one of two basic paradigms: *mixes* and *onions*:

Mix Servers. A mix [5] works as follows: a number of encoded messages enters a mix, they are recoded cryptographically by the mix and leave it. The recoding process guarantees that an external observer cannot link the incoming and the outgoing messages.

Usually, the messages are processed by many mixes forming a network of mixes (see for instance [9]). This provides more security, since some of the mixes might be corrupted. Another point is that in a large system capacity of a mix might be too low to process all messages simultaneously. In this case, many mixes applied in parallel overcome the processing bottleneck.

Onions. An idea of mix servers is used for anonymous communication with messages encoded as *onions* [15, 14, 17]. The main idea is that each message is sent along a random path; the route of each message is determined by the sender (and not by the network!), the nodes on the route are chosen independently at random. The messages are encapsulated cryptographically in a structure resembling an onion. Each layer contains the name of the next server on the path, the kernel of an onion contains a ciphertext of the message transmitted. When a node receives an onion it recodes the onion by *peeling off* one layer using its private key. Encoding guarantees that only the node determined by the sender may peel off and read the contents, which is the next node on the route and the sub-onion to be sent there.

Message Encoding. Traditionally, onions are encoded in a structure with many encapsulation layers. For instance, a message c that has to be sent through servers J_1, J_2, J_3 is encoded as $E_{J_1}(J_2, E_{J_2}(J_3, E_{J_3}(c)))$, where $E_{J_i}(-)$ denotes (probabilistic) encryption with a public key of J_i. In order to process such a message J_1 decrypts the message, finds the next destination J_2 and a "sub-onion" $E_{J_2}(J_3, E_{J_3}(c)))$ which has to be sent to J_2. Then upon receiving this sub-onion J_2 behaves in the same way.

Recently, universal re-encryption has been proposed [12]. Such an asymmetric encryption scheme has a remarkable property that one can re-encrypt a ciphertext so that a new ciphertext of the same message is obtained. The key point is that without the private decryption key one cannot derive any relationship between a ciphertext and its re-encrypted version – that is, one cannot detect that both ciphertexts are of the same plaintext or that they are encoded with the same public key. An important point is that for re-encryption we do not need the public key which was used to obtain the ciphertext. Let us remark that one can construct digital signatures that can be universally re-encrypted as well.

Adversary Model. We consider adversaries that cannot break encoding scheme, but can monitor the whole traffic. (A weaker model is considered in [3], where the adversary can see only a constant fraction of communication. Thanks to this modification security proofs of the onion protocol have been found.) Obviously, if a single message enters a server, no encoding scheme can hide the connection between a message that enters the server and the message that leaves the server immediately afterwards. Hiding effect occurs when at least two encoded messages are processed by a server at the same time. The key issue for the solutions such as networks of mixes and onions is to determine appropriate number of intermediate servers through which a message is to be processed in order to hide the link between the origin and destination of each message.

Mobile Mixing. We consider the following simple process (we describe it in terms of hobbits in order to escape some wrong intuitions coming from random walks on graphs). Given a regular graph G with n nodes. There are k hobbits that walk at random through G traversing one edge of G during one step; the edge to be followed is chosen by a hobbit uniformly at random from the set of outgoing edges of the current node. Each hobbit holds one ring. Among the rings there are m magic rings, each of them is different. When two or more hobbits meet at the same node they exchange their rings at random in the way that cannot be traced by an adversary. The adversary can see where the hobbits are, but cannot see which rings are they holding.

The goal of an adversary, who observes the hobbits is to determine the positions of the magic rings given their initial positions.

Anonymity Conditions. From the point of view of the adversary the mixing process is described by a probability distribution for the location of the magic rings (this distribution might be quite tedious to compute).

A naive approach to guarantee anonymity would be to show that the probabilities for different locations of a single ring are almost uniform. However, this does not suffice: in such a case it is still possible that there are strong dependencies between locations of rings, so that for instance revealing the position of one ring betrays much information about other rings. For this reason it is necessary to consider the probability distribution of locations of all magic rings and show that this distribution conditioned on the traffic information is almost uniform.

For measuring how far a given distribution μ is from the uniform distribution, we consider *variation distance* between μ and the uniform distribution. Recall that the variation distance between probability distributions μ_1, μ_2 defined over space X equals

$$d_{TV}(\mu_1, \mu_2) = \sum_{x \in X} |\mu_1(x) - \mu_2(x)| \ .$$

Another distance measures are considered in [3], among others also one based on mutual information. Somewhat surprisingly, all these measures turn to be equivalent to some extent. There have been also several weaker proposals in the literature (e.g. [8]), but they may provide only a limited degree of privacy.

Previous Results. As far as we know, mobile mixing was considered only for the case of a complete graph. Even then, the problem turns out to be hard. In papers [15, 7] it was proved that after a polylogarithmic number of steps probability distribution of magic rings is close to uniform. However, it was necessary to assume that $m = n$ and that the routes are confined in an artificial way. The execution is divided into $\log n$ phases, during phase i graph G is divided into $n/2^i$ complete subgraphs, each of cardinality 2^i, each subgraph of phase i contains nodes of exactly two subgraphs of phase $i - 1$. The number of steps necessary to achieve probability distribution of magic rings close to uniform distribution is $O(\log^2 n)$ [7].

There are recent results concerning this process in papers devoted to security of the onion protocol [3, 13]. Anonymity is already guaranteed with high probability after $O(\log m)$ steps provided that an adversary may trace only a constant fraction of the communication lines [13]. It is not required that the number of hobbits equals n.

No reasonable bound on the number of steps is known for the case when each hobbit holds a magic ring, G is a complete graph and the adversary monitors whole traffic. At least from a technical point of view, this is a challenging problem, since present proof techniques cannot cope with this case.

There is yet another line of research [2] for a different model of communication – "messengers" travel through a network and carry encoded messages that can be read by servers passed by the messenger. These solutions are not suitable for large scale or frequently changing networks, because each message is encoded in a data of size $O(n^2)$, where n is the number of nodes in the whole network. Delivery time in this model is proportional to the n.

New Results. We prove the following result (which we state here in a less formal way, see Theorem 3 for a precise statement):

Theorem 1. *Let G be a n-node regular graph with mixing time τ_G (for a definition of mixing time see Section 2). Assume that mobile mixing is executed by $m = \Theta(n)$ hobbits and k magic rings, where $k \leq \frac{1}{2}m$. Then after $\lambda = \Omega(\tau_G^3 \cdot \log^6 n \cdot \log k)$ steps the probability distribution of magic rings is close to the uniform distribution.*

We use graph mixing time as a crucial parameter in our estimation. This is useful, since mixing time of graphs has been intensively studied in the literature [1].

The point is that for d-regular graphs created in a random way mixing time is $O(\log n)$ with probability $1 - o(1)$ for $d \geq 3$. Let us also remark that assumption about regularity of the graph G can be removed under certain conditions. However, we skip these considerations, since our proof is quite technical anyway.

Our proof consists of two phases: first we consider mobile mixing with one magic ring (Section 2). Then we use path coupling arguments to generalize it to the general case (Section 3).

Applications. We mention here just three applications where mobile mixing is an essential component.

Anonymity in RFID-Systems. RFID tag is a powerless, small and low-cost device that may be activated by an external reader and communicates through a radio channel. There have been proposals to use these devices as identity tags, for instance on purchase items in supermarkets, library books, currency notes, and identity cards. Introducing RFID-tags as identification tags on a massive scale may bring many economical advantages. However, it has a dark side of losing privacy: anyone in the range of the tag can communicate with it. This makes it easy to trace the routes of the tags.

There has been a lot of concern about this problem (see for instance the story http://www.spychips.com/metro/scandal-payback.html).

In order to cope with this problem, universal re-encryption scheme can be used [12]. So in order to preserve privacy one can deploy a network of re-encryption points for the tags appearing in their proximity. Re-encryption guarantees that an unauthorized party (one without an appropriate private key) cannot link the tags before and after re-coding. Re-coding servers can be placed at *hot spots* such as street crossings. Then the tags perform a kind of a random walk on a graph, where the nodes of the graph denote re-encryption points.

As in the case of the onion protocol, effectiveness of the solution described depends not only on unbreakability of universal re-encryption, but also on resistance to traffic analysis of the underlying mobile mixing process.

Mobile Agents. One of major proposals how to use complex Web systems is to leave some tasks to mobile agents that migrate through the Web and try to collect information for the party that launched the agent. Even if it is a very appealing proposal for instance for information retrieval automation, implementing agents brings new security challenges. The first problem are dangers for the host machine from malicious agents. The second one are privacy problems for the owner of the agent.

In order to cope with the privacy problem we could encode information carried by the agent. However, the next privacy threat is to reveal the owner of the agent (for instance inspecting some specific stock exchange information in the Web). Even if re-coding of an agent is executed at each host before forwarding the agent, and re-coding is unbreakable cryptographically, the observer may perform traffic analysis. So one of the main questions here would be how effective is mixing of the agents, or in our terms, how fast mobile mixing protocol is approaching an almost uniform distribution.

Distributed Time Stamping. Implementing time stamping services on a single server is hard: one has to trust the server completely or the server has to use linking techniques. In the latter case, detecting forgery is possible, but it is painful and might occur too late. As a remedy, it has been proposed to fetch time-stamps from k pseudorandom *witnesses*: these time-stamping servers are determined from a hash value of the document to be time-stamped. However, this approach has a privacy weakness: at the same time many request are sent and time-stamping activity can be detected. Of course, one can make traffic analysis harder by sending each request encoded in an onion.

One can reduce the number of messages and make traffic analysis harder by implementing time-stamping by a single onion with a random route going through the network and back to the sender [11]. It is possible to design encoding so that each server processing an onion attaches an encrypted time-stamp to it. This has the advantage that most of the servers on the onion route do not know for whom they are issuing time-stamps, which makes the system more dependable and resistant against denial of service against certain users. The key point is that each hop on the route should have a small communication delay (otherwise long gaps in time stamping process would occur). This motivates considering connections that are described by a neighborhood graph rather than a complete graph.

A protocol of this type still faces some privacy risks: information that somebody is time-stamping a document should be protected from a party performing a traffic analysis. However, if time-stamping is hidden in ordinary onion traffic we have to do with a mobile mixing process.

2   On Hobbits and One Magic Ring

Consider mobile mixing process on a regular graph G with n nodes, $m = \Theta(n)$ hobbits and one magic ring. There are no restrictions about the initial placement of the hobbits (including Bilbo holding initially the only magic ring). For simplicity we assume that $m = n$, but as long as $m = \Omega(n)$ the proof below applies provided that we tune the constants accordingly. Also, for the sake of readability, in the estimations below we use somewhat arbitrary constants instead of general expressions.

2.1   Probability Distributions

Location of the Magic Ring. Unless otherwise specified P is a random variable. Let $\rho(P; i, t)$ denote the probability that after step t the ith hobbit has the magic ring, provided that the locations of the hobbits are given by pattern P.

Our goal is to consider the behavior of probability distribution $\rho(P; i, t)$. The random walk of hobbits and exchanges of the rings has the purpose of hiding who has the magic ring. For an overwhelming majority of all patterns P we shall estimate the time necessary so that an adversary observing a random walk of the hobbits cannot say much about the location of the magic ring. In order to state our result we need some more notation. Let

$$d_H(P; t) = \tfrac{1}{2} \sum_{i=1}^{m} |\rho(P; i, t) - \tfrac{1}{m}| = \max_{W \subseteq [m]} \sum_{i \in W} (\rho(P; i, t) - \tfrac{1}{m})$$

and $\tau_H(P) = \min_t \{d_H(P; t) \le 0.1\}$. That is, for given routes of hobbits described by P, number $\tau_H(P)$ denotes the first moment t such that the deviations of probabilities $\rho(P; i, t)$ from $\tfrac{1}{m}$ sum up to a small value. (Essentially, our ultimate goal is to find a moment such that the bound is very small.)

Location of Hobbits. By $p(i, j; t)$ we denote the probability that a hobbit who starts to walk from the vertex i will arrive to vertex j at time t (the probability is counted over all possible routes of the hobbit). Let

$$d_G(i, t) = \tfrac{1}{2} \sum_{j=1}^{n} |p(i, j; t) - \tfrac{1}{n}| = \max_{W \subseteq [n]} \sum_{i \in W} (p(i, j; t) - \tfrac{1}{n}) .$$

Let us recall that the *mixing time* τ_G for graph G is defined as

$$\tau_G = \min_t \max_i \{d_G(i; t) \le 0.1\} .$$

2.2   Key Technical Result

The key issue is that the values d_H and d_G concern stochastic processes of different nature, so similarity of notation is a little bit misleading. Estimating $\tau_H(P)$ is very much different from estimating τ_G. The main technical result in this section is the following theorem:

Theorem 2. *If $\tau_G = n^{o(1)}$, then $\tau_H(P) = O(\tau_G^3 \cdot \log^5 n)$ with probability at least $1 - \tfrac{1}{n}$ over the choices of P.*

For the proof of Theorem 2, we shall use the fact that typically we cannot have too many hobbits in one vertex.

Lemma 1. *Let $\tau_1 = \tau_G \cdot \log n$ and $\tau_2 = \tau_G^3 \cdot \log^5 n$. Then the probability that at some moment t, where $\tau_1 \le t \le \tau_2$, at least $\log n$ hobbits meet at some vertex of G is $o(1)$.*

Proof. Let $d_G(t) = \max_i d_G(i,t)$. It is well known and easy to derive (see also the appendix) that

$$d_G(s+t) \leq 2d_G(s) \cdot d_G(t) . \tag{1}$$

So for $t \geq \tau_G \log n$ we have $d_G(t) \leq \frac{1}{4n}$. Hence

$$\frac{1}{2n} \leq p(i,j;t) \leq \frac{2}{n} \quad \text{for every} \quad i,j \leq n. \tag{2}$$

Consequently, the probability that at some moment t, $\tau_1 \leq t \leq \tau_2$, at least $\log n$ hobbits meet at some vertex of G is bounded from above by

$$\tau_2 \cdot n \cdot \sum_{k=\log n}^{m} \binom{m}{k} \left(\frac{2}{n}\right)^k \leq n^3 \left(\frac{e}{\log n}\right)^{\log n} = o(1) . \qquad \Box$$

Potential Function g. Now we return to the proof of Theorem 2. We consider behavior of the following potential function

$$g(P;t) = \sum_{i=1}^{m} \rho^2(P;i,t) .$$

By Schwarz inequality, $g(P;t)$ is a non-increasing function of t. Furthermore, if a difference $|\rho(P;i,t+1) - \rho(P;i,t)|$ is large, then $g(P;t) - g(P;t+1)$ is significant as well. Namely, we have the following fact which can be easily deduced from the "defective" Schwarz inequality (see [16]):

Lemma 2. *Let* $a_1,\ldots,a_k > 0$ *and* $\mu = \frac{1}{k}\sum_{i=1}^{k} a_i$. *If* $|a_1 - \mu| \geq \epsilon$, *then*

$$\sum_{i=1}^{k} a_i^2 - k\mu^2 \geq \epsilon^2 .$$

The next lemma is the most tricky point in our considerations:

Lemma 3. *Let* $d_H(P;t_0) \geq 0.1$ *for a pattern* P *and* $t_0 \geq \tau_1$. *Then the expected change of potential* g *after time* τ_1 *is quite significant, namely*

$$\mathbf{E}[g(P;t_0) - g(P;t_0 + \tau_1 + 1)] \geq \Omega\big(g(P;t_0)/(\tau_G^2 \log^3 n)\big) .$$

Proof. We define the following sets of hobbits based on probability of holding the magic ring:

- set B of Baggins, which consists of all hobbits i for which $\rho(P;i,t_0) \geq 1.01/m$,
- set F of Proudfeet[1], which consists of all hobbits j for which $\rho(P;j,t_0) \leq 1/m$,
- set T of all other hobbits, i.e., those for which $1/m < \rho(P;j,t_0) < 1.01/m$.

It is easy to see that $|F| \geq 0.1 \cdot m$, namely:

$$0.1 \leq d_H(P;t_0) = \sup_{W\subseteq[m]} \sum_{j\in W} (\rho(P;j,t_0) - \tfrac{1}{m}) = \sum_{j\in F}(\tfrac{1}{m} - \rho(P;j,t_0)) \leq |F|/m .$$

---

[1] Proudfoots?

On the other hand

$$
\begin{aligned}
0.1 \le d_H(P;t_0) &= \sum_{i\in B}(\rho(P;i,t_0) - \tfrac{1}{m}) + \sum_{i\in T}(\rho(P;i,t_0) - \tfrac{1}{m}) \\
&\le \sum_{i\in B}(\rho(P;i,t_0) - 1/m) + 0.01 \le \sum_{i\in B}\rho(P;i,t_0) - |B|/m + 0.01 \ .
\end{aligned}
$$

So $\sum_{i\in B}\rho(P;i,t_0) \ge 0.09 + |B|/m$.

Hence by Lemma 2 we get

$$
\sum_{i\in B}\rho^2(P;i,t_0) \ge |B| \cdot \left(\tfrac{0.09}{|B|} + \tfrac{1}{m}\right)^2 \ge \tfrac{0.18}{m} \ .
$$

Since

$$
\sum_{i\in F\cup T}\rho^2(P;i,t_0) \le m \cdot \left(\tfrac{1.01}{m}\right)^2 \le 1.01/m \ ,
$$

we finally obtain

$$
\sum_{i\in B}\rho^2(P;i,t_0) \ge \tfrac{1}{7}\cdot g(P;t_0). \tag{3}
$$

Now we consider the configuration of the hobbits after τ_1 steps. By inequality (2), the probability (counted over the choice of P) that hobbits F will occupy less than $\frac{m}{200}$ vertices of G is bounded from above by

$$
\binom{n}{m/200}(0.01)^{|F|} = o(\tfrac{1}{n^2}) \ ,
$$

hence with high probability in time $t_0 + \tau_1$ Proudfeet are placed at vertices W of G, where $|W| \ge m/200$.

Now let us construct disjoint sets $B_i \subseteq B$, $F_i \subseteq F$, and $D_i \subseteq B \cup F$ recursively in the following way. Set $B_0 = B$, $F_0 = F$, $D_0 = \emptyset$. Take the Baggins b_1 for which the probability $\rho(P;b_1,t_0)$ is maximal. We consider the following cases:

Case 1: If $\rho(P;b_1,t_0 + \tau_1 + 1) \le \tfrac{1}{m} + \tfrac{1}{2}\cdot(\rho(P;b_1,t_0) - \tfrac{1}{m})$, then take the first t_1, $0 \le t_1 \le \tau_1$, for which

$$
\rho(P;b_1,t_0 + t_1) - \rho(P;b_1,t_0 + t_1 + 1) \ge \tfrac{1}{2\tau_1}(\rho(P;b_1,t_0) - \tfrac{1}{m}) \ ,
$$

put $D_1 := D_0 \cup \{b_1\}$, delete from B_0 and F_0 all the hobbits met by b_1 at time $t_0 + t_1$, and denote the sets obtained in this way by B_1, F_1, respectively.

Case 2: Now suppose that $\rho(P;b_1,t_0 + \tau_1 + 1) > \tfrac{1}{m} + \tfrac{1}{2}\cdot(\rho(P;b_1,t_0) - \tfrac{1}{m})$, but at time $t_0 + \tau_1$, the Baggins b_1 meets a Proudfoot f_1. Then,

$$
\rho(P;f_1,t_0 + \tau_1 + 1) = \rho(P;b_1,t_0 + \tau_1 + 1) \ge \tfrac{1}{m} + \tfrac{1}{2}\cdot(\rho(P;b_1,t_0) - \tfrac{1}{m}) \ .
$$

Take the first time t_1, $0 \le t_1 \le \tau_1$, for which

$$
\rho(P;f_1,t_0 + t_1 + 1) - \rho(P;f_1,t_0 + t_1) \ge \tfrac{1}{2\tau_1}(\rho(P;b_1,t_0) - \tfrac{1}{m}) \ ,
$$

put $D_1 := D_0 \cup \{f_1\}$, delete from $B_0 \setminus \{b_1\}$ and $F_0 \setminus \{f_1\}$ all the hobbits f_1 met at time $t_0 + t_1$ and denote the sets obtained in this way by B_1, F_1, respectively.

Case 3: If neither of the above two cases occurs, then do nothing, i.e., set $B_1 := B_0 \setminus \{b_1\}$, $F_1 := F_0$, $D_1 := D_0$.

Then take the element of $b_2 \in B_1$ which maximizes $\rho(P; b_2, t_0)$ and repeat the procedure until B_j becomes empty.

Note that, due to (2), a given Baggins visits a vertex from W with probability at least $\delta \geq \frac{1}{400}$. Hence, in particular, the probability that $|D_1| = 1$ is at least δ. So, by Lemma 2, the expected change of g due to the exchange of rings of b_1 or f_1 which took place at time t_1 is at least

$$\frac{\delta}{4\tau_1^2} \cdot \left(\rho(P; b_1, t_0) - \frac{1}{m}\right)^2 \geq \frac{\delta}{4\tau_1^2} \cdot \left(\frac{1}{101} \cdot \rho(P; b_1, t_0)\right)^2 = \Omega\left(\frac{1}{\tau_1^2} \cdot \rho^2(P; b_1, t_0)\right)$$

(for the first inequality recall that $\rho(P; b_1, t_0) > \frac{1.01}{m}$). So for the total expected change of g between t_0 and $t_0 + \tau_1 + 1$ we get for some constant ψ:

$$\mathbf{E}[g(P; t_0) - g(P; t_0 + \tau_1 + 1)] \geq \frac{1}{\log n} \frac{\psi}{\tau_1^2} \sum_{i \in B} \rho^2(P; i, t_0) \geq \frac{\psi}{7} \frac{1}{\tau_G^2 \log^3 n} \, g(P; t_0) \ (4)$$

(factor $\frac{1}{\log n}$ is due to the fact that at each step of our considerations we may remove up to $\log n$ Baggins from the set B_i). Inequality (4) has been derived under assumption that not too many hobbits meet in one node. Since probability of the opposite event is $o(1)$ and the values of g do not increase, we have

$$\mathbf{E}[g(P; t_0 + \tau_1 + 1)] \leq \left(1 - \frac{1}{w}\right) \cdot g(P; t_0) \qquad (5)$$

for some $w = \Theta(\tau_G^2 \log^3 n)$. This concludes the proof of Lemma 3.   □

Armed with the result of Lemma 3 we return to the proof of Theorem 2. Let $\rho(t) = t_0 + t(\tau_1 + 1)$. By X_t we denote random variable $g(P; \rho(t))$. Let S_t be the event that $d_H(P) > 0.1$ after step $\rho(t)$. We say that transition from step $\rho(t)$ to step $\rho(t + 1)$ has type A, if $\Pr[S_{t+1}|S_t] \leq 1 - \frac{1}{2w}$. Otherwise it has type B. Let us notice that in the latter case

$$\mathbf{E}[X_{t+1}|S_{i+1}] \leq \left(1 - \frac{1}{2w-1}\right) \cdot E[X_i|S_i] \ .$$

Indeed, it follows from the following inequality:

$$\mathbf{E}[X_{t+1}|S_{i+1}] \cdot \left(1 - \frac{1}{2w}\right) \leq \mathbf{E}[X_{t+1}|S_{i+1}] \cdot \Pr[S_{t+1}|S_t] \leq \mathbf{E}[X_{t+1}|S_t] \leq \left(1 - \frac{1}{w}\right) \mathbf{E}[X_i|S_i].$$

Since value of g may never drop below $1/m$ (the minimum value is obtained when ρ is constant and equals $\frac{1}{m}$), the number of transitions of type B leading to event S_T cannot exceed some $T_A = O(w \log m)$. So there are at least T_A transitions of type A when reaching event S_{2T_A}. Consequently, $\Pr[S_{2T_A}] < (1 - \frac{1}{2w})^{T_A} < \frac{1}{m}$.   □

After finishing the long and technical proof of Theorem 2 let us observe that we can easily derive a strong corollary concerning the behavior of d_H. Namely, analogously to (1), one can show easily that

$$d_H(P, s + t) \leq 2d_H(P, s) \cdot d_H(P', t) \ ,$$

where P' is the pattern P with the data for the first s steps removed. So, by Theorem 2:

Corollary 1. *If $\tau_G = n^{o(1)}$, then*

$$d_H(P,t) \leq \tfrac{1}{n^2} \quad for \quad t \geq t_H, \ where \ t_H = \Theta(\tau_G^3 \cdot \log^6 n)$$

with probability at least $1 - \tfrac{1}{n^2}$ (counted over the choices of P).

3   Many Magic Rings and Rapid Mixing

Process Definition. Let $m \leq n/2$. Now we consider a process such as in Section 2 but we assume that the number of magic rings is k instead of 1 and the number of hobbits is $m = \Theta(n)$.

We consider the assignment of rings to hobbits conditioned upon the event that the movements of hobbits are random but known. If P describes these movements, then let $\mathcal{L}^P$ stand for such a process. A configuration of $\mathcal{L}^P$ can be described by a mapping from the set of k magic rings into m hobbit positions predetermined for a given moment. Obviously, for each P the uniform distribution is a stationary distribution.

Path Coupling. We are interested how fast $\mathcal{L}^P$ approaches the uniform distribution for a random pattern P describing λ steps. For this purpose we use path coupling technique of Bubley and Dyer [4], and in fact a delayed version of path coupling [6].

For path coupling proof we need a distance function between configurations. Let $\Delta(L_1, L_2)$ be the minimum number of transpositions necessary to convert configuration L_1 to L_2, where by transposition we mean exchanging positions of a magic ring and an ordinary ring (the rings need not to be at the same node of G for this exchange). Obviously, Δ is a metric (over all configurations where the hobbits stay at the same nodes) with $\Delta(L_1, L_2) \leq 2k$. Let Γ denote the set of pairs of configurations that are at distance 1 according to metric Δ. Path coupling requires that for every pair $(L, L^\star)$ of configurations with $\Delta(L, L^\star) = r$, there exist a "path" $L = \Lambda_0, \Lambda_1, \ldots, \Lambda_r = L^\star$ such that $(\Lambda_{i-1}, \Lambda_i) \in \Gamma$ for $0 \leq i < r$. This condition obviously holds in our case.

In a path coupling proof we consider two configurations $(L, L^\star) \in \Gamma$ and define a pair of stochastic processes $(\mathcal{L}'^P_t, \mathcal{L}^{\star P}_t)_{t \in \mathbb{N}}$, where $(L, L^\star)$ is the initial state of the process, and processes $\mathcal{L}'^P$, $\mathcal{L}^{\star P}$ considered separately have the same transition functions as $\mathcal{L}^P$. However, we shall define some dependencies between $\mathcal{L}'^P$ and $\mathcal{L}^{\star P}$. According to delayed path coupling, the dependence may be built when considering a block of steps as a whole.

Let $\mu(\mathcal{L}^P_t)$ denote probability distribution of $\mathcal{L}^P_t$ and $\mu(I^P_t)$ uniform distribution on the same set of configurations.

Lemma 4 (Delayed Path Coupling Lemma). *Assume that the process $(\mathcal{L}'^P_t, \mathcal{L}^{\star P}_t)_{t \in \mathbb{N}}$ can be defined such that for some $\beta < 1$ and every t*

$$\mathbf{E}[\Delta(\mathcal{L}'^{P}_{t+\tau_H}, \mathcal{L}^{\star P}_{t+\tau_H})] \leq \beta \qquad (6)$$

for any pair of states of $\mathcal{L}'^{P}_{t}$ *and* $\mathcal{L}^{\star P}_{t}$ *from* Γ. *Then, after*

$$\tau_H \cdot \lceil \ln(k \cdot \varepsilon^{-1}) / \ln \beta^{-1} \rceil$$

steps of $\mathcal{L}^{P}$ *variation distance between the stationary (uniform) distribution and distribution of* $\mathcal{L}^{P}$ *is lower than* ε.

Convergence Proof. Now we apply path coupling to $\mathcal{L}^{P}$. Let Y_1, Y_2 be two configurations at step $t \cdot \tau_H$ that belong to Γ.

Now we assume that the random walk of hobbits is described by some pattern P. We may assume that P has the properties mentioned in Corollary 1, since almost all patterns P have this property. Let $\mathcal{R}$ be the only magic ring with different owner in configurations Y_1, Y_2. For the purpose of defining appropriate coupling let us fix the movements of all magic rings except $\mathcal{R}$ – these movements are the same for both processes. For a while we disregard all hobbits holding magic rings other than $\mathcal{R}$. So we are left with $m - k + 1 \geq m/2$ hobbits. Let us consider location of ring $\mathcal{R}$ when we start Y_1. Let $Z_1, \ldots, Z_w$, for $w \leq m - k + 1$, be all possible configurations at step $(t+1) \cdot \tau_H$ that agree with P and for that the positions of all magic rings except $\mathcal{R}$ are as revealed. Let $p_i^{(j)}$ denote probability of reaching Z_i from configuration Y_j. Let $\bar{p}_i = \min(p_i^{(1)}, p_i^{(2)})$ and $\bar{p} = \sum_{i \leq w} \bar{p}_i$. Then by Corollary 1

$$1 - \bar{p} = \sum_{i \leq w} (\tfrac{1}{w} - \bar{p}_i) \leq \sum_{i \leq w} |\tfrac{1}{w} - \bar{p}_i|$$
$$\leq \sum_{i \leq w} |\tfrac{1}{w} - p_i^{(1)}| + \sum_{i \leq w} |\tfrac{1}{w} - p_i^{(2)}| \leq \tfrac{2}{m^2}$$

Now we establish dependencies between both processes by defining the movements of $\mathcal{R}$. We may consider the first process as a process which chooses the position of $\mathcal{R}$ after τ_H steps in the following way:

step A: with probability $\bar{p}$ it goes to step B, otherwise it goes to step C,

step B: for $i \leq w$ with probability $\bar{p}_i/\bar{p}$ it changes its configuration to Z_i,

step C: for $i \leq w$ with probability $(p_i^{(1)} - \bar{p}_i)/(1 - \bar{p})$ it goes to configuration Z_i.

Obviously, in this way configuration Z_i is reached with probability $p_i^{(1)}$, as required. Then we may define the transitions of the second process:

- if case B has occurred for the first process, then the second process reaches the same configuration,
- if case C has occurred, then the second process changes its configuration to Z_i with probability $(p_i^{(2)} - \bar{p}_i)/(1 - \bar{p})$.

Obviously, in this way the second process reaches configuration Z_i with probability $p_i^{(2)}$ for $i \leq w$, as required. So it is defined correctly as a copy of $\mathcal{LL}^{P}$. It follows directly from the construction that with probability at least $\bar{p}$ both processes reach the same configuration. We conclude that due to path coupling lemma we have:

Theorem 3. *With high probability (over a random variable P)*

$$d(\mu(\mathcal{L}_t^P), \mu(I_t^P)) \leq \tfrac{1}{n^2}$$

for $t = \Omega(\tau_G^3 \cdot \log^6 n \cdot \log k)$.

4   Final Remarks and Conclusions

We have estimated how long the mobile mixing process must be so that traffic analysis cannot provide much information. As often in this field, **proving** privacy properties is technically quite involved. Our estimation is very rough; the main goal was to show a polylogarithmic bound. There is a lot of room to optimize the bound through a more involved and/or more detailed analysis.

We did some numerical experiments for certain simple graphs. These results indicate that coefficient d_H might converge much faster than it is stated in Theorem 3.

For practical applications, analysis should be performed for special classes of graphs – in the case of a proposed universal re-encryption of RFID tags this would not only involve analysis of a given network of re-encryption points, but also indications about necessary density of these points and their locations. Certainly, this is a new and technically challenging question.

References

1. Aldous, D., Fill, J.: Reversible Markov chains and random walks on graphs. In preparation, some chapters available at http://stat-www.berkeley.edu/pub/users/aldous/RWG/book.html
2. Beimel, A., Dolev, Sh.: Buses for Anonymous Message Delivery. Second International Conference on FUN with Algorithms, Carleton Scientific, 2001, 1–13
3. Berman, R., Fiat, A., Ta-Shma A.: Provable Unlinkability Against Traffic Analysis. Financial Cryptography'2004, Lecture Notes in Computer Science , Springer-Verlag
4. Bubley, B., Dyer, M.: Path Coupling: A Technique for Proving Rapid Mixing in Markov Chains. IEEE Symposium on Foundations of Computer Science (FOCS) '97, 223-231
5. Chaum, D.: Untraceable Electronic Mail, Return Addresses, and Digital Pseudonyms. Communication of the ACM 24(2) (1981) pp. 84-88
6. Czumaj, A., Kutyłowski, M.: Generating Random Permutations and Delayed Path Coupling Method for Mixing Time of Markov Chains. Random Structures and Algorithms 17 (2000), 238–259
7. Czumaj, A., Kanarek, P., Kutyłowski, M., Loryś, K.: Distributed Stochastic Processes for Generating Random Permutations. ACM-SIAM Symposium on Discrete Algorithms (SODA) '99, 271-280
8. Danezis, G., Serjantov, A.: Towards an Information Theoretic Metric for Anonymity. Privacy Enhancing Technologies Workshop (PET) '2002, Lecture Notes in Computer Science 2482, Springer-Verlag, 41-53

9. Dingledine, R., Shmatikov, V., Syverson, P.: Synchronous Batching: From Cascades to Free Routes. Workshop on Privacy Enhancing Technologies (PET) '2004, Lecture Notes in Computer Science, Springer-Verlag

10. Fairbrother, P.: An Improved Construction for Universal Re-encryption, Workshop on Privacy Enhancing Technologies '2004, Lecture Notes in Computer Science, Springer-Verlag

11. Gogolewski, M., Kutyłowski, M., Łuczak, T.: Distributed Time-Stamping with Boomerang Onions. Manuscript, 2004

12. Golle, P., Jakobsson, M., Juels, A., Syverson P.: Universal Re-encryption for Mixnets. Cryptographers' Track at the RSA Conference'2004

13. Gomułkiewicz, M., Klonowski, M., Kutyłowski, M.: Provable Unlinkability Against Traffic Analysis already after $\mathcal{O}(\log(n))$ steps! Information Security Conference (ISC) '2004, Lecture Notes in Computer Science 3225, Springer-Verlag, 354-366

14. Gülcü, C., Tsudik, G.: Mixing E-mail with BABEL. ISOC Symposium on Network and Distributed System Security, IEEE 1996, 2-16

15. Rackoff, C., Simon, D. R.: Cryptographic Defense Against Traffic Analysis. ACM Symposium on Theory of Computing (STOC) '93, 672–681

16. Szemerédi, E.: Regular Partitions of Graphs. "Problèmes Combinatoires et Théorie des Graphes, Proc. Colloque Inter. CNRS" (J.-C. Bermond, J.-C. Fournier, M. Las Vergnas, et D. Sotteau, Eds.), CNRS, Paris, 1978, 399-401

17. Syverson, P.F., Goldschlag, D., Reed, M.: Hiding Routing Information. Information Hiding '96, Lecture Notes in Computer Science 1174, Springer-Verlag, 137-150

Appendix

We prove here inequality (1). Let us use the following notation:

$$p(i, u; s) = \tfrac{1}{n} + \varepsilon_{i,u} \quad \text{and} \quad p(u, j; t) = \tfrac{1}{n} + \delta_{u,j} \ .$$

Obviously,

$$p(i, j; s + t) = \sum_{u \leq n} p(i, u; s) \cdot p(u, j; t) \ .$$

Hence,

$$
\begin{aligned}
2d_G(i, s+t) &= \sum_{j \leq n} \left| \sum_{u \leq n} (\tfrac{1}{n} + \varepsilon_{i,u}) \cdot (\tfrac{1}{n} + \delta_{u,j}) - \tfrac{1}{n} \right| \\
&= \sum_{j \leq n} \left| \sum_{u \leq n} (\tfrac{1}{n^2} + \tfrac{1}{n}\varepsilon_{i,u} + \tfrac{1}{n}\delta_{u,j} + \varepsilon_{i,u} \cdot \delta_{u,j}) - \tfrac{1}{n} \right| \\
&= \sum_{j \leq n} \left| \tfrac{1}{n} \cdot \sum_{u \leq n} \varepsilon_{i,u} + \tfrac{1}{n} \sum_{u \leq n} \delta_{u,j} + \sum_{u \leq n} \varepsilon_{i,u}\delta_{u,j} \right| \\
&= \sum_{j \leq n} \left| \sum_{u \leq n} \varepsilon_{i,u}\delta_{u,j} \right| \leq \sum_{j \leq n} \sum_{u \leq n} |\varepsilon_{i,u}| \cdot |\delta_{u,j}| \\
&= \sum_{u \leq n} \sum_{j \leq n} |\varepsilon_{i,u}| \cdot |\delta_{u,j}| = \sum_{u \leq n} |\varepsilon_{i,u}| \cdot 2d_G(j,t) \\
&= 2d_G(j,t) \cdot \sum_{u \leq n} |\varepsilon_{i,u}| = 4d_G(j,t) \cdot d_G(i,s) \leq 4d_G(t) \cdot d_G(s)
\end{aligned}
$$

A Location-Aware Secure Interworking Architecture Between 3GPP and WLAN Systems*

Minsoo Lee[1], Jintaek Kim[1], Sehyun Park[1,**],
Ohyoung Song[1], and Sungik Jun[2]

[1] School of Electrical and Electronics Engineering,
Chung-Ang University, Seoul 156-756, Korea
{lemins, groundiv}@ms.cau.ac.kr, {shpark, song}@cau.ac.kr
[2] Electronics and Telecommunications Research Institute,
161 Gajeong-dong, Yuseong-gu, Daejeon, 305-350, Korea
sijun@etri.re.kr

Abstract. Location-aware computing is widely discussed to be one attractor of the future ubiquitous networks for intelligent home and telematics services. Apart from technical difficulties in providing accurate position information there is also a lack of a clearly defined framework how to create innovative and secure services using location information. In this paper, we identify QoS, interworking and security problems in location-aware computing. We propose a number of concepts and infrastructures for location-aware security services including micro and macro mobility management in ubiquitous home networks and heterogeneous wireless networks. We further explain how the proposed approach can be applied to seamless secure interworking among wireless LANs and 3G, 4G cellular systems. Our approach might be helpful in the discussion for the frameworks especially highlighting evolutionary steps for the next ubiquitous networks.

Keywords: Location-awareness, ubiquitous computing, 4G, security, interworking, agents, home network.

1 Introduction

The concept of *context awareness*, sensing and reacting to dynamic environments and activities is in the heart of ubiquitous computing. Context information gathered from various sensors, networks, devices, user profiles, and other sources can

---

* This research was supported by the MIC(Ministry of Information and Communication), Korea, under the Chung-Ang University HNRC(Home Network Research Center)-ITRC support program supervised by the IITA(Institute of Information Technology Assessment).
** Corresponding author.

C. Park and S. Chee (Eds.): ICISC 2004, LNCS 3506, pp. 394–406, 2005.

enhance mobile applications' usability by letting them adapt to conditions that directly affect their operation [1].

Location information is a fundamental element of context. Determining the location of people and objects has been the focus of much research in ubiquitous computing for the future intelligent home and telematics services . Many location sensing technologies have been devised, resulting in systems which perform sensing using diverse physical media, such as ultrasonic, ultrawideband radio, visible light, RFID, infrared, bluetooth, GPS, wireless LANs (WLAN), and mobile phones. Much research in the past decade has focused on these location-sensing technologies, location-aware application support, and location-based applications. With numerous factors driving deployment of sensing technologies, location-aware computing may soon become a part of everyday life [2]. Examples of location-aware application include intelligent home services with person tracking, military training, asset tracking, conference assistants, environmental resource discovery and control, support systems for the elderly, tour guides, augmented reality, and mobile desktop control. Each demands different levels of service, for example in terms of location accuracy and update rate, infrastructure cost, deployment difficulty, robustness, and capacity for security and privacy guarantees.

Although studies have been made on location accuracy and robustness, little attention has been given to location interoperability. The widest existing deployments are based on GPS, which is particularly suited for outdoor applications. Although GPS offers near-worldwide coverage, its performance degrades in indoor environments, especially at home, in building and in high-rise urban areas. GPS receivers have a relatively long start-up time and high cost. As intentional interferences, GPS suffers from signal jamming attack and spoofing attack. Therefore we need fine-grained as well as robust positioning system.

Apart from technical difficulties in providing accurate position information there is also a lack of a clearly defined framework how to create innovative and security services by using location information. Location-based services (LBS) are emerging as the next killer application in personal wireless devices, but there are few safeguards on location privacy. In fact, the demand for improved public safety is pushing regulation in the opposite direction [3].

For providing a foundation of the flexibility and the interoperability of LBS, there has been a recent focus on location-aware platforms, which link data-gathering systems and the data-consuming applications in a flexible manner. Such work includes location representation, sensor fusion to combine location data from many sources, and software frameworks supporting the distributed nature of location-aware computing. Such abstractions are essential for the interoperability, usability and development of location-aware systems and applications.

To meet these various demands of location-aware services in the future wireless networks, this paper focused mainly on how security can be substantially improved through a new form of authentication based on location-aware architecture. First, we identify these problems of location-aware computing related to QoS, interworking and security. For the seamless QoS support of location-aware services, we introduce *fine-grained hybrid positioning* method as a network based approach, using cooperations of GPS, 3G systems and wireless LANs with the enhanced location accuracy and robustness.

As an abstraction of location-aware model, we designed *LBS Broker* that performs *location-aware authentication* for fast secure roaming. Carefully managed location information could be used as authentication information, such as who you are (biometrics), what you have (token, smart tag, certificate, smart card) and what you know (ID, password). Location information can be combined to enforce the authenticity of a mobile user by validating location history of the user. In wireless LANs, previous security association with old AP could enforces secure roaming with a new AP by proper validation of location information. Moreover, *LBS Broker* plays key role in protecting user's privacy from unauthorized LBS service providers or malicious users.

Finally we propose the infrastructure for location-aware security services in the heterogeneous wireless networks including micro and macro mobility management in ubiquitous home networks and heterogeneous wireless networks for global personal connectivity aspects. We further explain how the proposed approach can be applied to seamless secure interworking of wireless LANs and 3G, 4G cellular systems with evaluation of our testbed.

The rest of this paper is organized as follow. Section 2 gives related works about location-aware computing in wireless networks. Section 3 identifies the problems and requirements of location-aware security computing. In section 4 we introduce the fine-grained hybrid positioning method. In section 5 we propose the location-aware authentication for fast roaming with LBS Broker. Section 6 suggests our location-aware security architecture for 3G/WLAN interworking. Section 7 shows some of location-aware service scenarios applying the proposed architecture. In section 8, we discuss evaluation of our location-aware secure roaming. Finally, we conclude in Section 9.

2   Related Works

Location is one of the key contexts that determine which types of devices are available and how communication should be conducted to fit the user's needs. In the future of ubiquitous home networks and heterogeneous wireless networks, location information will be available from various types of network including sensor networks, WLAN, 3G including Universal Mobile Telecommunications System (UMTS) and code-division multiple access 2000 (CDMA2000), Global System for Mobile Communications (GSM) evolutions such as General Packet Radio Service (GPRS) and Enhanced Data for GSM Evolution (EDGE) [4]. According to development of WLAN, 3G and 4G networks with the enhanced accuracy of the positioning technology [5], many location-based services with efficiency and reliability will appear. Toward seamless security of the heterogeneous networks, fundamental features are required, such as smooth roaming techniques, QoS guarantee, data security, user authentication and authorization.

For smooth roaming, several studies have been made on a fast handover management in IPv6 Networks [6] and an integrated management that combines the strengths of Mobile IP Location Registers (MIP-LR) and Session Initiation Protocol (SIP) [7]. In the area of location security and privacy, there ware frameworks of a cryptographic approach of an authorized-anonymous-ID-based scheme [8] and algorithms for location discloser-control [9] and based on frequently changing pseudonyms [10]. But there are challenging issues with regard to location-aware seamless secure roaming as reconfigurable and adaptable features are needed in the future wireless networks. This paper concentrates mainly on location-aware schemes with agent technologies to enforcing security for seamless interworking of ubiquitous networks.

3   Motivations and Requirements

Completely new concepts and features will emerge for location-aware computing in ubiquitous environment where seamless interworking of self-organizing sensor networks, WLAN, 3G and 4G networks are possible. Location-aware computing includes many issues and requirements related to the user experience, privacy preservation, QoS guarantee, and security in the determination and transfer of personal data.

The following requirements should be considered to fulfill the promising services of the future ubiquitous networks.

- *Availability and Survivability*: The positioning technology like GPS is a single point of failure. If a single scheme with GPS is unavailable, location information for continuous location tracking services will not be served correctly to subscribers of the services.
- *Global Secure Roaming*: It will be a very common situation that the mobile node (MN) is served from foreign domain without any pre-established user authentication or authorization.
- *System Overhead*: Complexity of secure roaming and reauthentication process requires more computational power of the mobile device that usually has limited power and system resources.
- *Network Overhead*: There is also signaling overhead from secure roaming and reauthentication requests and replies.
- *Service delay or Handoff latency*: In addition to the latency related to handoff at the physical and link layers, secure roaming or reauthentication could add significant latency.
- *Privacy and Security*: It may cause an invasion of privacy by unwanted disclosure and commercial use of location information. Security is also a concern for wireless networks because connections passing over the air can be easily monitored.
- *Granularity*: The effectiveness of location-aware applications depends not only on the user population but also on the location sensing system's update

rate and spatial resolution. Several applications would not work at all if they got updates only on a scale of hours and kilometers, as opposed to seconds and meters.

The first major requirement for our secure mobility management scheme is that it handles mobility without reauthentication overhead. A second requirement is that continuous location tracking with real-time traffic must be handled with special care. For example, handoff latency is especially disruptive to continuous location tracking, even if most of the reauthentication during the handoff or roaming in different networks are not lost but delayed and eventually delivered to a user. A third requirement is that it must be survivable and robust in dynamically auto-configured ubiquitous networks.

4   Fine-Grained Location Positioning Systems

Local positioning will be one of the most exciting features of the next generation of wireless systems. Positioning systems fall into one of three categories. In the network-based approach, infrastructure receivers such as cell towers track cellular handsets or other mobile transmitting units. In the networked-assisted approach, location determination occurs in the network with the mobile device's active. In the client-based approach, mobile devices autonomously compute their own position, as is the case with a GPS unit. The widest existing deployments are based on GPS, which is particularly suited for outdoor applications. However, GPS suffers from signal jamming attack and spoofing attack as well as indoor usage. Therefore we need fine-grained as well as robust positioning system.

For ensuring consistent QoS support in location-aware services, we introduce fine-grained hybrid positioning method as a network based approach, using

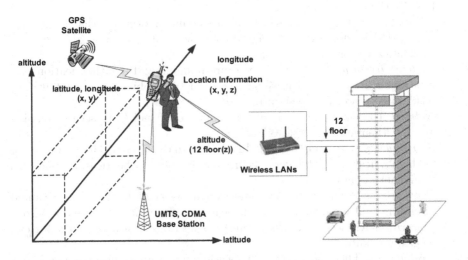

Fig. 1. A Hybrid Positioning for Fine-grained system

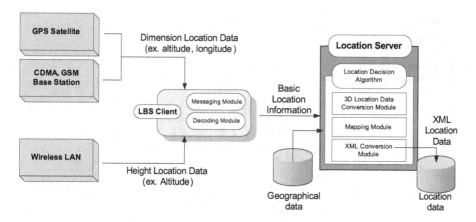

Fig. 2. 3D Location Decision

GPS, 3G cellular system and wireless LAN for enhancing location accuracy and robustness. Our positioning method also manages three-dimensional position information in the particular environments such as at home, in the office and the downtown area surrounded by high rise buildings. The positioning method could extend its coverage with sensor networks. If GPS is unavailable, the mobile device can interrogate a physically fixed reference sensor tag at its location to initialize the mobile device's position. Fig. 1 shows the hybrid-positioning method using GPS and wireless LAN for providing fine-grained three dimensional location information. Fig. 2 shows the module for getting 3D location information.

Due to the varying types of measurements from different sensors and the varying requirements of different applications, no dominant representation for location has emerged. But XML based representation will be the most interoperable method. We propose the scheme that Location Server converts 3D information to interoperable XML format for LBS Web services. The converted XML data is stored in the location DB of the Location Server for use of other LBS services. The Hybrid method mixing various networks helps mobile users gain more accurate location information and efficient LBS services. LBS with Web services will also become more effective by converting location data to XML under certain rules.

5   Location-Aware Secure Handoff

Location information can be combined to enforce the authenticity of a user by validating location history of the user. As an example in wireless LANs, previous security association with old AP enforces secure roaming with a new AP. In this section, we focus on how security can be substantially improved through a new form of authentication based on location-aware architecture. We also propose a LBS Broker using location information for secure interworking.

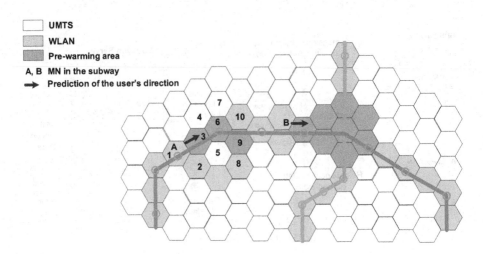

Fig. 3. Location-aware Roaming Service with Pre-warming area

5.1   Location-Aware Handoff

One of the difficulties in providing secure, seamless roaming service is how to promptly and securely exchange authentication information or security credentials during handoff. The exchange of authentication information in advance by tracking and predicting user's direction, so called pre-warming may helpful to simplify the authentication procedure during handoff and to support seamless roaming service [11]. For example, if the MN A in Fig. 3 is on the subway, it will obviously handoff from AP1 to AP3, even if AP2 is better reachable for a short period of time. Therefore, the authentication information of the MN needs to be delivered from AP1 to AP3 correctly in advance for seamless roaming before the handoff procedure.

5.2   Determining Paging Area

Handoff procedure using location information can be effectively completed if the user movements can be predicted accurately in a ubiquitous network environment. If the MN A in Fig. 3 is on a subway, its route may be constrained to areas along the one way track. In this case, location information for the area outside of the track is not necessary. In the case of the MN B, its paging area will have to include areas around the subway station with areas along the track. In the case of a pedestrian, its path may be effectively predicted by the history based location algorithms that has a record of the previous user movements and take into account the respective probability of movements together with factors such as the direction and the speed [12]. The concept of paging area can enhance the efficiency of the system from gathering location information only in the paging areas.

6   Location-Aware Security Architecture for 3GPP and WLAN Interworking

In this section, we propose our location-aware security architecture for ubiquitous networks. The architecture is designed to meet the location-aware computing requirements in section 3. First we describe the basic functionalities are described and we will show some practical scenarios in which location-aware fast roaming are provided while preserving user's location security and privacy. Fig. 4 shows proposed location-aware privacy and security enhanced interworking architecture.

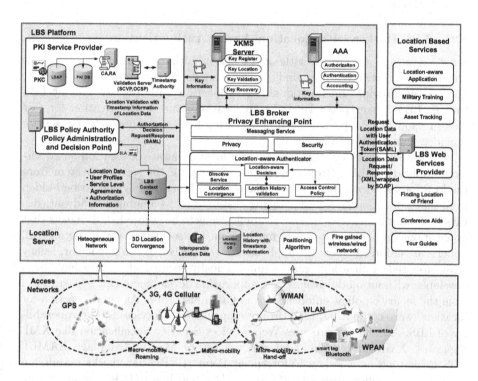

Fig. 4. The Proposed Location-aware Security Architecture

6.1   System Assumptions and Key Figures

– In the LBS platform of our architecture location information are managed by XML based protocol. Mobile Location Protocol (MLP) [13] is used to initiate, request and response of location information. This is a reasonable assumption to make, given the momentum XML based protocols have gained in recent years.
– Location server and mobile device support our fine-grained positioning methods using heterogeneous access networks in ubiquitous environments.

- Interdomain authentication, authorization, and accounting (AAA) are handled together with *LBS Broker* and AAA server.
- When a user moves into a foreign (different) network, it needs to obtain a reauthentication process in that network and LBS Broker and AAA take charge of reauthentication process.
- *LBS Broker* includes *location-aware authenticator* for fast secure roaming using the concepts of *direction of user* and *pre-warming zone* for reauthentication.
- *LBS Broker* enforce fast reauthentication as well as provide vertical handoff for ubiquitous network where loosely coupled and highly resource-constrained environments.

6.2   LBS Broker with Location-Aware Authentication

Sensed locations are useless without a location authority that gives a space of possible locations and can respond to queries on distances, routes, and proximity. To securely provide the functionalities of location authority we present LBS Broker. LBS broker plays key roles in location-aware authentication for fast roaming and also in protecting user's privacy from unauthorized LBS service provider or malicious users.

LBS Broker provides an abstraction for location-aware or location based services application. LBS Broker model could separate the security system from the location-based application (i.e., the security functionality should be hidden from the application layer). Supporting abstraction has a number of advantages. Firstly, it allows "security unaware" applications to be secured. This means that applications do not need to know much about the security features, because all policies are enforced below the applications by LBS Broker. In particular, legacy location-based applications can be automatically secured after they have been developed without modifications, and application development can be segregated from the security policy, enforcement, and administration. As an abstraction for location-aware computing, LBS Broker supports Web Services for interoperability of LBS. And it also provides Web Services Security Specification like XML Signature, XML Encryption and Security Assertion Markup Language (SAML). LBS Broker act as a Policy Enforcement Point (PEP) that checks permission with the LBS policy authority, the Policy Decision Point (PDP) by requesting SAML assertion before making decision and releasing the secured location data to the LBS service providers.

7   Location-Aware Access Control Model

In the location-aware secure roaming procedure, the privacy and security policies should be associated with an access control mechanism which refers to the characteristics of implementation and enforcement. Four different types of access control policies have been designed in literature. They include Mandatory Access Control (MAC), Discretionary Access Control (DAC), Double Discretionary Access Control (DDAC) and Role Based Access Control (RBAC). Among

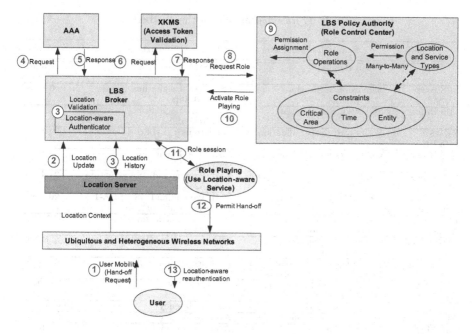

Fig. 5. The implementation of location-aware roaming with RBAC

these, RBAC [14] is the most extensible type of access control policy for various location-aware services. Fig. 5 shows the proposed RBAC model on our architecture.

8   Evaluation of Location-Aware Secure Roaming

8.1   The Testbed

We evaluate our location-aware secure roaming procedure over wireless LAN for micro-mobility. We also consider macro-mobility in our scheme with some of measurement parameters of interaction with CDMA, GPRS and Wireless LAN. We consider a single local correspondent communicating with our MNs. We focus on IEEE 802.11b and IEEE 802.11i [15] in order to have an overview of the real MN possibilities over wireless LANs. Table 1 summarizes the base parameter settings underlying the performance experiments. LBS Broker and LBS Policy Authority are running on server machines of Pentium III 933 MHz CPUs with Solaris 8 operating system (O/S). AAA Server is running on a server of Pentium III 800 MHz with Linux O/S and the modified FreeRADIUS library for RADIUS functionality. APs are working on a Pentium III 500 MHz machines with Linux O/S and MNs are running Pentium III 500 MHz machines with WindowsXP O/S and Lucent Orinoco 802.11b wireless LANs cards. The cryptographic library is OpenSSL 0.9.7a, and SAML Library is OpenSAML 0.9.1. Data size is 1KB in digital signature.

Table 1. Base parameters of the Testbed

Node	Simulation parameter	Remark	Value
MN	Token Request	RSA SHA-1 signature sign(512bits)	5.5 ms
LBS Broker	Signature verification	RSA SHA-1 signature verify(512bits)	0.1 ms
LBS Broker	SCVP(OCSP) Request	RSA SHA-1 signature sign(1024bits)	7.4 ms
XKMS with PKI	X.509 Certificate validation	Validate user certificate	30.3 ms
LBS Broker	OCSP Response validation	RSA SHA-1 signature verify(1024bits)	0.4 ms
LBS Broker	SAML Authorization Request	XML Parsing and RSA signature sign(1024bits)	27.4 ms
LBS Policy Authority	SAML Authorization Response	XML Parsing and RSA SHA-1 signature verify(1024bits)	20.4 ms
LBS Policy Authority	SAML Authentication Token generation	3DES Symmetric key encryption	7.702 MB/s
LBS Broker	Token Response with Location information	RSA encryption(512bits)	31.201 KB/s
LBS SP	Decrypt Token Response with Location Update Response	RSA decryption(512bits)	8.517 KB/s
MN-AAA	802.1X full authentication (EAP-TLS)	Average delay	1600ms
AP	802.11 scan (active)	Average latency	40∼300ms
AP	802.11 reassociation(IAPP)	Average latency	40ms
MN-AP	Fast Handoff (4-way handshake only)	Average latency	60ms
LBS Broker	Location history request to Location Server	Request location history	20∼100ms
LBS Broker	Location-aware authentication	Validation of Location history	80∼100ms
802.11/CDMA	TCP parameter adjustment	Average delay	5000ms
802.11/GPRS	TCP parameter adjustment	Average delay	20000ms

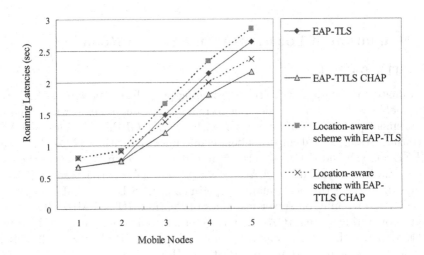

Fig. 6. Roaming Latency in wireless LAN

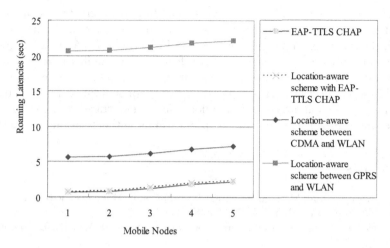

Fig. 7. Roaming Latency in wireless LAN, CDMA and GPRS

8.2   Analysis

For micro-mobility, the averages of our measurements in roaming between wireless LANs are shown in Fig. 6. For macro-mobility, the averages measurements in the roaming between wireless LANs and CDMA, wireless LANs and GPRS are shown in Fig. 7. The solid curves represent the measurements without our location-aware scheme and the dotted curves represent our location-aware case. We notice an important point between the existing roaming case and the location-aware roaming case. When the moving MNs are increasing, our location-aware scheme does not create much burden on the roaming as to selection of 802.1X authentication methods. In Fig. 6, our location-aware scheme with EAP-TTLS CHAP shows almost the same latency of the roaming without location-aware scheme with EAP-TLS. There are trade-off issues in selection of authentication methods. We should also consider trade-off among QoS of location-aware services with location update period and location precision, security and privacy policy, protection of location information. When more security features are introduced into the network, there are the ever-increasing computation, communication, and management overhead. In fact, both dimensions of security strength and network performance are equally important, and achieving a good trade-off between two extremes is the one of the fundamental challenges in security design for location-aware computing. In this context, our simulation results could be useful to provide guidelines as to how the security level is set to meet the user' needs.

9   Conclusions

In this paper, we analyze QoS, interworking and security issues in location-aware computing and give our view on the future prospects of ubiquitous networks. We introduce the different concepts of location-aware security and propose the solutions with several new and existing systems and applications. Our architecture integrates the location-aware authentication scheme for macro and micro mobility with the fine-grained positioning and LBS Broker with AAA. This integrated scheme could provide the desired security features and fulfill the requirements for survivable wireless networks. In the future work, we are going to evaluate our testbed in real world prototypes that collect outdoor movement traces through GPS and 3G cellular systems and indoor movement trace at home through WLANs and sensor networks with the pre-warming authentication mechanism.

References

1. Panu Korpip, et. al: Managing Context Information in Mobile Devices. IEEE PERVASIVE computing, JULY-SEPTEMBER 2003.
2. Mike Hazas, et. al: Location-Aware Computing Comes of Age. IEEE Computer, February 2004.
3. Bill Schilit, et. al: Wireless Location Privacy Protection. IEEE Computer, December 2003.
4. Petri Mähönen, et. al: Hop-by-Hop Toward Future Mobile Broadband IP. IEEE Communications Magazine, March 2004.
5. Martin Vossiek: Wireless Local Positioning. IEEE Microwave Magazine, December 2003.
6. Nicolas Montavont and Thomas Noël: Handover Management for Mobile Nodes in IPv6 Networks. IEEE Communications Magazine, August 2002.
7. K. Daniel Wong, et. al: Mobility Management Scheme for Auto-configured Wireless IP Networks. IEEE Wireless Communications, October 2003.
8. Qi He, et. al: The Quest for Personal Control Over Mobile Location Privacy. IEEE Communications Magazine, May 2004.
9. MARCO GRUTESER et. al: Protecting Privacy in Continuous Location-Tracking Applications. IEEE SECURITY & PRIVACY, March-April 2004.
10. Alastair R. Beresford and Frank Stajano: Location Privacy in Pervasive Computing. IEEE PERVASIVE computing, JANUARY-MARCH 2003.
11. Christian Prehofer and Qing Wei: Active Networks for 4G Mobile Communication: Motivation, Architecture and Application Scenarios. International Working Conference on Active Networks, 2002.
12. R. Chellappa, A. Jennings and N. Shenoy: User Mobility Prediction in Hybrid and Ad Hoc Wireless Networks. ATNAC, December 2003.
13. Location Inter-operability Forum (LIF): Mobile Location Protocol (MLP). TS 101 Specification Version 3.0.0 6, June 2002.
14. David F. Ferraiolo, et. al: Proposed NIST Standard for Role-Based Access Control. ACM Transactions on Information and System Security, Vol. 4, No.3, August 2001.
15. IEEE Draft 802.11i/D6.1, September 2003.
16. Minsoo Lee, Jintaek Kim, Sehyun Park, Jaeil Lee and Seoklae Lee: A Secure Web Services for Location Based Services in Wireless Networks. Lecture Notes in Computer Science, Vol 3042. Springer-Verlag, Berlin Heidelberg (2004), 332–344

ADWICE – Anomaly Detection
with Real-Time Incremental Clustering*

Kalle Burbeck and Simin Nadjm-Tehrani

Department of Computer and Information Science,
Linköpings universitet SE–581 83 Linköping, Sweden
{kalbu, simin}@ida.liu.se

Abstract. Anomaly detection, detection of deviations from what is considered normal, is an important complement to misuse detection based on attack signatures. Anomaly detection in real-time places hard requirements on the algorithms used, making many proposed data mining techniques less suitable. ADWICE (Anomaly Detection With fast Incremental Clustering) uses the first phase of the existing BIRCH clustering framework to implement fast, scalable and adaptive anomaly detection. We extend the original clustering algorithm and apply the resulting detection mechanism for analysis of data from IP networks. The performance is demonstrated on the KDD data set as well as on data from a test network at a telecom company. Our experiments show a good detection quality (95 %) and acceptable false positives rate (2.8 %) considering the online, real-time characteristics of the algorithm. The number of alarms is then further reduced by application of the aggregation techniques implemented in the Safeguard architecture.

Keywords: Intrusion Detection, Anomaly Detection, Adaptability, Real-time, Clustering.

1 Introduction

The threats to computer-based systems on which we are all dependent are ever increasing, thereby increasing the need for technology to handle those threats. One study[1] estimates that the number of intrusion attempts over the entire Internet is in the order of 25 billion each day and increasing. McHugh[2] claims that the attacks are getting more and more sophisticated while they get more automated and thus the skills needed to launch them are reduced. Intrusion Detection Systems (IDS) attempt to respond to this trend by applying knowledge-

* This work was supported by the European project Safeguard IST–2001–32685. We would like to thank Thomas Dagonnier, Tomas Lingvall and Stefan Burschka at Swisscom for fruitful discussions and their many months of work with the test network. The Safeguard agent architecture has been developed with the input from all the research nodes of the project, the cooperation of whom is gratefully acknowledged.

C. Park and S. Chee (Eds.): ICISC 2004, LNCS 3506, pp. 407–424, 2005.

based techniques (typically realised as signature-based misuse detection), or behaviour-based techniques (e.g. by applying machine learning for detection of anomalies).

Due to increasing complexity of the intrusion detection task, use of many IDS sensors to increase coverage and the need for improved usability of intrusion detection, a recent trend is alert or event correlation[3]. Correlation combines information from multiple sources to improve information quality. By correlation the strength of different types of detection schemes may be combined, and weaknesses compensated for.

The main detection scheme of most commercial intrusion detection systems is misuse detection, where known bad behaviour (attacks) are encoded into signatures. Misuse detection can only detect attacks that are well known and for which signatures have been written. In anomaly detection normal (good) behaviour of users or the protected system is modelled, often using machine learning or data mining techniques. During detection new data is matched against the normality model, and deviations are marked as anomalies. Since no knowledge of attacks is needed to train the normality model, anomaly detection may detect previously unknown attacks.

Anomaly detection still faces many challenges, where one of the most important is the relatively high rate of false alarms (false positives). We argue that the usefulness of anomaly detection is increased if combined with further aggregation, correlation and analysis of alarms[4], thereby minimizing the number of false alarms propagated to the administrator and helping to further diagnose the anomaly. In this paper we explain the role of anomaly detection in a distributed architecture for agents that has been developed within the European Safeguard project[5].

We apply clustering as the technique for training of the normality model, where similar data points are grouped together into clusters using a distance function. Clustering is suitable for anomaly detection, since no knowledge of the attack classes is needed whilst training. Contrast this to other learning approaches, e.g. classification, where the classification algorithm needs to be presented with both normal and known attack data to be able to separate those classes during detection. Our approach to anomaly detection, ADWICE (Anomaly Detection With fast Incremental Clustering), is an adaptive scheme based on the BIRCH clustering algorithm[6]. BIRCH has previously been used in applications such as web mining of user sessions on web-pages[7] but to our knowledge there has previously been no extensions of the algorithm for intrusion detection. We proceed by comparing our work with related research and point out the advantages of ADWICE. In section 3 we present the Safeguard agent architecture where the clustering based anomaly detection fits in. In section 4 we describe the anomaly detection algorithm. Evaluation of the algorithms is presented in section 5 followed by a concluding discussion in section 6.

2   Motivation

2.1   IDS Data Problems and Dependencies

One fundamental problem of intrusion detection research is the limited availability of good data to be used for evaluation. Producing intrusion detection data is a labour intensive and complex task involving generation of normal system data as well as attacks, and labelling the data to make evaluation possible. If a real network is used, the problem of producing good normal data is reduced, but then the data may be too sensitive to be released in public.

For learning based methods, good data is not only necessary for evaluation and testing, but also for training. Thus applying a learning based method in the real world, puts even harder requirements on the data. The data used for training need to be representative to the network where the learning based method will be applied, possibly requiring generation of new data for each deployment. *Classification based methods*[8, 9], or supervised learning, require training data that contains normal data as well as good representatives of those attacks that should be detected to be able to separate attacks from normality. Complete coverage of even known and recent attacks would be a daunting task indeed due to the abundance of attacks encountered globally. Even worse, the attacks in the training data set need to be labelled with the attack class or classes.

Clustering, or unsupervised learning, has attracted some interest[10, 11, 12, 13] in the context of intrusion detection. The interesting feature of clustering is the possibility to learn without knowledge of attack classes, thereby reducing training data requirement, and possibly making clustering based techniques more viable than classification-based techniques in a real world setting. There exist at least two approaches.

When doing *unsupervised anomaly detection* a model based on clusters of data is trained using unlabeled data, normal as well as attacks. The assumption is that the relative amount of attacks in the training data is very small compared to normal data, a reasonable assumption that may or may not hold in the real world context for which it is applied. If this assumption holds, anomalies and attacks may be detected based on cluster sizes. Large clusters correspond to normal data, and small clusters possibly correspond to attacks. A number of unsupervised detection schemes have been evaluated on the KDD network data set[10, 12] and command line sequences[11] with varying success. The accuracy is however relatively low which reduces the direct applicability in a real network.

In the second approach, which we denote simply *(pure) anomaly detection* in this paper, training data is assumed to consist only of normal data. Munson and Wimer[13] used a cluster based model (Watcher) to protect a real web server, proving anomaly detection based on clustering to be useful in real life.

Acceptable accuracy of the unsupervised anomaly detection scheme may be very hard to obtain, even though the idea is very attractive. Pure anomaly detection, with more knowledge of data used for training, may be able provide better accuracy than the unsupervised approach. Pure anomaly detection also avoids the coverage problem of classification techniques, and requires no labelling of training data similar to unsupervised anomaly detection. Generating training data in a highly controlled network now simply consists of generating normal data. This is the approach adopted in this paper, and the normality of the training data in our case is ensured by access to a large test network build specifically for experimental purposes in the Safeguard project.

In a real live network with connection to Internet, data can never be assumed to be free of attacks. Pure anomaly detection also works when some attacks are included in the training data, but those attacks will during detection be considered normal and therefore not detected. To increase detection coverage, attacks should be removed to as large an extent as possible, making coverage a trade-off with data cleaning effort. An efficient approach should be to use existing misuse detectors with updated rule-bases in the preparatory phase, to reduce costly human effort. Updated signature based systems should with high probability detect many of the currently known attacks, simplifying removal of most attacks in training data. A possibly complementary approach is to train temporary models on different data sets and let them vote on normality to decide what data to use for the final normality model.

Certain attacks, such as Denial of Service and scanning can produce large amounts of attack data. On the other hand, some normal types of system activities might produce limited amounts of data, but still be desirable to incorporate into the detection model. Those two cases falsify the assumption of unsupervised anomaly detection and need to be handled separately. Pure anomaly detection such as ADWICE does not have those problems since detection is not based on cluster sizes.

2.2   IDS Management Effort

One of the inherent problems of anomaly detection is the false positives rate. In most settings normality is not easy to capture. Normality changes constantly, due to changing user behaviour as well as hardware or software changes. An algorithm that can perfectly capture normality of static test data, will therefore not necessarily work well in a real life setting with changing normality. The anomaly detection model needs to be adaptable. When possible, and if security policy allows, it should be autonomously adaptive to minimize the human effort. In other cases an administrator needs to be able to update the anomaly model with simple means, without destroying what is already learnt. And the effort spent updating the model should be minimal compared to the effort of training the initial model. ADWICE is incremental, supporting easy adaptation and extension of the normality model.

2.3   Scalability and Performance Issues

For critical infrastructures or valuable company computer based assets it is important that intrusions are detected in real-time with minimal time-to-detection to minimize the consequences of the intrusion. An intrusion detection system in a real-time environment needs to be fast enough to cope with the information flow, have explicit limits on resource usage and also needs to adapt to changes in the protected network in real-time.

Many proposed clustering techniques require quadratic time for training[14], making real-time adaptation of a cluster-based model hard. This implies that most clustering-based approaches would require time consuming off-line training to update the model. They may also not be scalable, requiring all training data to be kept in main memory during training, limiting the size of the trained model.

- ADWICE is scalable, since only compact summeries of clusters are kept im memory rather then the complete data set.
- ADWICE has good performance due to local clustering and an integrated tree index for searching the model.

3   The Safeguard Context

Safeguard is a European research project aiming to enhance survivability of critical infrastructures by using agent technology. The Safeguard agent architecture is presented in Fig. 1. This architecture is evaluated in the context of telecommunication and energy distribution networks. The agents should improve survivability of those large complex critical infrastructures (LCCI:s), by detecting and handling intrusions as well as faults in the protected systems. The key to a generic solution applicable in many infrastructures is in the definition of roles for various agents. There may be several instances of each agent in each LCCI. The agents run on a platform (middleware) that provides generic services such as discovery and messaging services. These are believed to be common for

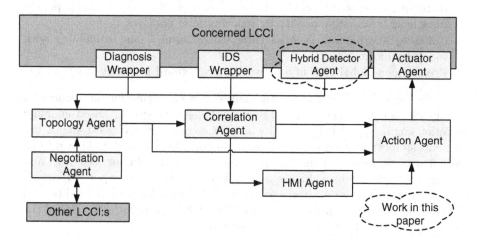

Fig. 1. The Safeguard agent architecture

the defence of many infrastructures, but should be instantiated to more specific roles in each domain. The generic roles can be described as follows:

- *Wrapper agents* wrap standard INFOSEC devices and existing LCCI diagnosis mechanisms, and provide their outputs after some filtering and normalisation for use by other agents.
- *Topology agents* gather dynamic network topology information, e.g. host types, operating system types, services provided, known vulnerabilities.
- *Hybrid detector agents* utilise domain knowledge for a given infrastructure, but combine it with behavioural detection mechanisms (e.g. anomaly detection with white lists).
- *Correlation agents* identify problems that are difficult to diagnose with one source of information in the network, by using several sources of information from wrapper, topology, or hybrid detector agents. They use the data sources to order, filter and focus on certain alarms, or predict reduced availability of network critical services. One type of correlation agent performs adaptive filtering and aggregation to further reduce the alarm rates.
- *Action agents* enable automatic and semi automatic responses when evaluation of a problem is finished.
- *Negotiation agents* communicate with agents in other LCCI:s to request services and pass on information about major security alarms.
- *HMI (Human-Machine Interface) agents* provide an appropriate interface, including overview, for one or many system operators.
- *Actuator agents* are wrappers for interacting with lower layer software and hardware (e.g. changing firewall rules).

In the context of a management network for telecom service providers we have identified the following needs:

- Reducing information overload ([4] and section 5.5 on aggregation)
- Increasing coverage by providing new sources of information (this paper)
- Increasing information quality by reducing false positives[4]
- Collating information, such as correlating alarms[4] and combining with topology information
- Presenting a global view of a network (in Safeguard Demonstrator)

In this paper we describe the anomaly detection engine for an instance of the Hybrid detection agent. The agent combines a clustering based anomaly detection engine (ADWICE) with a white-list engine. The white-list engine implements simple specification based intrusion detection where data known to be normal are described by manually constructed signatures. In our case hosts and services known to produce abnormal behaviour (e.g. DNS server port 53) are filtered away, but rules for arbitrary features can be used. Data considered normal by the white-list engine are not fed into ADWICE. This reduces the size of the normality model without decreasing detection coverage.

4   The Anomaly Detection Algorithm

This section describes how ADWICE handles training and detection. The basis of ADWICE, the BIRCH clustering algorithm, requires data to be numeric. Non-numeric data is therefore assumed to be transformed into numeric format by pre-processing.

4.1   Basic Concepts

The basic concepts are presented in the original BIRCH paper[6] and the relevant parts are summarized here.

Given n d-dimensional data vectors v_i in a cluster $CFj = \{v_i | i = 1 \ldots n\}$ the centroid v_0 and radius $R(CF_j)$ are defined as:

$$v_0 = \frac{\sum_{i=1}^{n} v_i}{n} \qquad R(CF_j) = \sqrt{\frac{\sum_{i=1}^{n} (v_i - v_0)^2}{n}} \qquad (1)$$

R is the average distance from member points in the cluster to the centroid and is a measure of the tightness of the cluster around the centroid.

A fundamental idea of BIRCH is to store only condensed information, denoted *cluster feature*, instead of all data points of a cluster. A cluster feature is a triple $CF = (n, S, SS)$ where n is the number of data points in the cluster, S is the linear sum of the n data points and SS is the square sum of all data points. Given the CF for one cluster, centroid v_0 and radius R may be computed. The distance between a data point v_i and a cluster CF_j is the Euclidian distance between v_i and the centroid, denoted $D(v_i, CF_j)$ while the distance between two clusters CF_i and CF_j is the Euclidian distance between their centroids, denoted $D(CF_i, CF_j)$. If two clusters $CF_i = (n_i, S_i, SS_i)$ and $CF_j = (n_j, S_j, SS_j)$ are merged, the CF of the resulting cluster may be computed as $(n_i + n_j, S_i + S_j, SS_i + SS_j)$. This also holds if one of the CF:s is only one data point making incremental update of CF:s possible.

A CF tree is a height balanced tree with three parameters, branching factor (B), threshold (T), and maximum number of clusters (M). A leaf node contains at most B entries, each of the form (CF_i) where $i \in \{1, \ldots, B\}$. Each CF_i of the leaf node must satisfy a threshold requirement (TR) with respect to the threshold value T. Two different threshold requirements have been evaluated with ADWICE. The first threshold requirement where $R(CF_i) \leq T$ corresponds to a threshold requirement suggested in the original paper and is therefore used as base line in this work (ADWICE–TRR). A large cluster may absorb a small group of data points located relatively far from the cluster centre. This small group of data points may be better represented by their own cluster since detection is based on distances. A second threshold requirement was therefore evaluated where $D(v_i, CF_i) \leq T$ was used as decision criteria (v_i is the new data point to be incorporated into the cluster). This version of the algorithm will be referred to as ADWICE–TRD.

Each non-leaf node contains at most B entries of the form (CF_i, child_i), where $i \in \{1, \ldots, B\}$ and child$_i$ is a pointer to the node's i-th child. Each CF at non-leaf level summarises all child CF:s in the level below.

4.2   Training

The CF tree is the normality model of the anomaly detection algorithm. During training, each data vector v is inserted into the CF-tree incrementally following the steps described below:

1. *Search for closest leaf:* Recursively descend from the root to the closest leaf, by in each step choosing child i such that $D(v, CF_i) < D(v, CF_j)$ for every other child j.

2. *Update the leaf:* Find closest CF_i by computing $D(v, CF_j)$ for all CF_j in the leaf. If CF_i may absorb v without violating the threshold requirement (TR) update CF_i to include v. If TR is violated, create a new CF_k entry in the leaf out of v. If the number of CF:s including CF_k is below B we are done. Otherwise the leaf needs to be split in two. During splitting, the two farthest CF:s of the leaf are selected as seeds and all other CF_k from the old leaf are distributed between the two new leafs. Each CF_k is merged with the leaf with the closest seed.

3. *Modify the path to the leaf:* After an insert, the tree needs to be updated. In the absence of split, the CF:s along the paths to the updated leaf need to be recomputed to include v by incrementally updating the CF:s. If a split occurred, we need to insert a new non-leaf entry in the parent node of the two new leafs and re-compute the CF summary for the new leafs. If there is free space in the parent node (i.e. the number of children is below B) the new non-leaf CF is inserted. Otherwise the parent is split in turn. Splitting may proceed all the way up to the root in which case the depth of the tree increases when a new root is inserted.

If the size of the tree increases so that the number of nodes is larger than M, the tree needs to be rebuilt. The threshold T is increased, all CF:s at leaf level are collected and inserted anew into the tree. Now it is not single data points that are inserted but rather CF:s. Since T has been increased, old clusters may be merged thereby reducing the size of the tree. If the increase of T is too small, a new rebuild of the tree may be needed to reduce the size below M again. A heuristic described in the original BIRCH paper may be used for increasing the threshold to minimize the number of rebuilds, but in this work we use a simple constant to increase T conservatively (to avoid influencing the result by the heuristic).

Of the three parameters T, B and M the threshold T is the simplest to set, as it may be initialised to zero. The branching factor B influences the training and detection time but may also influence detection accuracy. The original paper suggests using a branching factor of 15, but of course they do not consider anomaly detection accuracy since the original algorithm is not used for this purpose.

The M parameter needs to be decided using experiments. Since it is only an upper bound of the number of clusters produced by the algorithm it is easier to set than an exact number of clusters as required by other clustering algorithms. As M limits the size of the CF-tree it is an upper bound on the memory usage of

ADWICE. Note that in general M needs to be set much lower than the number of data represented by the normality model to avoid over-fitting (i.e. training a model which is very good when tested with the training data but fails to produce good results when tested with data that differs from the training data). M also needs to be set high enough so that the number of clusters is enough for representing normality.

4.3   Detection

When a normality model is trained, it may be used to detect anomalies in unknown data. When a new data point v arrives detection starts with a top down search from the root to find the closest cluster feature CF_i. This search is performed in the same way as during training. When the search is done, the distance $D(v, CF_i)$ from the centroid of the cluster to the new data point v is computed. Informally, if D is small, i.e. lower than a limit, v is similar to data included in the normality model and v should therefore be considered normal. If D is large, v is an anomaly.

Let the threshold T be the limit (L) used for detection. Using two parameters E_1 and E_2, $MaxL = E_1 * L$ and $MinL = E_2 * L$ may be computed. Then we compute the belief that v is anomalous using the formula below:

$$belief = \begin{cases} 0 & \text{if } D \leq MinL \\ 1 & \text{if } D \geq MaxL \\ \frac{D-MinL}{MaxL-MinL} & \text{if } MinL < D < MaxL \end{cases} \qquad (2)$$

A belief threshold (BT) is then used to make the final decision. If we consider v anomalous we raise an alarm. The belief threshold may be used by the administrator to change the sensitivity of the anomaly detection. For the rest of the paper to simplify the evaluation we set $E_1 = E_2 = E$ so that v is anomalous if and only if $D > MaxL$. Note that clusters are spherical but the area used for detection of multiple clusters may overlap, implying that the clusters may be used to represent also non-spherical regions of normality. Time complexity of testing as well as of training is in ordo $N \log C$ where N is the number of processed data and C is the number if clusters in the model.

4.4   Adaptation of the Normality Model

As described earlier, agents need to be adaptable in order to cope with varying LCCI conditions including changing normality. Here we describe two scenarios in which it is very useful to have an incremental algorithm in order to adapt to changing normality.

In some settings, it may be useful to let the normality model relearn autonomously. If normality drifts slowly, an incremental clustering algorithm may handle this in real-time during detection by incorporating every test data classified as normal with a certain confidence into the normality model. If slower drift of normality is required, a random subset of encountered data based on sampling could be incorporated into the normality model.

Even if autonomous relearning is not allowed in a specific network setting there is need for model adaptation. Imagine that the ADWICE normality model has been trained, and is producing good results for a specific network during detection. At this point in time the administrator recognizes that normality has changed and a new class of data needs to be included as normal. Otherwise this new normality class produces false positives. Due to the incremental property, the administrator can incorporate this new class without the need to relearn the existing working normality model. Note that there is no need for retraining the complete model or to take the model off-line. The administrator may interleave incremental training of data from the new normality class with detection.

5   Evaluation

In all following experiments ADWICE-TRD is used unless otherwise stated.

5.1   Data Set

Performing attacks in real networks to evaluate on-line anomaly detection is not realistic and our work therefore shares the weaknesses of evaluation in somewhat "unrealistic" settings with other published research work in the area. Our approach for dealing this somewhat synthetic situation is as follows. We use KD-DCUP99 data set[15] to test the real-time properties of the algorithm. Having a large number of attack types and a large number of features to consider can thus work as a proof of concept for the distinguishing attributes of the algorithm (unknown attacks, fast on-line, incremental model building). We then go on to evaluate the algorithm in a test network that has been specifically built with the aim of emulating a realistic telecom management network. While large number of future tests with different criteria are still possible on this test network, this initial set of tests illustrates the sort of problems that are detected by ADWICE and not covered by current commercial INFOSEC devices deployed on the emulated network.

Despite the shortcomings of the DARPA related datasets[15] (see also section 6) they have been used in at least twenty research papers and are unfortunately currently the only openly available data set commonly used data for comparison purposes. The original KDD training data set consists of almost five millions session records, where each session record consists of 41 fields (e.g. IP flags set, service, content based features, traffic statistics) summarizing a TCP session or UDP connection. Since ADWICE assumes all training data is normal, attack data are removed from the KDD training data set and only the resulting normal data (972 781 records) are used for training. All 41 fields of the normal data are considered by ADWICE to build the model.

The testing data set consists of 311 029 session records of which 60 593 is normal and the other 250 436 records belong to 37 different attack types ranging from IP sweeps to buffer overflow attacks. The use of the almost one million data records for training and more than 300 000 data for testing in the evaluation presented below illustrates the scalability of ADWICE.

Some features of KDD data are not numeric (e.g. service). Non-numeric features ranging over n values are made numeric by distributing the distinct values over the interval $[0, 1]$. However, two distinct values of the service feature (e.g. http, ftp) for example, should be considered equally close, regardless of where in the $[0, 1]$ interval they are placed. This intuition cannot be captured without extending the present ADWICE algorithm. Instead the non-numeric values with $n > 2$ distinct values are scaled with a weight w. In the KDD dataset $n_{protocol} = 3$, $n_{flag} = 11$ and $n_{service} = 70$. If $w/n > 1$ this forces the algorithm to place two sessions that differ in such non-numeric multi-valued attributes in different clusters. That is, assuming the threshold condition requiring distance to be less than 1 to insert data into a cluster ($M \gg$ distinct number of combinations of multi-valued non-numeric attributes). This should be enforced since numerical values are scaled to $[0, 1]$. Otherwise a large difference in numerical attributes will anyway cause data to end up in the same cluster, making the model too general. If multi-valued attributes are equal, naturally the difference in the numerical attributes decides wether two data items end up in the same cluster.

5.2   Determining Parameters

If the maximum number of clusters M is set too low, the normality model will be too general leading to lower detection rate and lower accuracy of the algorithm. This was confirmed in experiments where M was increased from 2 000 to 25 000 in steps of 1 000. When setting M above 10 000 clusters the accuracy reaches a stable level meaning that setting M at least in this range should be large enough to represent the one million normal data points in the training set. In the forthcoming experiment M is therefore set to 12 000.

Experiments where the branching factor was increased from 2 to 2 048 in small steps showed that the chance of finding the correct cluster increases with the branching factor (i.e. decreasing false positives rate). However, increasing the branching factor also increases the training and testing time. The extreme setting $B = M$ would flatten out the tree completely, making the algorithm linear as opposed to logarithmic in time. The experiments showed that the false positives rate stabilized when the branching factor is increased above 16. In the forthcoming experiments the branching factor is therefore set to 20. Experiments where the branching factor was changed from 20 to 10 improved testing time by roughly 16 % illustrating the time-quality trade-off.

5.3   Detection Rate Versus False Positives Rate

Figure 2 shows the trade-off between detection rate and false positive rate on an ROC diagram[16]. To highlight the result we also compare our algorithm ADWICE–TRD with ADWICE–TRR which is closer to the original BIRCH algorithm. The trade-off in this experiment is realized by changing the E-parameter from 5 (left-most part of the diagram) to 1 (right-most part of the diagram) increasing the detection space of the clusters, and therefore obtaining better detection rate while false positives rate also increases.

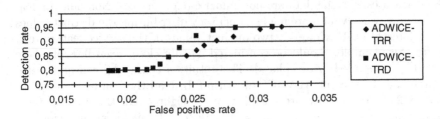

Fig. 2. Detection rate versus false positives when changing E from 5 to 1

The result confirms that ADWICE is useful for anomaly detection. With a false positives rate of 2.8 % the detection rate is 95 % when E = 2. While not conclusive evidence, due to short-comings of KDD data, this false positives rate is comparable to alternative approaches using unsupervised anomaly detection[10, 12]. On some subsets of KDD data Portnoy et al[10] produces 1–2 % false positives rate at 50–55 % detection rate, but unfortunately other subsets produces considerable inferior results. The significantly better detection rate of ADWICE is expected due to the fact that unsupervised anomaly detection is a harder problem then pure anomaly detection.

Since the KDDCUP data initially was created to compare classification schemes, many different classification schemes have been applied to the KDDCUP data set. Classification implies that the algorithms were trained using both normal and attack data contrasted to ADWICE which is only trained on the normal training data. The attack knowledge makes differentiating of attack and normal classes an easier problem, and it was expected that the results[9] of the winning entry (C5 decision trees) should be superior to ADWICE. This was also the case[1] regarding false positives (0,54 %), however detection rate was slightly lower, 91,8 %. Due to the importance of low false positives rate we indeed consider this result superior to that of ADWICE. We think the other advantages of ADWICE (section 2) make up for this. Also, we recall that ADWICE is one element in a larger scheme of other Safeguard agents for enhancing survivability.

The result shows that for values of E above 4.0 and values of E below 1.75 the false positives rate and detection rate respectively improve very slowly for ADWICE-TRD. The comparison with the base-line shows that using R in the threshold requirement (ADWICE–TRR) implies higher false positives rate. Section 5.5 describes further reduction of false positives in ADWICE–TRD by aggregation.

---

[1] In the original KDDCUP performance was measured using a confusion matrix where the result for each class is visible. Since ADWICE does not discern different attack classes, we could not compute our own matrix. Therefore overall false positives rates and detection rates of the classification scheme was computed out of the result for the individual classes.

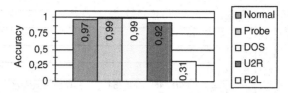

Fig. 3. The accuracy of ADWICE for individual attack classes and the normal class

5.4   Attack Class Results

The attacks in the test data can be divided into four categories:

- *Probe* - distinct attack types (e.g. IP sweep, vulnerability scanning) with 4 166 number of session records in total.
- *Denial of Service (DOS)* - 10 distinct attack types (e.g. mail bomb, UDP storm) with 229 853 number of session records in total.
- *User-to-root (U2R)* - 8 distinct attack types (e.g. buffer overflow attacks, root kits) with 228 number of session records in total.
- *Remote-to-local (R2L)* - 14 distinct attack types (e.g. password guessing, worm attack) with 16 189 number of session records in total.

Since the number of data in the probe and DOS classes is much larger than U2R and R2L, a detection strategy may produce very good overall detection quality without handling the U2R and R2L classes that well. Therefore it is interesting to study the attack classes separately. Note that since ADWICE is an anomaly detector and has no knowledge of attack types, it will give the same classification for every attack type unlike a classification scheme.

Figure 3 shows the four attack classes Probe, DOS, U2R and R2L as well as the normal class (leftmost column) for completeness.

The results for Probe, DOS and U2R are very good, with accuracy from 88 % (U2R) to 99 % (DOS). However, the fourth attack class R2L produces in comparison a very bad result with an accuracy of only 31 %. It should be noted that the U2R and R2L classes are in general less visible in data and a lower accuracy should therefore be expected. The best entries of the original KDD-cup competition had a low detection rate for U2R and R2L attacks, therefore also a low accuracy for those classes.

5.5   Aggregation for Decreasing Alarm Rate

While 2–3 percent false positives rate produced by ADWICE may appear to be a low false positive rate in other applications, in practice this is not acceptable for network security[16]. Most realistic network data is normal, and if a detection scheme with a small percent of false positives is applied to millions of data records a day, the number of false alarms will be overwhelming. In this section we show how the total number of alarms can be further reduced through aggregation.

An anomaly detector often produces many similar alarms. This is true for new normal data that is not yet part of the normality model as well as for attacks

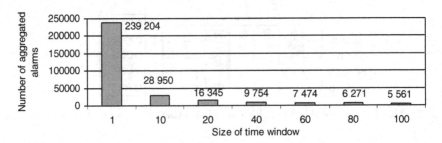

Fig. 4. The number of aggregated alarms for different time windows

types like DOS and network scans. Many similar alarms may be aggregated to one alarm, where the number of alarms is represented by a counter. In the Safeguard agent architecture aggregation is one important task of the Alert reduction correlation agent[4]. By aggregating similar alarms the information passed on to higher-level agents or human administrators becomes more compact and manageable. Here we evaluate how aggregation would affect the alarm rate produced from the KDD data set.

The KDD test data does not contain any notion of time. To illustrate the effect of aggregation we make the simplifying assumptions that one test data is presented to the anomaly detector each time unit. All alarms in which service, flag and protocol features are equal are aggregated during a time window of size 0 to 100. Of course aggregation of a subset of features also implies information loss. However, an aggregated alarm, referring to a certain service at a certain time facilitates the decision for narrowing down to individual alarms for further details (IP-address should have been included if present among KDD features). The result is shown in Fig. 4.

When a new alarm arrives, it is sent at once, to avoid increasing time to detection. When new alarms with the same signature arrive within the same time window, the first alarm is updated with a counter to represent the number of aggregated alarms. Without aggregation ADWICE produces 239 104 alarms during the 311 029 time units. Using a short time window of 10 time units, the number of aggregated alarms becomes 28 950. Increasing the time window to 100 will reduce the original number of alarms to 5 561, an impressive reduction of 97,7 %. The explanation is that many attacks (probes, DOS) lead to a large amount of similar alarms. Note that aggregation also reduces false positives, since normal sessions belonging to a certain subclass of normality may be very similar. While it might seem that aggregation only makes the alarms less visible (does not remove them) it is in fact a pragmatic solution that was appreciated by our industrial partners, since it significantly reduces the time/effort at higher (human-intensive) levels of investigation. The simple time slot based aggregation provides a flexible system in which time slots can be adaptively chosen as flexible 'knobs' in response to different requirements.

5.6   Usefulness of Incremental Training

To evaluate the incremental training of ADWICE we treat an arbitrary abnormal class as normal and pretend that the normality model for the KDD data should be updated with this class. Without loss of generality we choose the IP sweep attack type and call it 'normal-new'; thus, considering it a new normal class detected by the administrator. The model only trained on the original normal data will detect the normal-new class as attack, since it is not included in the model. This produces 'false positives'. The old incomplete normality model is then incrementally trained with the normal-new training data producing a new normality model that incorporates the normal-new class. Our evaluation showed that (without aggregation) the old model produced 300 false positives, whereas the new retrained model only three.

5.7   Evaluation in the Safeguard Test Network

One of the main efforts of the Safeguard project is the construction of the Safeguard telecom management test network, used for data generation for off-line use as well as full-scale on-line tests with the Safeguard agent architecture. At the time of evaluation of this work the network consisted of 50 machines (at present time about 100 machines) in multiple sub networks. Normal data can be generated by isolating the network from Internet and not running any internally generated attacks on the network.

Evaluation using data from the Safeguard test network is ongoing work. Here we present only some initial results from tests performed over a total time period of 36 hours. The ADWICE model was trained using data from a period known to contain only normal data. To keep parsing and feature computation time low to make real-time detection possible, features were only based on IP-packet headers, not on packet content (e.g. source and destination IP and ports, time, session length). This means of course that we at this stage can not detect events that are only visible by analyzing packet content. The purpose of this instance of the hybrid detection agent is to detect anomalies, outputting alarms that can be analysed by high level agents to identify time and place of attacks as well as failures or misconfigurations.

In Scenario 1 an attacker with physical access to the test network plugged in a new computer at time 15:33 and uploaded new scripts. In Scenario 2 those scripts are activated a few minutes later by the malicious user. The scripts are in this case harmless. They use HTTP on port 80 to browse Internet, but could just as well have been used for a distributed attack (e.g. Denial of Service) on an arbitrary port. The scripts are then active until midnight the first day, producing traffic considered anomalous for their respective hosts. During the night they stay passive. The following morning the scripts becomes active and execute until the test ends at 12:00 the second day.

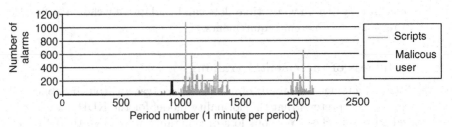

Fig. 5. Anomalous behaviour detected by ADWICE in the telecom management network. The figure shows the number of alarms for a certain time period and IP/port

Figure 5 illustrates the usefulness of the output of the hybrid detection agent. The 36 hours of testing was divided in periods of one minute and the number of alarms for each time period is counted.

For Scenario 1, all alarms relating to the new host (IP x.x.202.234) is shown. For Scenario 2 all alarms with source or destination port 80 are shown. The figure shows clearly how the malicious user connects at interval nr 902 (corresponding to 15:34), when the scripts executes, waits during the night, and then executes again. Some false alarms can also be noted, by the port 80 alarms occurring before the connection by the malicious user. This is possible since HTTP traffic was already present in the network before the malicious user connected.

6   Discussion and Future Work

The DARPA related data sets have been widely used but also criticized[15]. The normal traffic regularity as well as distribution of attacks compared to distribution of normality does not exactly correspond to network data in a real network. Generation of a new public reference data set for IDS evaluation without the identified weaknesses of the DARPA data remains therefore as an important task of the research community. With this in mind, our DARPA/KDD based evaluation still shows feasibility of ADWICE given the assumptions that relevant features are used and that those features separate normal data from attacks. Local clustering was, according to our knowledge, used for the first time in the intrusion detection setting, whereby an optimal global clustering of data is not necessary. The incremental property of ADWICE is important to provide flexible adaptation of the model. Future work includes further evaluation of the algorithm in the context of the Safeguard test network. If made available also other public data sets will be considered. Unfortunately the GCP data provided by DARPA's Cyber Panel program[3] is not currently released to researchers outside USA.

Current work includes using the incremental feature of ADWICE for autonomous normality adapation. Evaluation of such adapation, may require long periods of data to study the effect of adaptation over time.

Our experience with ADWICE indicates, as hinted in the original BIRCH paper, that the index is not perfect. Our on-going work includes full evaluation of an alternative grid-based index with initial indications of improvement of the false positive rate by 0,5–1 % at a similar detection rate.

References

1. Yegneswaran, V., Barford, P., Ullrich, J.: Internet intrusions: global characteristics and prevalence. In: Proceedings of the 2003 ACM SIGMETRICS international conference on Measurement and modeling of computer systems, San Diego, CA, USA, ACM Press (2003) 138–147
2. McHugh, J.: Intrusion and intrusion detection. International Journal of Information Security 1 (2001) 14–35
3. Haines, J., Kewley Ryder, D., Tinnel, L., Taylor, S.: Validation of sensor alert correlators. IEEE Security and Privacy 1 (2003) 46–56
4. Chyssler, T., Nadjm-Tehrani, S., Burschka, S., Burbeck, K.: Alarm reduction and correlation in defence of ip networks. In: Proceedings of International Workshops on Enabling Technologies: Infrastructures for Collaborative Enterprises (WETICE04), Modena, Italy, IEEE Computer Society (2004) 229–234
5. Safeguard: The safeguard project (2003) http://www.ist-safeguard.org/ Acc. May 2004.
6. Zhang, T., Ramakrishnan, R., Livny, M.: Birch: an efficient data clustering method for very large databases. SIGMOD Record 1996 ACM SIGMOD International Conference on Management of Data 25 (1996) 103–14
7. Fu, Y., Sandhu, K., Shih, M.Y.: A generalization-based approach to clustering of web usage sessions. In: Proceedings of the Web Usage Analysis and User Profiling. International WEBKDD'99 Workshop. Volume 1836 of Lecture Notes in Artificial Intelligence., San Diego, CA, USA, Springer-Verlag (2000) 21–38
8. Mukkamala, S., Janoski, G., Sung, A.: Intrusion detection using neural networks and support vector machines. In: Proceedings of the 2002 International Joint Conference on Neural Networks (IJCNN '02), Honolulu, HI, Institute of Electrical and Electronics Engineers Inc. (2002) 1702–1707
9. Elkan, C.: Results of the kdd'99 classifier learning. ACM SIGKDD Explorations 1 (2000) 63 – 64
10. Portnoy, L., Eskin, E., Stolfo, S.: Intrusion detection with unlabeled data using clustering. In: ACM Workshop on Data Mining Applied to Security. (2001)
11. Sequeira, K., Zaki, M.: Admit: Anomaly-based data mining for intrusions. In: Proceedings of the 8th ACM SIGKDD international conference on Knowledge discovery and data mining, Edmonton, Alberta, Canada, ACM Press (2002) 386–395
12. Guan, Y., Ghorbani, A.A., Belacel, N.: Y-means: A clustering method for intrusion detection. In: Canadian Conference on AI. Volume 2671 of Lecture Notes in Computer Science., Montreal, Canada, Springer (2003) 616–617
13. Munson, J., Wimer, S.: Watcher: the missing piece of the security puzzle. In: Proceedings of the 17th Annual Computer Security Applications Conference, New Orleans, LA, USA, IEEE Comput. Soc (2001) 230–9
14. Han, J., Kamber, M.: Data Mining - Concepts and Techniques. Morgan Kaufmann Publishers Inc., San Francisco, CA, USA (2001)

15. Mahoney, M.V., Chan, P.K.: An analysis of the 1999 darpa/lincoln laboratory evaluation data for network anomaly detection. In: Recent Advances in Intrusion Detection. Volume 2820 of Lecture Notes in Computer Science., Pittsburgh, PA, USA, Springer (2003) 220–237
16. Axelsson, S.: The base-rate fallacy and the difficulty of intrusion detection. ACM Transactions on Information and Systems Security **3** (2000) 186–205

Steganography for Executables and Code Transformation Signatures

Bertrand Anckaert, Bjorn De Sutter, Dominique Chanet,
and Koen De Bosschere

Ghent University, Electronics and Information Systems Department,
Sint-Pietersnieuwstraat 41 9000 Ghent, Belgium
{banckaer, brdsutte, dchanet, kdb}@elis.UGent.be
http://www.elis.UGent.be/paris

Abstract. Steganography embeds a secret message in an innocuous cover-object. This paper identifies three cover-specific redundancies of executable programs and presents steganographic techniques to exploit these redundancies. A general framework to evaluate the stealth of the proposed techniques is introduced and applied on an implementation for the IA-32 architecture. This evaluation proves that, whereas existing tools such as Hydan [1] are insecure, significant encoding rates can in fact be achieved at a high security level.

Keywords: code transformation signature, steganography, executables.

1   Introduction

Steganography embeds a secret message in a seemingly innocuous cover-object. Digital cover-objects most often are media, such as image and music files, that involve noise and are perceived by imperfect human senses. As a result, they contain many redundant bits, which can be modified to embed secret messages.

This paper explores the largely unexplored field of steganography for executable programs. This differs significantly from steganography for media because changing as little as a single bit of a program can cause it to fail entirely. Hence different techniques are required for embedding messages in executables.

With the exception of Hydan [1], little information on this subject is publicly available. While the related subjects of software watermarking and fingerprinting, which also involve information hiding, have received considerably more attention [2, 3], the results of that research are not applicable in the context of steganography. This follows from the fact that watermarking and fingerprinting typically deal with very short embedded messages (shorter than 1 Kb), and that those messages first of all need to be irremovable, rather than hidden. Moreover, some watermarking approaches also require knowledge of the embedded message in the detection phase, which is obviously not possible in steganography.

Rather than implementing ad-hoc techniques, as in Hydan [1], we present a thorough study of the available redundancy in compiled programs. Furthermore,

C. Park and S. Chee (Eds.): ICISC 2004, LNCS 3506, pp. 425–439, 2005.

we present a general framework for evaluating the stealthiness of the different program transformations that exploit the redundancies. Based on this framework, a number of countermeasures to prevent possible attacks are presented.

This paper is structured as follows: Section 2 presents the used model. The fitness of executables for steganography is explored in Section 3. A framework for the evaluation of statistical signatures of code transformations is discussed in Section 4. The concepts are then evaluated for the IA-32 architecture in Section 5. Related work is the topic of Section 7 and conclusions are drawn in Section 8.

2   The Prisoners' Problem

We will follow Simmons' [4] classic model, a.k.a. *the prisoners' problem* for invisible communication. Alice and Bob are two prisoners in different cells. Wendy, the warden, arbitrates all communication between them, and will not let them communicate through encryption or suspicious communication. Both prisoners therefore need to communicate invisibly about their escape plan.

Furthermore, we will assume that the mechanism in use is known to the warden (Kerkhoffs' principle [5]). Hence its security must depend solely on a secret key that Alice and Bob managed to share, possibly before their imprisonment.

The general principle of steganography is as follows. To share a secret message with Bob, Alice randomly chooses a harmless message, called a cover-object c, which can be transmitted to Bob without raising suspicion. The secret message m is then embedded in the cover-object using the secret key k, resulting in a stego-object s. This is to be done in such a way that Wendy, knowing only the apparently harmless message s, cannot detect the presence of the secret. Alice then transmits s to Bob via Wendy. Bob can reconstruct m since he knows the embedding method and has access to the key k. It should not be necessary for Bob to know the original cover c. The security of invisible communication lies mainly in the inability to distinguish cover-objects from stego-objects. The task of Wendy can be formalized as a statistical hypothesis-testing problem, for which she defines a test function on objects (of the set O) $f : O \rightarrow \{0, 1\}$:

$$f(o) = \begin{cases} 1 & if\ o\ contains\ a\ secret\ message \\ 0 & otherwise \end{cases}$$

This function can make two types of errors: detect a hidden message when there is none (false positive) and not detect the existence of a hidden message when there is one (false negative). In this paper we will further assume that the warden is passive, i.e. she will not modify the object, but only classify it. This is generally accepted in steganography [6].

3   Fitness of Executables as Cover-Objects

While changing a single bit in a program can cause it to fail, this does not imply a lack of redundancy for the purpose of steganography. Instead the specific characteristics of software indeed result in many forms of redundancy.

In theory, we can consider two programs extensionally equivalent if they produce identical output, given identical input. In practice, more stringent requirements for time, space and power consumption need to be taken into account. But even then a large number of equivalent executables exists. This has been exploited for several purposes including program optimization, program obfuscation, software watermarking and fingerprinting, and software diversity. It is thus generally accepted that the number of equivalent executables for any real-life application is large and that there is indeed a lot of redundancy in a program which, in this context, we would like to exploit to encode a secret message.

Besides being equivalent to the original program, any program with an embedded message also needs to pass the warden's test function described in the previous section. Since we believe useless (suboptimal) code added to a program will be easily detected, we will only allow embedding transformations that do not deoptimize a program. In other words, the message should be embedded in the code a (optimizing) compiler back-end has produced from its intermediate code representation of the program. Typically a compiler's back-end goes through 4 phases (in varying orders), each of which inserts a number of redundancies.

During *instruction selection*, the intermediate code operations are translated into assembly instructions. Often multiple instruction sequences can be chosen to implement an intermediate code operation. During the *register allocation*, architectural registers are chosen to store values temporarily. Usually there are multiple valid allocations. In the *instruction scheduling* phase, the selected instructions are put in their final order. Again, multiple orderings are often valid. Finally, multiple compiled files are combined into a program. During this *code layout*, multiple orderings can be chosen. These types of choices/redundancies and their exploitation are the topic of this section. As the target architecture, we have chosen the IA-32 architecture [7], because it is most commonly used.

3.1   Encoding Bits in a Choice

For each of the choices between equivalents, a number of bits can be encoded in the program. If there are n equivalent programs because of some type of choice, the number of bits that can be encoded can be computed as follows.

As $n \geq 2^{\lfloor \log_2(n) \rfloor}$, it is clear that at least $\lfloor \log_2(n) \rfloor$ bits can be encoded: it suffices to number each equivalent, and to take that equivalent who's (binary) number corresponds to the bit-string to be encoded. This simple approach may result in a significant decrease in encoding capabilities however: if $log_2(n) \notin \mathbb{N}$ for large n, many equivalents may not correspond to an encodable bit-string.

A more efficient scheme is as follows: If $\log_2(n) \notin \mathbb{N}$, then $\lfloor \log_2(n) \rfloor = \lceil \log_2(n) - 1 \rceil$. We can thus always embed $\lceil \log_2(n) - 1 \rceil$ bits. If we associate each of the remaining $n - 2^{\lceil \log_2(n) - 1 \rceil}$ equivalents with one of the $2^{\lceil \log_2(n) - 1 \rceil}$ already used ones, we can embed an additional bit by allowing the embedder to choose between one of the two associated equivalents, as illustrated for $n = 7$ in Figure 1. Therefore, we can embed an extra bit in $n - 2^{\lceil \log_2(n) - 1 \rceil}$ of the $2^{\lceil \log_2(n) - 1 \rceil}$ possibilities for the next $\lceil \log_2(n) - 1 \rceil$ bits.

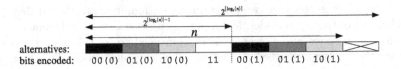

Fig. 1. Encoding bits in the choice of 7 equivalents

If the embedded message is encrypted with the secret key k, all bit-strings to be embedded have equal probability, and hence the average number of bits that can be encoded in the choice out of n valid equivalents is given by

$$b(n) = \lceil \log_2(n) - 1 \rceil + \frac{n - 2^{\lceil \log_2(n)-1 \rceil}}{2^{\lceil \log_2(n)-1 \rceil}}. \tag{1}$$

One can easily verify that equation (1) also holds if $\log_2(n) \in \mathbb{N}$.

3.2   Instruction Selection

To explore the steganographic potential of executables, we have developed a tool that is capable of exhaustively generating all possible instruction sequences for the IA-32 architecture. This tool operates in a similar manner as the so-called superoptimizer [8].

Its input consists of a code sequence, a set of output registers and a set of (scratch) registers whose value is no longer used after the sequence has been executed in a program. For all generated sequences, the tool checks whether they perform the same function as the original code sequence, by testing the output values for all possible input values. If the test succeeds an equivalent sequence is found.

Because of the halting problem, it is in general undecidable if a generated sequence will terminate. Hence the equivalence test can run forever. By restricting the set of instructions to the integer instructions, that do not include any control flow, we can assure that each tested sequence terminates. But even then the number of potential equivalent sequences is still too large. To make the problem tractable, and to terminate the exhaustive generations within reasonable time, we further limit the immediate operands (constants encoded in an instruction) that can be used to $\{-1, 0, 1, 31\}$. Finally, we restrict the length of the generated sequences.

Even with these restrictions we can still find many equivalent sequences that perform realistic computations. For the operation ECX= max(EAX,EDX), e.g., our tool was able to find 433 different encodings of three instructions. Similarly, for the computation EAX= (EAX/2), 3708 equivalent sequences of 4 instructions were generated. Note that the tool did not find shorter sequences because of the limited list of immediates that does not contain 2.

It should be noted that these examples are no exception. Moreover, the number of equivalents is exponential in the number of instructions: if we have n instructions which we can divide in groups of i instructions, of which each group has at least a equivalents then combined we have at least $a^{\frac{n}{4}}$ equivalents. Furthermore, many additional equivalents arise when considering the larger piece as a whole, in which instructions can be moved from one group to another.

While our tool thus shows great potential for encoding bits, it is too slow for a practical tool. Hence we had the tool generate a database of equivalence classes for the instructions that occur most often in our suite of training programs. During this process, we imposed the additional restriction that equivalent instructions can only read/write locations that are read/written in the original instruction. However, if liveness analysis [9] determines that certain status flags are dead, we allow them to be overwritten. Finally, the set of immediates is expanded with the immediates used in the original instruction and the negate thereof.

3.3   Register Allocation

On the IA-32, the number of registers is very limited, and most registers have fixed designations. Moreover, the calling conventions specify precisely how registers should be used. Hence the little choice that a compiler in theory has to choose a register allocation, is in practice unexploitable: any deviation from the calling conventions would be spotted by the ward. As a result, changing the allocated registers is not an option to embed secret messages in IA-32 programs.

3.4   Instruction Scheduling

Typically, instruction scheduling is performed per basic block. As two or more instructions that perform independent operations can be permuted within a basic block, we can encode bits in the instruction order within basic blocks.

To do so, we first determine all valid orderings by constructing a dependency graph of a block's instructions, in which dependent instructions are connected by directed edges. By iteratively removing instructions from this graph that do not depend on other instructions in it, a valid schedule can be determined. At each iteration, multiple instructions may be ready to be removed from the graph. They are, in other words, in the *ready-set* [9] of instructions. Using a branch and bound algorithm to select instructions from the ready-set, we can easily generate all the possible permutations. Supposing there are n possible schedules, the number of bits that can be encoded on average is given by equation (1).

Since finding valid instruction orderings using a dependency graph is time-consuming, and since the marginal gain of additional orderings decreases steadily when the number of orderings increases, it is useful to put an upper bound on the number of valid permutations that are considered. In our implementation this upper bound is 1024 orderings. As basic blocks are usually not longer than 4-5 instructions, this upper limit rarely is reached. Hence it has little influence on the amount of bits that can be encoded. For the rare, long basic blocks that offer billions of valid orders, setting an upper limit is absolutely necessary for obtaining practical execution times.

3.5   Code Layout

If there exists no fall-through path between two consecutive basic blocks, these blocks can be moved apart. Hence the order of the basic blocks in a program, i.e. the code layout, is to some degree free. More precisely, all basic block chains, i.e. lists of consecutive basic blocks with fall-through paths between them, can be positioned in any order. When there are c different chains, we have $c!$ possible orderings to choose from, and hence we can encode $b(c!)$ bits.

While the order of unique elements to encode a bit-string can be exploited with existing methods [10], all chains in a program are not necessarily unique. This follows from the fact that most compilers only compile one source code module at a time, and hence never have an overview over all the code that constitutes a final program. As a result, duplicated code ends up in programs [11].

This problem is aggravated for our purpose, since we need to number and qualify all chains independently of their position in the program. Hence we cannot base our differentiation between two chains on any contents of them that depends on their location. *In concreto*, this means that all relocatable addresses [12] encoded in the instructions in the chains need to be neglected when comparing chains. For the programs in our benchmark suite the thus computed number of sets of identical chains is only between 47 to 59% of the total number of chains.

With m chains divided in n sets of identical chains $s_1 \ldots s_n$, the theoretical average number of bits that can be encoded in their ordering is given by

$$b\left(\frac{m!}{\prod_{i=1}^{n}(|s_i|!)}\right). \tag{2}$$

We can approximate this number by iteratively selecting a chain for placement out of the n remaining sets of chains. The average number of bits that can be encoded in this selection is once again given by equation (1). Depending on whether the selected chain was the last of a set of identical chains or not, the number of sets will be $n - 1$, respectively n in the next iteration. The process is repeated until all chains have been placed.

3.6   Interactions Between the Techniques

The discussed techniques are not completely orthogonal. In order to combine them successfully, a couple of issues need to be addressed.

First, it is worth noting that the number of bits that can be encoded in instruction selection is dependent on the chosen ordering of instructions in the basic block, and vice versa. When orders change, liveness ranges change, and hence the condition flags and scratch registers that may be changed by equivalent instructions also change.

For the same reason, instruction selection influences the order in which an embedder or extractor will generate equivalent orderings, and hence how specific bit sequences are encoded in the ordering. Vice versa, if scheduling is applied first, it influences the order in which equivalent instructions are generated.

Moreover, if the embedder first encodes bits in the instruction selection of the instructions in their original order in the program, and subsequently reorders

```
55          push EBP         |   55          push EBP
89 e5       mov  ESP,EBP      |   89 e5       mov  ESP,EBP
83 ec 08    sub  0x8,ESP      |   83 c4 f8    add  0xfffffff8,ESP
```

Fig. 2. Two equivalent code sequences

the instructions, the extractor does not know the order in which the information embedded in the instruction selection needs to be extracted. Clearly, the extractor and the embedder need to depart from the same dependency graph in order for the extractor to obtain the correct embedded information.

Before the embedding and the extraction, all basic blocks in a program should therefore be transformed into a canonical form, in which both the instruction selection and their ordering are predetermined.

3.7 Practical Considerations for Extracting an Embedded Message

In order to extract embedded information from a program, an extractor needs to identify the basic blocks, and he needs to pinpoint relocated operands, since these should be neglected for the ordering of chains.

The necessary relocation information is available at the embedding phase, as the embedding is done at link-time, when the whole program is first available. This information is lost in the resulting executable however.

Fortunately most of the necessary information can be derived from a static analysis of the executable program itself. As a consequence, we only need to communicate the discrepancy between the derived information and the actual information to the decoder. To do so, we can store this information in the first instructions of the resulting binary, without taking liveness information into account. This is the only option since the decoder cannot identify basic blocks or chains and it cannot compute liveness information at this point.

4 Code Transformation Signatures

While Section 5 shows that the encoding rate achieved by the discussed techniques is fairly high, its security is obviously too low. The reason is that the techniques introduce very unusual code that will arise suspicion of the warden. Consider, e.g., the equivalent code sequences in Figure 2. Anyone somewhat familiar with assembly code will agree that the likelihood of a compiler generating the code on the right is extremely low. But this code is present in executables that have been put through Hydan or our tool (without countermeasures). In short, the application of our tool has left an obvious signature.

We define a *code transformation signature* (CTS) as a code property that results from that transformation. The security of the discussed embedding techniques depends by and large on the absence of such signatures. While this is obvious for steganography, it is also of importance for other embedding techniques such as watermarking and fingerprinting, as the distortion of a watermark or fingerprint is facilitated if an attacker can accurately locate it.

Despite the importance of the stealthiness of applied code transformations, almost all research efforts have targeted the development of new techniques. Little work has been done on the security evaluation of the techniques. So far, most claims for security have been ad hoc and often based on author's belief.

4.1   A Framework for Detecting Code Transformation Signatures

Because quantitative methods have proved so powerful in many other domains, we will first quantify unusual properties using quantitative software metrics. On these metrics, we build models of the expected behavior, after which we can compare the observed value of a metric to the expected behavior, and thus classify software into clean and suspect software.

Software Metric. A software metric summarizes and quantifies properties of a given piece of software, called a *unit*, in order to detect signatures. Hence the property to measure depends on the applied code transformations. Metrics can in general be classified along two axes: that of *aspects* and that of *granularities*.

The aspect identifies what type of software unit is inspected. This could, e.g., be the static code or the dynamically executed code. It could be the heap or the stack as well, as they result from the executed code. We should note that even a dynamic data watermark [2] may introduce a signature in the static code.

The granularity of a metric identifies the size of the unit that is the subject of measurement. Possible granularities are the instruction, the basic block, the procedure, the memory location, the graph structure, etc. Granularity is important for an attacker, because, the smaller the granularity, the more accurately the attacker can pinpoint the location of the suspicious software.

Statistical Code Model. In order to evaluate executables for the presence of suspicious units with respect to some metric, we need a model of what constitutes a "clean" unit. We will do so by means of statistical distributions that are constructed by evaluating a population of units for some metric. On such a distribution, a statistical test can then be postulated that decides on the behavior of a unit under investigation.

For each model, the population's *locality* identifies how closely related the units that make up the population are to the unit under investigation. If the granularity of the metric is, e.g., a basic block, then we could test each block by comparing it to the blocks in its own procedure or we could compare it to all the blocks in a training set of programs. In the former case, the locality of the model would be that of procedures, in the latter that of the software universe.

Based upon the postulated model of the clean behavior of a metric, we can then compute how unusual it is to observe a particular value for a metric. If we then define a threshold to differentiate suspect units from clean ones, we obtain a statistical test. In some cases, a single CTS will suffice to classify units, but in other cases several CTSs will need to be combined to increase the reliability.

Stealthy Code Transformations. Knowing that a warden uses such statistical models to detect CTSs of suspicious code, we need to defend against them.

This can either be done by elevating the false negative rate of a test, i.e. thwarting the recognition of the CTSs, or by elevating the false positive rate, i.e. transforming original code to contain the same CTSs.

Consider, e.g., a tamper-proofing mechanism that reads a piece of the program code, computes a checksum over it and compares it to some predefined value. Since programs rarely read their own code, an attacker trying to locate the detection mechanism may search for the CTS consisting of a (static) instruction that reads the code section. A countermeasure against this attack consists of hiding the fact that the instruction reads from the code section by obfuscating the involved address computation (increasing the false negative rate) or by transforming the original code to read constant values from the code section throughout (increasing the false positive rate).

4.2   Application of the Framework to the Steganographic Techniques

Presence of Unusual Instructions. When multiple instructions are available to perform a given operation, a compiler will deterministically select the (assumed) optimal equivalent during the instruction selection phase. A CTS can therefore be found in an unusual instruction in the static code.

To detect such CTSs, the aspect of the metric is static code, and the granularity is that of instructions. The locality of the model is that of a large, predefined set of clean programs. The test considers a program suspicious if it contains an instruction that never occurred in the clean programs.

As a countermeasure against this test, an embedder should only use alternative, possibly suboptimal, instructions when they could be the remnant of a choice based on information that is no longer available to the warden. This information could, e.g., be the source code or relocation information.

Unusual Relative Frequencies of Instructions. Replacing instructions by equivalent ones changes the frequencies with which the equivalents occur in a program. Therefore a suitable metric is the relative frequency of equivalents in the static code of the program. When an encrypted secret message is embedded in an executable using equivalent instructions (and without taking countermeasures), the equivalents for a certain instruction can be expected to have uniform frequencies. In clean programs, by contrast, they likely have skew distributions.

Suppose that there are n equivalents for some instruction, whose relative frequencies in a programs x are presented by $\mathbf{f}(x) = [f_1(x) \ ... \ f_n(x)]$. To measure how clean a particular program p is, the metric then consists of the euclidean distance from $\mathbf{f}(x)$ to $\mathbf{f}(P)$, in which P is large set of clean programs. The model of programs can be obtained by postulating that this distance follows a normal distribution, of which the parameters μ and σ^2 can be estimated by computing the average and variance of the distance in P.

To classify an executable e we compute the probability that a distance occurs that is larger than the observed distance between $\mathbf{f}(e)$ and $\mathbf{f}(P)$. A low likelihood indicates a high probability that a secret message is present. The reliability of this test can be increased by averaging the likelihoods of a number of sets of equivalent instructions. The classification could then classify programs with a score below, e.g., 0.5 as suspect.

To avoid the detection of the presence of an embedded message through the discussed CTS, we need to approximate the relative frequency of equivalents through a biased embedding. To do so the equivalents for a particular instruction are partitioned in two groups with relative frequency r_1 and $r_2 = 1 - r_1$ as close to 0.5 as possible. Without loss of generality we can assume that $r_1 \geq 0.5$. Using a random number generator seeded with a value derived from the secret key k (to assure that both the embedder and extractor follow the same scheme), numbers are then generated, on which a test is defined that succeeds with chance $p = (r_1 - 0.5) * 2$. Whenever the test succeeds for a generated number, we are forced to take an equivalent from group 1, and no bit is encoded. Otherwise, we encode a bit by choosing between group 1 and group 2. As a result of Bernoulli's theorem, the relative frequencies r_1 and r_2 will converge to the chance $p(g_1)$, that group 1 is chosen, resp. $p(g_2)$, that group 2 to is chosen, with increasing n. As it is equally likely that the next bit will be a 0 as a 1, $p(g_1) = (r_1 - 0.5) * 2 + \frac{1}{2} * (1 - (r_1 - 0.5) * 2) = r_1$ and $p(g_2) = \frac{1}{2} \times (1 - (r_1 - 0.5) * 2) = r_2$. Note that if a group is not a singleton, we iteratively partition it in subgroups on which the the same process is applied.

Unusual Jump Behavior of Program. Clean programs exhibit spatial locality as the basic block chains of procedures will usually be placed consecutively. This locality is further increased when cache-optimizing code layout techniques are applied [13]. When the code in a program is reordered randomly, the spatial locality of the code is hence likely to decrease. A poor locality may therefore indicate the presence of a secret message, and a suitable metric to test this consists of the average offset of executed jumps that cross chain boundaries.

In the execution of our population of training programs, the largest average jump offset during a program's execution was 3, 725. Respecting a considerable margin we therefore postulate the model that this average will be no higher than 10, 000 for clean programs. For our evaluation benchmark programs the largest average observed before embedding was 2, 575, while the smallest average observed after embedding (without countermeasures) was 59, 815. The classification thus has a perfect false positive and false negative rate for our benchmarks.

To avoid the presence of this CTS, we must retain the spatial locality present in the original code. To achieve this, it suffices to not reorder all of the different chains at once, but to take a three step approach. First, chains are reordered within their procedures. Secondly, all procedures are partitioned into equivalence classes, where two procedures are considered equivalent if they are (transitively) connected through inter-procedural jumps (that most often result from tail-call optimization by compilers), and procedures are reordered within their class. Finally, all classes are reordered.

Diversity in Scheduling. When instruction schedulers pick instructions from the ready-set (see Section 3.4), they try to optimize the generated schedules by taking instruction latencies, available processor resources, and other properties into account. As a result, any badly scheduled basic block will raise suspicion. Moreover, schedulers typically are deterministic, and hence two identical dependency graphs of instructions will most often be scheduled identically within a program. An important exception is where libraries and the main application are compiled independently of each other. Furthermore, there may be considerable differences between schedules in different programs. Different compilers may have been used, or different optimization levels, or different processor targets. Obviously, the warden does not know which combination has been used.

One attack on this type of CTS consists of comparing the basic blocks in a program, and to consider identical dependency graphs that are scheduled differently suspicious. For this type of model, the metric's aspect would be static code, and the granularity the basic block, while the locality is the equivalence class of the block's procedure within the program.

Since the warden does not know which properties the original scheduler has taken into account, he cannot know which graphs were identical according to the original scheduler. So while the warden may consider two graphs identical, the compiler may have seen them as different. Hence the warden needs to build a statistical model of clean code with a large set of programs.

The approach resembles the one to detect suspect instruction frequencies. Now $\mathbf{f}(x)$ indicates how often a compiler has generated i different schedules for (assumingly) identical graphs. Again, we can compute the euclidean distance and obtain a model by postulating that this distance follows a normal distribution. In practice, we observed that 95% of the graphs occurring more than once occurred with a single schedule, while approximately 5% occurred with two different schedules. More schedules are rare.

As a countermeasure to these attacks, we suggest the following approach: instead of choosing any instruction from the ready-set, limit this choice to the set of, within reasonable boundaries, good instructions to schedule. Furthermore, identical dependency graphs should result in i schedules with chance f_i. To implement this, it suffices to maintain a database of already scheduled blocks and enforce i different schedules with chance f_i.

Please note that making compilers non-deterministic, to increase the false positive rate, is not an option: besides the simple fact that one cannot control all compilers, making them non-deterministic would make debugging the compilers themselves and the compiled programs even more difficult than it is today.

5 Experimental Evaluation

To evaluate the presented concepts we have implemented Stilo, our steganographic tool for the IA-32 architecture, using the link-time rewriting framework Diablo [14], and applied it on 9 SPECint2000 benchmark programs to embed and extract "King Lear" by W. Shakespeare. The programs were compiled with GCC 3.2.2 and linked to glibc 2.3.2 for Linux. For each benchmark, the embedding and extraction took less than a minute on a 2.8GHz Pentium IV.

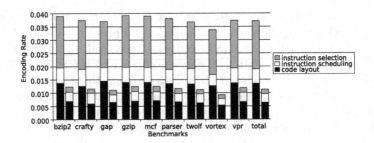

Fig. 3. Encoding rate before (left) and after (right) countermeasures for steganalysis

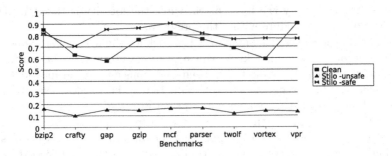

Fig. 4. Code transformation signature: unusual relative frequencies of instructions

The obtained encoding rates are presented in Figure 3. The distribution over the different techniques is also indicated. We achieve an encoding rate between 1/29.6 and 1/25.49 and a total encoding rate of 1/26.96 before countermeasures, four times the encoding rate of the previous prototype tool Hydan (1/110).

Without countermeasures, neither our tool nor Hydan, as noted by its authors, is stealthy. Most importantly, their encoding rate achieved through instruction selection is achieved by and large by choosing very unusual instructions. However, as the clean instruction distribution is uniform across executables [1], little information can be stealthily embedded this way. When we take countermeasures to address this problem, only 9% of the encoding rate due to instruction selection remains, as can be seen in Figure 3. This clearly illustrates the usual trade-off between data rate and stealth.

The need for countermeasures is acknowledged by the results of our attack on the unusual instruction frequency CTS as discussed in Section 4.2. These results are presented in Figure 4, and confirm that unlimited instruction selection freedom indeed results in big, easily detectable discrepancies from clean code.

Fortunately, there is a much greater variation in schedules across executables and 47% of the encoding rate due to instruction scheduling can be safely retained. Furthermore, as code layout is largely determined by the source code, a piece of information that is no longer available to the warden, 59% of the encoding rate due to code layout can be safely retained.

Combined, we thus achieve a stealthy encoding rate ranging from 1/108.59 to 1/80.1 and averaging 1/88.76. This is still higher than the unsafe rate of Hydan.

6   Future Work

The techniques discussed in this paper are portable to other architectures, as they exploit compiler choices common to all architectures. It would be interesting to see how the characteristics of an architecture influence the obtained data encoding rate. For example, a RISC architecture is unlikely to have the same redundancy in its instruction set as the IA-32 architecture. Therefore, the data rate due to instruction selection can be expected to be much lower on a RISC. On the other hand, RISC architectures typically have a larger set of registers, as a result of which register allocation might be a safe place to hide information.

While we have taken measures to prevent the detection of the presence of secret information in an executable in general, all executables generated by a single programmer are likely to be generated by the same compiler, with the same compiler flags, libraries, etc. If an embedder wants to use this tool repeatedly to defeat the same warden, his freedom of choice may need to be further reduced to assure that the attacker is not alarmed when different executables from the same programmer are unlikely to have been compiled with the same tool chain. This requires future research.

7   Related Work

Several types of cover-objects have been used to embed a secret message. The first reported occurrence is due to Herodotus. He tells of Histiæus, who shaved the head of his most trusted slave and tattooed it with a message that disappeared after his hair had regrown. Many other physical objects have since been used as cover-objects, e.g, earrings, written documents, and music scores.

Digital steganography has mainly been applied to media, such as images, sound and video. A large number of systems has been proposed [15, 16].

Steganography in the context of executables has, to the best of our knowledge, only been addressed by Hydan [1], a steganographic tool for IA-32 compatible executables.

Significantly more research has been conducted in the related field of software watermarking. The first one, proposed by Davidson and Myhrvold [17], encodes the watermark in the sequence of basic blocks. Pieprzyk [18] suggests assigning a unique identity to every copy in the choice of equivalent instructions. Another approach encodes the watermark in the frequency of groups of instructions [3]. All of these approaches change properties of the existing executable. Other techniques add a piece of data [19] or code [20] to the original program.

Whereas the mentioned work has mainly focused on the development of new techniques, more attention has recently gone into the evaluation of their security [21, 22, 23]. No general framework has been presented however.

8   Conclusion

This paper identified the redundancy present in executable programs and presented instruction selection, instruction scheduling and code layout as three techniques to exploit this redundancy for steganography. Combined, they resulted in encoding rates of approximately $\frac{1}{27}$, four times the rate of the previous approach by Hydan [1].

A framework for the evaluation of code transformation stealth was introduced and applied to the presented techniques, showing that our techniques can be made secure by the appropriate countermeasures, while still obtaining an encoding rate of $\frac{1}{89}$.

Acknowledgments

This work is supported by the Institute for the Promotion of Innovation through Science and Technology in Flanders (IWT-Vlaanderen), the Fund for Scientific Research - Belgium - Flanders (FWO) and Ghent University, member of the HiPEAC network.

References

1. El-Khalil, R., Keromytis, A.: Hydan: Hiding information in program binaries. In: International Conference on Information and Communications Security, LNCS. Volume 3269. (2004)
2. Collberg, C., Thomborson, C.: Software watermarking: Models and dynamic embeddings. In: ACM SIGPLAN-SIGACT symposium on Principles of Programming Languages, ACM Press (1999) 311–324
3. Stern, J., Hachez, G., Koeune, F., Quisquater, J.J.: Robust object watermarking: Application to code. In: Information Hiding, LNCS. Volume 1768. (1999) 368–378
4. Simmons, G.J.: The prisoners' problem and the subliminal channel. In: Advances in Cryptology. (1984) 51–67
5. Kerkhoffs, A.: La cryptographie militaire. Journal de Sciences Militaires **9** (1883) 5–38
6. Anderson, R.J., Petitcolas, F.A.: On the limits of steganography. I.E.E.E. Journal of Selected Areas in Communications (1998) 474–481
7. Intel: IA-32 Intel Architecture Software Developer's Manual. (2003)
8. Massalin, H.: Superoptimizer: a look at the smallest program. In: Architectual Support for Programming Languages and Operating Systems, IEEE Computer Society Press (1987) 122–126
9. Aho, A., Sethi, R., Ullman, J.: Compilers, Principles, Techniques and Tools. Addison-Wesley (1986)
10. Kwan, M.: Gifshuffle (1998) http://www.darkside.com.au/gifshuffle/.
11. De Sutter, B., De Bus, B., De Bosschere, K.: Sifting out the mud: Low level c++ code reuse. In: ACM SIGPLAN Conference on Object-Oriented Programming, Systems, Languages and Applications. (2002) 275–291
12. Levine, J.: Linkers & Loaders. Morgan Kaufmann Publishers (2000)

13. Gloy, N., Smith, M.D.: Procedure placement using temporal-ordering information. ACM Transactions on Programming Languages and Systems **21** (1999) 977–1027
14. De Bus, B., De Sutter, B., Van Put, L., Chanet, D., De Bosschere, K.: Link-time optimization of ARM binaries. In: ACM SIGPLAN/SIGBED Conference on Languages, Compilers, and Tools for Embedded Systems. (2004) 211–220
15. Cox, I., Miller, M., Bloom, J.: Digital watermarking. Morgan Kaufmann (2002)
16. Katzenbeisser, S., Petitcolas, F.: Information hiding techniques for steganography and digital watermarking. Artech House (2000)
17. Davidson, R., Myhrvold, N.: Method and system for generating and auditing a signature for a computer program (1996) Microsoft Corporation, US5559884.
18. Pieprzyk, J.: Fingerprints for copyright software protection. In: Information Security, LNCS 1729. (1999) 178–190
19. Holmes, K.: Computer software protection (1991) International Business Machines Corporation, US5287407.
20. Venkatesan, R., Vazirani, V., Sinha, S.: A graph theoretic approach to software watermarking. In: Information Hiding, LNCS. Volume 2137. (2001) 157–168
21. Collberg, C., Thomborson, C., Townsend, G.: Dynamic graph-based software watermarking. Technical report, Dept. of Computer Science, Univ. of Arizona (2004)
22. Curran, D., Cinneide, M.O., Hurley, N., Silvestre, G.: Dependency in software watermarking. In: Information and Communication Technologies: from Theory to Applications. (2004) 569–570
23. Sahoo, T.R., Collberg, C.: Software watermarking in the frequency domain: Implementation, analysis, and attacks. Technical report, Dept. of Computer Science, Univ. of Arizona (2004)

On Security Notions for Steganalysis
(Extended Abstract)

Kisik Chang[§], Robert H. Deng[†], Bao Feng[†], Sangjin Lee[‡],
Hyungjun Kim[‡], and Jongin Lim[‡]

[§,†]InfoComm Service Division(ICSD),
Institute for Infocomm Research(I²R),
21 Heng Mui Keng Terrace 119613 Singapore
{stusck, deng, baofeng}@i2r.a-star.edu.sg
[‡]Center for Information Security and Technologies(CIST),
Graduate School of Information Security,
Korea University, 136-701, Republic of Korea
honors@cist.korea.ac.kr
{sangjin, hyungjun, jilim}@korea.ac.kr

Abstract. There have been some achievements in steganalysis recently. Many people have been making strides in steganalysis. They have approached steganalysis from different angles; from information theory to complexity theory. Anderson gave a possibility that there is a provable secure steganographic system, but there had not been complexity theoretical approaches for years. In 2002, Katzenbeisser and Petitcolas defined the conditional security of steganography and gave a possibility for a practical, provable secure steganography for the first time, and Hopper *et al.* introduced a provable secure steganographic algorithm in the sense of complexity theory. Chang *et al.* also tried to define the complexity theoretical security and showed a practical, provable secure algorithm. Chang *et al.* presented chose-input attack model in a view of pseudoprocessingness for a steganographic system. In this paper, we try to improve this notion in detail. So we define chosen-cover attack model and chosen-message attack model. Moreover, we present the relation between them.

Keywords: steganography, steganalysis, pseudoprocessingness.

1 Introduction

Simmons proposed the *prisoners' problem* in [9] before many research papers related to information hiding techniques have been published. In this problem, Alice and Bob are accomplices in a crime and have been arrested, and then they are put in two different cells. After that time, they try to conspire to escape but their all communications are listened to by a guardian named Warden. Since Warden suspect that Alice and Bob want to collude an escape plan, he will only permit the exchanges through not an encrypted message but a plain text. Under this restriction, on the other hand, Alice and Bob attempt to deceive Warden

C. Park and S. Chee (Eds.): ICISC 2004, LNCS 3506, pp. 440–454, 2005.
© Springer-Verlag Berlin Heidelberg 2005

by finding a way of establishing an invisible communication channel between them in full view of the warden, even though the messages themselves contain no secret information.

A couple of solutions has been proposed for the prisoners' problem; subliminal channel, covert channel, etc.. There also have been introduced different approaches: steganography and digital watermarking. Especially, steganography is related to how Alice and Bob can construct a secret channel over a public channel without arousing Warden's suspicion. In other words, the goal of steganography is to conceal the existence of secret communication between Alice and Bob without being suspected by Warden. So the security of steganography depends on that Warden can not distinguish between a cover-object and a stego-object, and also the embedding algorithm has not to leave any special feature on a stego-object.

Although many researchers have studied the steganography in theoretic perspective and in practical, relatively little researchers have approached the security of steganography theoretically. First work was produced by Anderson[1]. He introduced some explanation of information hiding and present a couple of approaches to the theoretical security of steganography. And he also mentioned the computational security using polynomial-time turing machine. After publishing his paper, Cachin[2] and Zöllner et al.[11] proposed information theoretic security models separately. Cachin defined the security of steganography using *relative entropy*, while Zöllner et al. gave a different definition using *mutual information*. During that time, many theoretical and practical steganographic algorithms had been proposed and published, but almost all of them are identified as insecure[7, 6]. Katzenbeisser and Petitcolas defined the conditional security of steganography and gave a possibility for a provable secure steganography for the first time[8]. Hopper et al. introduced a provable secure steganographic algorithm in the sense of complexity theory[5]. However, the security of their algorithm just depends on not the algorithm itself but the security of the encryption algorithm used in their algorithm. In addition, it is different from a practical one because it is not efficient. Chang et al. defined some novel notions, *pseudoprocessing function(PPF)*, *chosen-input-attack*, *chosen-stego-attack*[3]. They also constructed a more practical, provable steganographic system using them.

In this paper, we will develop their chose-input-attack model in detail, *chosen-cover-attack* and *chosen-message-attack* in the sense of complexity theory. And we will also construct a practical, provable secure steganographic algorithm that is secure under the adversary models.

2   Security Notions

2.1   Security Requirement

We can regard the steganographic algorithm as a postcard on which the message is written openly. The postcard should be such that only legal senders can write down a just plain text on it and hide a secret message in the text, but anyone including an adversary can read only the plain text itself. So the adversary can also examine it in polynomial time. Even worse, we assume that the adversary can query a polynomial number of stego-objects to extract messages hidden in them for himself. This is called a *passive* adversary. On the other hand, there may be a more powerful adversary who can intercept stego-objects being transmitted and either bar the channel or modify them in some way. This is called an *active* adversary. In designing process, adversaries must have been considered at first stage and steganographic algorithms which are secure against them have been designed. Defining the security against powerful adversaries is more complicated than for passive adversaries. So in this paper, we consider only a passive adversary who knows the probability distribution over the message space.

When can we say that steganography is *secure* or *broken*? It's difficult to answer the question, but it is really clear that a minimal requirement of security would be that any adversary who is capable of watching the stego-object and knows which steganographic algorithm is being used, can not identify whether a given object is a stego-object or not. However, we need more desirable properties. We propose the principle of designing steganographic algorithm.

Condition 1. It should be hard to distinguish a stego-object from a given-object when a message and a cover-object are drawn from arbitrary probability distributions defined on a message space and a cover-object space. But we must assume that these spaces are known to the adversary. (Definition 9)

Condition 2. It should be hard to say whether a given-object has a message or not even though the adversary obtain the embedding algorithm and use it. In other words, it should be difficult to say whether a given-object has a message or a random string. Even more it should be also difficult find some useful information that can be used for detecting. (Definition 12)

Condition 3. It should be hard to extract information about the embedded message from the given-object where the adversary knows that it is a stego-object. Even more it should be also difficult to compute partial information about the message or to find some useful information that can be used for detecting. (Definition 13)

Condition 4. It should be hard to obtain the key information from the stego-object.

Condition 5. The above properties should be hold with high probability.

In above requirements, the most important thing is the first requirement. Condition 1 incorporates the essence of the traditional security of steganography. The other requirements are needed after Condition 1. is broken. To identify whether a give-object is a stego-object, an adversary should be able to make out objects. So we can consider the first requirement as the notion of indistinguishability between a cover-object and a stego-object.

Let us now consider the computing resources needed when an adversary tried to break system. As cryptography, it is widely accepted that it is impractical assumption that the adversary has infinite computing resources[2, 11]. Instead, we suppose that the computing resources who the adversary can use are bounded in some reasonable way. Especially, in this paper we will assume that the adversary is a probabilistic algorithm who runs in polynomial time. So the steganographic algorithms also should be probabilistic and run in polynomial time.

On the other hand, let us look at the problem of measuring the running time of algorithms including adversary. It should be calculated as a function of a security parameter which is fixed at the time the steganography is setup. Thus, the adversary algorithm runs in polynomial time means that time is bounded by some polynomial function of a security parameter[4]. Steganography should be designed based on a gap between the efficiency of embedding algorithms for the senders and the computational infeasibility of detecting task for the adversary so that it requires that one has available primitives with certain special kinds of computational hardness properties. As the most basic function is an one-way function to build secure encryption scheme in cryptography, Chang *et al.* showed that pseudoprocessing function can be used as a building block in steganography[3]. We will describe their definitions briefly and then, introduce the security of steganography in the sense of indistinguishability.

2.2   Basic Notions

In this section, we introduce some basic notions used for analyzing a steganographic system. We begin with the definition of Channel. Hopper et al. assumed that there exists an oracle that can draws objects from the channel.

Definition 1 (Channel; Hopper *et al.*(2002)). A *channel* $\mathcal{N}$ is a distribution on bit sequences where each bit is also time-stamped with monotonically increasing time value. Formally, a channel is a distribution with support $(\{0,1\}, t_1), (\{0,1\}, t_2), \ldots$, where for all $i > 0$ and $t_{i+1} > t_i$.

An easy computation is one which can be carried out by a probabilistic polynomial time algorithm and a function is *negligible* if it vanishes faster than the inverse of any polynomial[4].

Definition 2 (Negligible Function). A *function* $\nu : \mathbb{N} \to \mathbb{R}$ *is a negligible if for every constant* $c \geq 0$, *there exits an integer* k_c *such that* $\nu(k) < k^{-c}$ *for all* $k \geq k_c$.

A symmetric-key steganographic system is defined as follows;

Definition 3 (Symmetric-Key Steganographic System). A *symmetric-key (or secret-key) steganographic system* $\mathcal{SKS} = (\mathsf{Keys}, \mathsf{Enc}, \mathsf{Ext})$ consists of three polynomial-time algorithms.

Key Generation. On input 1^k (the *security parameter*), the *key generation* algorithm $\mathsf{Keys}(\cdot)$ outputs produces a string k, we write $K \xleftarrow{R} \mathsf{Keys}(1^k)$.

Embedding. The *embedding* algorithm $\mathsf{Emb}(\cdot, \cdot, \cdot)$ takes the key $K \in \mathsf{Keys}(1^k)$, an *object* $c \in \mathcal{N}$, and a message $m \in \mathcal{M}$ to return a *stego-object* $s \in \mathcal{S}$, we write $s \longleftarrow \mathsf{Emb}(K, c, m)$. We can consider an embedding algorithm is a function, $\mathsf{Emb} : \mathsf{Keys} \times \mathcal{N} \times \mathcal{M} \longrightarrow \mathcal{S}$.

Extracting. The *extracting* algorithm $\mathsf{Ext}(\cdot, \cdot)$ takes the key $K \in \mathsf{Keys}(1^k)$ and a stego-object s to return a message $m \in \mathcal{M}(\cup\{\bot\})$, we write $m \longleftarrow \mathsf{Ext}(K, s)$. We require that for any key $K \in \mathsf{Keys}(1^k)$, any object $c \in \mathcal{N}$, and any message $m \in \mathcal{M}$, if $\mathsf{Emb}(K, c, m)$ returns a stego-object s, that is $s \longleftarrow \mathsf{Emb}(K, c, m)$, then $\mathbf{Pr}[s \longleftarrow \mathsf{Emb}(K, c, m)] = 1$ and $\mathbf{Pr}[\mathsf{Ext}(K, s) \neq m]$ is *negligible*. We can consider an embedding algorithm is a function, $\mathsf{Ext} : \mathsf{Keys} \times \mathcal{S} \longrightarrow \mathcal{M}$.

To provide the security for the steganographic system, Chang *et al.* presented the Steganographic Decision Problem(SDP)[3].

Definition 4 (Steganographic Decision Problem(SDP)). *Given an object $s \in \mathcal{C}$ to Warden, he must determine whether s has a message $m \in \{0,1\}^*$ or not.*

However, this definition is not enough to apply itself to steganalysis directly because the definition is quite vague. Thus Chang *et al.* defined a processing function and a pseudoprocessing function additionally to provide a solution in the sense of complexity theory[3]. Nevertheless, their idea is not sufficient, but imperfect to develop theory. So we need more additional definitions to adopt complexity theory on steganalysis. To do this, we define a processing function and a criterion as follows.

Definition 5 (Processing and Criterion Function). A *processing function* is a map from a channel $\mathcal{N}$ to a cover-object set $\mathcal{C}$. That is, $\mathsf{Proc} : \mathcal{N} \longrightarrow \mathcal{C}$. And a *criterion function* is a map from an object set $\mathcal{U}$ to $\{0,1\}$. That is, $\mathsf{Crit} : \mathcal{U} \longrightarrow \{0,1\}$.

A processing function, for example, is a compression algorithm or an enhancement. And a criterion function plays a role to classify into two object sets. If a domain is a union of a cover-object set and a stego-object set, we can separate stego-objects from cover-objects using steganalytic methods, such as χ^2-test, as a criterion function.

Definition 6 (Distinguishability and Indistinguishability). Let $\mathcal{U}$ be a union of $\mathcal{S}$ and $\mathcal{C}$. If there is a criterion function f such that $f(\mathcal{S})$ and $f(\mathcal{C})$, be a partition of $f(\mathcal{U})$, i.e., $f(\mathcal{S}) \cap f(\mathcal{C}) = \varnothing$, then $\mathcal{S}$ is said to be *distinguishable* from $\mathcal{C}$, denoted by $\mathcal{S} \not\cong \mathcal{C}$. If all criterion functions on $\mathcal{U}$ are biased, that is $f(\mathcal{S}) = f(\mathcal{C})$, then $\mathcal{S}$ is said to be *indistinguishable* from $\mathcal{C}$, denoted by $\mathcal{S} \cong \mathcal{C}$.

The ideal steganalytic method must separate stego-objects from cover-objects completely. But all analyses introduced until now can distinguish stego-objects from cover-objects only partially. So we need a definition for partial distinguishability.

Definition 7 (Computational Distinguishability and Indistinguishability). Let $\mathcal{U}$ be a union of $\mathcal{S}$ and $\mathcal{C}$. When the intersection of $f(\mathcal{S})$ and $f(\mathcal{C})$ is negligible, $\mathcal{S}$ is said to be *computationally distinguishable* from $\mathcal{C}$, denoted by $\mathcal{S} \not\cong_c \mathcal{C}$. In the case that the difference of $f(\mathcal{S})$ and $f(\mathcal{C})$ is negligible, $\mathcal{S}$ is said to be *computationally indistinguishable* from $\mathcal{C}$, denoted by $\mathcal{S} \cong_c \mathcal{C}$.

For example, let the embedding be EzStego algorithm and a processing function be GIF compression algorithm. Since EzStego embeds a encrypted message into the sorted palette sequentially, visual test or χ^2-test can discriminate between stego-images produced by EzStego and cover-images with high probability[10]. So the stego-image set and the cover-image set are computationally distinguishable. From Definition 7, we get the following equivalent directly.

Proposition 1. *The following statements are equivalent;*

(1) $\mathcal{S}$ *and* $\mathcal{C}$ *are computationally indistinguishable,* $\mathcal{S} \cong_c \mathcal{C}$.
(2) *There exists no criterion function such that* $f(\mathcal{S}) \cap f(\mathcal{C})$ *is negligible.*

Proof. (1)$\Rightarrow$(2) Assume that there exist a criterion function f such that $|f(\mathcal{S}) \cap f(\mathcal{C})|$ be negligible. Then $\mathcal{S} \ncong_c \mathcal{C}$. This is contradiction.

(2)$\Rightarrow$(1) Suppose that $\mathcal{S} \ncong_c \mathcal{C}$. Then there exists a criterion function f such that $|f(\mathcal{S}) \cap f(\mathcal{C})|$ be negligible. This is contradiction. $\square$

If stego-objects produced by an embedding function could be separated from cover-objects computationally, we can consider the embedding function to be like a processing function. In other words, if a stego-object set is computationally indistinguishable from a cover-object set, it is difficult for Warden who uses a function like a black box to decide whether a given function is an embedding function or a processing function. We define the such function as a pseudoprocessing function.

Definition 8 (Pseudoprocessing Function). Let $e :$ Keys $\times \mathcal{M} \times \mathcal{N} \longrightarrow \mathcal{S}$ be an embedding function, and $p : \mathcal{N} \longrightarrow \mathcal{C}$ be a processing function. If $\mathcal{S} \cong_c \mathcal{C}$, then e is said to be a *pseudoprocessing function*, denoted by $e \cong_c p$. If $\mathcal{S} \ncong_c \mathcal{C}$, then e is said to be *computationally distinguishable* from p, denoted by $e \ncong_c p$. That is,

$$\mathcal{S} \cong_c \mathcal{C} \Longrightarrow e \cong_c p,$$
$$\mathcal{S} \ncong_c \mathcal{C} \Longrightarrow e \ncong_c p.$$

And the *pseudoprocessingness* as the degree of the similarity between an embedding function and a processing function.

Now we get an important theorem for the provable security from Definition 8.

Theorem 1. *Let* $e :$ Keys $\times \mathcal{M} \times \mathcal{N} \longrightarrow \mathcal{S}$ *be an embedding function and* $p : \mathcal{N} \longrightarrow \mathcal{C}$ *be a processing function. If* e *be a pseudoprocessing function, then* $\mathcal{S}$ *be computationally indistinguishable. That is,*

$$e \cong_c p \Longrightarrow \mathcal{S} \cong_c \mathcal{C}.$$

Theorem 1 says that the security of steganography relies on the pseudoprocessingness of an embedding function. We also get a corollary from the above theorem.

Corollary 1.

$$\mathcal{S} \cong_c \mathcal{C} \Longleftrightarrow \mathsf{Enc} \cong_c \mathsf{Proc}$$

2.3   Pseudoprocessing Function Experiment

Now we have a question for the pseudoprocessingness. How can we measure the pseudoprocessingness? Actually assessing the amount of pseudoprocessingness is not so easy work. To estimate the pseudoprocessingness, Chang *et al.* formulated the pseudoprocessing function experiment mathematically[3]. In this experiment, a test function is given to Warden and what can he only see is input-output behavior of the function. After experiment, Warden should tell whether a test function is an embedding function or a processing. The action of Warden is could be considered as the notion of a *distinguisher*. The measure is computed as the probability that the distinguisher or adversary say a given-object is a stego-object. The experiment is defined as follows.

Definition 9 (Pseudoprocessing Function Experiment). Let Emb : Keys$\times$ $\mathcal{N} \times \mathcal{M} \to \mathcal{S}$ be an embedding functions, and let $\mathcal{W}$ be an algorithm that takes an oracle for a function $p : \mathcal{N} \to \mathcal{C}$, and returns a bit d. We consider following two experiments:

Experiment $\mathbf{Exp}_{\mathsf{Emb},\mathcal{W}}^{\text{PPF-1}}$	Experiment $\mathbf{Exp}_{\mathsf{Emb},\mathcal{W}}^{\text{PPF-0}}$
$K \xleftarrow{r} \mathsf{Keys}$	$p \longleftarrow \mathsf{Proc}^{\mathcal{C} \times \mathcal{M} \to \mathcal{S}}$
$d \longleftarrow \mathcal{W}^{\mathsf{Emb}(K,\cdot,\cdot)}$	$d \longleftarrow \mathcal{W}^p$
Return d	Return d

The advantage for a pseudoprocessing function of a distinguisher $\mathcal{W}$, *PPF-advantage*, is defined as

$$\mathbf{Adv}_{\mathsf{Emb},\mathcal{W}}^{\text{PPF}} = \left| \mathbf{Pr}\left[\mathbf{Exp}_{\mathsf{Emb},\mathcal{W}}^{\text{PPF-1}} = 1 \right] - \mathbf{Pr}\left[\mathbf{Exp}_{\mathsf{Emb},\mathcal{W}}^{\text{PPF-0}} = 1 \right] \right|. \qquad (1)$$

For any t, q, μ, we define the *PPF-insecurity of an embedding function* Emb,

$$\mathbf{InSec}_{\mathsf{Emb}}^{\text{PPF}}(t, q, \mu) = \max_{\mathcal{W}} \left\{ \mathbf{Adv}_{\mathsf{Emb},\mathcal{W}}^{\text{PPF}} \right\}$$

where the maximum is over all distinguishers $\mathcal{W}$ having time-complexity t and making at most q oracle queries, the sum of the lengths of these queries being at most μ bits. An embedding function Emb is (t, q, μ, ε)-*pseudoprocessing function* or (t, q, μ, ε)-*pseudoprocessing* when $\mathbf{InSec}_{\mathsf{Emb}}^{\text{PPF}}(t, q, \mu) \leqslant \varepsilon$, where ε is negligible.

To solve the SDP, adversaries can use some resources: a cover-object, a message and a stego-object. Depending on the available resources, there are several kinds of adversary models for the steganographic systems[7], and Hopper *et al.* introduced several adversary models close to cryptographic adversary model[5]. Chang *et al.* also presented independently two kinds of adversary models, Chosen-Input-Attack(CIA) and Chosen-Stego-Attack(CSA)[3]. Their classification is based on the algorithms which an adversary can use as an oracle. In CIA model, the adversary is accessible to an embedding oracle, so he can inquire of an embedding oracle with different cover-objects for distinct messages. So, in the sense of PPF-CIA, the security is defined as follows;

Definition 10 (PPF-CIA Security). Let Emb : Keys $\times \mathcal{N} \times \mathcal{M} \to \mathcal{S}$ be an embedding functions, and let $\mathcal{W}$ be an algorithm that takes an oracle for a function $p : \mathcal{N} \to \mathcal{C}$, and returns a bit d. We consider following two experiments:

$$
\begin{array}{c|c}
\text{Experiment } \mathbf{Exp}_{\mathsf{Emb},\mathcal{W}}^{\text{PPF-CIA-1}} & \text{Experiment } \mathbf{Exp}_{\mathsf{Emb},\mathcal{W}}^{\text{PPF-CIA-0}} \\
K \xleftarrow{r} \mathsf{Keys} & p \longleftarrow \mathsf{Proc}^{\mathcal{N} \to \mathcal{C}} \\
d \longleftarrow \mathcal{W}^{\mathsf{Emb}(K,\cdot,\cdot)} & d \longleftarrow \mathcal{W}^p \\
\text{Return } d & \text{Return } d
\end{array}
$$

The advantage for a pseudoprocessing function of a distinguisher $\mathcal{W}$, *PPF-CIA-advantage*, is defined as

$$
\mathbf{Adv}_{\mathsf{Emb},\mathcal{W}}^{\text{PPF-CIA}} = \left| \Pr\left[\mathbf{Exp}_{\mathsf{Emb},\mathcal{W}}^{\text{PPF-CIA-1}} = 1 \right] - \Pr\left[\mathbf{Exp}_{\mathsf{Emb},\mathcal{W}}^{\text{PPF-CIA-0}} = 1 \right] \right|.
$$

For any t, q, μ, we define the *PPF-insecurity of an embedding function* Emb

$$
\mathbf{InSec}_{\mathsf{Emb}}^{\text{PPF-CIA}}(t, q, \mu) = \max_{\mathcal{W}} \left\{ \mathbf{Adv}_{\mathsf{Emb},\mathcal{W}}^{\text{PPF-CIA}} \right\}
$$

where the maximum is over all distinguishers $\mathcal{W}$ having time-complexity t and making at most q oracle queries, the sum of the lengths of these queries being at most μ bits. An embedding function Emb is (t, q, μ, ε)-*pseudoprocessing function under Chosen-Input-Attack(CIA)* or (t, q, μ, ε)-*pseudoprocessing under CIA* when $\mathbf{InSec}_{\mathsf{Emb}}^{\text{PPF-CIA}}(t, q, \mu) \leqslant \varepsilon$, where ε is negligible.

In CIA setting, an adversary can use two resources, a message and a cover-object for a query. He is capable of fixing a cover-object and then varying a message, or the other way. On the other hand, an adversary can access a stego-object besides the inputs. So PPF-CSA is defined as follows.

Definition 11 (PPF-CSA Experiment). Let Emb : Keys $\times \mathcal{N} \times \mathcal{M} \to \mathcal{S}$ be an embedding functions, and let $\mathcal{W}$ be an algorithm that takes an oracle for a function $p : \mathcal{N} \to \mathcal{C}$, and returns a bit d. We consider following two experiments:

$$
\begin{array}{c|c}
\text{Experiment } \mathbf{Exp}_{\mathsf{Emb},\mathcal{W}}^{\text{PPF-CSA-1}} & \text{Experiment } \mathbf{Exp}_{\mathsf{Emb},\mathcal{W}}^{\text{PPF-CSA-0}} \\
K \xleftarrow{r} \mathsf{Keys} & p \longleftarrow \mathsf{Proc}^{\mathcal{N} \to \mathcal{C}} \\
d \longleftarrow \mathcal{W}^{\mathsf{Emb}(K,\cdot,\cdot),\mathsf{Ext}(K,\cdot)} & d \longleftarrow \mathcal{W}^{p,p^{-1}} \\
\text{Return } d & \text{Return } d
\end{array}
$$

The advantage for a pseudoprocessing function of a distinguisher $\mathcal{W}$, *PPF-CSA-advantage*, is defined as

$$
\mathbf{Adv}_{\mathsf{Emb},\mathcal{W}}^{\text{PPF-CSA}} = \left| \Pr\left[\mathbf{Exp}_{\mathsf{Emb},\mathcal{W}}^{\text{PPF-CSA-1}} = 1 \right] - \Pr\left[\mathbf{Exp}_{\mathsf{Emb},\mathcal{W}}^{\text{PPF-CSA-0}} = 1 \right] \right|.
$$

For any $t, q_c, \mu_c, q_s, \mu_s$, we define the *PPF-insecurity of an embedding function* Emb

$$
\mathbf{InSec}_{\mathsf{Emb}}^{\text{PPF-CSA}}(t, q_c, \mu_c, q_s, \mu_s) = \max_{\mathcal{W}} \left\{ \mathbf{Adv}_{\mathsf{Emb},\mathcal{W}}^{\text{PPF-CSA}} \right\}
$$

where the maximum is over all distinguishers $\mathcal{W}$ having time-complexity t and making at most q_c oracle queries to the p oracle, the sum of the lengths of these queries being at most μ_c bits, and also making at most q_s queries to the p^{-1} oracle, the sum of the lengths of these queries being at most μ_s bits. An embedding function Emb is $(t, q_c, \mu_c, q_s, \mu_s, \varepsilon)$-*pseudoprocessing function under Chosen-Stego-Attack(CSA)* or $(t, q_c, \mu_c, q_s, \mu_s, \varepsilon)$-*pseudoprocessing under CSA*, when $\mathbf{InSec}_{\mathsf{Emb}}^{\mathrm{PPF\text{-}CSA}}(t, q_c, \mu_c, q_s, \mu_s) \leqslant \varepsilon$, where ε is negligible.

Definition 10 is too loose, so we will define the security for each adversary models corresponding to the accessible embedding oracle in the experiment in the next section.

3   Indistinguishability Under Chosen-Object Attacks

In this section, we discuss chosen-input attack in detail. An embedding function takes three inputs; a key, a cover-object, and a message. It is reasonable for us to consider two inputs, a cover-object and a message, as variables. So we will introduce chosen-cover attack and chosen-message attack in the sense of indistinguishability. We want to show that if an embedding algorithm is not secure against each attack, then the algorithm is not a pseudoprocessing function. We formulate this idea in the next section.

3.1   Chosen-Cover Attack Model

First, assume that the adversary who knows a message and a cover-object; in fact he is allowed to select cover-objects and a message on his own. There is a sequence of couples of a pair of cover-objects and a message; $\{(c_{1,0}, c_{1,1}), m_1\}, \ldots, \{(c_{q,0}, c_{q,1}), m_q\}$. This sequence is known to the adversary. Now, a *challenge* bit b is chosen randomly, and a sequence of a pair of objects $(s_{1,0}, s_{1,1}), \ldots, (s_{q,0}, s_{q,1})$ is produced for a message m_i and a random string m_i^r, $|m_i^r| = |m_i|$, where

$$s_{i,b} \longleftarrow \begin{cases} \mathsf{Emb}_K(c_{i,0}, m_i), & \text{for } b = 0, \\ \mathsf{Emb}_K(c_{i,1}, m_i), & \text{for } b = 1. \end{cases}$$

$$s_{i,1-b} \longleftarrow \begin{cases} \mathsf{Emb}_K(c_{i,1}, m_i^r), & \text{for } b = 0, \\ \mathsf{Emb}_K(c_{i,0}, m_i^r), & \text{for } b = 1. \end{cases}$$

In each stage, the embedding algorithm uses fresh internal coins each time. The adversary gets the sequence of challenge-objects $(s_{1,0}, s_{1,1}), \ldots, (s_{q,0}, s_{q,1})$ and must guess which side object has a message m_i. This means that the adversary should determine whether the sender sent $s_{1,0}, \ldots, s_{q,0}$ or $s_{1,1}, \ldots, s_{q,1}$. In this environment, we say that the embedding function is secure under *chosen-cover attack* if it is hard for the adversary to tell whether $s_{i,0}$ or $s_{i,1}$ has a message m_i. We will formalize this idea. Let us fix a steganographic system $\mathcal{SKS} = (\mathsf{Keys}, \mathsf{Emb}, \mathsf{Ext})$ and assume that a warden is $\mathcal{W}$, which is a program accessible to an oracle to which it can inquire as input any pair $\{(c_0, c_1), m\}$.

Then the oracle returns an object (s_0, s_1) for the query. There are two possible ways in which these 'challenge' stego-objects are computed by the oracles corresponding to two sets. To do this, we define the *Two-Sides embedding oracle* $\mathrm{TS}(\mathsf{Emb}_K(\cdot, \cdot), \mathsf{Emb}_K(\cdot, \cdot), b)$ as follows:

$$\text{Oracle } \mathrm{TS}(\mathsf{Emb}_K(c_b, m), \mathsf{Emb}_K(c_{1-b}, m^r), b), \text{ where } b \in \{0, 1\}$$

$$(s_0, s_1) \longleftarrow \begin{cases} (\mathsf{Emb}_K(c_0, m), \mathsf{Emb}_K(c_1, m^r)), \text{ for } b = 0, \\ (\mathsf{Emb}_K(c_0, m^r), \mathsf{Emb}_K(c_1, m)), \text{ for } b = 1. \end{cases}$$

$$\text{Retrun } (s_0, s_1)$$

The oracle embeds the message m and a random string m^r into the given cover-objects each according to the bit b. Now two sets are defined as follows:

Set 0: The oracle $\mathrm{TS}(\mathsf{Emb}_K(c_0, m), \mathsf{Emb}_K(c_1, m^r), 0)$ is given to the adversary. So, whenever the adversary makes a query $\{(c_0, c_1), m\}$ to its oracle, the former embeds the message m into c_0, $s_0 \leftarrow \mathsf{Emb}_K(c_0, m)$, and the later does a random string m^r into c_1, $s_1 \leftarrow \mathsf{Emb}_K(c_1, m^r)$ and returns (s_0, s_1) as the answer.

Set 1: The oracle $\mathrm{TS}(\mathsf{Emb}_K(c_0, m^r), \mathsf{Emb}_K(c_1, m), 1)$ is given to the adversary. So, whenever the adversary makes a query $\{(c_0, c_1), m\}$ to its oracle, the former embeds a random string m^r into c_0, $s_0 \leftarrow \mathsf{Emb}_K(c_0, m^r)$, and the later does the message m into c_1, $s_1 \leftarrow \mathsf{Emb}_K(c_1, m)$ and returns (s_0, s_1) as the answer.

We call the first set or oracle, the 'left-side' set or oracle, and the second the 'right-side' set or oracle. Now the adversary should tell which side oracle hides the message after making queries to its oracle for some time. We define an embedding function is secure against chosen-cover attack if an adversary can not get significant advantage in distinguishing the cases $b = 0$ and $b = 1$ given access to the oracle, where the adversary can use reasonable resources. This security notion is also called indistinguishability under chose-cover attack, denoted IND-CCA as following definition.

There is a thing to discuss about the above experiment. There are certain queries that an adversary can make to it TS-embedding oracle which will give information about the bit b, but we assume that those queries are illegitimate. For example, an adversary may make query $\{(c_0, c_1), m\}$ such that $\mathsf{Emb}_K(c_b, m) = \perp$ and $\mathsf{Emb}_K(c_b, m^r) \neq \perp$, or vice versa for $b \in \{0, 1\}$. So we simply suppose that an adversary is prohibited from making such queries. In this paper, we will have only legitimate adversary. Now we define the IND-CCA security as follows.

Definition 12 (IND-CCA Security). Let $\mathsf{Emb} : \mathsf{Keys} \times \mathcal{N} \times \mathcal{M} \to \mathcal{S}$ be an embedding functions, let $b \in \{0, 1\}$, and let $\mathcal{W}$ be an algorithm that takes an oracle for a function Emb, and returns a bit d. We consider following experiment:

$$\text{Experiment } \mathbf{Exp}_{\mathsf{Emb}, \mathcal{W}}^{\mathrm{IND\text{-}CCA}\text{-}b}$$
$$K \xleftarrow{R} \mathsf{Keys}$$
$$d \longleftarrow \mathcal{W}^{\mathrm{TS}(\mathsf{Emb}_K(c_b, m), \mathsf{Emb}_K(c_{1-b}, m^r), b)}$$
$$\text{Return } d$$

The *IND-CCA-advantage* of $\mathcal{W}$ is defined as

$$\mathbf{Adv}_{\mathsf{Emb},\mathcal{W}}^{\text{IND-CCA}} = \left| \mathbf{Pr} \left[\mathbf{Exp}_{\mathsf{Emb},\mathcal{W}}^{\text{IND-CCA-1}} = 1 \right] - \mathbf{Pr} \left[\mathbf{Exp}_{\mathsf{Emb},\mathcal{W}}^{\text{IND-CCA-0}} = 1 \right] \right|.$$

For any t, q, μ, we define the *IND-CCA advantage of* Emb via

$$\mathbf{InSec}_{\mathsf{Emb}}^{\text{IND-CCA}}(t, q, \mu) = \max_{\mathcal{W}} \left\{ \mathbf{Adv}_{\mathsf{Emb},\mathcal{W}}^{\text{IND-CCA}} \right\},$$

where the maximum is over all legitimate $\mathcal{W}$ having time-complexity t and making at most q oracle queries, the sum of the lengths of these queries being at most μ bits. An embedding function Emb is (t, q, μ, ε)-*indistinguishable under Chosen-Cover-Attack(CCA)* or (t, q, μ, ε)-indistinguishable under CCA when $\mathbf{InSec}_{\mathsf{Emb}}^{\text{IND-CCA}}(t, q, \mu) \leqslant \varepsilon$, where ε is negligible.

Definition 12 says that $\mathcal{W}$ is returning 1 just about as often in set 0 as in set 1 if $\mathbf{Adv}_{\mathsf{Emb},\mathcal{W}}^{\text{IND-CCA}}$ is small, namely the adversary $\mathcal{W}$ does not tell which set is selected. But if if $\mathbf{Adv}_{\mathsf{Emb},\mathcal{W}}^{\text{IND-CCA}}$ is large, then the adversary does well meaning the embedding algorithm Emb is insecure.

3.2   Chosen-Message Attack Model

Next, we discuss the other input, message. Let us assume that the adversary who knows a pair of a cover-object and a message; in fact he is allowed to select a cover-object and a pair of messages on his own. There is a sequence of couples of a cover-object and a pair of messages; $\{c_1, (m_{1,0}, m_{1,1})\}, \ldots, \{c_q, (m_{q,0}, m_{q,1})\}$. This sequence is known to the adversary. Now, a *challenge* bit b is chosen randomly, and a sequence of stego-objects $s_1, \ldots, s_q$ is produced for a cover-object c_i and a pair of messages $m_{i,0}$ and $m_{i,1}$, $|m_{i,0}| = |m_{i,1}|$, where

$$s_i^b \longleftarrow \begin{cases} \mathsf{Emb}_K(c_{i,0}, m_i^r), \text{ for } b = 0, \\ \mathsf{Emb}_K(c_{i,1}, m_i), \text{ for } b = 1. \end{cases}$$

In each stage, the embedding function uses fresh internal coins each time. The adversary gets the sequence of challenge-objects $s_1, \ldots, s_q$ and must guess which message is embedded into the cover-object c. This means that the adversary should determine whether the sender sent $m_{1,0}, \ldots, m_{q,0}$ or $m_{1,1}, \ldots, m_{q,1}$. In this environment, we say that the embedding function is secure under *chosen-message attack* if it is hard for the adversary to tell whether s_i has a message $m_{i,0}$ or $m_{i,1}$. We will formalize this idea. Let us fix a steganographic system $\mathcal{SKS} = (\mathsf{Keys}, \mathsf{Emb}, \mathsf{Ext})$ and assume that a warden is $\mathcal{W}$, which is a program accessible to an oracle to which it can inquire as input any pair $\{c, (m_0, m_1)\}$. Then the oracle returns an object s for the query. There are two possible ways in which the 'challenge' stego-object is computed by the oracles corresponding to two sets. To do this, we define the *Left-or-Right embedding oracle* $\mathsf{Emb}_K(\cdot, \mathrm{LR}(\cdot, \cdot, b))$ as follows:

Oracle $\mathsf{Emb}_K(c, \mathrm{LR}(m_0, m_1, b))$, where $b \in \{0, 1\}$

$\quad s \longleftarrow \mathsf{Emb}_K(c, m_b)$

$\quad$ Retrun s

The oracle embeds the message m_0 or m_1 into the given cover-object c according to the bit b. Now two sets are defined as follows:

Set 0: The oracle $\mathsf{Emb}_K(\cdot, \mathrm{LR}(\cdot, \cdot, 0))$ is given to the adversary. So, whenever the adversary makes a query $\{c, (m_0, m_1)\}$ to its oracle, it embeds the message m_0 into c, $s \leftarrow \mathsf{Emb}_K(c, m_0)$, and returns s as the answer.

Set 1: The oracle $\mathsf{Emb}_K(\cdot, \mathrm{LR}(\cdot, \cdot, 1))$ is given to the adversary. So, whenever the adversary makes a query $\{c, (m_0, m_1)\}$ to its oracle, it embeds the message m_1 into c, $s \leftarrow \mathsf{Emb}_K(c, m_1)$, and returns s as the answer.

We call the first set or oracle, the 'left' set or oracle, and the second the 'right' set or oracle. Now the adversary should tell which oracle hides the message after making queries to its oracle for some time. We define an embedding function is secure against chosen-message attack if an adversary can not get significant advantage in distinguishing the cases $b = 0$ and $b = 1$ given access to the oracle, where the adversary can use reasonable resources. This security notion is also called indistinguishability under chose-message attack, denoted IND-CMA as following definition.

There is a thing to discuss about the above experiment. There are certain queries that an adversary can make to it LR-embedding oracle which will give information about the bit b, but we assume that those queries are illegitimate. For example, an adversary may make query $\{c, (m_0, m_1)\}$ such that $\mathsf{Emb}_K(c, m_0) = \bot$ and $\mathsf{Emb}_K(c, m_1) \neq \bot$, or vice versa. So we simply suppose that an adversary is prohibited from making such queries. In this paper, we will have only legitimate adversary. Now we define the IND-CMA security.

Definition 13 (IND-CMA Security). Let $\mathsf{Emb} : \mathsf{Keys} \times \mathcal{N} \times \mathcal{M} \to \mathcal{S}$ be an embedding functions, let $b \in \{0, 1\}$, and let $\mathcal{W}$ be an algorithm that takes an oracle for a function $\mathsf{Emb} : \mathcal{C} \times \mathcal{M} \to \mathcal{S}$, and returns a bit d. We consider following experiment:

$$\text{Experiment } \mathbf{Exp}_{\mathsf{Emb}, \mathcal{W}}^{\text{IND-CMA-}b}$$
$$K \xleftarrow{R} \mathsf{Keys}$$
$$d \longleftarrow \mathcal{W}^{\mathsf{Emb}_K(\cdot, \mathrm{LR}(\cdot, \cdot, b))}$$
$$\text{Return } d$$

The *IND-CMA-advantage* of $\mathcal{W}$ is defined as

$$\mathbf{Adv}_{\mathsf{Emb}, \mathcal{W}}^{\text{IND-CMA}} = \left| \mathbf{Pr}\left[\mathbf{Exp}_{\mathsf{Emb}, \mathcal{W}}^{\text{IND-CMA-}1} = 1 \right] - \mathbf{Pr}\left[\mathbf{Exp}_{\mathsf{Emb}, \mathcal{W}}^{\text{IND-CMA-}0} = 1 \right] \right|.$$

For any t, q, μ, we define the *IND-CMA advantage of* Emb via

$$\mathbf{InSec}_{\mathsf{Emb}}^{\text{IND-CMA}}(t, q, \mu) = \max_{\mathcal{W}} \left\{ \mathbf{Adv}_{\mathsf{Emb}, \mathcal{W}}^{\text{IND-CMA}} \right\},$$

where the maximum is over all legitimate $\mathcal{W}$ having time-complexity t and making at most q oracle queries, the sum of the lengths of these queries being at most μ bits. An embedding function Emb is (t, q, μ, ε)-*indistinguishable under Chosen-Message-Attack(CMA)* or (t, q, μ, ε)-*indistinguishable under CMA* when $\mathbf{InSec}_{\mathsf{Emb}}^{\mathrm{IND\text{-}CMA}}(t, q, \mu) \leqslant \varepsilon$, where ε is negligible.

Definition 13 says that $\mathcal{W}$ is returning 1 just about as often in set 0 as in set 1 if $\mathbf{Adv}_{\mathsf{Emb}, \mathcal{W}}^{\mathrm{IND\text{-}CMA}}$ is small, namely the adversary $\mathcal{W}$ does not tell which set is selected. But if if $\mathbf{Adv}_{\mathsf{Emb}, \mathcal{W}}^{\mathrm{IND\text{-}CMA}}$ is large, then the adversary does well meaning the embedding function Emb is insecure.

Statistical analysis[10] is a good example of this definition. Westfeld and Pfitzmann noticed that the message length affects the randomness of the least significant bits of the cover-image. So LSB-based steganographic algorithms such as EzStego, Jsteg, Steganos, S-Tools are insecure under chosen-message attack. To analysis these algorithms, they had chosen messages that have different length for a fixed image and tested them. They found the fact that stego-images of those algorithms are detected with a probability near 1 if the length of embedded messages is around the half size of the capacity of cover-images.

3.3   Chosen-Stego Attack Model

In previous sections, we described concealment of algorithm under chosen-cover attack and chosen-message attack. In addition, we can assume that adversary is capable of attacking with more stronger way, chosen-stego attack. In this setting, an adversary has access to a extracting oracle. The adversary gives a stego-object for a query to the extracting oracle and gets back the corresponding message.

4   Security Against Chosen-Input Attacks

Let Emb: $\mathsf{Keys} \times \mathcal{N} \times \mathcal{M} \longrightarrow \mathcal{S}$ be a family of embedding functions. Then we can guarantee the security of Emb under chosen-cover and chosen-message attack when it is a pseudoprocessing function family. In other words, if an adversary can tell that Emb is computationally distinguishable from Proc, then he could distinguish the oracles used in the experiment. We will show these in the following theorems. For the chosen-cover adversary we have:

Theorem 2. *Let* Emb : $\mathsf{Keys} \times \mathcal{N} \times \mathcal{M} \to \mathcal{S}$ *be a family of embedding function, where* $\mathsf{Keys} = \{0,1\}^k$, $\mathcal{C} = \{0,1\}^\ell$, $\mathcal{M} = \{0,1\}^l$, *and* $\mathcal{S} = \{0,1\}^L$. *Then for any* t, q *we have*

$$\mathbf{InSec}_{\mathsf{Emb}}^{\mathrm{IND\text{-}CCA}}\Big(t, q, q(2\ell + l)\Big) \leqslant 2 \cdot \mathbf{InSec}_{\mathsf{Emb}}^{\mathrm{PPF}}\Big(t, 2q, 2q(\ell + l)\Big).$$

And for the chosen-message adversary:

Theorem 3. *Let* Emb : $\mathsf{Keys} \times \mathcal{N} \times \mathcal{M} \to \mathcal{S}$ *be a family of embedding function, where* $\mathsf{Keys} = \{0,1\}^k$, $\mathcal{C} = \{0,1\}^\ell$, $\mathcal{M} = \{0,1\}^l$, *and* $\mathcal{S} = \{0,1\}^L$. *Then for any* t, q *we have*

$$\mathbf{InSec}_{\mathsf{Emb}}^{\mathrm{IND\text{-}CMA}}\Big(t, q, q(2\ell + l)\Big) \leqslant 2 \cdot \mathbf{InSec}_{\mathsf{Emb}}^{\mathrm{PPF}}\Big(t, 2q, 2q(\ell + l)\Big).$$

These theorem say that we can give provable assurances of security of a steganographic system based on the assumption that an embedding function is a pseudoprocessing function. Before proving the theorems, we consider the random processing function $\mathsf{Proc} : \mathcal{N} \longrightarrow \mathcal{C}$. Then, we stress the probability that an adversary breaks Proc is zero. For the chose-cover adversary we have:

Lemma 1. *Let* $\mathsf{Proc} : \mathcal{N} \longrightarrow \mathcal{S}$ *be a processing function. Then, for any IND-CCA-adversary attacking*

$$\mathbf{InSec}_{\mathsf{Proc}}^{\mathrm{IND\text{-}CCA}}\left(t, q, \mu\right) = 0.$$

And for the chosen-message adversary:

Lemma 2. *Let* $\mathsf{Proc} : \mathcal{N} \longrightarrow \mathcal{S}$ *be a processing function. Then, for any IND-CMA-adversary attacking*

$$\mathbf{InSec}_{\mathsf{Proc}}^{\mathrm{IND\text{-}CMA}}\left(t, q, \mu\right) = 0.$$

We will prove the lemmas in the full version of this paper. Instead we first see how to use them to prove the theorems. These lemmas say that the steganographic scheme is perfect secure when the embedding function is a processing function. But Emb may be not secure even though Proc is secure. Namely, there may be an adversary having large IND-CMA or IND-CCA advantage in attacking Emb, even though the advantage in attacking Proc is zero. But we will show that this is impossible when Emb is PPF. To prove this, assume that $\mathcal{W}_{\mathrm{CCA}}$ be a IND-CCA-adversary and $\mathcal{W}_{\mathrm{CMA}}$ be a IND-CMA-adversary attacking Emb. We associate them with $\mathcal{W}_{\mathrm{PPF}}$ that is given oracle access to a function $p \in \mathsf{Proc}^{\mathcal{N} \to \mathcal{C}}$ and is trying to determine which set is chosen. $\mathcal{W}_{\mathrm{PPF}}$'s strategy is as follows. First, it runs $\mathcal{W}_{\mathrm{CCA}}$ or $\mathcal{W}_{\mathrm{CMA}}$, and replies to their oracle queries in such a way that $\mathcal{W}_{\mathrm{CCA}}$ and $\mathcal{W}_{\mathrm{CMA}}$ are attacking Proc in $\mathcal{W}_{\mathrm{PPF}}$'s set 0, and they are attacking Proc in $\mathcal{W}_{\mathrm{PPF}}$'s set 1. If $\mathcal{W}_{\mathrm{CCA}}$ or $\mathcal{W}_{\mathrm{CMA}}$ can distinguish the oracles, $\mathcal{W}_{\mathrm{PPF}}$ certainly ensure that p is an instance of Emb, and otherwise $\mathcal{W}_{\mathrm{PPF}}$ declare that p is an instance of Proc. This is the key point of the proof. We give the full proofs of the following theorems in the full version of this paper.

Theorem 4. *Let* $\mathsf{Emb} : \mathsf{Keys} \times \mathcal{N} \times \mathcal{M} \to \mathcal{S}$ *be a family of an embedding function, where* $\mathsf{Keys} = \{0,1\}^k$, $\mathcal{N} = \{0,1\}^\ell$, $\mathcal{M} = \{0,1\}^l$, *and* $\mathcal{S} = \{0,1\}^L$. *Then for any* t, q *we have*

$$\mathbf{InSec}_{\mathsf{Emb}}^{\mathrm{IND\text{-}CCA}}\left(t, q, q(2\ell + l)\right) \leqslant \mathbf{InSec}_{\mathsf{Emb}}^{\mathrm{IND\text{-}CMA}}\left(t, 2q, 2q(\ell + 2l)\right).$$

5   Conclusion

Many practical steganographic algorithms have been introduced to construct a secret communication channel as a part of solutions for the prisoners' problem, however almost all of them are revealed as insecure. Recently, Hopper *et al.* and Chang *et al.* introduced a provable secure steganographic algorithm in the sense of complexity theory. But Chang's works looked more reasonable, but they only defined some broadly. Therefore, we have introduced more detail security notions in Chang's works and the relation between them. With these notions, we may construct an efficient algorithm which has a post-processing algorithm as a subroutine, and proved that it is secure against CCA and CIA. We also showed that some steganographic algorithms are insecure under our adversary models.

Acknowledgement

This work was supported in part by the Ministry of Information & Communications, Korea, under the Information Technology Research Center (ITRC) Support Program.

References

1. R.J. Anderson, "Stretching the Limits of Steganography," in *Information Hiding: First International Workshop*, Lecture Notes in Computer Science 1174, Cambridge, U.K, May 30–June 1, Proceedings: Springer-Verlag, 1996, pp.39–48.
2. C. Cachin, "An Information-Theoretic Model for Steganography," in *Information Hiding - Second International Workshop, IH'98*, Lecture Notes in Computer Science 1525, Portland, Oregon, USA, April 14–17, 1998, Proceedings: Springer-Verlag, 1998, pp.306–318.
3. K. Chang, R Deng, B. Feng, S. Lee, and H. Kim, "On Security Notions of Steganographic Systems," accepted in *International Workshop on Digital Watermarking 2004*, 2004.
4. G. Goldwasser and M. Bellare, Lecture Notes on Modern Cryptography, August, 2001. Available at http://www-cse.ucsd.edu/users/mihir/papers/gb.html.
5. N.J. Hopper, J. Langford, and L. von Ahn, "Provably Secure Steganography," in *Proceedings of Crypto'02*, Lecture Notes in Computer Science 2442, Springer-Verlag, 2002, pp.77–92.
6. N.F. Johnson, Z. Duric, and S. Jajodia, *Information Hiding: Steganography and Watermarking Attacks and Contermeasures*, Kluwer Academic Publishers, 2001.
7. S. Katzen Beisser and F.A.P. Petitcolas, *Information Hiding - techniques for steganography and digital watermarking*, Artech House Books, 1999.
8. S. Katzenbeisser and F.A.P Petitcolas, "Defining Security in Steganographic Systems," in *Proceeding of the SPIE*, vol.4675, Security and Watermarking of Multimedia Contents IV, 2002, pp.50–56.
9. G.J. Simmons, "The Prisoners' Problem and the Subliminal Channe," in *Advances in Cryptology: Proceedings of Crypto'83*, Plenum Press, New York, 1984, pp.51–67.
10. A. Westfeld and A. Pfitzmann, "Attacks on Steganographic Systems," in *Information Hiding - Third International Workshop, IH'99*, Lecture Notes in Computer Science 1768, Springer-Verlag, 1999, pp.61–76.
11. J. Zöllner, H. Federrath, H. Klimant, A. Pfitzmann, R. Piotraschke, A. Westfeld, G. Wicke, and G. Wolf, "Modeling the Security of Steganographic Systems," in in *Information Hiding - Second International Workshop*, Lecture Notes in Computer Science 1525, Portland, Oregon, USA, April 14–17, 1998, Proceedings: Springer-Verlag, 1998, pp.344–354.

A Block Oriented Fingerprinting Scheme in Relational Database

Siyuan Liu[1,3], Shuhong Wang[2,4], Robert H. Deng[4], and Weizhong Shao[1]

[1] Institute of Electronics Engineering & Computer Science,
Peking University (PKU), China 100871
iceice@cs.pku.edu.cn, wzshao@pku.edu.cn
[2] School of Mathematical Sciences, PKU, China 100871
wshong@math.pku.edu.cn
[3] Institute for Infocomm Research, Singapore 119613
[4] School of Information Systems,
Singapore Management University, Singapore 259756
robertdeng@smu.edu.sg

Abstract. The need for protecting rights over relational data is of ever increasing concern. There have recently been some pioneering works in this area. In this paper, we propose an effective fingerprinting scheme based on the idea of block method in the area of multimedia fingerprinting. The scheme ensures that certain bit positions of the data contain specific values. The bit positions are determined by the keys known only to the owner of the data and different buyers of the database have different bit positions and different specific values for those bit positions. The detection of the fingerprint can be completed even with a small subset of a marked relation in case that the sample contains the fingerprint. Our extensive analysis shows that the proposed scheme is robust against various forms of attacks, including adding, deleting, shuffling or modifying tuples or attributes and colluding with other recipients of a relation, and ensures the integrity of relation at the same time.

Keywords: Fingerprinting, Scheme, Block, Security.

1 Introduction

1.1 Background

Due to the rapid development and widespread use of digital assets, such as software, images, video, audio and text, protection of ownership of digital content is increasingly being a matter of great concern. There are many methods to prevent piracy of digital content and fingerprinting, a special type of information hiding technique, is a very promising one. Consider a scenario where merchants sell digital data to buyers. Some dishonest buyers may redistribute the data to others without permission from the merchants. A merchant may use a fingerprint scheme to embed a buyer-specific mark into a data copy and subsequently detect the mark in pirated data and use the mark to identify the traitor who distributed

C. Park and S. Chee (Eds.): ICISC 2004, LNCS 3506, pp. 455–466, 2005.
© Springer-Verlag Berlin Heidelberg 2005

the data. Fingerprinting is often discussed in comparison or extension to watermarking. Watermarking is another type of information hiding technique whose purpose is to identify the sources of data. A merchant may use a watermarking scheme to embed a merchant-specific mark into her data and assert ownership of the data by detecting the watermark. Thus, watermarking is used to embed marks that identify the merchant while fingerprinting is used to embed marks that identify legitimate buyers.

Till now, the study of fingerprinting and watermarking has focused mainly on multimedia content which includes digital images, audio and video. There has recently been some pioneering research presented in [1,2,3] and more recent work of [7,8] in the area of protecting relational database. In [1], the authors present the first known database watermarking technique that marks the numeric attributes of relational data. The algorithm uses a hash function depending on a private key known only to the owner. The hash function decides the tuples, attributes within a tuple, and bit positions within an attribute to be marked. Only when the attackers have access to the private key, can they detect the watermark with a high probability. The technique survives several attacks and preserves mean and variance of all numerical attributes. In [2], the authors generalized the watermarking technique in [1] to enable the fingerprinting of relational data. The fingerprinting technique enables a buyer-specific bit string to be embedded and extracted from a relational database, as compared to the watermarking technique which enables a single watermark bit to be embedded and extracted from a relational database. In [3], the authors extend the technique in [1,2] which is dependent on primary keys and construct a virtual primary key scheme for relational databases which do not have primary keys. The schemes in [1,2,3] are robust against various attacks including flipping bits, adding or deleting tuples and guessing secret keys.

1.2   Related Work

In [4], the authors propose a block oriented fingerprinting scheme in spatial domain, which inspires us very much. The scheme first produces one fingerprint for every buyer and then divides the image to be fingerprinted into a number of blocks of size $\beta \times \beta$ and the number of blocks m is equal to $\frac{ht(I) \times wd(I)}{\beta \times \beta}$ ($ht(I)$ and $wd(I)$ are the height and width of the image in pixels, respectively). Then the scheme permutes the blocks in an order which is specific for every buyer. The permutation and the information of the buyer are both stored in a database known to the merchant only. Then for every block the scheme calculates the minimum and maximum intensities of the pixels in the block, and according to the corresponding bit of the fingerprint, increases intensities of the pixels in the block if the bit is 1 or decreases the intensities if the bit is 0. So every buyer will get one marked image which is different for everyone.

1.3   Our Contribution

In this paper, we propose a novel and flexible scheme to fingerprint a relational database based on the block method for fingerprinting an image as described above. Compared to the previous works, our scheme has three novel features. First, it is based on the block method so that the owner can change the size of the blocks and change the degree of distortion to the database. Secondly, our scheme has low distortion introduced to the data values without compromising the integrity (mean and variance) of the data. As will be seen later, our analysis show that our scheme is resistant to many attacks including collusion attacks. Third, our scheme can be applied to databases with primary keys and with a little extension to databases without primary keys.

The scheme we propose is inspired by the method described above, but is very different from it. Because there exist some fundamental differences between the characteristics of multimedia data and relational data so that we can not carry the multimedia techniques over to the realm of relational database. The differences include:

- It generally does not cause perceptual changes in the object to drop or replace portions of a multimedia object. However, the pirate of a relation can frequently delete, insert or replace the database tuples.
- Multimedia objects consist of a large number of bits, with considerable redundancy, thus providing a larger cover to hide information. A database relation consists of tuples, each of which represents a separate object. The fingerprint needs to be spread over these separate objects.
- The relative spatial/temporal positions of various pieces of a multimedia object is often fixed, but it is not the case for the tuples of a relation database.
- There are many psycho-physical phenomena based on the human visual system and human auditory system which can be exploited for mark embedding. However, one can not exploit such phenomena in relational databases.

Due to the differences, our fingerprinting scheme for relational database is only inspired by the block fingerprinting method for images, and is very different from it.

The paper is structured as follows. Section 2 describes an effective fingerprinting scheme based on the block method. Section 3 analyzes the security of the proposed scheme. The conclusion is given in Section 4.

2   An Effective Fingerprinting Scheme

Let's consider the following scenario. Alice is the owner of a relational database which is sold to many buyers. Later Alice found that someone owned the database but she never sold it to him. She needs certain methods to detect who distributed the database illegally. In this section, we will describe an effective fingerprinting scheme which can be used to embed a fingerprint into the database and detect it when necessary. We assume that the relational database has a primary key. The scheme can be extended for relational databases without primary keys based on the technique in [3].

2.1   Requirements

A fingerprinting scheme should satisfy the following properties.

- Detectability: Alice should be able to detect the fingerprint by examining limited tuples from a suspicious database. The suspicious database may be only a small part of the fingerprinted database or a modified version of the fingerprinted database.
- Imperceptibility: Modifications caused by fingerprinting should not reduce the usefulness of the database. In addition, commonly used statistical measures such as mean and variance of the numerical attributes should not be significantly affected.
- Robustness: A fingerprint scheme should be robust against benign database operations and malicious attacks that may destroy or modify embedded fingerprints. Benign operations include adding tuples, deleting tuples, and updating tuples in database relations. Malicious attacks include selective modifications of fingerprinted relations, taking subsets of relations, and modifying or erasing the embedded fingerprint. These common attacks have been identified in [1,2] and described in the following.

  1. Randomization attacks: Certain bits of a fingerprinted database are assigned random values so that some fingerprint bits may not be detected.
  2. Zero out attacks: Values of some bits of fingerprinted database are changed to zero which results in that the fingerprint can not be detected correctly.
  3. Bit flipping attacks: Values of some bits of fingerprinted database are inverted thus the fingerprint can not be detected correctly.
  4. Rounding attacks: Some bits of fingerprinted database are deleted due to the rounding of numerical values so that the fingerprint may not be detected correctly.
  5. Subset attacks: A subset of tuples or attributes of a fingerprinted relation appear in a pirated database so that the fingerprint can not be detected correctly.
  6. Superset attacks: Some new tuples or attributes are added to a fingerprinted database, which can affect the correct detection of the fingerprint.
  7. Additive attacks: Adding an additional fingerprint to a pirated copy thus to confuse a third party.
  8. Invertibility attacks: Discovering a fictitious fingerprint in a relation thus confusing the owner.
  9. Majority attacks: Creating a new relation with the same schema as the copies but with each bit value computed as the majority function of the corresponding bit values in all copies so that the owner can not detect the fingerprint.
  10. Mix and match attacks: Creating a pirated copy by combining subsets of tuples and attributes from each fingerprinted copy so that the owner can not detect the fingerprint.

The first four types of attacks reduce the accuracy of data. The following two classes of attacks modify relations but didn't reduce accuracy. The seventh and eighth types of attacks seek to provide a traitor or pirate with evidence that raises doubts about a merchant's claims. The last two types of attacks are collusion attacks which require attackers to have access to multiple fingerprinted copies of the same relation but with different embedded fingerprints.

2.2 Notation and Parameters

Consider a database relation R that has a single primary key attribute P and v numerical attributes $A_0, \ldots, A_{v-1}$. Without loss of generality, let the schema of R be $R(P, A_0, \ldots, A_{v-1})$ and let the database has η tuples. For each attribute value $r.A_i$ of tuple $r \in R$, one of its $\xi(r.A_i)$ least significant bits could be used to embed a mark bit. $\xi(r.A_i)$ could depend on the number of bits in a standard binary representation of $r.A_i$, or it could be a constant number that is independent on the value $r.A_i$. To be simple, we use ξ for $\xi(r.A_i)$ unless otherwise stated.

Let n be the number of users (or buyers) to whom the data is being distributed. A fingerprint $\Gamma = (f_0, \ldots, f_{L-1})$ is a binary string with length $L \geqslant logn$. Each user is assigned a unique fingerprint of the same length L. A fingerprint is embedded into each copy of R and the fingerprinted data is then distributed to the corresponding user.

User i's fingerprint is computed by a cryptographic hash function H_0 whose input is the concatenation of a secret key K(known by the merchant only) and user identifier ID_i. The output of H_0 is a binary string of length L. We shall assume that this results in a unique fingerprint for each user $i = 0, \ldots, n-1$. This is usually the case when $L > logn$ because of the collision-free property of the hash function. If collisions do exist, we may use a larger L, reserve the user identifiers that cause collision. We use one pseudo-random producer, which can be the BBS producer, to produce a series of random numbers, and every user have one different threshold for the pseudo-random producer. We also use one cryptographic hash function H_1:

$$H_1(K, ID_i) = H(K\|H(K\|ID_i)) \tag{1}$$

where H is a standard hash function(e.g., MD5 or SHA), and $\|$ denotes concatenation. Table 1 gives the notations we use.

2.3 Insertion Stage

At the stage of fingerprint insertion, we first regard the bits of the attributes that can be used to embed the fingerprint bits as a two-dimension image. For example, Table 2 gives a small part of a relational database. P is the primary key, the last three bits of A_1 and A_2 can be used to embed fingerprint bits. We first extract the three least significant bits of A_1 and A_2 and combine them together as shown in Table 3. Then we divide Table 3 into 6 parts each of size $\beta \times \beta$ (here $\beta = 2$), as given in Table 4.

Table 1. The notions

Notations	Meaning
ν	The number of attributes in the relational database that can be marked
η	The number of tuples in the relational database
ξ	The number of the least significant bits that can be used for marking in an attribute
β	The size of every block
n	The number of users

Table 2. Part of a relational database

P	A_1	A_2
1	01100011	00001001
2	10000010	00100111
3	01001111	10010001
4	00000000	00000101

Table 3. The bits available for fingerprinting

011001
010111
111001
000101

Table 4. The 6 2×2 blocks

01	10	01
01	01	11
11	10	01
00	01	01

Table 5. The insertion algorithm

```
1. for one buyer
2.     produce the fingerprint Γ for the buyer
3.     choose one threshold r₀ for the pseudo random number generator
4.     divide the database attributes bits into blocks of size β × β
5.     i=0,j=0
6.     for each block Bᵢ
7.         r₁ =random(r₀)
8.         x = H₁(r₁, ID) mod β
9.         r₂ =random(r₁)
10.        y = H₁(r₂, ID) mod β
11.        Bᵢ(x, y) = Bᵢ(x, y) ⊕ fⱼ
12.        r₀ = r₂
13.        i++,j++ if j==L then j=0
14.    end for
15. end for
```

Now we use the pseudo random generator to produce a random number r, and according to the result of $H_1(r, ID) \bmod \beta$ to decide where the fingerprint bit should be embedded. Table 5 gives the insertion algorithm.

Table 6. The detection algorithm

---

1. sort S to S' according to the primary key

2. divide bits in S' into blocks of size $\beta \times \beta$

3. for each buyer, retrieve the corresponding r_0:

4.   for each block B_i

5. $r_1 =$ random(r_0)

6. $x = H_1(r_1, ID) \bmod \beta$

7. $r_2 =$ random(r_1)

8. $y = H(r_2, ID) \bmod \beta$

9. $F_i = S'(B_i(x, y)) \oplus R(B_i(x, y))$ if $S'(B_i(x, y))$ is in S

10. $r_0 =$ random(r_2)

11.   i++

12.   end for

13. end for

14. for each buyer, retrieve his fingerprint $\Gamma = (f_0, \ldots, f_{L-1})$

15.   define $f'_i = 1, 0 \leq i \leq L - 1$, if $\frac{\#\{k| F_{i+kL}=1,0\leq i+kL\leq m-1\}}{\omega} \geq \tau$, otherwise define $f'_i = 0$.

16.   if $\Gamma = \Gamma' = (f'_0, \ldots, f'_{L-1})$, the data is said to had distributed by this buyer.

Line 7 and line 9 use a pseudo random number generator to produce a random number, respectively. Line 8 and line 10 determine the position in a block the fingerprint bit should be embedded. Line 13 means that the next fingerprint bit is ready to be embedded and the next block is ready to be marked. When all the fingerprint bits have been embedded and the blocks are not over the fingerprint bits will be embedded again until all the blocks have been used.

2.4   Detection Stage

In the fingerprint detection stage, the merchant first sorts the suspicious database S according to the primary key. If there are some tuples deleted, they are added as in the unfingerprinted original data according to the primary key. Then divide the bits that can be used to embed fingerprint bits into blocks of size $\beta \times \beta$ and mark the blocks which are included in R but not in S. Comparison to the original blocks which don't contain fingerprint, the merchant decides whether the suspected relational database is pirated or not. Table 6 gives the fingerprint detection algorithm.

Line 9 means that for this buyer, the i-th block is detected being marked with F_i. $S(B_i)$ is the i-th block in the suspicious database and $R(B_i)$ is the i-th block in the original database. In line 15, m is the number of blocks, ω times is the number of times a fingerprint has been embedded in the database. If $\lceil \tau\omega(\tau \in [0.5, 1]) \rceil$ detected bits are 1, the detected fingerprint bit is said to be 1, otherwise is said to be 0.

2.5   Fingerprinting Relational Databases Without Primary Keys

The fingerprinting scheme described above is predicated on the assumption that the relational database have a primary key. And our scheme can be easily extended to be used for relations that are without primary keys based on the techniques proposed in [3].

3   Robustness

In this section, we analyze the robustness of our fingerprint algorithm against representative attacks under the assumption that the attackers don't change the values of primary keys. In addition, we also investigate *false hit rate*, the probability of failing to detect an embedded codeword correctly. Let R be a fingerprinted relation with the embedded fingerprint $\Gamma = (f_0, \ldots, f_{L-1})$.

3.1   Some Discussion

An important parameter in our scheme is β, which determines the size of the block. If β is large, the number of blocks reduces and the number of times a fingerprint being embedded also reduces. But if β is too small, and the number of buyers is more than the size of the block, the situation will happen that different buyers' fingerprint is embedded into the same bit position in the same block. To avoid the situation, the size of the block should be more than the number of buyers, meaning that β should be at least more than $\sqrt{n}$. On the other hand, if β is too big, the fingerprint can be embedded for only a small number of times, the correctness of the detection may be impaired. The times that the fingerprint can be embedded can be computed as $\omega = \lfloor \frac{\xi \eta \nu}{\beta \times \beta \times L} \rfloor$. It is to say that every fingerprint bit f_l is embedded ω times

3.2   Cumulative Binomial Probability

We use Bernoulli trials in our robustness analyze. Repeated independent trails are called Bernoulli trials if there are only two possible outcomes for each trial and their probabilities remain the same throughout the trials. Let $b(k; n, p)$ be the probability and n Bernoulli trials with probabilities p for success and $1 = 1-p$ for failure result in k successes and $n - k$ failures. Then

$$b(k; n, p) = \binom{n}{k} p^k q^{n-k} \tag{2}$$

$$\binom{n}{k} = \frac{n!}{k!(n-k)!} \tag{3}$$

Denote the number of successes in n trials as S_n. The probability of having at least k successes in n trials, the cumulative binomial probability, can be written as

$$P\{S_n \geqslant k\} = \sum_{i=k}^{n} b(i; n, p) \tag{4}$$

For brevity, define

$$B(k; n, p) = \sum_{i=k}^{n} b(i; n, p) \tag{5}$$

3.3   Bit-Flipping Attacks

In a bit-flipping attack, an attacker selects some bits and toggles their values. In our scheme, the bit positions used for fingerprinting are computed by a pseudo-random generator which has a threshold known only to the merchant and a cryptographic hash function. We assume that the attacker does not know the threshold so that he has no information about the values or positions of embedded bits. We also assume that the attacker possesses a single fingerprinted copy of the data. Now, let the attacker examine each bit available for fingerprinting independently and select it for flipping with probability p. Let $q = 1 - p$. We model bit flipping as Bernoulli trials with probability p of success and q of failure. Let the attacker apply the attack to a fingerprinted relation. Consider the probability p_l that one particular fingerprint codeword bit f_l is destroyed. Each fingerprint bit f_l is actually embedded w times as discussed above. For the detection algorithm to fail to recover the correct fingerprint bit, at least $(1 - \tau)w$ embedded bits that correspond to the fingerprint bit must be changed. It also can be said that more than $w - \lceil \tau w \rceil + 1$ bits must be changed. So $p_l = B(w - \lceil \tau w \rceil + 1; w, p)$. The probability that the codeword bit is recovered is $q_l = 1 - p_l$. Then the probability that the entire codeword is recovered correctly is $\prod_l q_l = (1 - p_l)^L$. And *the false hit rate is* $1 - (1 - p_l)^L$. Table 7 describes the probability of a successful attack for different parameter values. Here we set $\eta = 100000, \xi = 3, \nu = 3$, $\tau = 0.5$ and $L = 100$. We can see that when β is increasing and p is less than 40% the probability for a successful bit-flipping attack is also increasing for the same p. And when p is more than 40%, the probability is decreasing while β is increasing. So we can choose the appropriate β. For example, $\beta = 30$ is adapt to the situation when the length of fingerprint is 100 bits. Comparison to the scheme in [2], there is some development in our scheme on robustness against bit-flipping attacks.

Table 7. The probability for a successful bit-flipping attack for different block sizes

	$p = 10\%$	$p = 20\%$	$p = 30\%$	$p = 40\%$	$p = 50\%$
$\beta = 5$	0	0	0	0	0
$w = 360$	0	0	0	0	0
$\beta = 10$	0	5.0610×10^{-9}	2.2564×10^{-3}	0.8837	1.0000
$w = 90$	0	8.3235×10^{-8}	1.2954×10^{-2}	0.9962	1.0000
$\beta = 15$	1.9597×10^{-9}	5.0260×10^{-4}	0.2151	0.9996	1.0000
$w = 40$	1.6237×10^{-7}	8.4818×10^{-3}	0.7743	0.9999	1.0000
$\beta = 20$	2.4620×10^{-5}	3.4324×10^{-2}	0.7567	0.9999	1.0000
$w = 22$	2.0701×10^{-3}	0.4599	0.9999	1.0000	1.0000

Table 7 also gives the comparison result on the probability for a successful bit-flipping attack for different w based on the scheme in [2]. For every cell, the first line is the probability for our scheme and the second line is the probability for the scheme in [2].

3.4   Subset Attacks

Consider a subset attack where the pirated data is a subset of tuples of a finger-printed relation. Note that a relation has η tuples and that an attacker examines each tuple independently and selects it with probability q' for inclusion in the pirated relation. The pirated relation will thus have $\zeta = q'\eta$ tuples on average. The probability that a tuple is deleted is $p' = 1 - q'$. Suppose that a subset attack is applied to a fingerprinted relation and that there is no other attack or benign update on the data. Then, for the attack to be successful, it must delete at least $\lfloor \omega - \tau\omega \rfloor$ embedded bits for some codeword bit. Now, each codeword bit f_l is embedded ω times in the original relation, so the probability μ_l that a codeword bit f_l is erased completely is $\mu_l = B(\lfloor \omega - \tau\omega \rfloor; \omega, p')$. Then, $v_l = 1 - \mu_l$ is the probability that a codeword bit f_l is detected, $\prod_l v_l$ is the probability that the entire codeword is detected correctly, and $1 - \prod_l v_l$ is the false miss rate.

Table 8 shows the probability of a successful attack for different parameter values. Here we set $\eta = 100000, \xi = 3, \nu = 3, \tau = 0.5$ and $L = 100$. We can see that when β is increasing the probability for a successful subset attack is also increasing for the same p'. So we can change β to adapt to different needs.

Table 8. The probability for a successful subset attack for different block sizes

	$p' = 10\%$	$p' = 20\%$	$p' = 30\%$	$p' = 40\%$	$p' = 50\%$
$\beta = 5$	0	0	0	0	0
$\beta = 10$	0	2.09722×10^{-8}	5.5264×10^{-3}	0.9706	1.0000
$\beta = 15$	1.8719×10^{-8}	2.1669×10^{-3}	0.4660	0.9999	1.0000
$\beta = 20$	2.4596×10^{-4}	0.1471	0.9808	0.9999	1.0000

3.5   Attribute Attacks

If an attacker adds one new attribute into a fingerprinted relation. Because our detection algorithm first reassorts the suspected relation, the new attribute can be omitted.

If an attacker delete some attributes from a fingerprinted relation, tuples in which the deleted attributes were marked can be regarded as deleted. The situation can be analyzed as tuple deletion described in section 3.4.

If an attacker modifies some attributes from a fingerprinted relation, the situation can be analyzed as the bit-flipping attacks.

3.6   Collusion Attacks

Fingerprinting schemes are susceptible to collusion attacks by coalitions with access to multiple fingerprinted copies of the relation but with different embedded fingerprints. The attackers can create a useful data copy that does not implicate anyone of the attackers. During fingerprint detection, the copy may yield the fingerprint of an innocent buyer, or it may not yield a valid fingerprint at all.

There are many solutions to the collusion problem, a well-known of which was proposed by Boneh and Shaw [5], and many others have been proposed such as [6]. These solutions focus on the choice of codewords used by a fingerprinting scheme. They show that by a proper choice of codewords, fingerprinting schemes can be made collusion secure. Most of the solutions need the fingerprinting satisfy two properties, one is that an attacker can only detect that a bit position was used during fingerprint insertion if the attacker has data copies that differ in value at that position and the other is that although an attacker may determine that a particular data bit was used to embed some codeword bit, the attacker cannot determine which codeword bit it represents. The fingerprinting schemes described in the paper satisfy these properties because the positions the codeword is embedded are hidden using a pseudo random function. And because β is known to the owner, the attacker do not know the relationship between the embedded bits and the codeword bits. On the other hand, different buyers' codewords are embedded in different locations, which also increase the difficulties an attacker can destroy the fingerprint. Thus, we can use any of those collusion-secure codeword schemes by replacing the hash function H_0 in the fingerprinting algorithms. This has been demonstrated in [2] using an adapted version of Boneh and Shaw's algorithm. It's the same for our schemes.

3.7   Additive Attacks

In an additive attack, an attacker may insert another mark before distributing a pirated database. A traitor may insert a watermark to claim ownership of the database and he may insert a fingerprint to claim that the database was provided to a user legitimately. This type of attack is discussed in [1] in the context of watermarking, the solution they propose is applicable to our fingerprinting schemes as well.

3.8   Distortion to Integrity and Consistency

In our scheme, the bits ready for embedding fingerprint bits are the least significant bits of the candidate attributes. The bits are randomly chosen for embedding fingerprint bits, which means that it is impossible for the bits are mostly embedded in only one or few attributes. So the changes of the mean and variance of the candidate attributes are almost imperceptive, which ensure the integrity and consistency of the relation.

4   Conclusion

In this paper, we have presented the scheme for embedding and detecting fingerprints in relational databases based on a block method. In addition, we have presented security analysis to show the robustness of our technique against various attacks.

For future work, we would like to optimize the detection process and investigate the possibility of extending our embedding scheme for both non-numeric and numeric attributes.

References

1. R.Agrawal and J.Kiernan. Watermarking relational databases. Proc. 28th International Conference on Very Large Data Bases (VLDB 2002), 2002.
2. Y.Li, V.Swarup, S.Jajodia. Fingerprint relational databases. Technical report, Center for secure Information Systems, George Mason University, Fairfax, VA, May 2003.
3. Yingjiu Li, Vipin Swarup, Sushil Jajodia. Construting a virtual primary key for fingerprinting relational data. DRM 03.
4. Tanmoy Kanti Das, Subhamoy Maitra. A robust block oriented watermarking scheme in spatial domain. 4the International Conference(ICICS 2002),2002.
5. D.Boneth and J.Shaw. Collusion secure fingerprint for digital data. IEEE Transactions on Information Theory, 44(5):1897-1905,1998.
6. H. Guth and B. Pfitzmannn. Error and collusion secure fingerprinting for digital data. In Information Hiding '99, LnCS 1768, Springer-Verlag, pages 134-145,2000.
7. R.Sion, M.Atallah, and S. Prabhakar. On watermarking numeric datasets. In Proc. First International Workshop on Digital Watermarking(IWDW 2002), 2002.
8. R.Sion, M.Atallah, and S. Prabhakar. Rights protection for relational data. In Proc. ACM International Coference on Management of Data(SIGMOD 2003), 2002.
9. E.Bertino, S. Jajodia, and P. Samarati. A flexible authorization merchanism for relational data management systems. ACM Transactions on Information Systems, 17(2):101-140, 1999.
10. D. Gross-Amblard. Query-preserving watermarking of relational databases and xml documents. In Proc. of the Twenty-second ACM SIGMOD-SIGACT-SIGART Symposium on Pringciples of Database Systems(PODS 2003), 2003.
11. S. Katzenbeisser and E.F. Petitcolas. Information Hiding Techniques for Steganography and Watermarrking. Artech House, Boston MA, 2000.
12. S. Khanna and F.Zane. Watermarking maps: hiding information in structured data. In Proc. Symposium on Discrete Algorithms(SPDA),2000.
13. W. Stallings. Cryptographic Techniques and Data Security. Third Edition. Prentice Hall, 2002.

A Study on Evaluating the Uniqueness of Fingerprints Using Statistical Analysis

Y. Han[1], C. Ryu[1], J. Moon[1], H. Kim[1], and H. Choi[2]

[1] School of Information and Communication Eng., Inha University,
#253 Yonghyun-Dong, Nam-Ku, Incheon, Korea
anilyca@ieee.org
cwryu, jhmoon @vision.inha.ac.kr
[2] Dept. of Information Eng., Myongji University,
San #38-2, Namdong, Yongin, Kyungkido, Korea
hschoi@mju.ac.kr

Abstract. Biometrics-based methods for personal authentication assume that the biometric characteristics used for the verification of an individual's identity are unique from person to person. The purpose of this study is to verify the uniqueness of fingerprints, by analyzing the similarity of live-scanned fingerprints for all ten fingers of twins and family members. In order to maintain the consistency and to guarantee the repeatability of the analysis, we established an evaluation framework, and studied the uniqueness of fingerprints from two points of view, namely the similarity between fingerprint types and the distribution of the similarity scores produced by a minutia-based matching algorithm. Preliminary experiments were carried out using the live-scanned ten-finger fingerprints of sixty-six twins and fifty-two families consisting of the parents and two children. The results demonstrate that fingerprints are sufficiently unique to distinguish one person from another, with an insignificant decrease in the recognition accuracy for identical twins.

Keywords: Uniqueness, Contingency Table, Personal authentication, Twin, Family, Fingerprint.

1 Introduction

Traditional methods of personal identification, such as passwords and access cards, are prone to be deceived, because tokens may be stolen and passwords lost or forgotten. On the other hand, the biological characteristics of human beings cannot be forgotten, easily shared or misplaced. Moreover, biometrics-based authentication requires that the person to be authenticated be present at the point of authentication to provide his biometric measurements [1]. Biometrics consists of a series of processes including the extraction of the unique physical or behavioral properties from a human body and their authentication against a set of previously enrolled properties, so called, the enrolled template. It assumes that the biometric traits of an individual are sufficiently unique to distinguish one person from another.

C. Park and S. Chee (Eds.): ICISC 2004, LNCS 3506, pp. 467–477, 2005.

Fingerprints are part of an individual's phenotype, which arise from the interaction of the individual's genes and the developmental environment in the uterus [2], and are fully formed when the fetus has developed for about seven months. Furthermore, it is known that the finger ridge configurations remain unchanged throughout a person's lifetime, except in the case of accidents, such as bruises and cuts on the finger tips. The flow of amniotic fluids and the position of the fetus in the uterus change the growth patterns of the cells on the fingertip and determine the structure of the fingerprints. While the differences in the microenvironment between the fingers are small and subtle, their effect is amplified by the differentiation of the cells, and this produces macroscopic differences that enable the fingerprints of twins to be differentiated [3].

The purpose of this study is to verify the uniqueness of fingerprints, by analyzing the similarity between the live-scanned fingerprints for all ten fingers of twins and family members. Jain, et al.[1] quantitatively determined the similarity between the fingerprints of identical twins collected from the National Heart, Lung and Blood Institute (NHLBI), and assessed the impact of this similarity on the performance of automatic fingerprint-based verification systems. However, they used only ink-rolled-scanned thumb fingerprints for their similarity analysis of the twin fingerprints and, in their study, the similarity of family members' fingerprints was not considered.

To the best of our knowledge, there have been no studies in which a statistical analysis was performed for the purpose of verifying the uniqueness of fingerprints. In this paper, we propose a framework for evaluating the uniqueness of fingerprints and provide empirical methods by which family or identical twin's fingerprints can be distinguished from each other with slightly lower accuracy than those of non-twins or the members of different families, from the standpoint of the score distribution. Furthermore, in this paper, we analyze the correlation between different fingerprint types, using the mirror effect and a contingency table. For the purpose of this study, we constructed a ten-finger live-scan fingerprint database from the fingerprints of 66 pairs of twins (51 identical and 15 fraternal twins) and 52 families.

The remainder of this paper is organized as follows: Section 2 explains the framework used for evaluating the uniqueness of fingerprints. Section 3 briefly describes the algorithm used for automatic fingerprint verification. Section 4 presents the experimental procedures and results. Finally, Section 5 ends this paper by drawing various conclusions.

2   Evaluation Framework

As in the case of other technologies, the uniqueness evaluation for fingerprint recognition technology requires generality, expertise, fairness, and reliability [4-8], that is to say:

i) Generality: the method employed must be applicable to various types of technology involving different biometric modes.

ii) Expertise: the method must be designed and carried out by people who have a good knowledge of the technology in question.

iii) Fairness : the method used and the evaluation indices must be fairly designed.

iv) Reliability : the results of the evaluation must be reliable.

The proposed framework used for the uniqueness evaluation of twins and families' fingerprints consists of seven elements, as shown in Fig. 1, viz. the evaluation target, evaluation type, evaluation criteria, evaluation method, evaluation environments and construction of the database, test, and reporting of the results. In our framework, the evaluation target is the uniqueness of the twins and families' fingerprints. A fingerprint is a biometric piece of information used in personal authentication, however its properties are inherited by the chromosome. The purpose of this work is to analyze the uniqueness of the fingerprint from the viewpoint of the genetic relationship, using the fingerprints of twins and families as the evaluation target.

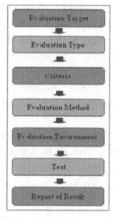

(a) Evaluation Procedure

(b) Evaluation framework

Fig. 1. Procedure and Framework used for uniqueness evaluation

The evaluation type is established using a technological evaluation and offline test. It is difficult to perform an operational or scenario test, because a suitable group of volunteers is not always available for such an evaluation. In technological evaluations, the test data should be collected in the environment that is to be tested, and it should be neither too difficult nor too easy for this data to match the enrollment templates. Also, the time interval between the enrollment templates and the test data should be considered. Long time intervals generally make it more difficult to match the search and file templates, due to template aging. Presentation and channel effects are either uniform or vary randomly across volunteers. Systematic variations caused by presentation and channel effects between the enrollment and test data will inevitably lead to the results being distorted by these factors. The advantage of using a technological evaluation is that the database that is compiled can be used to evaluate the algorithm being developed and to tune it through the process of feedback.

For the evaluation criteria, three points of view are considered, viz. the class correlation, the similarity of the fingerprints and the system performance. For the class correlation, six fingerprints classes are used, viz. the whorl (W), double loop (DL), left loop (LL), right loop (RL), arch (A) and tented arch (TA). The class correlation is a comparison of the frequency of occurrence of an identical fingerprint class between two sets of fingerprints. The score distribution is produced by a proprietary minutiae-based fingerprint algorithm, which outputs the similarity score between the enrollment and test data. Systematically speaking, the False Match rate(FMR) and False Non-match rate(FNMR) are considered by this algorithm.

In the evaluation method, the class correlation step first involves the classification of the entire fingerprint database. This procedure was performed by expert, who is well versed in this particular issue. Then, various analyses were carried out, such as the probability of the same type occurring for two fingerprints obtained from a pair of corresponding fingers of a child and one of parents, the probability of the two fingerprints having the same type when both the father and mother have the same type for each corresponding fingers, the hereditary trend from father and mother, separately, to the children. The score distribution is generated by a genuine and imposter matching process, using the FVC2000 protocol. In order to avoid the possible correlation among fingerprints of ten fingers of each individual, within-individual comparisons are not included in the set of imposter transactions.

The database was constructed by scanning the fingerprints in a normal office environment using the fingerprints of 66 pairs of twins and 52 families, and then tests and reports were carried out, in order to evaluate the uniqueness of the fingerprints contained in the constructed database. To reduce the amount of errors in the database, a checklist was established.

a) Use the system correctly
b) Use the correct PIN(Personal Identification Number)
c) Collect good quality images
d) Use the correct body parts
e) Check for blank and corrupted images
f) Minimize the amount of data requiring keyboard entry
g) Allow the volunteers to visualize the input image instantly
h) In case of need, the volunteers should be able to obtain help from a supervisor
i) The supervisor must be fully aware of how to use the data collection software
j) Perform three impressions of each finger for all ten fingers

3   Minutiae Based Fingerprint Recognition Algorithm

An automatic fingerprint recognition algorithm consists of two basic steps, namely feature extraction and matching. Feature extraction produces a set of

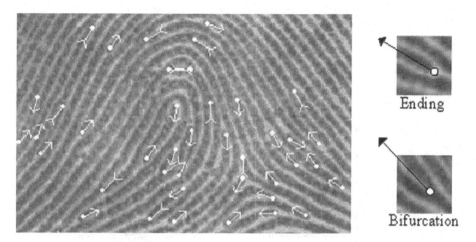

Fig. 2. Definition of fingerprint minutiae types

local ridge characteristics or minutiae, as shown in Fig 2., which together constitute the so called feature template. A feature template $T = \{m_1, m_2, ..., m_3\}$ consists of multiple feature vectors whose components are its type, position and orientation.

A minutiae-based fingerprint matching algorithm is generally decomposed into three stages:

i) The alignment stage: the parameters used for the transformation between the two fingerprint images are estimated, and the input minutiae are aligned with the enrolled minutiae according to the estimated parameters.

ii) The matching stage: the corresponding minutiae are determined and the difference in the angle and position of each pair of corresponding minutiae are computed.

iii) The scoring stage: a similarity measure between the two fingerprints is calculated using a decision strategy.

The overall matching process for the algorithm used in this study is depicted in Fig. 3. This algorithm produces triangular minutiae structures named cliques from the minutiae of the search and file fingerprints, where the file fingerprint is the fingerprint registered in the enrollment process and the search fingerprint is the fingerprint being authenticated. The geometry of a clique is shown in Fig. 4. A pair of cliques from the file and search fingerprints is said to be identical if all the elements of the cliques are within an allowable range of similarity. The amounts of translation and rotation are calculated for all pairs of cliques [9-10].

Two cliques are said to be identical if the following conditions are satisfied. Note that, in equations (1) through (4), the superscript F denotes the file fingerprint, while the superscript S denotes the search fingerprint.

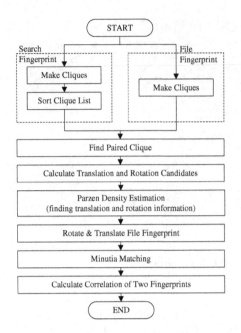

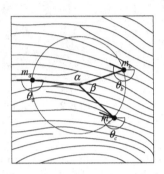

Fig. 3. Overview of the proposed algorithm

Fig. 4. Geometry of cliques over thinned fingerprint image

a) The radius of the cliques r must be similar.

$$\frac{\min(r^F, r^S)}{\max(r^F, r^S)} \geq r_{th} \tag{1}$$

b) The angles α and β must be similar.

$$|\alpha^F - \alpha^S| \leq \alpha_{th}, |\beta^F - \beta^S| \leq \beta_{th} \tag{2}$$

c) The minutiae angles $\theta_a, \theta_b, \theta_c$ must be similar.

$$|\theta_a^F - \theta_a^S| \leq \theta_{th}, |\theta_b^F - \theta_b^S| \leq \theta_{th}, |\theta_c^F - \theta_c^S| \leq \theta_{th} \tag{3}$$

d) The minutiae types $\zeta_a, \zeta_b, \zeta_c$ must be the same.

$$T(\zeta_a^F, \zeta_a^S) = \begin{cases} 1 & \text{if } \zeta_a^F \equiv \zeta_a^S \\ 0 & otherwise \end{cases} \tag{4}$$
$$T(\zeta_a^F, \zeta_a^S) = T(\zeta_b^F, \zeta_b^S) = T(\zeta_c^F, \zeta_c^S) = 1$$

4   Experimental Results

For this study, we constructed a live-scan fingerprint database consisting of the fingerprints obtained from 66 pairs of twins and 52 families(comprising 4 members : the father, mother and two children). Each fingerprint was scanned three times using an optical sensor in 256 level grayscale mode at a resolution of 500 dpi in a normal office environment. In each case, this process was repeated for all ten fingers. The resulting twins database contained 3,960 (66 pairs × 3 times × 2 members × 10 fingers) images with a size of 248 × 292, while the family database contained 6,240 (52 families × 3 times × 4 members × 10 fingers) images.

Next, we will discuss the uniqueness of the twins and family members' fingerprints from the following points of view: class correlation, fingerprint similarity and system performance.

We begin by examining how much correlation exists between the fingerprints. Firstly, all of the fingerprints were manually classified by an expert in the field of fingerprint recognition area into six classes, viz. the left loop(LL), right loop(RL), double loop(DL), whorl(W), arch(A) and tented arch(TA).

Table 1. Mirror effects of the fingerprint database

	Index Mir.	Little Mir.	Middle Mir.	Ring Mir.	Thumb Mir.	Total Mir.	Extra Mir.	#of Comp. Mir.
Twin-Gen.	54	32	32	70	46	234	3726	3960
Ratio divide by Finger	1.4	0.8	0.8	1.8	1.2	5.9	94.1	100.0
Twin-Impo.	43	27	23	65	40	198	3762	3960
Ratio divide by Finger	1.1	0.7	0.6	1.6	1.0	5.0	95.0	100.0
Family-Gen.	80	48	64	126	78	396	5844	6240
Ratio divide by Finger	1.3	0.8	1.0	2.0	1.3	6.3	93.7	100.0
Family-Imp.-Parent-Child	104	64	79	198	82	527	11935	12480
Ratio divide by Finger	0.8	0.5	0.6	1.6	0.7	4.2	95.8	100.0
Family-Impo.-Father-Mother	23	15	19	46	16	119	3001	3120
Ratio divide by Finger	0.7	0.5	0.6	1.5	0.5	3.8	96.2	100.0

We investigated the mirror effect, which is used to indicate whether a particular finger in one hand has the same pattern type as the corresponding finger in the other hand [6]. Table 1 summarizes the results of the analysis of the mirror effect. In the comparison of the 3,960 images, index mirroring occurred 54 times in the case of twin-genuine mirroring, representing a 1.4% ratio, while total mirroring occurred 234 times, representing a 5.9% ratio. The result shown in Table 1 led us to the conclusion that the frequency of genuine mirroring is higher than that of imposter mirroring and that the frequency of twin imposter mirroring is higher than that of parent-child or father-mother imposter mirroring. From this, we can formulate a contingency table of twin class correlations, as shown in Table 2. Table 2 shows that class correlations are not independent between twins. We can see from Table 3 that in the case of the RL class, the child's class is not independent of the parent's class. That is, the child tends to belong to the RL class when both the father and mother belong to the same class.

Secondly, we evaluated the similarity between the fingerprints using a proprietary minutiae-based fingerprint recognition algorithm based on the FVC2000 protocol [7]. In the imposter matching as shown in Fig. 5(a) and (b), a higher score distribution was observed for the matching of twin siblings in the twin

Table 2. Contingency Table for twin-twin imposter

Frequency Expected Percentage Row% Colume%	A	DL	LL	RL	TA	W	Total
A	6	0	1	1	0	0	8
	0.1939	1.0667	1.6848	1.9273	0.1576	2.9697	
	0.91	0	0.15	0.15	0	0	1.21
	75	0	12.5	12.5	0	0	
	37.5	0	0.72	0.63	0	0	
DL	0	39	7	6	0	23	75
	1.8182	10	15.795	18.068	1.4773	27.841	
	0	5.91	1.06	0.91	0	3.48	11.36
	0	52	9.33	8	0	30.67	
	0	44.32	5.04	3.77	0	9.39	
LL	2	19	108	14	5	19	167
	4.0485	22.267	35.171	40.232	3.2894	61.992	
	0.3	2.88	16.36	2.12	0.76	2.88	25.3
	1.2	11.38	64.67	8.38	2.99	11.38	
	12.5	21.59	77.7	8.81	38.46	7.76	
RL	7	4	2	122	6	25	166
	4.0242	22.133	34.961	39.991	3.2697	61.621	
	1.06	0.61	0.3	18.48	0.91	3.79	25.15
	4.22	2.41	1.2	73.49	3.61	15.06	
	43.75	4.55	1.44	76.73	46.15	10.2	
TA	1	0	4	3	2	0	10
	0.2424	1.3333	2.1061	2.4091	0.197	3.7121	
	0.15	0	0.61	0.45	0.3	0	1.52
	10	0	40	30	20	0	
	6.25	0	2.88	1.89	15.38	0	
W	0	26	17	13	0	178	234
	5.6727	31.2	49.282	56.373	4.6091	86.864	
	0	3.94	2.58	1.97	0	26.97	35.45
	0	11.11	7.26	5.56	0	76.07	
	0	29.55	12.23	8.18	0	72.65	
Total	16	88	139	159	13	245	660
Total	2.42	13.33	21.06	24.09	1.97	37.12	100

database and the matching of non-twin siblings in the family database, respectively. The twin-twin imposter score distribution had the highest score distribution, followed by that between the non-twin siblings, that between the parents and children, and finally the father-mother imposter score distribution, in that order.

Finally, we examined the matching algorithm from a system performance point of view. The experimental results show that the Equal Error rate (EER) is generally 1−2% higher than that of the twin-nontwin or father-mother imposter matching (Fig. 6).

Table 3. Contingency Table for family imposter(Parent-Child)

Frequency Expected Percentage Row% Colume%	A	DL	LL	RL	TA	W	Total
A	0	2	0	2	0	4	8
	0.5211	0.7356	0.1533	4.7203	0.0613	1.8084	
	0	0.77	0	0.77	0	1.53	3.07
	0	25	0	25	0	50	
	0	8.33	0	1.3	0	6.78	
DL	0	3	0	7	0	7	17
	1.1073	1.5632	0.3257	10.031	0.1303	3.8429	
	0	1.15	0	2.68	0	2.68	6.51
	0	17.65	0	41.18	0	41.18	
	0	12.5	0	4.55	0	11.86	
LL	0	4	2	2	0	3	11
	0.7165	1.0115	0.2107	6.4904	0.0843	2.4866	
	0	1.53	0.77	0.77	0	1.15	4.21
	0	36.36	18.18	18.18	0	27.27	
	0	16.67	40	1.3	0	5.08	
RL	8	10	2	103	2	24	149
	9.705	13.701	2.8544	87.916	1.1418	33.682	
	3.07	3.83	0.77	39.46	0.77	9.2	57.09
	5.37	6.71	1.34	69.13	1.34	16.11	
	47.06	41.67	40	66.88	100	40.68	
TA	0	0	0	0	0	2	2
	0.1303	0.1839	0.0383	1.1801	0.0153	0.4521	
	0	0	0	0	0	0.77	0.77
	0	0	0	0	0	100	
	0	0	0	0	0	3.39	
W	9	5	1	40	0	19	74
	4.8199	6.8046	1.4176	43.663	0.567	16.728	
	3.45	1.92	0.38	15.33	0	7.28	28.35
	12.16	6.76	1.35	54.05	0	25.68	
	52.94	20.83	20	25.97	0	32.2	
Total	17	24	5	154	2	59	261
Total	6.51	9.2	1.92	59	0.77	22.61	100

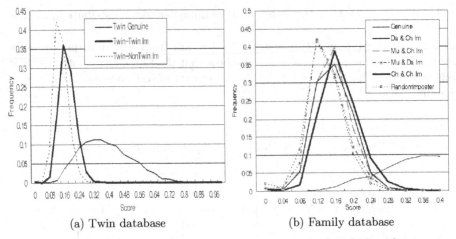

(a) Twin database   (b) Family database

Fig. 5. Similarity score distribution using minutiae−based algorithm

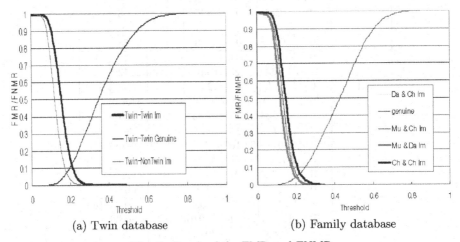

(a) Twin database   (b) Family database

Fig. 6. Graph of the FMR and FNMR

5   Conclusion

In this work, we proposed a framework for the evaluation of the uniqueness of
human fingerprints and implemented this in the form of an algorithm based on
the FVC2000 protocol. This algorithm was tested on an experimental database,
and the experimental results were analyzed from the point of view of several
evaluation criteria.

The results of the experiments show that there is some correlation between
both the class and minutiae-based similarity between the fingerprints of parents
and their children, and the same pattern was also observed for identical twins.
The similarity between the fingerprints of siblings was found to be higher than
that between those of parents and their children.

In conclusion, we attempted to establish the impact of similarity on the performance of the minutiae based fingerprint algorithm, particularly in the case of twins and sets of parents and children. Even though the fingerprints of twins exhibit a high class correlation, they can still be distinguished using the minutiae-based fingerprint recognition algorithm due to the uniqueness of the fingerprints, although with a slightly higher error rate than those of non-twins.

More research is necessary to develop a class independent matching algorithm. We also intend to statistically analyze the genetic relations and inherited pattern conditions based on twin and family fingerprints. This will form the foundation for the development of a fingerprint algorithm for the analysis of a database containing genetically intimate relations, namely relations having high class similarity.

Acknowledgements

This work was supported in part by the Biometrics Engineering Research Center, KOSEF.

References

1. Anil Jain, Salil Prabhakar, Sharath Pankanti, "On the similarity of identical twin fingerprints," *Pattern Recognition*, Volume 35, Issue 11, pp. 2653-2663, 2002.
2. E.P. Richards, "Phenotype vs. Genotype: Why Identical Twins Have Different Fingerprint?", http://www.forensic−evidence.com/site/ID_Twins.html
3. Anil K. Jain, Salil Prabhakar and Sharath Pankanti, "Twin Test: On Discriminability of fingerprints", *Lecture Notes in Computer Science on AVBPA*, Vol.1, No.2091, pp. 211-216, 2001.
4. CESG/BWG "Biometric Test Program", online at, http://www.cesg.gov.uk/biometrics/pdfs/BiometricTest Reportpt1.pdf.
5. "BioIS Project", online at, http://www.igd.fhg.de/igd−a8/projects/biois/biois_de.html
6. James S. McCANN, "A Family Fingerprint Project", *IDENTIFICATION NEWS* May 1975:7-11.
7. "FVC 2000(Fingerprint Verification Competition 2000)", online at, http://bias.csr.unibo.it/fvc2000
8. "FVC 2002(Fingerprint Verification Competition 2002)", online at, http://bias.csr.unibo.it/fvc2002
9. Choongwoo Ryu and Hakil Kim, "A Fast Fingerprint Matching Algorithm Using Parzen Density Estimation", *Lecture Notes in Computer Science on ICISC*, No. 2587, pp.525-533, 2002.
10. Hyosup Kang, Hakil Kim, Dae-cheol Shin, and Jae-sung Kim, "Test and Evaluation of Liveness Detection for Various Fingerprint Recognition Systems", *Lecture Notes in Computer Science on AVBPA*, Vol.1, No.2774, pp. 1245-1253, 2003.
11. Gaye Shahan, "Heredity in Fingerprints", *International Association for Identification*, Vol. XX, April 1970, pp. 9-14, 2001.

Profile-Based 3D Face Registration
and Recognition

Chao Li[1] and Armando Barreto[1,2]

[1] Electrical and Computer Engineering Department,
Florida International University, 33174 Miami, USA
{cli006, barretoa}@fiu.edu
http://dsplab.eng.fiu.edu
[2] Biomedical Engineering Department,
Florida International University, 33174 Miami, USA

Abstract. With the rapid development of 3D imaging technology, face recognition using 3D range data has become another alternative in the field of biometrics. Unlike face recognition using 2D intensity images, which has been studied intensively by many researchers since the 1960's, 3D range data records the exact geometry of a person and it is invariant with respect to illumination changes of the environment and orientation changes of the person. This paper proposes a new algorithm to register and identify 3D range faces. Profiles and contours are extracted for the matching of a probe face with available gallery faces. Different combinations of profiles are tried for the purpose of face recognition using a set of 27 subjects. Our results show that the central vertical profile is one of the most powerful profiles to characterize individual faces and that the contour is also a potentially useful feature for face recognition.

Keywords: 2D, 3D, biometrics, contour, face, intensity, moment, profile, range, recognition, registration.

1   Introduction

Face recognition has been widely studied during the last two decades. It is a branch of biometrics, which studies the process of automatically associating an identity with an individual by means of some inherent personal characteristics [1]. Biometric characteristics include something that a person is or produces. Examples of the former are fingerprints, the iris, the face, the hand/finger geometry or the palm print, etc. The latter include voice, handwriting, signature, etc. [2]. Compared with other biometric characteristics, the face is considered to be the most immediate and transparent biometric modality for physical authentication applications. Despite its intrinsic complexity, face- based authentication still remains of particular interest because it is perceived psychologically and physically as noninvasive. Significant motivations for its use include the following [2]:

– Face recognition is a modality that humans largely depend on to authenticate other humans.

C. Park and S. Chee (Eds.): ICISC 2004, LNCS 3506, pp. 478–488, 2005.

- Face recognition is a modality that requires no or only weak cooperation to be useful.
- Face authentication can be advantageously included in multimodal systems, not only for authentication purposes but also to confirm the aliveness of the signal source of fingerprints, voice, etc.

The definition of face recognition was formulated in [3] as: "Given an image of a scene, identify one or more persons in the scene using a stored database of faces." This is called the 'one to many' problem or identification problem in face recognition. Another kind of problem is 'one to one', i.e., the authentication problem. This kind of problem is to determine whether the input face of a person is really the person he or she claims to be or not. In this paper, we deal with face recognition in the first scenario. The potential field of the application of face recognition is very wide, mostly in areas such as authentication, security and access control, which include the physical access control and logical access control. Especially in recent years, anti-terrorism has been a big issue throughout the world. Face recognition will play a more and more important role in its efforts.

In the last ten years, most of the research work in the area of face recognition used two-dimensional images, that is, gray level images taken by a camera. Many new techniques emerged in this field and achieved good recognition rates. A number of these techniques are outlined in survey publications, such as [5]. However, most of the 2D face recognition systems are sensitive to the illumination changes or orientation changes of the subjects. All these problems result from the incomplete information contained in a 2D image about a face. On the other hand, a 3D scan of a subject's face has complete geometric information about the face, even including texture information, in the case of some scanners. It is believed that, on average, 3D face recognition methods will achieve higher recognition rates than their 2D counterparts. With the rapid development of 3D imaging technology, 3D face recognition will attract more and more attention.

In [6], Bowyer provides a survey of 3D face recognition technology. Some of the techniques are derived from 2D face recognition, such as Principal Component Analysis(PCA) used in [7, 8] to extract features from faces. Some of the techniques are unique to 3D face recognition, such as the geometry matching method in [9], the profile matching proposed in [10, 11] and the isometric transformation method presented in [4].

This paper outlines a new algorithm used to register 3D face images automatically. Specific profiles are defined in the registered faces and these are used for matching against the faces on a database including 27 subjects. The impact of using different types of profiles for matching is studied. Also the possibility of using the contour of a face as a feature for face recognition is explored.

The structure of the paper is as follows: Section 2 describes the database used for this research. Section 3 presents the registration algorithm and Section 4 outlines the matching procedure using different profiles and contours and gives the results of the experiments. Section 5 is the conclusion.

2   3D Face Database

Unlike 2D face recognition research, for which there are numerous databases available in the Internet, there are only a few 3D face databases available to researchers. Examples are the Biometrics Database from the University of Notre Dame [12] and the University of South Florida(USF) face database[13]. In our experiment, the USF database is used.

The USF database of human 3D face images is maintained by researchers in the department of Computer Science at the University of South Florida, and sponsored by the Defense Advanced Research Projects Agency (DARPA). The USF database has a total number of 111 subjects (74 male; 37 female). All subjects have a neutral facial expression. Some of the subjects were scanned multiple times. In our experiment, the 3D faces of the subjects who were scanned multiple times are considered, so that one scan can be used as a gallery image, i.e., one of the faces that are assumed to be prerecorded, and the remaining scans from the same subject can be used as probe images, i.e., faces to be identified. A subset of 27 subjects is used in this research, with 27 faces in the gallery and 27 scans to be identified (probe faces).

 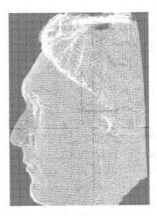

Fig. 1. Rendered 3D face image(Left) and triangulated mesh 3D face image(Right)

The 3D scans in the USF database were acquired using a Cyberware 3030 scanner. This scanner incorporates a rugged, self-contained optical range-finding system, whose dynamic range accommodates varying lighting conditions and surface properties [14] .

The faces in the database were converted into Stereolitography (STL) format. Each face has an average of 18,000 vertices and 36,000 triangles. Figure 1 shows a face from the database in its rendered and triangulated mesh forms.

3   Registration and Preprocessing

In 3D face recognition, registration is a key pre-processing step. Registering may be crucial to the efficiency of some matching methods. Earlier work used Principal Curvature and Gaussian Curvature to segment the face surface and register it, such as the methods in [9, 10, 15]. The disadvantage of using curvatures to register faces is that this process is very computationally intensive and requires very accurate range data [16].

Another method often used involves choosing several user-selected landmark locations on the face, such as the tip of the nose, the inner and outer corners of the eyes,etc., and then using the affine transformation to register the face to a standard position[7, 8, 11].

A third method performs registration by using moments. The matrix(Equation 1) constituted by the six second moments of the face surface: m200,m020, m002, m110,m101,m011,contains the rotational information of the face[17].

$$M = \begin{bmatrix} m200 & m110 & m101 \\ m110 & m020 & m011 \\ m101 & m011 & m002 \end{bmatrix} \tag{1}$$

$$U \Delta U' = SVD(M) \tag{2}$$

By applying the Singular Value Decomposition (Equation 2), the unitary matrix U represents the rotation and the diagonal matrix Δ represents the scale, for the three axes. U can be used as an affine transformation matrix on the original face surface. The problem with this method is that during repeated scans for the same subject, besides the changes in the face area, there are also some changes outside the face area, such as the different caps worn by the subjects during the scanning process (Fig. 1). These additional changes will also impact the registration of the face surface, causing the registration for different instances of the same subject not to be the same. This limitation constrains this approach to only the early stages of registration.

Figure 2 is an example of a scanned face rendered in a Cartesian coordinate system, with the X axis corresponding to the depth direction of the face, the Y axis corresponding to the length direction of the face and the Z axis corresponding to the width direction of the face. In the registration process, we assume that each subject kept his head upright during scanning, so that the face orientation around the X axis does not need to be corrected, but the orientation changes in the Y and Z axes need to be compensated for.

The registration algorithm proposed does not require user-defined landmark locations and can be done automatically.

First, the tip of the nose is found by looking for the point with the maximum value in the X direction. Then a 'cutting plane', parallel to the XZ plane is set to contain the tip of the nose (Fig. 3). The intersection of this cutting plane with the face defines the horizontal profile curve. In effect, the result is a discretized curve with a spacing of 1.8 mm between samples (Fig. 4).

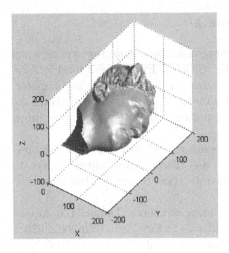

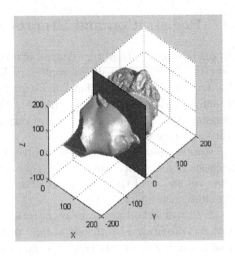

Fig. 2. Face surface in a Cartesian coordinate system(the units in the three axes are mm)

Fig. 3. Illustration of the extraction of the horizontal profile

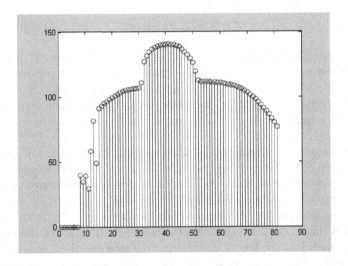

Fig. 4. Discrete horizontal profile before registration

A trilinear interpolation method is used to find the value of each point in this profile. (Fig 5). The point P is in the YZ plane. P' is the intersection between the triangle ABC and the straight line PP', which is normal to the YZ plane. The length of PP' is the profile value corresponding to point P.

Next, the following cost function is minimized with respect to α, where I is the index of the maximum point of X.

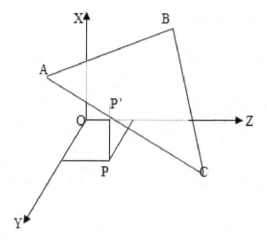

Fig. 5. Trilinear interpolation to get exact values of profile elevations

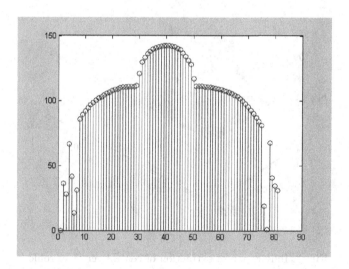

Fig. 6. Horizontal profile after registration around Y axis

$$E = \sum_{i=1}^{15} [(X(I+i) - X(I-i)]^2 \tag{3}$$

For every α, the affine transformation is applied to the face surface using the following transformation matrix, and the horizontal profile is found, as illustrated before.

$$T = \begin{bmatrix} \cos\alpha & 0 & -\sin\alpha \\ 0 & 1 & 0 \\ \sin\alpha & 0 & \cos\alpha \end{bmatrix} \tag{4}$$

$$\alpha = arg\ min[\sum_{i=1}^{15}[(X(I+i)-X(I-i)]^2]\qquad(5)$$

The final value of α represents the orientation change around the Y axis required for the registration.

Figure 6 shows the horizontal profile seen in Figure 4, after the Y axis adjustment has been performed:

Typically, a rotational adjustment around the Z axis will also be required. Analogous to Figure 3, Figure 7 shows the intersection of the face surface with a cutting plane, which is parallel to the XY plane and passes through the tip of the nose. This intersection is the central vertical profile. Similar to Figure 4, Figure 8 shows the discretized central vertical profile, before adjustment.

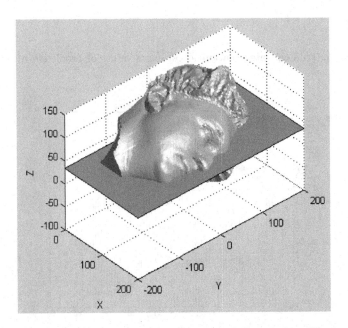

Fig. 7. Illustration of extraction of central vertical profile

The cost function to be minimized in this case is the following,

$$E = abs(X(I-50)-X(I+40))\qquad(6)$$

Minimization is with respect to α. I is the index of profile point with the largest value of X.

$$\alpha = arg\ [min(abs(X(I-50)-X(I+40)))]\qquad(7)$$

For every α, the affine transformation is applied, using the following transformation matrix.

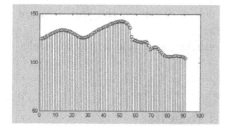

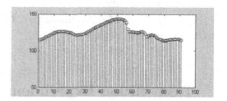

Fig. 8. Discretized central profile before registration

Fig. 9. Central vertical profile after registration

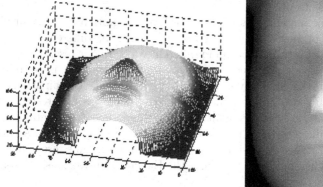

Fig. 10. Mesh plot of the range image(left) and gray level image plot of range data(right)

$$T = \begin{bmatrix} \sin\alpha & 0 & \cos\alpha \\ \cos\alpha & 0 & -\sin\alpha \\ 0 & 1 & 0 \end{bmatrix} \quad (8)$$

The aim is to equalize the X coordinates of two critical points contained in the central vertical profile: the end point on the forehead side and the end point on the chin side. Figure 9 is the central vertical profile after adjustment around the Z axis.

To complete the registration process, a grid of 91 by 81 points is prepared that corresponds to pairs of (y, z) coordinates. The point (51, 41) of the grid is made to coincide with the tip of the nose in the adjusted face surface. This grid assumes a spacing of 1.8 mm in both the Y and Z directions, with 91 points in the length direction and 81 points in the width direction. The value associated to each point in the grid is the distance between the point in the face surface and the corresponding location on the YZ plane, calculated by trilinear interpolation (Fig. 5). The values are offset so that the value corresponding to the tip of the nose is normalized to 100 mm. Values below 20 mm in the grid area are thresholded to 20 mm.

Figure 10 is a Matlab mesh plot of the resulting grid, and a gray level plot of the same range image.

4   Recognition Experiments and Results

For the experiments described here, a gallery database of 27 range images of 27 subjects (one for each subject) and a probe database of 27 different scans of the same 27 subject were used. The time interval between the acquisition of the gallery image and the corresponding probe image for a given subject ranges from several months to one year.

The use of profile matching as a means for face recognition is a very intuitive idea that has been proposed in the past. In [10, 11, 18, 19], different researchers explored the profile matching method in different ways. In our research, because the range image has already been obtained, profile extraction is simple. We have, in fact, tested the efficiency of several potential profile combinations used for identification. Besides profiles, the contour of a face was also tested for its potential applicability for face recognition. In our experiment, a frontal contour defined 30 mm behind the tip of the nose was extracted for each scan. Although in computing the distance or dissimilarity between profiles, some researchers [19] used the Hausdoff distance, we found that the Euclidean distance is suitable for the context of our experiment.

The following six different feature combinations and direct range image matching variations were tested with the experimental data described above:

(a) Central vertical profile alone.
(b) Central horizontal profile alone.
(c) Contour, which is 30 mm behind the tip of the nose.
(d) Central vertical profile and two horizontal profiles. The two horizontal profiles are defined at 18 mm and 36 mm above the tip of the nose. The distance between central profiles is given the weight of 0.7; the two horizontal profile distances are given the weight of 0.15 each, towards the overall matching score for identification.
(e) Central vertical profile and two more vertical profiles, one passing 18 mm to the left of the central profile, the other passing 18mm to the right of the central profile. The distance between central profiles is given the weight of 0.7; the other two vertical profile distances are given a weight of 0.15 each.
(f) Using the entire range image.

From the results in Figure 11, we can see that scheme (a), i.e., matching the central vertical profile alone, has the highest recognition rate. On the other hand, using the whole range image for matching yields the lowest recognition rate. Because the probe image was taken several months to one year after the gallery image was taken, we have sufficient reason to assume there were changes in the face for every subject. The high recognition rate using the central vertical profile suggests that this profile has the most distinctive properties between different subjects and is the most consistent through time for the same subject. These observations concur with a similar analysis, presented in [11]. Besides the central vertical profile, the contour of a face also shows its potential as a feature to be used in face recognition.

5   Conclusion

In this paper, a new registration algorithm for 3D face range data was proposed. This algorithm is valid under some constrains; i.e., only orientation changes along the width direction and length direction of the face need to be compensated. But this algorithm can also be extended to register arbitrarily oriented face surfaces in 3D space, combined with simple registration algorithms that use the six second moments of the face surface.

Also in this paper, face identification based on profile matching was explored. Different combinations of profiles for matching were compared. It was found that the central vertical profile is the feature that best represented the intrinsic characteristics of each face and had the highest identification value among all the profile combinations tested. The contour of a face also has the potential to be used as one of the features in face recognition.

Acknowledgments

This work was sponsored by NSF grants IIS-0308155 and HRD-0317692. The participation of Mr. Chao Li in this research was made possible through the support of his Florida International University Presidential Fellowship.

References

1. A.K.Jian, R.Bolle and S.Pankanti, Biometrics-Personal Identification in Networked Society. 1999, Norwell, MA: Kluwer
2. J. Ortega-Garcia, J.Bigun, D.Reynolds and J.Gonzales-Rodriguez, Authtication gets personal with biometrics. IEEE Signal Processing, 2004. 21(No.2): p. 50-61
3. R. Chellappa, C.Wilson and S. Sirohey, Human and Machine Recognition of Faces: A Survey. Proceedings of the IEEE, 1995. 83(5): p. 705-740
4. A. Bronstein, M.Bronstein and R.Kimmel, Expression-invariant 3D face recognition. In the Proceedings of Audio and video-based Biometric Person Authentication (AVBPA),2003: p. 62-69
5. W.Zhao, R.Chellappa and A.Rosenfeld, Face recognition:a literature survey. ACM Computing Survey, 2003. 35: p. 399-458
6. K. Bowyer, K.Chang and P. Flynn, A Survey of Approaches to 3D and Multi-Modal 3D+2D Face Recognition. In Proceedings of IEEE International Conference on Pattern Recognition. 2004: p. 358-361

7. K. Chang, K. Bowyer and P. Flynn, Multimodal 2D and 3D biometrics for face recognition. In the Proceedings of ACM workshop on Analysis and Modeling of Faces and Gestures. 2003: p. 25-32

8. C.Hesher, A.Srivastava and G.Erlebacher, A novel technique for face recognition using range images. In the Proceedings of Seventh International Symposium on Signal Processing and Its Application. 2003

9. G.Gordon, Face recognition based on depth maps and surface curvature. In Geometric Methods in Computer Vision, SPIE. July 1991: p. 1-12

10. J.Y.Cartoux, J.T.LaPreste and M.Richetin, Face authentication or recognition by profile extraction from range images. In the Proceedings of the Workshop on Interp. of 3D Scenes. 1989: p. 194-199

11. T.Nagamine, T.Uemura and I.Masuda, 3D facial image analysis for human identification. In the Proceedings of Internaitonal Conference on Pattern Recognition (ICPR,1992): p. 324-327

12. K. Bowyer, University of Notre Dame Biometrics Database. http://www.nd.edu/%7Ecvrl/UNDBiometricsDatabase.html

13. K. Bowyer, S. Sarkar, USF 3D Face Database. http://marathon.csee.usf.edu/HumanID/ 2001.

14. www.cyberware.com

15. H.Tanaka, T. Ikeda, Curvature-based face surface recognition using spherical correlation-principal directions for curved object recognition. In the Proceedings of the 13th International Conference on Pattern Recognition, 1996: p. 638-642

16. Y. Wu , G.Pan and Z. Wu, Face Authentication Based on Multiple Profiles Extracted from Range Data. In the Proceedings of 4th International Conference on audio- and video-based biometric person authentication. 2003: p. 515-522

17. M. Elad, A.Tal and S. Ar, Content Based Retrieval of VRML Objects-An Iterative and Interactive Approach. In the Proceedings of the 6th Eurographics Workshop in Multimedia. 2001.

18. C. Beumier, C.Acheroy, Face verification from 3d and grey level clues. Pattern Recognition Letters, 2001. 22(12): p. 1321-1329

19. G. Pan, Y.Wu and Z. Wu, Investigating Profile Extraction from Range Data for 3D Face Recognition. In the Proceedings of 2003 IEEE International Conferrence on Systems Man and Cybernetics: p. 1396-1399

Author Index

Lecture Notes in Computer Science

For information about Vols. 1–3427

please contact your bookseller or Springer